Principles of Public Health Microbiology

Robert S. Burlage, PhD

Professor

College of Health Sciences

School of Public Health

University of Wisconsin–Milwaukee

Milwaukee, Wisconsin

BARTLETT
LEARNING

World Headquarters
Jones & Bartlett Learning
40 Tall Pine Drive
Sudbury, MA 01776
978-443-5000
info@jblearning.com
www.jblearning.com

Jones & Bartlett Learning
Canada
6339 Ormindale Way
Mississauga, Ontario L5V 1J2
Canada

Jones & Bartlett Learning
International
Barb House, Barb Mews
London W6 7PA
United Kingdom

Jones & Bartlett Learning books and products are available through most bookstores and online booksellers. To contact Jones & Bartlett Learning directly, call 800-832-0034, fax 978-443-8000, or visit our website, www.jblearning.com.

Substantial discounts on bulk quantities of Jones & Bartlett Learning publications are available to corporations, professional associations, and other qualified organizations. For details and specific discount information, contact the special sales department at Jones & Bartlett Learning via the above contact information or send an email to specialsales@jblearning.com.

Production Credits
Publisher: Michael Brown
Editorial Assistant: Teresa Reilly
Associate Production Editor: Kate Stein
Senior Marketing Manager: Sophie Fleck
Associate Photo Researcher: Jessica Elias
Manufacturing and Inventory Control Supervisor: Amy Bacus
Composition: Achorn International
Art: diacriTech
Cover Design: Scott Moden
Cover and Interior Image: © Gabor Racz/ShutterStock, Inc.
Printing and Binding: Malloy, Inc.
Cover Printing: Malloy, Inc.

Library of Congress Cataloging-in-Publication Data
Burlage, Robert S.
Principles of public health microbiology / Robert S. Burlage.
p. ; cm.
Includes bibliographical references and index.
ISBN-13: 978-0-7637-7982-5 (pbk.)
ISBN-10: 0-7637-7982-2 (pbk.)
1. Microbial ecology. 2. Public health. I. Title.
[DNLM: 1. Environmental Microbiology. 2. Public Health. QW 55]

QR100.B87 2012
579'.17—dc22

2011000575

6048

Printed in the United States of America
15 14 13 12 11 10 9 8 7 6 5 4 3 2 1

Dedication

To John and Wilma

CONTENTS

Preface

People are naturally interested in the things that can harm them. We are far more interested in learning about a dangerous animal, like a cobra, than an ordinary kingsnake, even though the latter is of more benefit to us. We are more interested in toxins that kill in small amounts than we are in vitamins that we require in small amounts. And we are more interested in *Staphylococcus aureus*, the pathogen, than in *Staphylococcus epidermidis*, the helpful commensal microbe. Much is known about infectious disease, although it is far better to avoid it.

Of course, avoidance is not always possible. For example, there was an unfortunate incident at our department's annual Christmas party. It was held on a Wednesday, and on Thursday and Friday we had a number of people out sick. Even though colds and flu are frequent in the campus community, a cluster like this one seemed unusual. One professor, rather pale and heading home to recuperate, suggested that it was food poisoning and even suggested what she thought was the culprit.

A quick survey of the department found that, of 20 attendees, 8 had become ill with gastrointestinal symptoms (vomiting and/or diarrhea). No one suggested that anything tasted "funny." Comments from the sufferers suggested that the contaminant was in one of three items: the pasta salad, the spinach salad, or the vegetable dip. I calculated the odds ratio (covered in this text) and found that the odds ratio for spinach salad was 1.5. With such a small sample size, this number was hardly significant. The odds ratio for the pasta salad was 3.8, which certainly looked suspicious. The odds ratio for the vegetable dip was 12. This was far too great to ignore.

So what was the source? An epidemiological approach can only point you in the right direction; it does not provide substantive proof of causation. However, I did not eat either the dip or the pasta salad, and I was fine. It was one of those items. The real lesson (at least for this book) is that foodborne illness can strike any group, even an educated bunch of Health Sciences faculty. While we will never know what the ultimate source of contamination was in this case, it is certain that there was a breakdown in the food preparation, and any breakdown should be avoidable if we just analyze it, understand it, and engineer our way around it.

Some people take a fatalistic approach to illness. People have been getting sick from all manner of sources for a long, long time. And, of course, no one died from our office party. They were sick for a while, and everyone is better now. But that is too parochial a vision. Our faculty is a group of otherwise strong, healthy people, but they will not always be. Everyone gets old and prone to illness, and there are many, many other people who, through no fault of their own, are at risk for major illness from microorganisms from which the rest of us would easily recover.

The measure of a great society is the care it takes with the most vulnerable members of that society. We care about these things because we, as a society, wish to reduce suffering and sadness. Industrialized societies have been quite successful at reducing infectious disease, and we have a happier and healthier society for it. The mechanisms that we have put in place have been successful in avoiding a huge amount of illness. Knowing how this was accomplished, and knowing the problems that still remain, are requisite for knowing what to preserve and what to concentrate on next.

Acknowledgments

I greatly appreciate the patience of my family and the faculty and staff of my department as I wrote this text. It is a time-consuming task, and many other things often had to wait.

Special acknowledgment is due to some of the wonderful people who went out of their way to help me with this text. Pam Akers and Thomas Achankunju, University of Wisconsin–Milwaukee, helped greatly with data collection and presentation. Stanley Kucharski, Fox River Water Pollution Control Center, was very generous with his time in explaining the science of wastewater treatment. Bill Graffin, Milwaukee Metropolitan Sewerage District, was another valuable resource. Joe Kelly of Global Water, Joan Maxwell of ThermoFisher, Ron Kean of the University of Wisconsin, and Carrie Lewis of the Milwaukee Water Works for their generous contributions. John Burlage of Cherry Hill, New Jersey, and Dr. Thomas Tang of Brookfield, Wisconsin, for their unique photographs. Michael Peters gave me a terrific tour of the University of Wisconsin Dairy Science Center. Special thanks to Dr. Steven Gradus, Milwaukee Health Department, who not only supplied some of the interesting material found in the text but has been the model of what a good public health official should be.

Additionally, I would like to acknowledge the reviewer of this text, Michael Pentella, PhD, D(ABMM), who is a professor in the College of Public Health at the University of Iowa, Iowa City.

About the Author

Robert S. Burlage, PhD, is a professor in the Health Sciences Department at the College of Health Sciences, University of Wisconsin–Milwaukee. He received his doctoral degree from the University of Tennessee and performed post-doctoral work at Oak Ridge National Laboratory in Tennessee. He continued there until 2001, working on bioremediation of hazardous wastes, including detection of specific chemical species using bacterial bioreporter systems.

His current research involves gene expression of bacterial pathogens in natural settings, such as rivers. He has received one U.S. patent and has coedited another text, *Techniques in Microbial Ecology*. He served as chairman of the Health Sciences Department from 2007 to 2010 and as graduate coordinator from 2001 to 2007.

He is the graduate coordinator of the Zilber School of Public Health at UWM. He was instrumental in starting the school and was the principal developer of the Environmental and Occupational Health doctoral program at the school. Dr. Burlage continues to serve on their executive committee.

CHAPTER 1

The Scope of Public Health Microbiology

LEARNING OBJECTIVES

- Describe how public health microbiology fits into the world of public health.
- Contrast medical microbiology with public health microbiology.
- Describe how the development of civilization has both helped and hindered the health of populations.

KEY TERMS

- Biostatistics
- Division of labor
- Environmental health
- Epidemiology
- Health services administration
- Social and behavioral sciences

INTRODUCTION TO THE TOPIC

Public health microbiology is a subdiscipline of environmental health, which itself is one of the five principal areas of public health (**Box 1-1**). Because public health is concerned with the causes of human morbidity and mortality, the scope of its practice is practically universal. It includes the environment in which people live, the attitudes and beliefs that people hold, and the methods that are used to deal with perceived problems. It can include subjects as diverse as infant mortality, folk medicine traditions, blood pressure screenings, irrational fears, lead paint removal, handling of health data, design of health laws and their implementation, and water purification.

The five areas of public health are interconnected and interdependent. **Biostatistics** and

BOX 1-1 The Areas of Knowledge Basic to Public Health

Biostatistics
Epidemiology
Environmental health sciences
Health services administration
Social and behavioral sciences

epidemiology are methods to understand the data that are collected by researchers in the other areas. Epidemiology is used to discern the trends in health and disease and the probable factors involved in these trends, and biostatistics are used to analyze whether those data pass the test of statistical significance. This information allows researchers to pursue cause-and-effect relationships that can then be used to change our environment for the better. **Social and behavioral sciences** concern the human element of public health: people form social organizations (e.g., families, neighborhoods, clubs, etc.) and practices (e.g., habits, addictions, preferences, customs, and traditions) that may affect health and well-being. These stand apart from the more obvious requirements of life—food, water, shelter, warmth—but are every bit as important. Humans are social animals and require personal interactions. It is essential to determine why people do what they do, whether what they do is helpful or deleterious, and how to effectively promote good social and behavioral practices. **Health services administration** concerns the public health policies of a group of people, as expressed through their government and the rules of society. A good example is the campaign against cigarette smoking. While smoking is a legal activity, it is also an unhealthy one. Legal requirements have been put into place to protect nonsmokers from other people's cigarette smoke, and high taxes on cigarettes have been levied to discourage the activity.

Because of its emphasis on the products that people use and the environment in which people live, **environmental health** is the area that is most closely associated with public health microbiology. In essence, environmental health is concerned with the dangers posed to humans from a variety of sources, such as acute chemical exposure, which may lead to acute poisoning, or chronic chemical exposure, which may lead to cancer. Other dangers include injuries from physical dangers, such as high energy sources (e.g., electricity, radiation) and accidents (e.g., falls, motor vehicle accidents, gunfire), and from biological agents (e.g., dog bites, falls from horses, and even bullfighting injuries). Environmental health concerns also include infectious disease and the means by which people encounter pathogens, the agents of infectious disease. Infectious disease comprises such a vast element of environmental health that the topic is most often treated separately.

Public health microbiology is dependent on an understanding of concepts from both environmental health and from microbiology, and it also requires an understanding of concepts from other disciplines such as hydrology, engineering, law, and food science. Throughout this study of public health microbiology, therefore, data from different disciplines will be woven into a core presentation of microbiology. This is the true value of this topic: it not only describes the pathogens and their means of infection but also how we avoid encountering them on individual and societal bases.

WHY STUDY PUBLIC HEALTH MICROBIOLOGY SEPARATELY?

Infectious disease is still a major health problem in the United States (**Table 1-1**) and worldwide (**Table 1-2**). In the United States, the biggest killers are, by far, cancer and heart disease. Both of these diseases can debilitate the patient and create opportunities for infectious disease to do further harm. For example, treatment for cancer involves radiation or chemotherapy, both of which severely weaken the immune system. Lengthy hospital stays, as often happen with cardiovascular disease, often involve invasive procedures and place the patient in proximity to antibiotic-resistant bacteria. Nosocomial (i.e., hospital-acquired) infections remain a huge problem in modern society.

Although industrialized countries have made major strides in preventing and treating infectious diseases, pathogens are still responsible for tens of thousands of deaths and millions of illnesses every year. Deaths attributable to influenza (often accompanied by a bacterial pneumonia), to antibiotic-resistant bacterial strains, and to both bacterial and viral meningitis are not uncommon. Illnesses caused by contaminated water and food number in the millions, and, although the vast majority are mild, a significant percentage require hospitalization. In tropical areas of the world a variety of other pathogens remain as dangerous as ever, including malaria and dengue fever, both of which are transmitted by mosquitoes. Malaria causes nearly a million deaths around the world annually, mostly in

TABLE 1-1 The Top Ten Causes of Death in the United States in 2006

Cause of Death	Number	Percentage
Heart disease	631,636	26.0
Cancer	559,888	23.1
Stroke (cerebrovascular diseases)	137,119	5.7
Chronic lower respiratory diseases	124,583	5.1
Accidents (unintentional injuries)	121,599	5.0
Diabetes	72,449	3.0
Alzheimer's disease	72,432	3.0
Influenza and Pneumonia	56,326	2.3
Nephritis, nephrotic syndrome, and nephrosis	45,344	1.9
Septicemia	34,234	1.4

NOTE: Percentage sums to 76.5% because all other causes of death equal the other 23.5%. No other cause of death exceeds 1.4% of total deaths.

SOURCE: http://www.cdc.gov/nchs/fastats/lcod.htm

TABLE 1-2 The Top Ten Causes of Death Worldwide in 2004

Cause of Death	Number (in millions)	Percentage
Coronary heart disease	7.20	12.2
Stroke and other cerebrovascular diseases	5.71	9.7
Lower respiratory infections	4.18	7.1
Chronic obstructive pulmonary disease	3.02	5.1
Diarrhoeal diseases	2.16	3.7
HIV/AIDS	2.04	3.5
Tuberculosis	1.46	2.5
Trachea, bronchus, lung cancers	1.32	2.3
Road traffic accidents	1.27	2.2
Prematurity and low birth weight	1.18	2.0

NOTE: Percentage adds to 50.3% because all other causes of death equal the other 49.7%. No other cause of death exceeds 2.0% of total deaths.

SOURCE: http://www.who.int/mediacentre/factsheets/fs310_2008.pdf

countries where the income level is low. The prevalence of dual infection with the HIV virus and *Mycobacterium tuberculosis* is also a major cause of mortality in many countries (Table 1-2).

A course in microbiology is a prerequisite to a full understanding of this topic. Microorganisms constitute a diverse group of organisms, which are central to the functioning of our ecosystem. Therefore, the student must understand the various groups of microorganisms (**Box 1-2**), how they acquire carbon and energy, what their growth requirements generally are, and how they exist in the greater ecosystem. A vital component is knowing what autotrophs and anaerobes are. Some understanding of disease is also necessary because much of public health microbiology is concerned with avoiding or destroying pathogens. Other vital components include knowing which microbes are pathogenic, understanding the diseases they cause, communicating the importance of each disease to the medical community, and establishing access to the treatment of each disease. However, a course in medical microbiology (the study of pathogenic microorganisms) is not essential. Much of medical microbiology is devoted to the progression of disease after the host has encountered the pathogen and to the ability of the host to defeat the invaders using either the immune system or through medical intervention (e.g., antibiotics). Although these concepts are addressed, the emphasis here is on avoidance of pathogens and infection or on the destruction of pathogens before they can infect the population.

BOX **1-2** **The Classes of Microorganisms**

Bacteria
Protozoa
Viruses
Fungi (yeast, mold)
Algae

AN OVERVIEW OF THE DISEASE PROCESS

A person may encounter infectious disease in a number of ways. Contaminated drinking water is a major source of exposure, and great efforts are expended in society to reduce or eliminate this particular threat. The chapters on water treatment and waterborne diseases are extensive, reflecting the multiplicities of dangers and the monitoring and treatment methods that are used to counter them. Contaminated food is another obvious threat, especially considering the great variety of foods that people eat. Diseases that are spread through close contact include casual contact (e.g., being the recipient of airborne pathogens through a cough or sneeze), direct contact (e.g., through an insect bite or an inanimate object), and the much more intimate contact required to pass sexually transmitted diseases. Finally, deliberate exposure to pathogens is possible as a result of bioterrorism.

Because public health is concerned mainly with populations rather than individuals, issues of scale are relevant. While the practice of medicine for infectious disease is concerned with the individual patient and the specific disease involved, public health is more concerned with all pathogens, their prevalence, and the methods used to detect their presence and to limit their opportunities to infect people. While individual illness or a small outbreak of cases may be unfortunate, it does not pose a problem for society as a whole. For this reason, the examples shown in the text often describe large outbreaks or citywide problems. The potential for disaster is much greater when a large number of people are dependent on a common factor.

A HISTORICAL PERSPECTIVE

Epidemic illness has been with us since the start of civilization. The Greek historian Thucydides wrote of a plague in Athens in the year 430 AD. By his account nearly a third of the city's population died from this plague, the source and identity of which remains unclear. However, it must have been terrifying, given his description of the symptoms:

> *. . . people in good health were suddenly attacked by violent heats of the head, redness*

and inflammation of the eyes, with the inward parts, such as the throat and tongue, becoming bloody and emitting an unnatural and fetid breath. These symptoms were followed by sneezing and hoarseness, after which the pain soon reached the chest, producing a hard cough. When it fixed in the stomach it upset it, causing discharges of bile of every kind named by physicians, and accompanied by very great distress. In most cases also an ineffectual retching followed, producing violent spasms, which in some cases ceased soon after, in others much later.

We live in a modern, industrialized society, and it can be challenging to acquire a perspective on how our built environment is sometimes less than ideal in terms of health promotion. We have created structures and societies that serve certain purposes, and those purposes sometimes conflict with other purposes, such as avoiding disease. A good example of this concept comes from the time of the Roman Empire, when water pipes and goblets were constructed from lead. The pipes served a very useful purpose in delivering water to people who would otherwise have to carry it great distances, but lead is a toxic metal which, over time, accumulated in the bodies of the citizens. The Romans surely suffered from its effects. Nothing was known about the toxic effects of lead at that time, and some historians have speculated that it was a significant contributing factor in the decline of their empire. Their example makes us cautious today about effects that we may not fully appreciate.

Communities can create their own problems. To grasp the challenge of public health in the modern world, it is helpful to reflect on the beginnings of civilization, approximately 10,000 years ago. In small clans of hunter-gatherers, people searched for food and water and the other necessities of life. This is not a particularly bad way to survive because their diets were usually adequate and varied (and healthy), at least for small groups passing through a region. Such tribes have survived for thousands of years, and nomadic tribes still exist today. However, this lifestyle has distinct disadvantages: sometimes the game disappears and the local plants are not abundant, making starvation an occasional reality. Also, hunting can be a dangerous occupation, and hunters died from infected injuries that we would consider relatively minor.

Another often overlooked advantage of nomadic hunting–gathering was that food was under the direct observation and control of the consumers. If food appeared to be spoiled, they decided whether or not to eat it. If they made the wrong choice, they got sick and perhaps died. But the ultimate responsibility for food and water quality resided with the individual consuming them.

The adoption of agriculture somewhere between 8,000 and 10,000 years ago was a major advance in human development. People still hunted local game, and humans would eventually domesticate animals, but cultivation of food allowed a more stable food supply, while fiber crops (e.g., cotton and flax) became important for clothing. The foods that were selected could usually be stored for extended periods of time, making famine less likely. Occasionally, however, famine did occur due to climactic events such as a drought or a hailstorm or due to a swarm of locusts that destroyed the crops.

People settled in an area because it was amenable to cultivation and because it was near a reliable water supply, such as a river. The variety of foods in the diet declined because a farmer could not grow every plant that existed in the wild, and some crops were simply easier to cultivate than others. But the resulting food supply was usually greater. Early farming settlements were communities, and population density increased.

In these communities the person doing the planting and harvesting was responsible for the quality of the food consumed. But a greater advance in civilization accompanied the concept of **division of labor**. This was a very simple concept that had enormous benefits. For example, in a hunter–gatherer clan each member of the hunting party made his own weapons and used them to kill game. But not everyone excelled at hunting; some were exceptionally skilled at making flint arrowheads and spear points. At some point those making spear points decided to offer them to another hunter—perhaps someone lacking the patience to produce a good spear point—in exchange for part of the kill. This simple barter arrangement represents a division of labor because the two people involved each did something at which they excelled. This method allows each person to get the things he wants, which is another way of saying that it creates a more efficient economy. In fact, our spear-point maker may produce spear points so fast and so well that he created leisure time for himself. Alternatively, he might have started trading them to other clans for their specialty goods (e.g., clothing or jewelry). Hereby the creation of wealth emerged (at least by Neolithic standards). Although this arrangement has many benefits, it has a major drawback. The spear-point maker had given a large measure of responsibility for the quality of his food to the hunter. This simple arrangement should pose few problems; if the hunter traded a piece of spoiled meat the spear-point maker would know immediately and would complain loudly. A wise hunter would trade only good meat for spear points, or he might not get any in the future. Even after the advent of agriculture, this type of trade predominated. One farmer might have abundant wheat that he would trade for a butchered hog. Again, each party could easily determine whether the food was wholesome. Barter arrangements persist to the present day, and actually work quite well, even if they are ultimately limited in scope.

As societies became more complex, other means of supporting workers were needed. The Biblical book of Daniel describes how the young Daniel was educated as a servant of the royal court of King Nebuchadnezzar of Babylonia. The king ". . . appointed them a daily provision of the king's meat, and of the wine which he drank" (King James Version: Daniel chapter 1), indicating that room and board for the servants was provided directly by the king. Daniel and his friends refused the rich food as unhealthy, and they worked out an agreement to eat only vegetables and drink water. After a trial period they were found to be healthier than the other servants. This story suggests the risk of allowing others to prepare food: consumers are never completely certain of its quality.

CIVILIZATION HAS BECOME COMPLEX

A more robust economy required a common medium of exchange, i.e., money. Money promoted a further division of labor because it was storable, portable, and allowed the valuation of certain types of labor that were not easily traded. How could the work of judges, scribes, teachers, and other learned individuals be valued in an ancient society? The earliest recognizable money was probably in the form of precious metals such as gold and silver. While useful in themselves for creating jewelry and decorative objects, they could not be eaten or worn as clothing. Money became a means of exchange for such essential things. With money, a person could not only buy a measure of wheat but he could also let someone else do the milling of the wheat and just buy the flour. Or he could completely skip the steps in food production and buy a meal prepared by someone else. That is, he could have a restaurant meal. Then and now, this is very convenient, and a tremendous number of restaurant meals are consumed every day. However, the consumer of the meal has entirely surrendered control of the quality of the food to someone else, which can be a large disadvantage. In fact, the person responsible for his safety may be a complete stranger. The diner has no idea whether the food preparer uses proper hygiene, is passing off questionable food as fresh, or is even using the ingredients that are claimed. For example, it is not unusual today to find a cheaper grade of fish offered as a more expensive variety to unsuspecting shoppers.

As civilization progressed, specialization of labor increased. The gap between producer and consumer is now vast, and the variety of products offered has become immense. Potential health problems have been magnified by improvements in transportation that allow people and goods to travel thousands of miles in a matter of hours. Not only does modern transportation bring exotic products from all over the world, it also brings exotic diseases. With the advent of air freight, fresh produce can be delivered from South America to North America during the winter months, greatly improving the diets of American citizens and making their meals more varied and interesting.

In a modern, industrialized society, it is simply impractical for individuals to monitor the quality of their food. Complexity requires that an impartial authority—the government—must set appropriate rules for food handling and must monitor the process for safety. We give the government the authority to establish safe practices, to monitor production, and to punish those who put people at risk by breaking the rules. A major element of trust is involved with this authority; we expect them to be effective and impartial sentinels. The study of public health microbiology requires an understanding of the authorities (i.e., federal, state, and local) that affect processes and the most important rules and regulations at work in the food industry, i.e., effective regulation does not interrupt the supply of food and drink. In a complex, integrated world every nation is under threat from infectious disease, and our awareness of just how many pathogens exist and how they can become epidemic problems encourages us to adopt preventative measures (Fauci et al., 2005).

A simple analogy is helpful in thinking about public health microbiology: a community is much like the body of a single organism. The body needs to consume wholesome food and water and to remove waste products so that they no longer pose a health threat. The body also needs uncontaminated air to breathe, and it must take active measures to avoid other sources of infection, such as sexually transmitted diseases, exotic (unexpected) pathogens, and deliberate exposures. A healthy body—and a healthy community—is aware of the sources of pathogens and the risks of exposure. It takes active measures to avoid unnecessary exposure and to incorporate effective procedures to stay healthy. The active measures that we take to protect our citizens are both complex and effective; skilled workers are necessary to maintain that level of safety. Most people never think about the people who purify their water, treat their wastewater, and monitor their food supply, but without them we would all be at greater risk of disease.

QUESTIONS FOR DISCUSSION

1. Which innovations of the past century have contributed to public health? Which of them also have significant unintended deleterious outcomes?
2. How are the various fields of public health interdependent?
3. How do public health and medicine complement each other?
4. Can you foresee a public health problem arising in the future based on your experience today? How could it be avoided?

Reference

Fauci, A.S., N.A. Touchette, and G.K. Folkers. 2005. Emerging infectious diseases: a 10-year perspective from the National Institute of Allergy and Infectious Diseases. Emerg. Infect. Dis. 11:519–525.

CHAPTER 2

THE IMPORTANCE OF WATER

LEARNING OBJECTIVES

- Describe the hydrological cycle and where humans interact directly with it.
- List the major uses of water in the household.
- Define the saturated and vadose zones of groundwater.
- Identify the characteristics of surface water and groundwater, and why the distinctions are important.
- Learn how water is obtained.

KEY TERMS

- Artesian well
- Community water
- Groundwater
- Lentic
- Lotic
- Noncommunity water system
- Nontransient noncommunity water system
- Public water system
- Run-off
- Saturated zone
- Surface water
- Transient noncommunity water systems
- Unsaturated
- Vadose zone
- Water table

EVERYONE NEEDS WATER

Water is essential for life. Approximately 70% of the Earth's surface is covered with water, yet little of it is potable. Most is seawater and undrinkable (**Figure 2-1**). However, about 3% of the water on Earth is freshwater found in lakes, rivers, underground, as rainwater, and in the polar ice caps. An adequate supply of freshwater is necessary for the health of the population, and therefore great care is taken in the stewardship of this natural resource. In the past this was not always the case, and water sources were often polluted with many kinds of waste. However, changes in laws and in human attitudes have decreased water pollution greatly. While there is currently sufficient water in the United States, there is reason to be concerned about the future. Demand for water has increased, while supplies are finite. More effective use of water and cleaning of water once it has been used, will be needed.

The importance of water is self-evident. Without it life cannot exist. NASA has devoted major projects to the detection of water elsewhere in our solar system, notably on Mars. If water is detected, then perhaps life exists there as well, but we are reasonably certain that without water there is no possibility of life. Human beings have always intrinsically understood the value of water (**Figure 2-2**). Settlements were created near a stable supply of water, such as a river or a natural spring. Ancient cities created cisterns to hold water against times of

FIGURE 2-1 A lighthouse.

drought (and especially while besieged by invaders). The city of Rome was founded on the banks of the Tiber River, but it gradually outgrew this water supply. The Romans constructed a vast aqueduct system to supply the city with fresh water from nearby mountains. By some estimates the population of Rome during this time exceeded one million. Many other Roman cities constructed aqueducts as well.

As precious as water is, national conflicts over water rights are relatively rare. It is not at all unusual to see low-intensity conflict between different claimants to the same water supply, and in westernized countries this often presents as legal challenges, but full-scale wars are rare. Perhaps this is because everyone understands the importance of fresh water and that the issue can quickly get out of control. For instance, the Israelis and the Palestinians must share water resources, and they have endured longstanding animosity. But their conflict over water remains far down on the list of grievances. In practice, nations adapt to water issues by promoting industries that correspond to realistic water availability and use trade to make up the water difference. This has proved to be a workable approach so far, but in the future water shortage may become even more acute. Rivers are good sources of water, especially if the water is moving swiftly. Egypt became one of the first great nations because of the Nile River and its supply of fresh water. The important Chinese cities of Wuhan, Nanjing, and Shanghai are all located along the Yangtze River. Paris is located along the Seine River; London along the Thames. In North America major

FIGURE 2-2 **Water is essential every day. It is used for (a) drinking, (b) cooking, (c) recreation, and (d) decorative fountains, among other uses.**

cities have grown up alongside rivers (**Figure 2-3**): Boston on the Charles River, Montreal at the juncture of the St. Lawrence and Ottawa rivers, Pittsburgh at the meeting point of three rivers (the Allegheny River, the Monongahela River, and the Ohio River).

Quantity of water was obviously important, but quality of water was also a major concern. While the link between fresh water and health was not appreciated for many years, the idea of different levels of acceptable water based on taste and smell was quite obvious. Water from mountain streams was generally considered to be of high quality and cold as well. River water or well water (the ancients could dig shallow wells by hand) was also considered potable. Lake water was usually suitable only for cattle feeding, although various methods were proposed that would reduce the turbidity of the water and make it drinkable. For example, there is evidence that the Egyptians used alum to clarify water.

But water treatment is really a relatively modern invention, in common use only within the last 200 to 300 years. The use of sand filters for water purification appears to have started in the 1700s. By 1804, the city of Paisley, Scotland used sand filters to purify water for the entire town. Additional cities followed suit over the next half-century. Usually these developments were implemented to improve the aesthetic qualities of the water. In 1854 a substantial outbreak of cholera hit London, one of many such outbreaks over the years. Dr. John Snow analyzed the affected population and concluded that the water supply was to blame. He identified the public water pump at Broad Street as the likely source and stopped the outbreak by removing the handle of the pump. However, Snow based his conclusions on an analysis of illness (and thus became the father of epidemiology) rather than describing the cause of the illness. It was not until 1885 that Robert Koch identified *Vibrio cholerae* as the causative agent of cholera (**Figure 2-4**).

Although it seems obvious to us today, the idea of small living creatures causing disease is a modern concept. Two hundred years ago it was more common to believe that "bad air" or divine judgment was the cause of disease. One of the pioneers of microbiology, Robert Koch, probably did more than anyone else to prove the germ theory of disease—the idea that specific microorganisms caused specific illnesses. Koch's Postulates established the rules for assigning disease to microbes, and they

FIGURE **2-3** **The city of Milwaukee, Wisconsin, was founded at the confluence of three rivers, the Milwaukee, the Menomonee, and the Kinnickinnic.**

FIGURE 2-4 **An electron micrograph of *Vibrio cholerae*, causative agent of Cholera. For many years it was one of the most feared infectious diseases.**

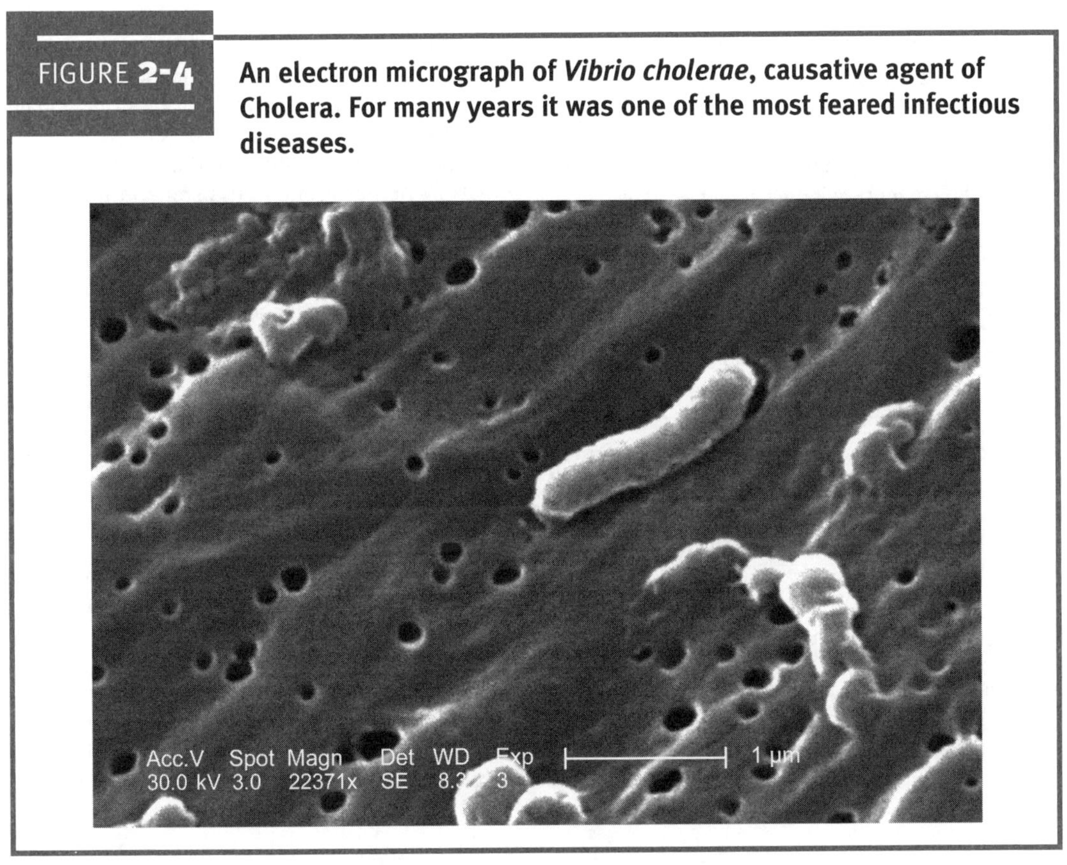

SOURCE: Photo courtesy of the Centers for Disease Control and Prevention Public Health Image Library.

are still very much in use today. After the general acceptance of the germ theory of disease, many more researchers examined water as a vector for communicable disease, and public authorities began to address these threats with engineering techniques. During this time it was also discovered that the sand filters that had been used were also good at removing bacteria. Of course, killing the bacteria is just as good as removing them. In 1902 Belgium instituted chlorine addition specifically for the treatment of biological threats in the water supply. These techniques are still in use.

According to the U.S. Environmental Protection Agency (EPA), the average American family of four uses about 400 gallons of water per day, or about 146,000 gallons annually. Household uses of water include personal hygiene (e.g., brushing teeth, showering, and bathing), drinking, cooking, clothes washing, dishwashing, and outdoor activities such as lawn sprinkling, car washing, and water used in swimming pools (**Box 2-1**). However, household water use is a small fraction of the total fresh water that is needed. Nonhousehold uses include irrigation of farm fields (by far the largest use, requiring about 80% of freshwater use in the United States), industrial and commercial use, and livestock watering. Other uses of water include cooling water for thermoelectric generation of power, navigation, recreational use, and support of the fishing industry (**Figure 2-5**).

BOX 2-1 Water Quality

We demand high-quality household water, even though most of it does not need to be drinking-water quality. It is arguable whether water that is used for drinking and food preparation and water that is in close contact with the body should be potable. But water that is used in the toilets, for washing clothes, or for outdoor use does not need to be as clean. By some estimates toilets account for over 40% of indoor household water use, so toilet water represents a large use of water that does not need to be potable.

SOURCES OF WATER

Water is a renewable resource (**Box 2-2**). Water falls to the earth as precipitation (i.e., rain or snow), and then gravity guides it downhill where it seeks the lowest point and forms streams and rivers, eventually spilling into lakes and oceans. Sunlight warms the water and leads to water evaporation. When sufficient water vapor collects in the atmosphere, clouds form, and the rain falls once again (**Figure 2-6**).

Some water percolates into the soil, also due to gravity. Whether it enters the soil and how far it penetrates is dependent on soil porosity. Water that collects below the ground surface will pool at a low level called the **saturated zone**. In this zone the soil has acquired as much water as its porosity will permit. A conventional well bores into the ground in search of the saturated zone of water (**Figure 2-7**). The empty hole of the well then fills with water and can be pumped out. Many communities still rely on conventional wells; drilling costs are a significant cost of this resource, as is the cost of pumping the water out of the well.

Above the saturated zone, and proceeding to the ground surface is soil that is not saturated with water. This is the **unsaturated** or **vadose zone**. The soil in this layer has some moisture content, but it is usually unsuitable for wells. The line between the saturated and unsaturated zones is referred to as the **water table**. The water table changes constantly as water drains away or as more water percolates through the soil. In times of

FIGURE 2-5 **Water is needed for every area of our lives.**

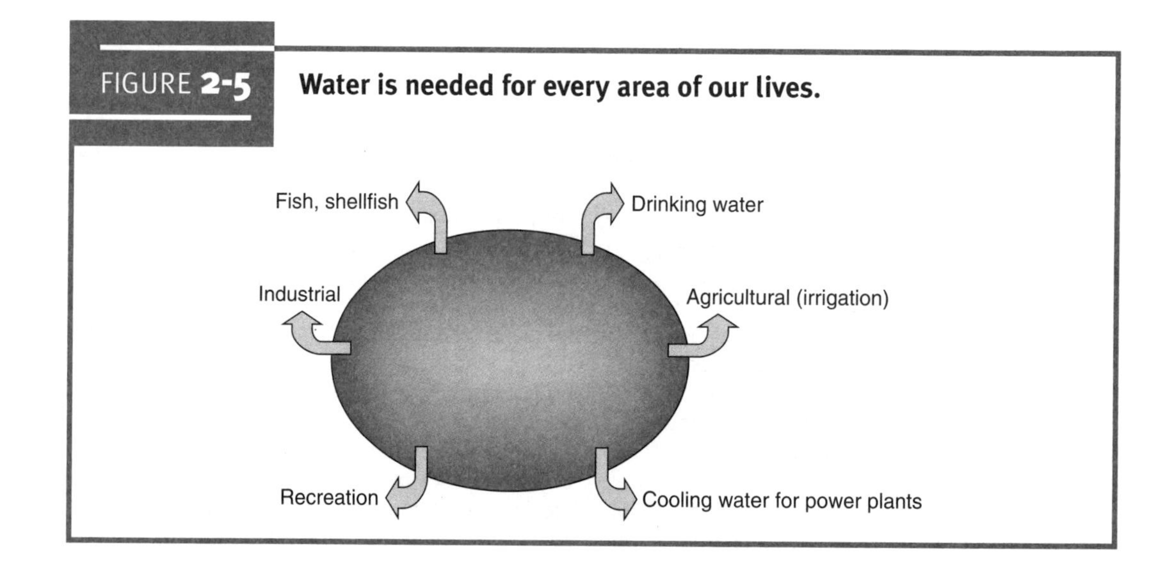

BOX 2-2 All Rivers Run to the Sea . . .

"All the rivers run to the sea, yet the seas are not filled up. From the place where the water flows, there the waters return again."
Ecclesiastes 1:7

The above quotation was written long ago, perhaps 1000 BCE, and is usually attributed to King Solomon. In a simple statement the author accurately describes the hydrologic cycle. That is, the idea that water falls as rain, collects in streams that run downhill, finally emptying into a large body of water (in this case, the Mediterranean) and then, through the process we call evaporation, returns to the atmosphere and falls as rain again.

To the ancients, knowledge of water was absolutely critical. They had to know where to find drinkable (potable) water. Wells were good sources of water, and occasionally disputes over the ownership of a well or access to a well would lead to warfare. Rivers were also good sources of water, especially if the source of the river was a nearby mountain range. A mountain stream is typically a good place to get fresh, cold water.

drought the water table may drop below the bottom of the well, causing the well to run dry.

Water sources are typically divided into two types, depending on where they are located: groundwater and surface water (**Table 2-1**). This is a useful distinction because the two types of water have different rules associated with their use as drinking-water sources. **Groundwater** is defined as water that is not open to the atmosphere and is not subject to run-off. Water **run-off** occurs when rain falls or snow and ice melt and the water runs over the land surface. Some of it will eventually percolate through the soil, and some will follow the easiest downhill course until it meets a receiving water supply, such as a lake. The fact that groundwater is not typically affected by run-off is important because run-off can carry contaminants. Although some run-off eventually reaches the groundwater, the soil typically filters out the contaminants. Notably, groundwater contains a microbial community, but typically in low concentrations of bacteria and viruses. Rarely are pathogens found. Sources of groundwater include wells and aquifers (**Figure 2-8**). One example of groundwater is natural springs, in which the water emerges from the ground at a specific point.

The advantage of using groundwater as a water supply is that the water is often very clean. It will pick up mineral content as it percolates through rock, giving the water "hardness." Notably, however, water that is completely free of dissolved minerals (such as distilled water) is not particularly appealing, having a "flat" flavor. Many bottled waters, including expensive boutique varieties, are natural types of groundwater that impart a distinctive taste. One disadvantage of groundwater is its limited availability. Groundwater recharges as rain and snow seep into the land above and percolate through the soil to reach the water table. This is a slow process, and over-pumping of wells eventually drains the resource. Many aquifers in the United States are in danger of overuse (Alley et al., 1999). This source of water should be considered as nonrenewable for any meaningful time frame (**Figure 2-9**).

Below the saturated zone is a layer of nonporous rock, and for many years it was considered untenable to drill through this very hard rock. However, another source of groundwater is found here—aquifer water. This water has been trapped between confining layers of rock for a very long time, in many cases hundreds of thousands of years. The water is not found in the rock itself but within gravel, sand, or other unconsolidated materials. That is, a material with a significant porosity and free volume.

The advantage of aquifer water is that it is often under pressure, which means that it will be forced upward toward the surface when the aquifer is

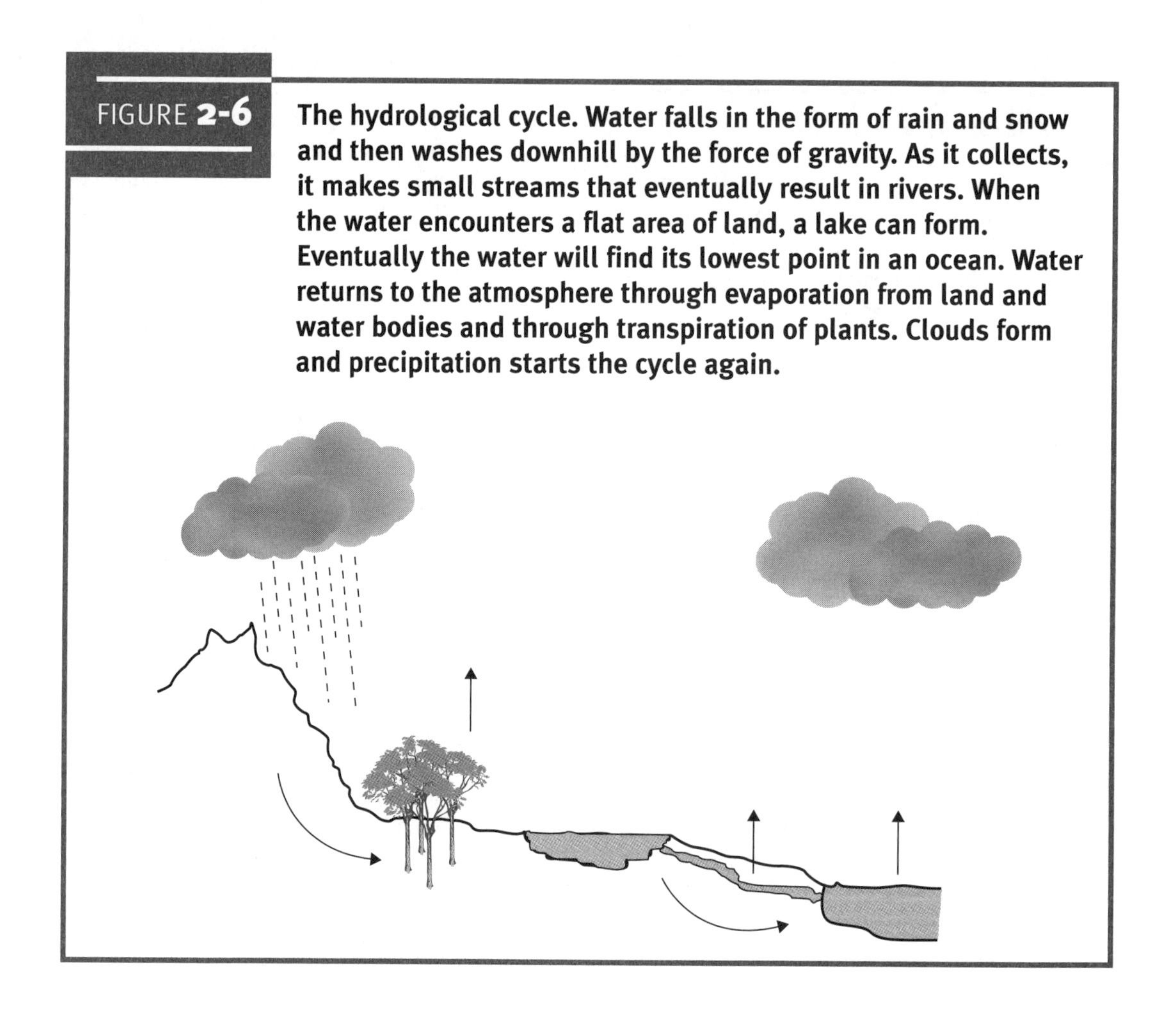

FIGURE **2-6** **The hydrological cycle. Water falls in the form of rain and snow and then washes downhill by the force of gravity. As it collects, it makes small streams that eventually result in rivers. When the water encounters a flat area of land, a lake can form. Eventually the water will find its lowest point in an ocean. Water returns to the atmosphere through evaporation from land and water bodies and through transpiration of plants. Clouds form and precipitation starts the cycle again.**

tapped. This is then known as an **artesian well**. The disadvantage is that the water must be pumped a long distance if it is not under pressure and, more importantly, the water is typically recharged very, very slowly.

Surface waters are very visible and varied, ranging from fast-moving mountain streams to flowing rivers to quiet ponds, lakes, and reservoirs. Not surprisingly, the quality of the water and the dissolved and suspended material in it also vary. The area comprising the water source—the rivers which carry it and its final destination—are referred to as the watershed. Some watersheds have a vast geographic extent.

Flowing water is referred to as a **lotic** system, while still water is referred to as a **lentic** system. Different ecological factors, biotic and abiotic, predominate in each. **Surface water** is defined as water that is open to the atmosphere and subject to water run-off. As the water moves, it can pick up many kinds of contaminants, which makes this distinction for surface water very important. Run-off water that crosses agricultural fields may pick up pesticide residues and fecal material from animals.

FIGURE 2-7 **A well takes advantage of the saturated zone of soil, in which water fills up the available soil pore space. Above this is the vadose, or unsaturated zone of soil. The line between them is the water table. As water is pumped from the well a "cone of depression" is formed, in which the unsaturated zone is below the water table. If too much water is pumped from the well, or if the water table drops, the well will go dry.**

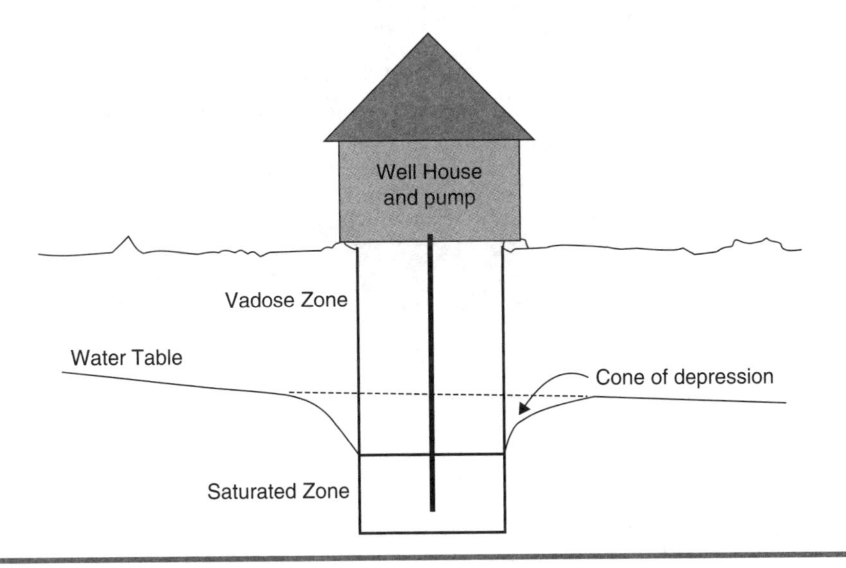

TABLE 2-1 **Characteristics of Surface Water and Groundwater**

Surface Water	Groundwater
Open to the atmosphere	Not open to the atmosphere
Subject to run-off	Not subject to run-off
Easy to contaminate	Hard to contaminate
Easy to clean	Hard to clean

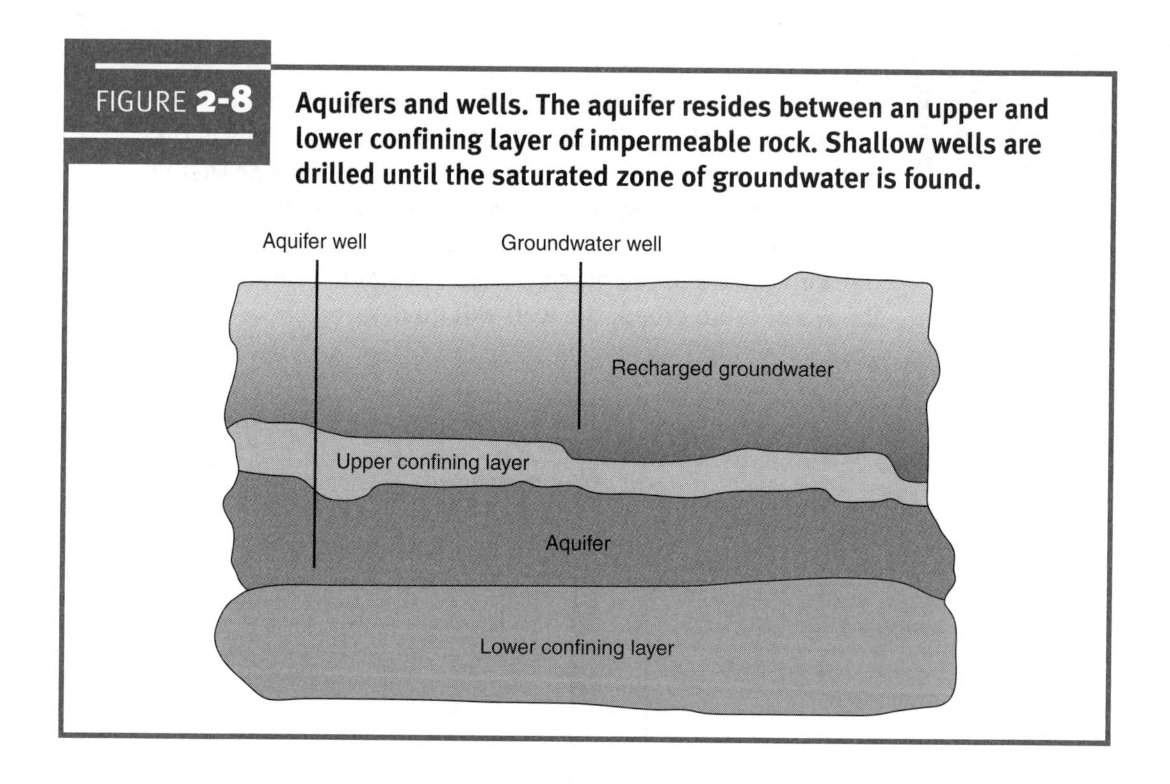

FIGURE 2-8 **Aquifers and wells. The aquifer resides between an upper and lower confining layer of impermeable rock. Shallow wells are drilled until the saturated zone of groundwater is found.**

Run-off in urban areas may pick up antifreeze, oils or other petroleum products, and litter. This is especially important in urban areas where asphalt and concrete cover much of the land, stopping rain and snowmelt from seeping into the ground. Run-off may travel for long distances before reaching another water source and may thereby pick up significant amounts of contaminants. The biota of surface water are typically very robust, and may include fish, crustaceans, amphibians, algae, protozoa, fungi and yeasts, and many, many species of bacteria and viruses. Other animal species (e.g., mammals, reptiles, and birds) may visit the surface water to drink and leave behind their waste products. Pathogens (e.g., bacteria, viruses, and parasites) are a risk in these water supplies. Determining flow of nutrients in such a complex web of life becomes difficult, and the picture changes with seasons and daily weather conditions in this dynamic environment. The water surface is the habitat for photosynthetic microorganisms (i.e., true algae and the cyanobacteria). The depth of this niche is dependent on the penetrating power of sunlight and, because different wavelengths of light penetrate to different depths, the photosynthetic communities vary with depth of the water column. Solid structures within the water are home to biofilms of bacteria, which are typically highly differentiated and home to a variety of microorganisms, and also support many types of stalked and appendaged bacteria (e.g., *Caulobacter* and *Hyphomicrobium*) that depend on the movement of water to bring nutrients to them. The sediments at the bottom of the water supply are also teeming with life due to the nutrients that settle here. Anaerobic zones in the sediment can support many different communities, such as the sulfate-reducing bacteria and, in swampy regions, the methanogens. Convection currents begin with the warming of the surface of the water by sunlight and lead to the mixing of surface and deep

The Ogallala Aquifer is massive, part of 8 states west of the Mississippi River. It has been used for many years as a water resource in the nation's farmland, and is directly responsible for the high productivity of the area. However, it is a nonrenewable resource, and water levels within the aquifer have been declining. The two views in this figure show different aspects of the decline.

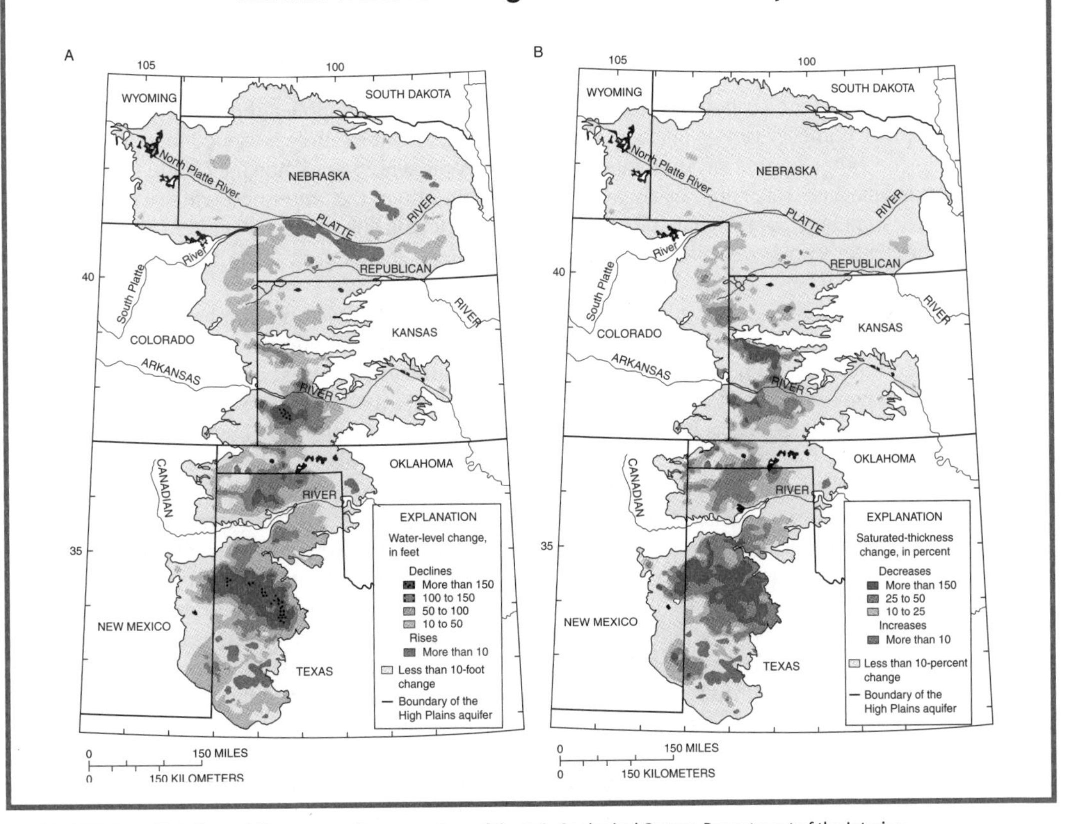

SOURCE: From McGuire and Sharp, 1997. Figure courtesy of the U.S. Geological Survey, Department of the Interior.

waters and the nutrients in each. If the surface water is a river, then rapid mixing of the water and a high degree of oxygenation results as the water is transported downhill.

Another distinction between groundwater and surface water is that surface water is easy to contaminate but relatively easy to clean. The extensive biota and exposure to the environment provides the means for degradation of contaminants. The availability of oxygen and sunlight helps in the degradation process. Groundwater is difficult to contaminate, but once it is contaminated it is very difficult to decontaminate. Past practices of dumping toxic wastes led to infiltration of the groundwater supply. This created cubic miles of contaminated soil and contaminated the water held there. There is no practical way to dig up and treat so much soil. The best options employ a variety of techniques to treat the soil in place, often through bioremediation and the pretreatment of water that is pumped from the site. This can be expensive. Surface water often acquires contaminants through illegal dumping of waste materials or from run-off. This is particularly true in areas where the surface of the land is rendered impermeable to water by structures such as houses and roads.

About 60% of the water distributed through public water supplies comes from a surface water source, while the other 40% is groundwater. But groundwater and surface water are not always mutually exclusive; sometimes one affects the other. In practice groundwater can become contaminated with run-off, as occurs when local flooding infiltrates a wellhead. Because groundwater is typically cleaner than surface water, a special designation is reserved for groundwater that is (or can be) infiltrated by surface water: groundwater under the direct influence (GWUDI) of surface water. Such sources can sometimes be easily seen if the groundwater supply is contaminated by insects, algae, pathogens like *Giardia*, or if the physical characteristics of the water (e.g., pH, turbidity, temperature, or conductivity) change rapidly and in apparent correlation with changes in local surface-water sources.

Nearby surface-water supply might recharge the groundwater supply. If the bedrock is sufficiently fractured, or if the underlying material is coarse sand, gravel, or boulders, water has a greater opportunity to travel to the groundwater source. Rain and snow recharges groundwater by seeping through the soil, which acts as a filter for most contaminants. However, improper disposal of hazardous wastes has been known to contaminate the vast reservoir of groundwater, which can result in cubic miles of contaminated soil and water. Finally, if there is evidence that the aquifer is unconfined, or that the vadose zone is highly permeable, or if the well is poorly designed or maintained (cracks and holes in the well casing), there is the suspicion of water mixing.

WATER SYSTEMS

If you draw water from a well on your property or if you collect rainwater in a cistern, you have a private water system. Very few people are served by these systems, although in rural areas it is not uncommon to have a household well. The **public water system** supplies the water to most people. That is, a water source is tapped for a large number of people who pay a small fee for the treatment and distribution of the water by a water utility. Typically the system is linked to wastewater removal, and the two items are charged to the customer on the same bill.

Public water systems can be either community water systems or noncommunity water systems. A **community water** system provides water to a distinct population (at least 25 people) throughout the year. The typical city or suburban system is a community water system. The vast majority of Americans receive their drinking water from (often very large) community water systems.

A **noncommunity water system** is further divided into transient and nontransient systems. **Transient noncommunity water systems** typically involve businesses, such as campgrounds, gas stations, or restaurants that are not on a community water system and that serve individuals for a short time rather than on a continuous basis. They may be open year-round but serve different people during that time. A **nontransient noncommunity water system** supplies water to a specific population, but only for part of the year. Schools, which are typically open only 9 months per year, would be a typical example. The distinction with noncommunity water systems is that they either serve a variable group of people or are not available year-round.

QUESTIONS FOR DISCUSSION

1. If you wanted freshwater, where would you be likely to find it?
2. Define each of these terms:
   a. Groundwater
   b. Surface water
   c. Vadose zone
   d. Water table
   e. Saturated zone
   f. Aquifer
   g. Community water system
3. Why are wells dug so that the bottom is in the saturated zone? Why can these wells sometimes go dry?
4. Which contaminants would you expect to find in run-off from a city? From a farm?
5. What is GWUDI and how is it treated?
6. What is the source for the water in your own home?
7. Where are issues of water availability likely to emerge in the near future?

References

Alley, W.M., T.E. Reilly, and O.L. Franke. 1999. Sustainability of Ground-Water Resources. U.S. Geological Survey Circular 1186. http://pubs.usgs.gov/circ/circ1186/pdf/circ1186.pdf

McGuire, V.L. and J.B. Sharpe. 1997. Water-level changes in the High Plains Aquifer—predevelopment to 1995. US Geological Survey Water-Resources Investigations Report Number 97-4081. http://pubs.er.usgs.gov/usgspubs/wri/wri974081

CHAPTER 3

WATER TREATMENT

LEARNING OBJECTIVES

- List the major steps in water purification.
- Describe the role of microorganisms in water purification.
- List acceptable water purification alternatives.
- Explain the theory of disinfection.
- Describe how fluoridation inhibits tooth decay and how it is applied.
- Define chain of custody and explain where and when it is needed.
- Define coliform bacteria and describe their use.
- Identify the presumptive, confirmed, and completed tests.
- Describe the other essential monitoring techniques and when they are appropriate.
- Identify the major laws relating to water and wastewater treatment and quality.

KEY TERMS

- Accuracy
- Available chlorine residual
- Biofilm
- Biofilter
- Chain of custody
- Chlorine breakpoint
- Clear well
- Coliform bacteria
- Completed
- Confirmed
- Conventional filtration
- Dental fluorosis
- Disinfection
- Fecal coliforms
- Fecal–oral route of transmission
- Filtration
- Flocculation
- Membrane filter technique
- Multiple-tube fermentation technique
- Precision
- Presumptive
- Quality assurance
- Sedimentation
- Skeletal fluorosis
- Sensitivity
- Specificity
- Total coliforms
- Total Coliform Rule
- Viable but not culturable (VBNC)

TREATMENT GOALS

The goal of water treatment is to produce potable water, that is, water that is safe to drink. Natural water sources are not pure water; they always contain dissolved chemical compounds. Some of these compounds contain minerals that give the water a distinct and pleasant taste, but high concentrations of some minerals (e.g., calcium and magnesium) are unwanted because they give water "hardness." Water with a high degree of hardness leaves mineral stains on kitchen and bathroom fixtures and interferes with the proper action of soaps. In these cases a water softener using ion exchange columns can be used to remove the unwanted minerals within the household or commercial settings. In some industrial settings it is extremely important to remove impurities. For instance, manufacture of biomedical components requires very pure water. Both anion and cation exchange columns are used to create cleaner water at these production sites.

Acceptability of water for consumption is evaluated in short-term objectives, such as avoiding toxic amounts of chemicals and quantities of microorganisms that are high enough to cause disease (i.e., an infectious dose). Finding a toxic dose of chemicals in a public water supply is a very rare event, but finding microorganisms in the water is entirely possible. Safety is also measured in long-term objectives. While some chemical compounds are likely to be dissolved in the water, we wish to keep their concentrations low enough so that they do not, over time, have a deleterious effect. That is, we want to severely limit their potential to cause cancer over a lifetime. Drinking water quality is thus based on several factors: concentrations of microorganisms (especially pathogens), concentrations of a variety of chemical compounds (e.g., organics, inorganics, and heavy metals), pH, physical attributes (e.g., color and turbidity), and subjective (or organoleptic) qualities such as taste and smell.

One means of achieving the goal of potable water would be to remove all constituents from water and to sterilize it. This is impractical for several reasons. First, it is needlessly expensive, and cost is always a major consideration. Second, it is ultimately futile: water distribution systems are vast, and contaminants are inevitable. Third, few people really want sterile, distilled water. It tastes flat; people consistently choose "tasty" water that contains dissolved minerals. In fact, sales of mineral water in the supermarkets are quite brisk. So the objective is to remove the pathogens in the water supply and to either remove the toxic chemicals or to reduce their concentrations to a level that is safe for people to drink over a long period of time (i.e., a lifetime).

This task is compounded by the vast scale of this work. Cities may use hundreds of millions of gallons of potable water every day, and an interruption in water supply would be devastating (**Figure 3-1**). Water must be available constantly, and the people being served must have confidence that water quality will not vary significantly. If you were confident that the drinking water was safe only 10% of the time, would you drink it at all? Providing a safe and continuous water supply requires significant use of engineering, hydrology, chemistry, and microbiology.

Effective maintenance of the watershed that contributes to a surface water supply is another facet of the job. In earlier times the purity of water was only considered once the water entered the treatment plant. However, it has become clear that attention to the quality of water in the watershed contributes to better water quality and a cheaper treatment process. Therefore, purity of surface water requires effective management of large areas of land. For instance, agricultural areas may be responsible for run-off that contains large amounts of fertilizer, which can create algal blooms. Management of fertilizer deposition provides the needed field treatment but leaves little extra for run-off to carry away (Hapeman et al., 2002). Another critical task is monitoring of the finished water product. Effective monitoring detects water problems before the water is consumed, sparing the population from illness. In addition, the involvement of all stakeholder groups allows different perspectives to be addressed before they become politically difficult.

The major problem for production of potable water is the mixing of wastewater with fresh water, which is referred to as the **fecal–oral route of transmission**. This describes the contamination of drinking water with fecal material. People who drink the water have a high risk of becoming ill, and infected people will excrete the infecting pathogen with their fecal waste, which may then reenter the water supply. Although it may be unpleasant to

FIGURE 3-1 **The Milwaukee (Wisconsin) Water Works purifies millions of gallons every day using water from Lake Michigan.**

SOURCE: Courtesy of the Milwaukee Water Works. These photos may not be used to infer or imply Milwaukee Water Works or City of Milwaukee endorsement of any product, company, or position.

think about, the fact is that wastewater is mostly water and flows downhill until it meets another water source. At that point it contaminates the water supply. This can happen, for example, when septic tanks overflow and infiltrate well water. Some of the routes of contamination are shown in **Figure 3-2**. In the past, wastewater was discharged to streams and rivers, where it contaminated the water supply of people downstream. In many parts of the world this practice continues. Breaking the fecal–oral route is absolutely essential to public health; safe water must be available, or all other attempts to improve health will be inadequate.

In communities where the sanitary sewer system is connected with the storm sewers, there is a risk that heavy storms (rain or snowmelt) will flood the sanitary sewer and cause backups (**Figure 3-3**). Rather than allowing the fecal material back up into homes, it is preferable to release the wastewater into the receiving water system (e.g., lake or river). This is the lesser of two evils and may lead to contamination of the drinking water supply if another (downstream) community uses that receiving water as a source of fresh water. The water and/or sewage infrastructure is always at risk. Mistakes can occur at wastewater or water treatment facilities. Broken pipes can easily lead to cross contamination of the water supply, and the extent of the distribution system is so vast that continual vigilance is needed to replace old pipes and to spot trouble before a problem leads to disease.

Unlike the chemical contaminants, which are detected individually, routine monitoring of the

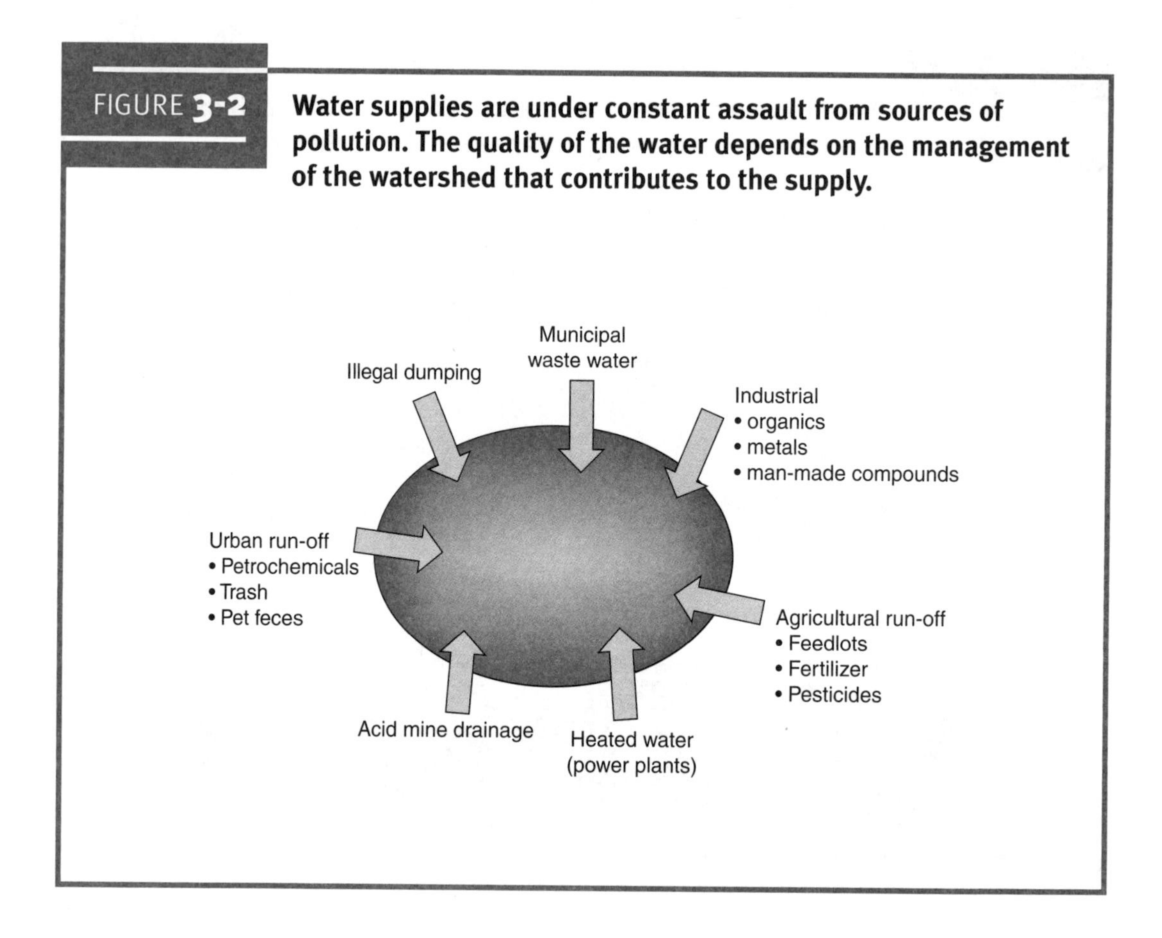

FIGURE 3-2 Water supplies are under constant assault from sources of pollution. The quality of the water depends on the management of the watershed that contributes to the supply.

pathogens is not performed in the course of water treatment. One reason is that it is very difficult to cultivate and positively identify many pathogenic species. Some true pathogens look much like the nonpathogenic species, and some pathogens cannot really be identified with any assurance in a timely and cost-effective manner. For instance, detection of Norovirus requires a molecular method, such as polymerase chain reaction. This technique requires a highly skilled technician, and enumeration is dependent on a well-controlled assay. This is beyond the capacity of most water utilities.

To avoid the difficulty of finding and enumerating all pathogens, two methods are used. The first is a Treatment Technique (TT). The TT is dependent on published research in which the technique has been evaluated for its ability to remove pathogens from the water supply. If this technique is then used by the water utility, it is assumed with a high degree of confidence that the same efficiency of removal will occur. The second technique is to use indicator microorganisms as surrogates for all other pathogens. The use of coliform bacteria is described later in this chapter.

THE PROCESS OF WATER TREATMENT

Water purification is not especially difficult. Even the most challenging water purification problem can be handled in a continuous-flow system at a reasonable cost. In fact, the average cost of household water is approximately four cents per gallon. This chapter provides an overview of water treatment details similar to those that might be used for

FIGURE 3-3 **Sanitary sewer overflows are a menace in some communities. Having separate sanitary and storm sewers is preferable to a combined system.**

SOURCE: © Mikhail Malyshev/ShutterStock, Inc.

a surface water supply. A schematic is shown in **Figure 3-4**. Depending on the water supply and the history of contaminants in that water, the public water system (PWS) might not use certain steps. Therefore, a full description of the individual steps and alternative approaches to water treatment are presented.

The first consideration is the source of the water. Aquifer water is often very clean. If it is under pressure, no pumping is needed. Surface water may come from a river system; water quality is variable. If this water is confined in a reservoir, then the water will run downhill into the PWS. If the water source is a lake, then there is often sufficient water for use, although substantial pumping is required to raise the water to a level at which it will flow to the community. This is significant because pumping costs may be a major factor in the cost of water purification. Groundwater supplies are frequently very clean and might need minimal processing. Soil has a tremendous ability to filter out microorganisms as water percolates through it. The only major exception is viral particles, which do seem to get through to groundwater supplies. Bacteria stuck in the soil gradually die or are consumed by protozoa. Chemical contaminants can bind to soil particles, especially clay particles.

When water enters the treatment facility, it passes through a rough screen to remove large objects that might have been pulled in with the

FIGURE 3-4 A schematic of water purification. A water source (here, surface water) is initially screened to remove large objects, then subjected to coagulation and flocculation. The sludge is removed through sedimentation and passed through a fast sand filter. The disinfection step usually involves chlorination or chloramination. The final step is the addition of fluoride and then the water can be stored in a clearwell for extended Ct time or sent through the distribution system.

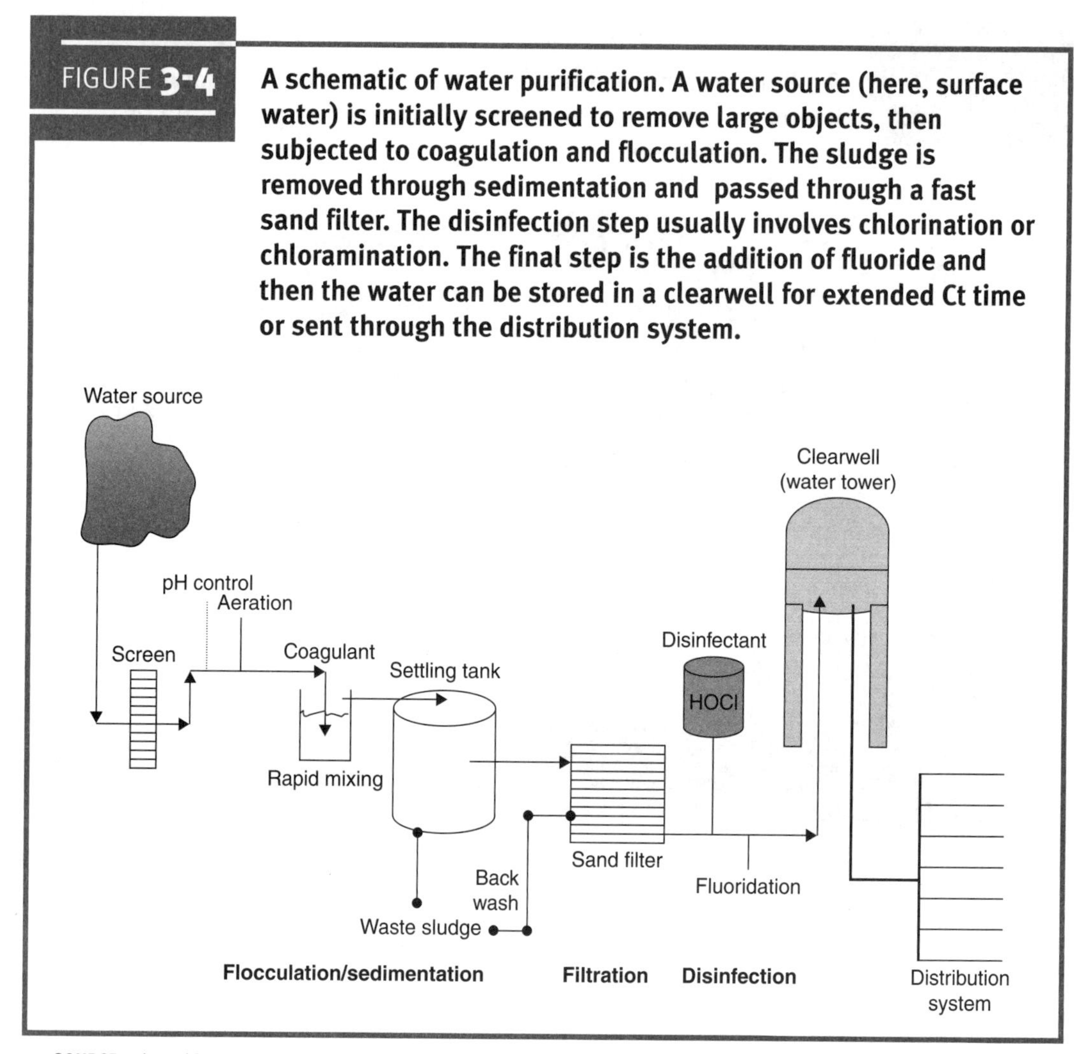

SOURCE: Adapted from Hammer and Hammer, 1996.

water (e.g., rocks, branches, plastic, and other debris). If carbon dioxide or hydrogen sulfide concentrations in the water are too high, then air stripping is used to expel these gases. In the unlikely event that the pH is outside of normal range, an adjustment can be made to this factor as well. Water treatment proceeds in three main phases: flocculation/sedimentation, filtration, and disinfection. After the disinfectant is added, the water is held in a storage tank called a **clear well** until it enters the distribution system; at this stage the water is said to be "polished." The water may also be fluoridated at this step.

Flocculation/Sedimentation

Raw water often contains many suspended solids. Some of these, like sand particles, are expected to sediment fairly quickly and are of little concern. However, many fine particles and colloids sediment very slowly. In order to accelerate this process, a coagulant is added to the water (the **flocculation** step). The coagulant creates larger particles (flocs), typically having a higher density that sediment faster. The most common coagulant is alum (hydrated aluminum potassium sulfate or $KAl(SO_4)_2 \cdot 12H_2O$), although other aluminum or iron salts can be used, and synthetic coagulants are now available. The coagulant is added to water as it enters a large holding tank. It is mixed rapidly and then exits to another tank where the water slowly mixes. The floc settles to the bottom of the tank in this **sedimentation** step. The sediment or "sludge" must be periodically removed and discarded. Some modern facilities have a means to rake the bottom while still in operation. Otherwise, the tank must be drained and the sediment must be shoveled out.

A large percentage (perhaps 99%) of bacteria in the water can be removed in this manner because bacteria cling to the solid surfaces that are removed in this step. Free-swimming cells, including algal cells, are not removed to the same extent. Algae can be a particular problem for water treatment because the cells tend to clog the filtration system.

Filtration

The next step is **conventional filtration**. The water passes through a column of sand or other permeable material—a rapid sand filter (**Figure 3-5**). The sand is rather coarse. The Effective Size of the sand particles is at least 0.5 mm, and the Uniformity Coefficient is less than 1.5 mm (**Box 3-1**). This is a large pore size, allowing water to travel through

FIGURE **3-5** **Filtration beds for water purification.**

SOURCE: Courtesy of the Milwaukee Water Works. These photos may not be used to infer or imply Milwaukee Water Works or City of Milwaukee endorsement of any product, company, or position.

BOX 3-1 Terminology

Sand particle size is important in these processes because it determines pore size. Pore size determines whether biological activity will be effective (whether the nutrients will contact the biofilms) and how fast the water will travel through the filter. In order to define sand sizes, two parameters are determined.

Effective Size refers to the diameter of the sand particles. A stated Effective Size means that 90% of the sand particles are greater than this diameter, while 10% are smaller. It is sometimes referred to as D10. This is conveniently measured using a sieve.

The Uniformity Coefficient refers to the distribution of sizes of a sand sample. This number is calculated as the Effective Size for passage of 60% of the sand particles (the D60) divided by the D10 value.

while catching the flocculent material. Throughput is high, with rates that can reach 15 $m^3/m^2/h$ (compared to slow sand filtration, described later). This is roughly 15,000 liters of water per hour, or at least 3,500 gallons per hour. In practice, the rapid sand filter can remove as much as 90% of coliform bacteria and 90% of parasitic oocysts and spores. Occasionally the filter becomes clogged with sludge or algae and must be cleared by backwashing, and the accumulated sludge is discarded. Because of the higher flow rate through these filters, the biologically active zone can be half a meter deep in the sand layer. During backwashing, the sand bed may become fluidized, creating a risk that the sediment bed will be disrupted. To counteract the backwashing problem, a layer of crushed anthracite is often placed on top of the sand.

Some biological activity can be seen in the conventional filter, but it is a minor component of the method. (This contrasts with slow sand filtration described in the following section.) In facilities using this purification method, a disinfection step is necessary. Conventional sand filtration takes up little space and is an excellent choice for large and very large PWSs. However, it is ineffective in dealing with issues of taste, odor, and color of the water.

Flow rate is determined according to Darcy's Law, which can be rendered in a simple form:

$$Q = \frac{AK\Delta h}{L}$$

where Q = flow rate in m^3/h

A = cross-sectional area of the filter (m^2)

K = hydraulic conductivity (m/h), which is dependent on fluid viscosity and density

Δh = head loss (m), which is a loss of energy of the water pressure due to frictional forces

L = distance through the sand medium that the water must travel (m)

Note that all length measurements are in meters and that 1 m^3 = 1000 l. Other equivalent units (i.e., English measurements) can also be used. In addition, flow rates may be so slow that a time unit of days is used. When the flow rate declines and the efficiency of water purification decreases, it is time to backwash the system. This equation is used to calculate optimum flow conditions and, more importantly, to determine when flow is no longer optimal because of clogging.

Disinfection

The third and final step in water system purification is **disinfection**. In this step a disinfectant, often a chlorine compound, is added to the water to kill any harmful microorganisms that may remain (**Box 3-2**). Usually these substances are both oxidizers and disinfectants. A disinfectant reduces the total number of microorganisms in the water, but does not typically sterilize the sample. The disinfectant does not discriminate between live cells and mere detritus in the water. For that reason it is important to purify the water as much as possible before adding disinfectant, otherwise much more disinfectant is needed, and additional problems with disinfection by-products (DBPs) are created (**Box 3-3**). In other words, if microscopic particles remain in the water, the disinfection process is more difficult. Bacteria and viruses can hide within small crevices of these

BOX 3-2 Emergency Iodination of Water

Halides other than chlorine can be used. Bromination has been used to disinfect water for swimming pool use, but it is not a suitable method for treating drinking water. Iodination is another method that is often used for drinking water purification in the field. Camping supply stores carry iodine tablets that can be added to a canteen of water from a field site such as a mountain stream (not common pond water). After about 30 minutes, the water is safe to drink. This is effective because the iodine concentration combined with the time factor is sufficient to kill the microbes, and in this case a very high concentration of disinfectant is used. If the same concentration of chlorine was used, the water would taste like swimming pool water—completely unpalatable.

particles, affording them some protection from disinfection compounds. This is a major reason that coagulation and sedimentation are needed for effective water treatment.

The Theory of Disinfection

Many people are familiar with disinfection through their use of chlorine tablets in swimming pool maintenance. However, there is much more to disinfection than simply adding chlorine to a water sample. The effectiveness of a disinfectant is measured by how well it kills pathogens of interest (**Table 3-1**). This is dependent on the concentration (*C*) of the disinfectant and the contact time (*t*) of the disinfectant in the water sample. Together they create the *Ct* value. A high concentration within a short time is thus equivalent to a low concentration over a longer period. Often this value is expressed in terms of treatment goals, such as Ct_{99}, which indicates the concentration and time needed to reduce the pathogen concentration by 99% (that is, two orders of magnitude). $Ct_{99.9}$, then, indicates a reduction of three orders of magnitude.

BOX 3-3 Disinfection Byproducts

Disinfection byproducts (DBPs) are a great concern whenever chlorine compounds are used. Their concentrations in the water supply are regulated by federal law. These include the trihalomethanes (e.g., chloroform, bromodichloromethane, dibromochloromethane, and bromoform), the haloacetic acids (e.g., monochloroacetic acid, dichloroacetic acid, trichloroacetic acid, monobromoacetic acid, and dibromoacetic acid), bromate (BrO_3), and chlorite (ClO_2).

trichloroacetic acid

DBPs are created when the very reactive chlorine combines with other compounds in the water supply. Removing dissolved and suspended organic matter as much as possible before the chlorination step is advantageous for this reason. DBPs are associated with an unacceptably high risk of carcinogenicity (e.g., bladder cancer). There may also be concerns regarding reproductive and developmental effects.

If DBPs are a problem, they can be removed from the water supply, but this is an expensive process. Removing more organic compounds before the disinfection step can usually be accomplished with simple engineering. Use of alternative disinfectants, such as chlorine dioxide or ozone, can be effective.

TABLE 3-1 Disinfectant

Pathogen	Disinfectant	Ct (mg · min/liter) (or range)	Log reduction	pH, temperature	How assayed	Reference
Helicobacter pylori	Na hypochlorite	0.12	2	6, 5°C	Agar cultivation	Johnson et al.
Norovirus	Na hypochlorite	0.66	2	6, 5°C	RT-PCR	Shin and Sobsey
Poliovirus	Na hypochlorite	10	2	6, 5°C	RT-PCR	Shin and Sobsey
Cryptosporidium parvum	Chlorine dioxide	75–1000	2	8, 21°C	MPN cell culture	Chauret et al.
Burkholderia pseudomallei	Na hypochlorite	0.7	2	8, 5°C	Agar cultivation	O'Connell et al.
Burkholderia pseudomallei	Monochloramine	204	2	8, 5°C	Agar cultivation	O'Connell et al.
Adenovirus	Chlorine dioxide	0.49 – 0.74	4	6, 15°C	TC infectivity	Thurston-Enriquez et al.
Feline Calicivirus*	Chlorine dioxide	4.2 – 6.7	4	6, 15°C	TC infectivity	Thurston-Enriquez et al.
Cryptosporidium parvum (Maine)	Na hypochlorite	11100	2	7.5, 22°C	TC infectivity	Shields et al.
Cryptosporidium parvum (Iowa)	Na hypochlorite	6900	2	7.5, 23°C	TC infectivity	Shields et al.
Adenovirus 2	Na hypochlorite	0.02	2	7, 5°C	TC infectivity	Cromeans et al.
Adenovirus 2	Monochloramine	600	2	7, 5°C	TC infectivity	Cromeans et al.
Coxsackie virus B5	Na hypochlorite	3.6	2	7, 5°C	TC infectivity	Cromeans et al.
Coxsackie virus B5	Monochloramine	270	2	7, 5°C	TC infectivity	Cromeans et al.
Echovirus 1	Na hypochlorite	0.96	2	7, 5°C	TC infectivity	Cromeans et al.
Echovirus 1	Monochloramine	8	2	7, 5°C	TC infectivity	Cromeans et al.
Murine Norovirus*	Na hypochlorite	< 0.02	2	7, 5°C	TC infectivity	Cromeans et al.
Murine Norovirus*	Monochloramine	26	2	7, 5°C	TC infectivity	Cromeans et al.

NOTE: TC = tissue culture.

* A surrogate for Norovirus.

A high concentration with a short time period is equivalent to a low concentration with a long time period. This is important because disinfectants are not particularly healthy to ingest and their taste is unpleasant; therefore, it is best to keep the concentration as low as possible while still accomplishing the goal. However, there are practical limitations. Enough disinfectant must be present to kill the pathogens. Using just enough to do the job is also not practical because the last molecule of disinfectant cannot be specifically measured to kill the last pathogen in the sample. At a certain concentration the disinfectant is in such low supply that its chance of encountering the pathogen is vanishingly small. The time factor also has limits. Perhaps a low concentration of disinfectant would be sufficient if the time factor was one month. But in most cases this is far too long. The greatest practical length of time is probably overnight, perhaps 12 hours. In practice, the PWS will calculate how long, on average, the water stays in the clear well and how long, also on average, the water stays in the distribution system before being utilized by a consumer (**Figure 3-6**). Water distribution systems

FIGURE **3-6** **A water tower is useful for two reasons. It is a clearwell that holds a water supply for an extended period, allowing longer times for disinfection to work. It also holds the water column above the distribution system and therefore provides water pressure household use. Even if the power fails the water pressure is usually present. Taller buildings would need their own pumps to force water higher than the water tower.**

SOURCE: Photo courtesy of John W. Burlage.

are not free of bacteria; they certainly grow in the water pipes and they can, over time, become a problem. Keeping the residual chlorine level at 0.2 ppm is helpful in keeping bacterial numbers low. Routine monitoring of the water distribution system is also required.

Within the water system pipes, the bacteria are not merely free-swimming, or "planktonic." They attach to the walls of the pipe and create a **biofilm**. A biofilm is a physical structure attached to a hard surface that is composed of microbial species and their extracellular products. Practically any hard surface is suitable for colonization by bacteria. In the natural environment, bacteria stick to soil particles, vegetation, rocks, or any other surface. Practically any bacterial species can become part of a biofilm, and bacteria may trap virus particles as well. Because biofilms are somewhat porous and are certainly a rich food source, protozoa are likely to be associated with biofilms, even if only along the surface.

Biofilms probably begin with a few pioneering bacteria that attach to a surface (**Figure 3-7**). If conditions are favorable, a microcolony will form and the bacteria will produce an extracellular polymeric substance (EPS), which is a thick layer of carbohydrates in polymer form. Like the capsule that surrounds individual cells, this layer protects the bacteria from predation and from toxic chemicals. The residual chlorine content in water pipes is usually insufficient to kill bacteria in biofilms. The presence of these microcolonies and their EPSs attracts other species of bacteria, which may simply become trapped in the EPS and thus add to the colony. Eventually the structure grows into a mature biofilm with many species and a diverse EPS composition. From this structure some bacterial cells will break off and become planktonic, eligible to colonize other sites.

The complexity of these biosystems makes them extremely difficult to model. The conditions that shape the biofilm structure are difficult to replicate in the laboratory with any degree of precision. Previously, the biofilm was thought to be rather solid, a simple slime layer. The current model is more porous, with globs of EPS separated by fluid-filled spaces. This porosity allows nutrients to flow through the EPS and feed bacteria throughout the biofilm. In some cases the biofilm is so thick and complex that different species achieve a selective advantage. Anaerobes may take over near the hard surface and away from the oxygenated fluid. A community of different final electron-accepting organisms may become established that efficiently utilizes any nutrients in the bulk fluid. Here again, the complexity and the range of environmental conditions are vast.

Biofilms are a medical problem because bacteria forms on both biotic and abiotic surfaces, such as foreign objects in the body (e.g., wood splinters that get stuck in the skin) or medical devices (e.g., catheters, artificial joints, or stents) that are inserted into the body. On these objects bacteria have the advantage of being somewhat removed from the immune system and therefore free from destruction. Although the immune system eventually destroys something as small as a wood splinter, a biofilm on a metal hip joint presents a greater problem. Antibiotics have a hard time reaching the biofilm on a foreign object inside the body. Over time some cells break free and start a systemic infection all over again. The best treatment in this case is probably the removal of the contaminated metal structure. Biofilms can form on biotic surfaces as well, leading to persistent infections. Here again, attack by the immune system or with antibiotics is more difficult.

Biofilms have practical benefits. Free swimming uses energy, and if the bacteria only drift, they will quickly utilize any nutrients within reach. Predation is also a concern because protozoa have little trouble consuming free-swimming bacteria. However, within a biofilm, bacteria are protected from predation and from toxic chemicals. In addition, the environment constantly changes for biofilm bacteria. The fluid that passes the biofilm may bring with it nutrients, which diffuse through the surface and feed the bacteria. Little effort is required on the part of the bacteria.

The most common disinfectant for water systems is chlorine, which is both a strong oxidizer and a disinfectant. It is available as a solid, calcium hypochlorite [$Ca(OCl)_2$], and as a liquid sodium hypochlorite [$NaOCl$], which is shipped in a concentration of 5–15%. When chlorine gas reacts with water, it forms hypochlorous acid, HOCl:

$$H_2O + Cl_2 \rightarrow HOCl + HCl$$

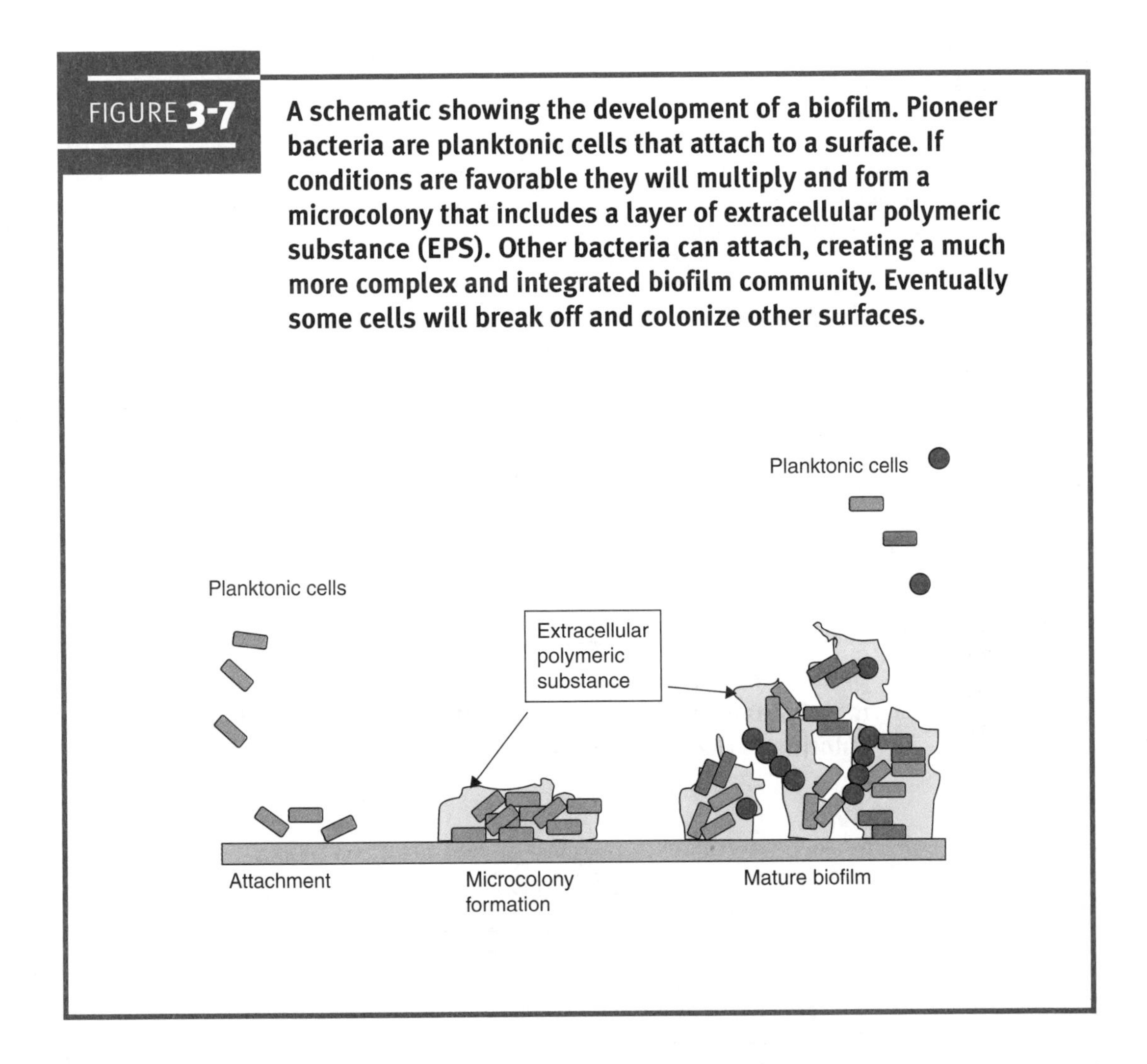

FIGURE 3-7 A schematic showing the development of a biofilm. Pioneer bacteria are planktonic cells that attach to a surface. If conditions are favorable they will multiply and form a microcolony that includes a layer of extracellular polymeric substance (EPS). Other bacteria can attach, creating a much more complex and integrated biofilm community. Eventually some cells will break off and colonize other surfaces.

The stability of hypochlorous acid is dependent on the pH of the water. At pH < 7 it is mostly HOCl, which is preferred. At pH > 8 it is mostly ^{-}OCl (hypochlorite ion) which is a much weaker oxidizer than HOCl. High pH greatly increases the *Ct*. As a neutral compound, HOCl probably has an easier time crossing the membranes of cells, which allows it to be more effective against the proteins of the cell. In contrast, ^{-}OCl is negatively charged and crosses the membrane with more difficulty.

Sodium hypochlorite combines with water:

$$NaOCl + H_2O \rightarrow HOCl + NaOH$$

The use of sodium hypochlorite here creates sodium hydroxide, a powerful alkali. This raises the pH, which is a disadvantage to the disinfection process. Therefore, appropriate control of the final pH is necessary.

How much chlorine to add to a system is dependent on many factors (Wang, 2006). One factor is the amount of reactive material in the water. Organic matter, sulfides and ferrous iron, and ammonia are all in this category. Most should have been removed in the purification process, but some concentration always remains. Remaining organic material is also capable of interacting with the chlorine disinfectant. Therefore, the chlorine demand of the system must first be determined, that is, how much chlorine is needed to oxidize reactive components in the water. Initially the added chlorine will result in the creation of chloramine compounds, but subsequent reactions will then break down these chloramines, giving the available chlorine graph a curved look. As shown in **Figure 3-8**, the initial addition of chlorine results in no residual chlorine (Zone 1); this is the area of immediate demand for chlorine, where it reacts with reducing agents. Further addition of chlorine results in chlorine residual (Zone 2) as chlororganics and chloramines are formed. This is followed by a decrease (Zone 3) in which the newly formed compounds are destroyed. Next the **chlorine breakpoint** occurs, in which free chlorine is available in the water supply as the **available chlorine residual**. Chlorine residual then accumulates (Zone 4). This factor is composed of chlorite, the chloramines, and related reactive chlorine-containing molecules. It is desirable to leave a small concentration of chlorine in the water supply in order to guard against the regrowth of bacteria within the pipes. This residual concentration should be at least 0.2 ppm (mg/l) even in the farthest point of the water distribution system. Leaving larger amounts of chlorine might seem like a good precaution, but too much chlorine makes the water unpalatable.

The objective of disinfection is the killing of target microorganisms over a period of time with a specific disinfectant. Conditions of the environment

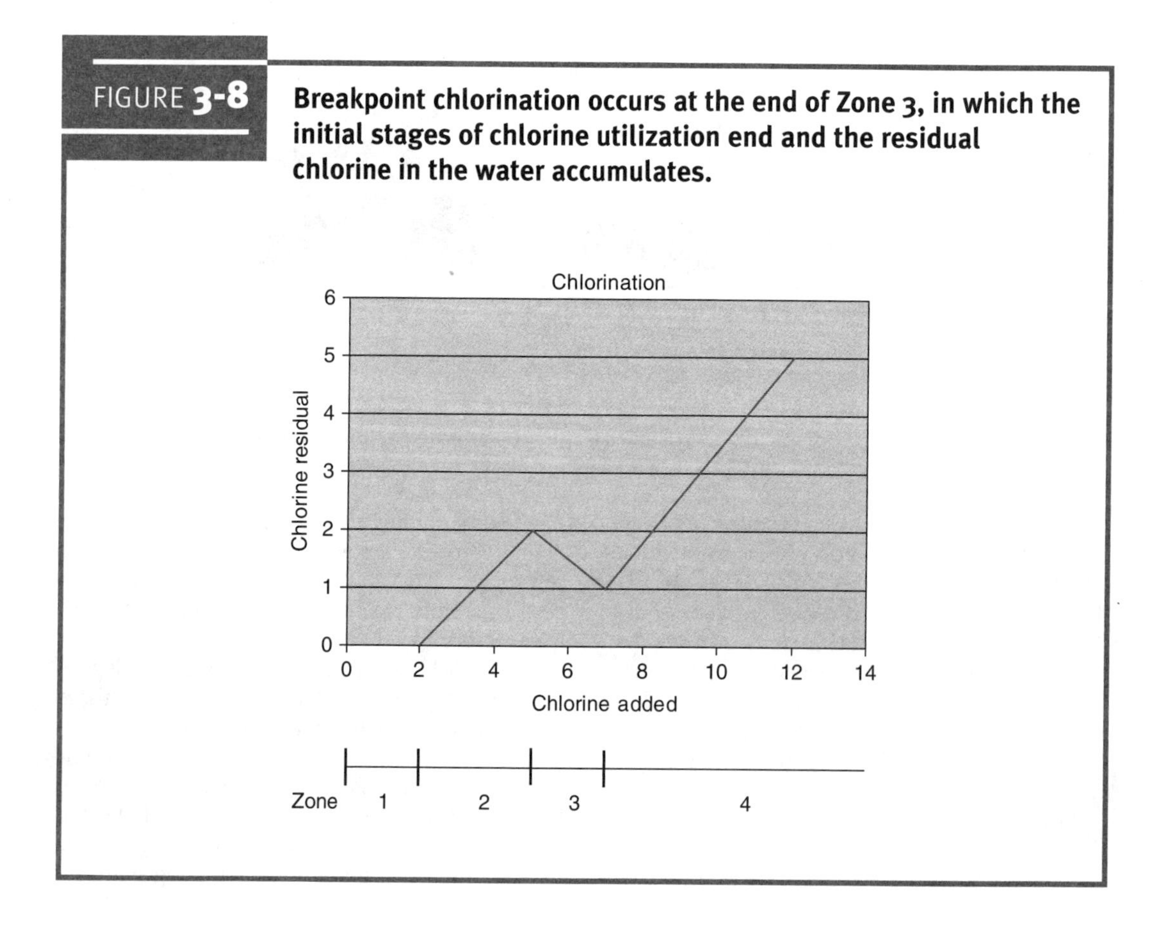

FIGURE 3-8 **Breakpoint chlorination occurs at the end of Zone 3, in which the initial stages of chlorine utilization end and the residual chlorine in the water accumulates.**

may greatly affect this process, such as presence of other substances (turbidity), temperature, and pH.

The general theory for disinfection was first proposed by Harriet Chick (1908) and refined by Watson (1908). The general law (simply called Chick's Law) is

$$-k[D] = \frac{\ln \frac{N_t}{N_0}}{t}$$

where

N_0 = number of microorganisms at time = 0
N_t = number of microorganisms at time = t
t = contact time
$[D]$ = concentration of disinfectant
k = coefficient of lethality, a constant for the system

Note that, as the disinfectant concentration increases, the time decreases and vice versa. The bacterial factors produce a unitless number; either numbers or concentrations can be used. The k value must be derived empirically. It changes with the disinfectant and the target microorganism, and perhaps with the environmental conditions. It carries units of liter · mol^{-1} · min^{-1} (or equivalent values in keeping with the $[D]$ and t values). Typical values range from 10^3 to 10^6.

SIGNIFICANT PROCESS ALTERNATIVES

Direct Filtration

In conventional filtration a sedimentation step precedes filtration. Most flocculent material settles in the sedimentation tank as sludge. In direct filtration the coagulant is mixed with the water, but the sedimentation phase is skipped; the water goes directly to a rapid sand filter. This is possible if the source water has a low turbidity (i.e., little suspended material) at the outset. The sand filter eventually clogs, but it can be backwashed to remove most of the sediments.

Slow Sand Filtration

Slow sand filters are probably the oldest technique for purifying large amounts of water; the process is now about 200 years old. Its construction is quite simple. The bottom of a tank is fitted with a drainage system for the water that gets through the filter. This is overlaid with large gravel, and then 2–5 feet of coarse sand are placed on top. This sand is finer than the sand used in rapid sand filtration, with an Effective Size between 0.15 and 3.0 mm and a Uniformity Coefficient less than 2. The water is allowed to trickle over the top of the filter, and gravity pulls it through. If there is a problem with dissolved toxic chemicals (e.g., pesticides, organics, or halogenated compounds), then a thin layer of granular activated carbon (GAC) can be sandwiched between sand layers. Chemicals that get through the sand are adsorbed into the GAC. Notably, bacteria, including pathogens, can also stick to the GAC and create biofilms. However, because this step is usually followed by a disinfection step, the problem can be avoided.

These filters work on the basis of two technologies: physical removal of particles that are caught in the pores of the filter, and biological removal of dissolved and suspended materials. This latter step is a key feature. Sand alone traps particles, and, as they do so, the bacteria and protozoa in the input water get trapped there as well. Bacteria form biofilms on the surfaces of the sand particles and are then able to efficiently remove nutrients from the water as it passes. In effect, this is a **biofilter**, commonly called the *Schmutzdecke* (German, literally "dirt blanket"). Protozoa eat many of the bacterial cells. Other microorganisms, such as parasite cysts and viral particles, can get caught in the biofilm matrix. Biofilms typically create large masses of EPS, which is a sticky or slimy material. If the pathogens are not consumed by some other microorganism, they may stay in the filter until they die. Either way, they are removed from the water, which flows through to the next step. It is not unusual to reduce bacteria and viruses by three orders of magnitude with this method. In colder weather the efficiency is reduced because bacterial physiology is slower at lower temperatures.

A water column is always placed above the filter so that the biofilter does not dry out. If the biofilter is utilized too long, the organic matter penetrates further into the sand column. This overgrowth of microorganisms tends to use all the dissolved oxygen, creating a nitrifying environment. This is not optimal; the water should have some degree of oxygenation. If maintained properly, the water that exits the filter is usually of good quality, although disinfection is still needed. Issues of color, odor, and taste are usually handled well with this system. Reports of 95% removal of coliform bacteria and 99% of parasitic spores and cysts show that this is a very good system for water purification.

Slow sand filters depend solely on gravity to pull water through the system. If too much suspended material is not removed during the sedimentation step, it will get caught in the filter. If maintained correctly the filter will still occasionally clog, and the growth might not be removed adequately by backwashing. This is evident by the decreased volume of water exiting the column. However, because the bulk of the biofilm is in the top few centimeters of sand, this layer can be simply scraped off, and the filter is thus reactivated. When either event occurs, the top layer can be skimmed off or the filter is backwashed to remove excess material. Removal of layers eventually reduces the sand column and thus requires replacement, but several layers can typically be withdrawn before this becomes necessary.

The major and predictable disadvantage of the slow sand filter is that it is slow. Water output is typically on the order of 0.1 $m^3/m^2/h$, which is 0.1 cubic meter of water per square meter of sand filter surface area per hour. That is, about 100 liters of water per hour, or about 25 gallons. If the system runs continuously, it processes 2,400 liters per day, or about 600 gallons. If a medium-sized city needed 100 million gallons of water per day, then it would need a slow sand filter bed in excess of 160,000 m^2, or the equivalent of 30 football fields. The area required is too large for most cities. For smaller towns this may be an acceptable tradeoff, especially because the slow sand filter is very economical.

The biofilm feature of the filter reveals a key weakness of this technology. If toxic chemicals get into the system, the biofilm can be destroyed, leading to a huge loss of efficiency of the system. For this reason (as well as others) people are discouraged from dumping harmful chemicals into the water supply, such as motor oil, antifreeze, pesticides, etc. When this occurs, it takes a long time to regrow the biofilter.

Other Disinfectants

Ozone (O_3)

Ozone is a very effective oxidizing agent (**Figure 3-9**). It has an oxidation potential of –2.07 V, making it even more reactive than chlorine. *Ct* values are thus correspondingly lower for pathogens, which is especially important for the recalcitrant pathogens like *Cryptosporidium*. The solubility of ozone in water is highly dependent on temperature, with warmer temperatures yielding lower solubility. This must be factored into the engineering of the system. Ozone is created by high-voltage arcing of electricity in an oxygen atmosphere, which is an expensive process. The advantage of using ozone is that it is very effective, even against difficult pathogens like *Cryptosporidium*. It is particularly good at preserving the so-called organoleptic factors (e.g., taste, smell, and color) of water, and it creates few DBPs. It is more difficult to determine a *Ct* value for ozone, but this has been achieved by measuring the time required for a residual amount of ozone in the water sample. That is, if too little ozone is added to the sample, the ozone quickly dissipates, and no residual is detected. At a certain higher concentration some ozone is still detectable at a specific time, and this can be quantified (Glaze, 1987). This time period is short, however, which highlights the next point.

The disadvantage of ozone treatment is that it is very unstable, and therefore leaves no residual long-term protection of the water supply. It may be advantageous to use an initial ozone treatment and then follow it with addition of a small concentration of chlorine (for its residual value). However, ozone treatment does not create trihalomethane compounds as chlorine will, but it can create nitric acids that corrode pipes and containers. Ozone is also produced on site, which means that significant capital must be invested in an ozone system.

Chloramine

If ammonia compounds are present in the water, free chlorine combines with it and forms chloramine molecules: chloramide (NH_2Cl), chlorimide ($NHCl_2$), and nitrogen trichloride (NCl_3). Often these are called monochloramine, dichloramine, and trichloramine, respectively. Chloramide (monochlo-

FIGURE 3-9 **Industrial ozone generators are used to provide disinfectant for the fresh water supply.**

SOURCE: Courtesy of the Milwaukee Water Works. These photos may not be used to infer or imply Milwaukee Water Works or City of Milwaukee endorsement of any product, company, or position.

ramine) is the disinfectant of choice in water treatment; it is not as strong as free chlorine. To create chloramine, a 3:1 ratio of chlorine to ammonia is used. Monochloramine is generally favored, with lesser amounts of dichloramine.

Chlorine dioxide (ClO_2)

This compound works well at higher pH values (8.0 and higher), so it may be very useful in some water system applications. It produces a residual effect for longer periods of protection in the water distribution system. Like ozone, it eliminates the organoleptic factors, such as those caused by organic halides. Chlorine dioxide does not create DBPs to the same extent that chlorine does, which may be a deciding factor in its use.

Ultraviolet light

UV light has been demonstrated to be germicidal by causing damage to DNA. This form of nonionizing radiation does not cause the severe damage of strand breakage that ionizing radiation (gamma rays) causes. UV light is particularly effective at creating pyrimidine dimers in DNA. For instance, two contiguous thymine molecules will dimerize and hinder the typical molecular processes of the cell (i.e., DNA replication and transcription). Besides the well-researched thymine–thymine dimers, cytosine–thymine and cytosine–cytosine dimers have been demonstrated to occur. In RNA viruses it is probable that uracil dimers will form; UV light is also effective at killing these viruses. Cells of all types have mechanisms to fix these dimers because they are probably common in nature. However, extensive DNA damage will interfere with their ability to replicate, which is an effective means of killing many pathogens, including some parasitic cysts. UV lights have many applications where their killing power is available in a "line-of-sight" with the target; however, particles in the water may shield bacteria from the UV light and therefore hinder the efficiency of the light. Short-wave UV light (200–300 nm) is especially damaging. However, it is not very penetrating and leaves no residual disinfectant. Short-wave UV light also damages human tissue, so care must be taken in positioning the lamp and in its maintenance. As lamps age, their power output is affected, and therefore monitoring the system is an ongoing activity. It is also difficult to establish a *Ct* value for a wavelength of light.

Optional Processes

Ion Exchange Systems

Ion exchange systems reduce the concentrations of dissolved ionic species (anions and cations) in the water. They are fairly well known as household water softeners, in which water is typically treated to remove iron, calcium, and magnesium. This method removes cations; anions such as chloride can be removed with a different column. The same technology is used for large-scale water softening, where the concentration of cations or anions is too high according to regulatory statutes. The ion exchange system should be located after the point where most of the suspended solids have been removed, otherwise they will clog the columns. Attaching a pre-filter to this unit is also generally recommended because some solids always remain in the water.

Reverse Osmosis (RO)

RO depends on the use of a semipermeable membrane to separate water from the dissolved chemical species that it may contain. This occurs through the application of high pressure to one side of the membrane, which forces the water to the other side of the membrane. RO is not typically used for large-scale water purification, but it has been used to treat saltwater and brackish water to make fresh water. RO is also used extensively in laboratories to make very pure water.

FLUORIDATION

The only bones that are exposed to the environment are the teeth. The oral environment is very challenging in terms of body defenses. A variety of different foods passes through the mouth, and during the act of chewing some food invariably is caught between the teeth. There, some of the many bacterial species found in the oropharynx ferment the food particles, producing acids. Acids erode the enamel of the teeth, pitting them and creating cavities (known more accurately as dental caries). *Streptococcus mutans* is often cited as a major contributor to tooth decay, and much is known about this microorganism (Russell, 2008). However, the biofilms that form on the teeth are very complex and involve many microbial species. In one study of oral flora some 509 species were identified in a set of volunteers (Moore and Moore, 1994), and this number is certainly much greater when the large number of unculturable bacteria is included. Genetic approaches are now being used to model these complex communities and their effects on tooth enamel (Kuramitsu, 2003). Dental hygiene effectively reduces dental caries because it removes the food particles and excess bacteria. However, it has been estimated that 98% of all people worldwide suffer from some degree of tooth decay, making it the most prevalent disease on Earth (**Figure 3-10**).

One of the major public health victories of the last century was the introduction of fluoride into municipal water supplies. With the right concentration, the inclusion of fluoride in the diet retards tooth decay. Some water supplies naturally contain fluoride, which has dissolved from minerals in groundwater sources. Many other communities add fluoride as the last step before sending the water to individual households. Despite the effectiveness of fluoridation, only part of the U.S. population receives fluoridated water (**Figure 3-11**). Even excluding private wells that probably do not have fluoride, only 60% of the population that is served by public water systems has fluoridated water.

How Did Fluoride Become Part of Our Water Supply?

The history of this development illustrates some important concepts in epidemiology and the introduction of new technology. Fluoride is a relatively common element, and it can be found at low con-

FIGURE 3-10 **Dental caries are common all over the world. Bacterial action can be slowed by the addition of fluoride in the water supply.**

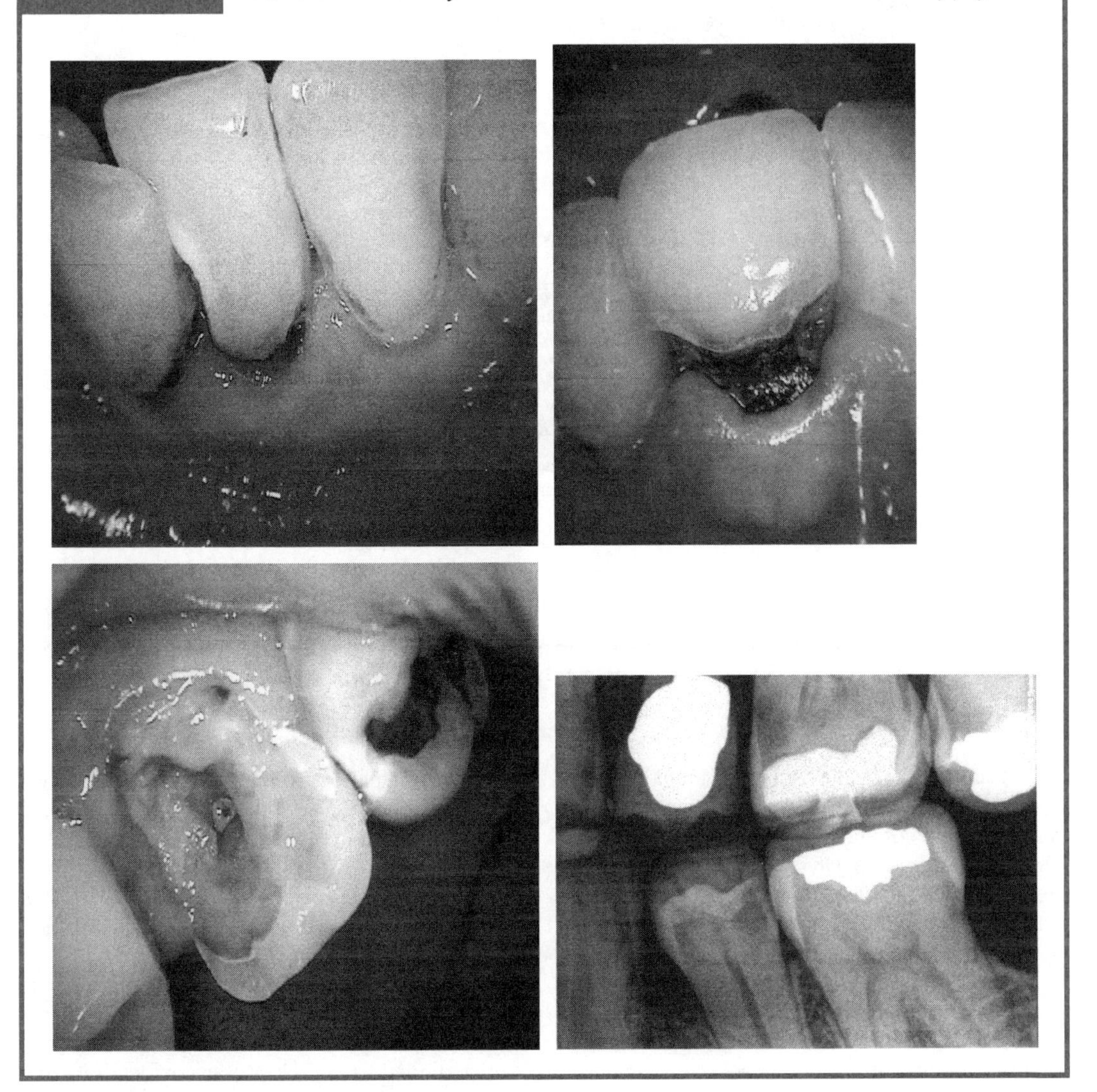

SOURCE: Photos courtesy of Dr. Thomas Tang.

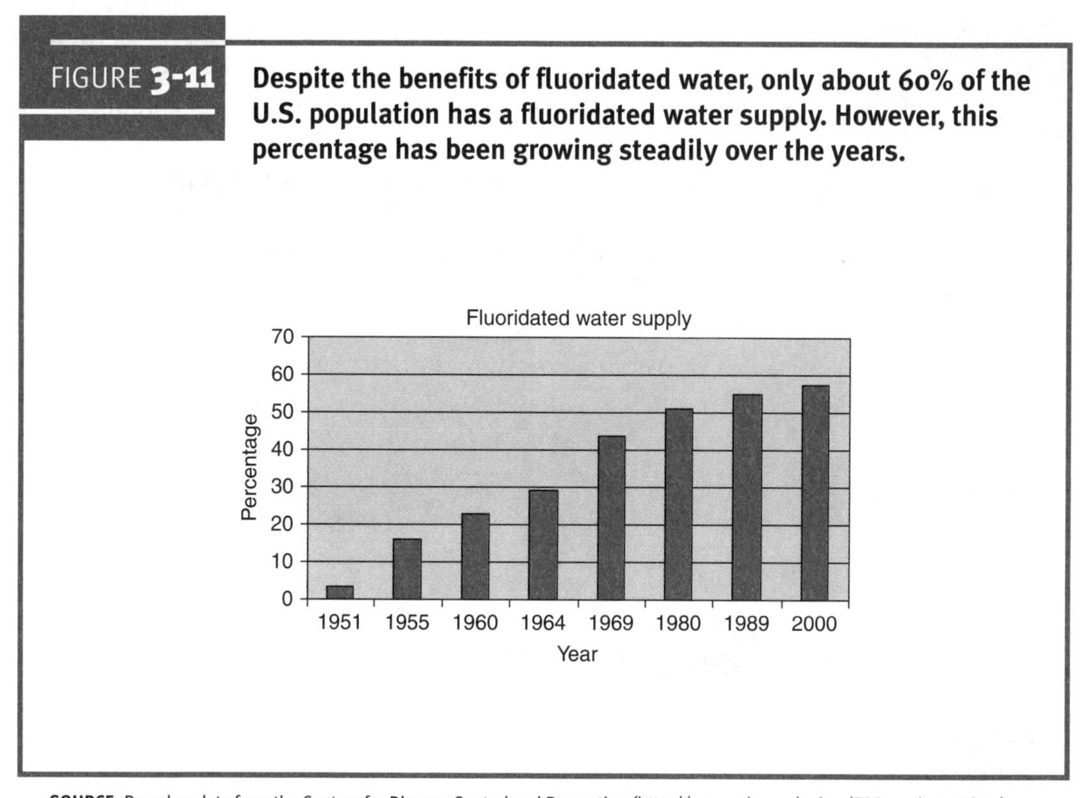

SOURCE: Based on data from the Centers for Disease Control and Prevention (http://www.cdc.gov/nohss/FSGrowth_text.htm).

centrations in most water supplies, including seawater (Fawell et al., 2006). Fluoride is a component of a number of common minerals, such as apatite, feldspar, and cryolite. Groundwater that contacts these minerals can dissolve the minerals, although their solubility is typically quite low. Fluoride can also be found in the air, and high concentrations have been detected in some industrial settings. For example, cryolite (Na_3AlF_6) is used in the production of aluminum, and airborne fluoride concentrations can be high around production facilities. Deposition in local water supplies is possible, although a substantial dilution effect may avoid any problems. However, in some areas the concentrations in fresh water supplies (chiefly groundwater) are much higher.

In 1901 a U.S. Public Health Service employee working in Italy named John M. Eager reported that residents of Naples had a high incidence of severely blackened teeth. He attributed the discoloration to the local water supply, which he speculated was contaminated because the natural springs were located in volcanic rock. However, after this initial description, Eager did not pursue his investigation of the problem.

As described in the following paragraphs, fluoride levels that are high enough to cause blackening are very destructive to teeth. Around the same time, a dentist named Frederick McKay made the same observation in a large majority of his patients in Colorado Springs, CO. The phenomenon was known locally as "Colorado Brown Stain" and by his reckoning some 80% of his patients of school age had it to some degree (Freeze and Lehr, 2009). The brown tooth stains were so predominant in some communities that the cause had to be found in a very common source, and here again the local water supply was suspected. Analysis of the water supply in Colorado Springs and in other affected communities revealed higher levels of fluoride than levels found in other water supplies. Today Colorado Brown Stain is referred to as **dental fluorosis**; this is the gray or brown mottled appearance on teeth. In higher amounts the teeth blacken and decay. In very high concentrations a condition known as **skeletal fluorosis** occurs. In these cases the other bones are affected, leading to brittle bones and osteoporosis.

At the time of McKay's study, fluoride was considered a hazard, and thus attracted the attention of the National Institutes of Health. A study in 1931 led by Dr. H. Trendley Dean examined whether specific levels of fluoride were, in fact, the cause of tooth discoloration (**Box 3-4**). His study, published in 1935, provided an extensive analysis of water supplies throughout western Texas, and provided a handy reference scale for the severity of the staining (Dean et al., 1935). Dr. Dean was a very careful researcher and a very observant scientist. He noticed that patients with Colorado Brown Stain had otherwise healthier teeth than other patients. He also suggested that the optimum fluoride level for the benefits of fluoride without the discoloration was 1 part per million (ppm, equivalent to 1 mg/l). Interestingly, McKay had made much the same observation about the decrease in tooth decay in his patients, although he preferred to pursue the destructive characteristics of fluoride rather than its possible benefits.

Dean's observation prompted a more extensive epidemiological analysis of fluoride for the prevention of tooth decay. Initial investigations made use of existing data in communities that used water with various concentrations of fluoride (Jones et al., 2005). The definitive epidemiological study involved two cities in Michigan—Grand Rapids and Muskegon—and a water supply to which fluoride was added after conventional treatment (Arnold et al., 1953). The results of this multi-year study showed that the city with fluoridated water had a 60% reduction in the prevalence of dental caries among 6-year-old children. This age group was significant because these children had been under the study conditions for their entire lifetimes. In addition, improvements in dental health were also seen with other age groups. Fluoride treatment was found to be optimal if it began early in life, but benefits still accrued if it was begun during childhood or even as a teenager (Arnold et al., 1956).

Many communities moved quickly to add fluoride to their drinking water supplies. This was not without controversy. Everyone depends upon the water supply, and nearly everyone agrees that the government has a legitimate interest in removing harmful substances from that water supply. However, in this

BOX 3-4 H. Trendley Dean's Scale for Dental Fluorosis

- Normal—Teeth are smooth, glossy, and usually of a pale, creamy white color.
- Questionable—Slight aberrations in the translucency of normal enamel ranging from a few white flecks to occasional white spots.
- Very Mild—Small, opaque, paper-white areas are scattered irregularly or streaked over the tooth surface.
- Mild—The white opaque areas in the enamel of the teeth involve at least half of the tooth surface. Light brown stains are sometimes apparent.
- Moderate—All tooth surfaces are involved. Minute pitting is often present. Brown stain is frequently a disfiguring complication.
- Moderately Severe—A greater depth of enamel appears to be involved. Pitting is more frequent and generally observed on all tooth surfaces. Brown stain is deeper in hue.
- Severe—The form of the teeth is at times affected. The pits are deep and very often confluent. The outer surface of the enamel is lost in places and the tooth often presents a corroded appearance. Stains are widespread and range in color from chocolate brown to almost black.

SOURCE: Adapted from Dean et al., 1935.

case the government was adding (or at least declining to remove) a substance that could reasonably be referred to as a medicine. In fact, it was a substance that had preventative powers at one dose, but at higher doses (mottled teeth start to appear at 2.5 ppm fluoride) had already been proven to be harmful.

The current optimum for fluoride in drinking water is 1.0 ppm. Once the concentration rises above 1.5 ppm, there are no additional benefits (Leone et al., 1954). The maximum contaminant level (MCL) is 4 ppm. In areas where the water supply naturally has an overabundance of fluoride, the excess can be removed chemically, but the preferred removal method is simply dilution with water that does not have fluoride.

On a community water quality report (typically sent to households once a year), it is not unusual to see fluoride levels that frequently exceed recommended levels. There are two reasons for this. First, it is difficult to keep fluoride concentrations within a narrow range, and occasionally a sample is taken from a fluoride spike. Second, an occasional spike in fluoride levels does not necessarily indicate adverse health effects. Only regular and continually high concentrations of fluoride will produce the characteristic dental fluorosis.

In addition to the deliberate fluoridation of water, other consumer products can be fluoridated in order to provide the protection to teeth. Fluoridated milk and fluoridated salt have been used successfully (Mullen, 2005). Many toothpastes and mouth rinses are fluoridated, and lozenges are also available. Dentists may prescribe a more potent fluoride solution for some patients. In communities where fluoride is added exogenously, it is supplied as a concentrated solution that dissolves easily. Sodium fluoride, cryolite, fluorosilicic acid, or similar mineral solutions are used. The fluoride is metered into the polished water before sending it into the distribution system. The presence of excess cations interferes with the optimum level of fluoride. While most communities in the United States have fluoridated water, some do not. The only way to be sure is to contact the water provider and ask. In addition, there is no requirement for fluoridation in bottled water.

Precisely how fluoride works to retard tooth decay remains unknown. The current theory is that tooth enamel, which consists of a calcium phosphate mineral called hydroxyapatite, is strengthened by adding fluoride during early development of the teeth, which explains why early intervention is more effective than later introduction (i.e., after the adult teeth appear). The fluorohydroxyapatite is more resistant to the acid produced by bacteria, which slows the process of demineralization. Fluoride may also have a direct inhibitory action on oral microflora.

BOTTLED WATER

The single most popular bottled beverage in the United States today is water. Supermarkets typically have large displays devoted to a variety of bottled water products of all types. The source of these products varies considerably. Some come from artesian wells, and the purity of these products is promoted. Others come from mountain streams or melting glaciers, and still others come from ordinary surface waters that are filled by a municipal distributor (**Box 3-5**).

For several years bottled water did not receive substantial regulatory oversight. However, as a product that is consumed and that is typically found in supermarkets, it falls under the authority of the Food and Drug Administration (FDA); bottled water is a food. The new regulations are very similar to the regulations for municipal water supplies and were designed to ensure that the purity of bottled water is comparable to public drinking water (i.e., "tap water"). The key regulations for bottled water definition and quality are found in 21 CFR 165.110. See, for example, http://edocket.access.gpo.gov/cfr_2006/aprqtr/pdf/21cfr165.110.pdf. The navigation of the Code of Federal Regulations (CFR) is presented in Chapter 10.

When testing for coliform bacteria, the water must not exceed 2.2 coliform bacteria per 100 ml as measured by the multiple-tube MPN test (see MPN tests in the following section), or 4 coliform bacteria per 100 ml as measured by the membrane filtration method. Standard chemical composition is measured (i.e., for iron, chloride, manganese, and total dissolved solids), but these are considered normal constituents and may be quite high in mineral waters.

BOX 3-5 Acceptable Names for Bottled Water

Bottled water—generic term.

Drinking water—generic term.

Artesian well water or artesian water—obtained from a confined aquifer.

Ground water—from a subsurface saturated zone; not under the direct influence of surface water.

Mineral water—containing at least 250 ppm of total dissolved solids from a natural underground source.

Purified/distilled/deionized/reverse osmosis/demineralized water—can be used if the water has been treated with a suitable process that meets that description.

Sparkling bottled water—contains carbon dioxide at the same level as found under natural circumstances.

Spring water—coming from an underground source that is collected at the site of emergence at the land surface.

Sterile or sterilized water—meets the definition for sterility found in the United States Pharmacopeia.

Well water—from a standard well.

Note: On many food products the ingredients will include one or more references to water, such as "water," "carbonated water," "disinfected water," "filtered water," "seltzer water," "soda water," "sparkling water," and "tonic water." These refer only to an ingredient, not specifically to bottled water.

Source water for bottling must be monitored periodically—in some cases as infrequently as once per year—for a number of chemical contaminants. In keeping with the idea of producing bottled water that is comparable to tap water, the list includes known contaminants: 19 metals (e.g., barium, lead, nickel, and nitrate), 21 volatile organic chemicals (VOCs) (e.g., carbon tetrachloride, toluene, and trichloroethylene), and 29 pesticides and synthetic organics (e.g., endrin, glyphoste, and PCBs). Radiological monitoring must be performed for alpha particle emission (like those found with radium) and for uranium concentration. In addition, bottled water must be monitored for disinfectants (e.g., chloramine, chlorine, and chlorine dioxide) and for DBPs (e.g., bromate, chlorite, trihalomethane, and haloacetic acids). Physical measurements (e.g., turbidity, color, odor) are also used.

Bottlers may use a variety of techniques for monitoring for these contaminants, but if the FDA wishes to test a water sample, it will use only previously approved techniques. Some antimicrobial agents (e.g., ozone) can be added to the bottled water, as well as fluoride within standard limits (no more than 1.4 ppm). Individual states may make regulations that are more stringent than the federal regulations.

MONITORING

Overview: The Need for Monitoring

As has been mentioned, potable water is not sterile, nor is it free from all contaminants. Even though a residual chlorine concentration is found in water at virtually every household faucet, some microbes remain. Some bacteria come from biofilms found in the water distribution system and slough off the biofilm and are carried along in the water. The most important factor is that there are no pathogens in the water.

Methods of finding all known pathogens are not routinely performed; procedures are limited to specific species. Often pathogens are present at low concentrations and, at very unpredictable intervals, are occasionally found in infectious dose concentrations. Other pathogens are detected only with very complex, time-consuming, and expensive procedures. For these reasons the bioindicator bacteria are still used to determine whether there is fecal contamination of the water supply. These are the

coliform bacteria, which will be described in the following section.

The search for coliform bacteria is divided into three phases: presumptive, confirmed, and completed (**Figure 3-12**). Several techniques can be used in these phases, all of which are explained in detail in the *Standard Methods for the Examination of Water and Wastewater* (Eaton et al., 2005). In addition, separate tests for specific pathogens can be used.

Successful monitoring is not a matter of haphazard sample collection and analysis; it is part of a developed plan that takes into account several factors: the regulatory limits (concentrations) of the microorganisms that are analyzed, the goals of the treatment facility, the frequency of the monitoring, the size of the sample that is analyzed, the methods of analysis (along with their known limitations), and the physical space of the system being analyzed. Each of these factors must be considered as part of the whole. Regulatory limits are fairly straightforward: the rules tell water utility officials what to search for, how often, and how to do it. The goals of the treatment facility may augment the regulatory limits. For example, the concentration of algae or of *Pseudomonas* may be of interest.

Sample size and sampling frequency are issues. Sample analyses are costly, and most treatment facilities would prefer to perform tests only as often as absolutely necessary. It may be more cost-efficient to sample more sites less often or fewer sites more often. Testing in an on-site laboratory may be more efficient than sending samples to a central laboratory.

Acceptable analytical techniques are usually defined very specifically. It is important to use reli-

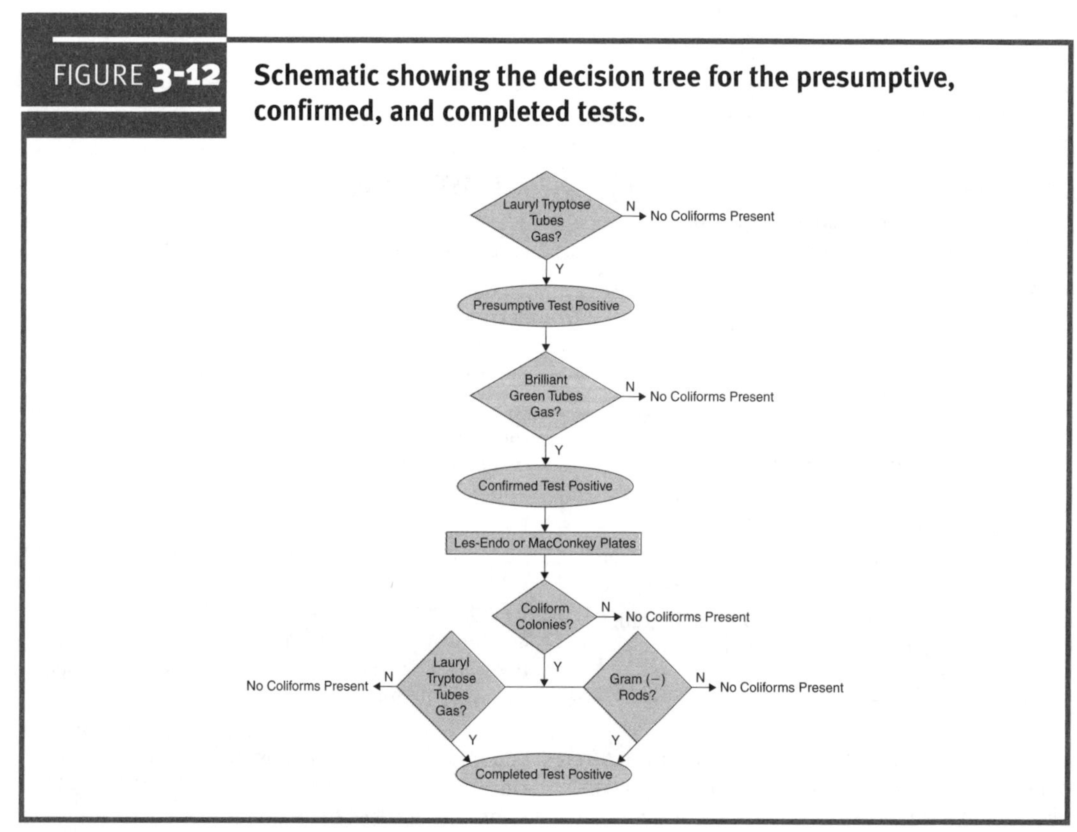

FIGURE 3-12 **Schematic showing the decision tree for the presumptive, confirmed, and completed tests.**

SOURCE: Based on information from Eaton et al., 2005.

able techniques that give data that can be reasonably compared to other treatment facility reports. New molecular and immunological techniques are developed all the time, and some of them are quite promising, especially for unusual pathogens. However, acceptance of a new technique requires a large data collection and substantial patience in getting the technique approved by the right authorities.

The physical space of the system under analysis may be very confined, such as the process stream of a water treatment plant, or it may be huge, such as a watershed. Obviously, the time between sample collection and processing can vary greatly under such diverse conditions. Bacteria may multiply in a water sample at room temperature and skew the true concentrations. Immediate processing (within an hour of collection) is best; keeping samples cold (below 8°C)—but not frozen—is acceptable when a lag between collection and processing is unavoidable. A simple ice chest and ordinary cold packs are sufficient.

It has long been recognized that microorganisms, especially bacteria, can be injured during treatment processes yet not be killed. They remain alive in such a damaged condition that they do not grow readily on artificial media. In the parlance of the trade, they are "stressed," and they are referred to as **viable but not culturable (VBNC)** microorganisms. They are important because they may recover after a certain period and either cause illness or become part of a biofilm and cause illness later. Techniques of searching for pathogens and bioindicators have been developed to rescue these VBNC bacteria and detect them. Often this is done using techniques that are physically gentler than standard spread plating, or using pre-incubation techniques under more agreeable conditions before subjecting them to incubation on selective media.

Laboratory Maintenance

It is vital that laboratories performing the testing maintain a high degree of professionalism and confidence that they are performing the tests correctly. Testing laboratories must promote a different mindset from research laboratories. Research labs are maintained by creative thinkers, people whose job it is to think originally. Often they create procedures that will be developed later for testing labs, but these procedures have a long way to go before they gain general acceptance. It is not unusual for a researcher to experiment with a procedure to see if it can be optimized or to gain an additional benefit.

Testing labs, on the other hand, are maintained by careful and conscientious workers—people that are precise and exact. The lab workers perform the testing procedures exactly as prescribed, and they maintain all of the laboratory equipment in suitable condition as well. Record-keeping is essential to demonstrate that all procedures were done precisely and with no variation. Labs qualify for certification based on their ability to demonstrate these high-quality test results.

Testing labs receive samples from other labs and must perform specific tests on those samples. Often the testing has a legal aspect: a water utility must demonstrate results to the Environmental Protection Agency (EPA), or a public health department might be called upon to provide evidence (i.e., in court) that a restaurant was not following proper hygienic conditions. Here again, the maintenance of a proper testing lab creates confidence that the information presented is accurate.

One of the systems utilized in the testing of samples is **chain of custody**. This means that the samples under analysis can be proven to have an origin (i.e., when they were collected, who collected them, and from where they were collected) and a known custodian from the point of sampling until testing or as long as legally necessary. The samples constitute evidence in a legal sense, so a chain of custody is necessary, as in a police investigation. For example, in a court of law a defendant might blame the public health authorities for mishandling the evidence. In the case of a biological sample, defendants might argue that the sample was not processed for many hours and/or that it was held under high temperatures that allowed the bacteria to grow to much higher concentrations than were originally present. The public health authority must provide evidence that the sample was handled appropriately: it was kept at the right temperature and processed in a timely manner using the appropriate techniques.

Jurors may be asked to hear evidence along these lines as well. Suppose that the Defense alleges that the sample was taken at 9 A.M. and was left out on the counter until the next day, when it was processed. Then the Prosecution presents a chain of custody log that states that the sample was collected by Person A at 9 A.M., that it was transported to the lab in a refrigerated container, that it was removed at 10:35 A.M. by Person B for processing. The procedures are all documented, and the results are positive. An additional list of everyone who had access to the sample is provided. Who will the jury believe?

Another system for the testing lab is a **quality assurance** (QA) plan. This is a general analysis of everything the lab does and how it does it. QA procedures ensure that the lab is working optimally when samples are analyzed using standard conditions. Central to QA is ongoing monitoring of the lab itself: checking the refrigeration temperature, checking the quality of the distilled water, running tests on the equipment, etc. Personnel must be trained on a regular basis. And, of course, documentation of all of these activities is essential. QA is time-consuming, but it is a necessary and valuable element of the overall water treatment system.

Techniques for Water Monitoring

Using artificial media for the detection and enumeration of bacteria may seem archaic, but there are still many good reasons for doing so. Testing laboratories have a great deal of experience working with these techniques, so errors are rare. A tremendous amount of data are available from the past 100 years of testing that can be easily accessed and used because the testing techniques remain largely unchanged. This method is also less expensive than molecular techniques, and most laboratories can perform these tests without difficulty. This series of tests depends on the detection of coliform bacteria as surrogates for pathogens of gastrointestinal origin. The term "coliform" was first suggested by Breed and Norton (1937). **Coliform bacteria** are facultative anaerobic, Gram-negative, non-spore-forming, rod-shaped bacteria that ferment lactose with gas and acid formation within 48 hours at 35°C (**Box 3-6**). This definition includes several genera of the family Enterobacteriaceae, including *Escherichia*, *Klebsiella*, *Citrobacter*, and *Enterobacter*. **Fecal coliforms** are a subset of total coliforms. These coliforms grow and produce gas at 44.5°C within 24 hours. They do not have a growth optimum at this temperature but are instead thermotolerant at this temperature. The coliform bacteria genera are listed in **Table 3-2**, with the fecal coliforms shown in bold. All of these genera have been isolated from clinical samples, although some opportunistic pathogens are more important than others.

BOX 3-6 Description of Coliforms (adapted from Eaton et al., 2005)

Coliform bacteria—facultative, anaerobic, Gram-negative, non-spore-forming, rod-shaped bacteria that ferment lactose with gas and acid formation within 48 h at 35°C.

Fecal coliforms—Coliform bacteria that also grow at 44.5°C.

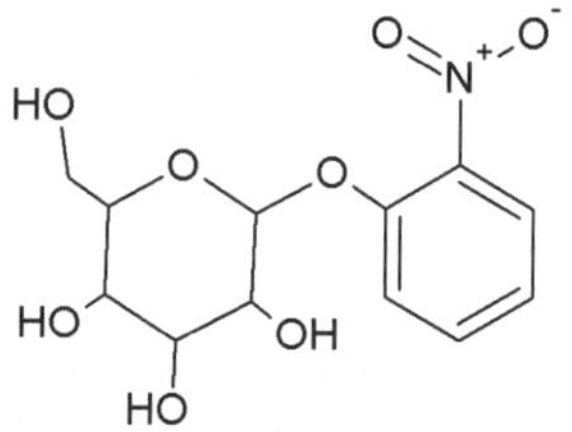

The coliform group is now defined as members of the *Enterobacteriaceae* that are positive for reaction with ONPG (a synthetic substrate).

HO, O, O, N^+, O^-, HO, OH, HO

6-(2-nitrophenoxy)tetrahydro-2*H*-pyran-2,3,4,5-tetrol

Coliform bacteria are usually associated with the fecal contents of warm-blooded animals (e.g., mammals and birds). The facultative anaerobes found in human feces are a minority of the total microbial count, but they still occur at levels of 10^8 per gram. Several species are of nonfecal origin and are free-living in the environment, notably *Enterobacter*, which is easily found on plants and in soil. *E. coli*

TABLE 3-2 Genera of Coliform Bacteria

Arsenophonus	***Klebsiella***
Budvicia	*Kluyvera*
Buttiauxella	*Lecleria*
Cedecea	***Moellerella***
Citrobacter	*Pantoea*
Enterobacter	*Rahnella*
Erwinia	*Serratia*
Escherichia	*Trabulsiella*
Ewingella	***Yersinia***
Hafnia	*Yokenella*

NOTE: Based on information from Leclerc et al., 2001.

is frequently considered the prototypical fecal coliform. Both fecal and nonfecal coliforms comprise the **total coliforms**.

Coliform or Total Coliform Test

The basis for much of water monitoring (and monitoring of many other food products) is the enumeration of coliform bacteria. As a starting point, we consider the Maximum Contaminant Level (MCL) for coliforms in finished water, which is <1 per 100 ml. Because of its importance, a number of techniques have been designed to evaluate the same type of sample. These include a multiple-tube fermentation test (based on the Most Probable Number concept), a membrane filter test (for concentrating large volumes), and the newer enzymatic substrate tests.

Obvious confounding factors are anticipated. Water samples that have chlorine can be treated with sodium thiosulfate ($Na_2S_2O_3$) to dechlorinate the water. Otherwise, the chlorine exposure may kill microorganisms that must be monitored. If heavy metal contamination is a concern, then a chelating agent like Na_2EDTA can be added. Water samples should be taken from a site that is representative of the water source. Sampling downstream from an outflow for street run-off is probably going to show more contaminants than a site slightly upstream. While the added contaminants are important, they might not be characteristic of the stream in general. Usually it is best to take the sample from the point at which water is removed for use because this is the most relevant location in terms of microbial load that must be treated. In some situations a sample of sediment may be useful, and published methods for obtaining a representative sample are available.

The Presence–Absence Test

The presence–absence test (P–A test) is a rapid means of determining whether a water sample meets the MCL. The idea is very simple: a 100-ml water sample is mixed with a differential medium. If even a single coliform bacterium is in the sample, it causes a color change and the sample will fail the test. If no coliform bacteria are present, the sample color will remain the same, and the sample passes. The P–A medium is a rich medium containing lactose (the signature substrate for the coliforms) and bromcresol purple as a pH indicator. The broth is typically produced in a triple-strength (3×) batch with 50-ml aliquots. When 100 ml of water is mixed in, the strength then becomes 1×. Coliforms utilize the lactose and produce acid, which lowers the pH and changes the green color to yellow. In addition, gas is produced, and gentle shaking yields bubble formation in the broth.

P–A tests are simple, reliable, and inexpensive. For finished water samples, the vast majority are negative anyway. As a broth test, this is a gentler means of handling damaged cells such as the VBNC cells. The disadvantage of the test is that it is not quantitative. A single coliform bacterium will cause a positive reaction, but a million will do the same in the same volume of water.

The Standard Tests for Coliform Presence and Quantification

Quantification of coliforms in a water sample is generally a good idea. It tells you whether you have a slight contamination problem that may require a minor adjustment or a major failure of your treatment system that, for example, would trigger a "boil water" alert. The test is comprised of three parts: the **presumptive** test, the **confirmed** test, and the **completed** test. Each is described in more detail in the next section, but first there is a need to explain the types of tests for which they can be used. Quantification of coliforms is initially performed either using the multiple-tube fermentation technique or the membrane filtration technique. In general they constitute a liquid-based assay and a solid-based assay. Each has advantages and disadvantages.

The **membrane filter technique** is very easy to perform. A nylon membrane (pore size: 0.45 μm) is inserted into a filter apparatus. The water sample is then passed through the membrane using a vacuum, and the bacteria are caught on the filter. The filter is then placed on a selective and differential medium, and the samples are incubated overnight at 37°C. The coliform colonies are then counted and are expressed as the number of coliforms per 100 ml of water.

The **multiple-tube fermentation technique** is based on the random distribution of coliform bacteria in a water sample. It can be used as the first part of the presumptive test and is used as the confirmed test. The basis for the test is a little challenging from a conceptual standpoint. Here is a simplified illustration.

Consider a sample of 1000 ml of water (1 liter) that contains 11 coliform bacteria. Because this concentration is greater than the 1 per 100 ml limit, the sample fails the test. If the coliform count were only nine, then the concentration would be below the 1 per 100 ml limit and the sample would pass. To examine this water sample, 100 ml could be used for the P–A test. However, with 11 total coliforms in the sample, it is quite possible that none of them would be in the 100-ml sample. The same is true of filtering 100 ml of water and using the membrane filtration test. In both cases the sample would pass, when technically it should fail.

The multiple-tube fermentation technique does not reveal the concentration of bacteria in a sample, but it does present the most probable number of bacteria in the sample (hence MPN), with a range to account for the values that fall within a 95% confidence interval for that test. It depends on the analysis of many smaller-sized samples of the water and on the probability of finding a positive result in any of them. That is, this test has a foundation in statistical probability (**Table 3-3**).

At each decade level (e.g., 10 ml, 1.0 ml, 0.1 ml) at least three tubes should be tested. Five tubes are better; the more tubes used, the more accurate the results. Larger volumes overly dilute the medium, so concentrated medium solutions are used. The pattern of positive tubes is provided per volume decade. If four out of five 10-ml tubes are positive, and two out of five 1.0-ml tubes are positive, and none of the 0.1-ml tubes are positive, the result would be given as 5-2-0. This result is then interpreted by a standard table of statistical values, which includes the MPN of coliforms per 100 ml as well as the confidence limit (95% of the possible values).

In this example, even a single positive tube would be enough to cause the sample to fail the test. However, even if all tubes are negative, the statistical possibility exists (as indicated by the confidence limits) of coliforms in the water. Another sampling scheme can be utilized that is more definitive for cases like this one: 100 ml, 10 ml, 1 ml.

The Tests

Presumptive Test

In the presumptive test the water sample is examined with the multiple-tube fermentation test or the membrane filtration test. If the multiple-tube fermentation test is used, then a series of water samples that are delineated by decades of volume (e.g.,

TABLE 3-3 **Selected MPN Patterns**

Positive tubes	MPN/100 ml	95% Confidence interval	
		Low	High
0-0-0	<1.8	–	6.8
0-0-1	1.8	0.09	6.8
0-2-0	3.7	0.7	10
1-0-0	2	0.1	10
1-0-1	4	0.7	10
1-1-0	4	0.71	12
1-1-1	6.1	1.8	15
1-2-1	8.2	3.4	22
1-4-0	10	3.5	22
2-0-0	4.5	0.79	15
2-0-2	9.1	3.4	22
2-1-2	12	4.1	26
2-3-0	12	4.1	26
2-3-1	14	5.9	36
2-4-0	15	5.9	36
3-0-2	13	5.6	35
3-1-2	17	6	36
3-2-2	20	6.8	40
3-3-2	24	9.8	70
3-5-0	25	9.8	70
4-0-1	17	5.9	36
4-1-2	26	9.8	70
4-2-1	26	9.8	70
4-3-0	27	9.9	70
4-4-0	34	14	100
4-5-0	41	14	100
5-0-0	23	6.8	70
5-0-3	58	22	150
5-1-1	46	14	120
5-1-3	84	34	220
5-2-2	94	34	230
5-2-4	150	58	400
5-3-0	79	22	220
5-3-2	140	52	400
5-4-1	170	58	400
5-4-4	350	100	710
5-5-2	540	150	1700
5-5-5	>1600	700	–

SOURCE: Adapted from Eaton et al., 2005.

10 ml, 1 ml, and 0.1 ml) are used to inoculate lauryl tryptose tubes. Tubes that show acid or gas production are considered positive and are used in the confirmed test. The pattern of positive tubes can be used to determine an MPN for the original water sample.

The medium that is used for the test is lauryl tryptose broth. It contains sodium lauryl sulphate as a selective agent for coliforms and lactose as the nutrient source for coliforms. If coliforms are present, they will ferment the lactose, creating acid and gas and decreasing the pH. This can be detected in either of two ways. First, the inclusion of bromcresol purple in the medium at 0.01 g/l provides a pH indicator much like that seen for the P–A broth.Second, a small tube, known as a Durham tube, is inserted into the larger tube of medium. The Durham tube is inverted. During autoclaving the tube expels the air inside it and fills with medium. During fermentation the tube will refill with the produced gas, which is easily seen. The tube does not need to refill completely to be positive; any bubble in the tube counts as positive. Occasionally the autoclaving will leave a very small bubble, so the prior presence of a slight bubble must be ruled out. Any more than a tiny bubble, and the entire tube should probably be discarded. Any positive tube may be the result of more than one species, but the important fact of the test is not the identity of the species but the fact of coliform detection. The positive tubes are then subjected to the confirmed test.

If the membrane filtration technique is used, the typical and atypical coliform colonies are counted and used to calculate the coliform concentration in the original water sample. The coliforms are then examined in the confirmed test.

The membrane is transferred, bacteria up, to either m-Endo agar or LES Endo agar plates, or to an absorbent pad soaked in m-Endo broth (**Figure 3-13**). On these media, typical coliforms appear red with a metallic sheen. However, atypical coliforms appear as dark red, or mucoid, nucleated colonies that do not have a metallic sheen. These are also counted as coliforms. Noncoliform colonies can also grow on these media, appearing as blue, white, pink, or colorless. They do not demonstrate the metallic sheen.

This technique is considered very accurate for enumeration of coliforms, but in some instances it is ambiguous. It works well for clear waters, but the filters will clog easily if suspended solids are present. If too many noncoliforms are present, there is a possibility that they will overgrow the plate and that the coliform colonies will be hidden from view. It is permissible to examine microcolonies on the filter through a microscope; this avoids the overgrowth problem. Both typical and atypical coliform colonies can be subjected to the next phase of identification, the confirmed test.

Confirmed Test

For this part of the procedure, Brilliant Green Lactose Bile broth is used. The medium contains lactose and oxgall, which is dehydrated fresh bile. It is used to select for coliform bacteria. As with the lauryl tryptose tubes, these contain small Durham tubes as well.

Tubes of Brilliant Green are inoculated with the coliform colonies (typical and atypical) from the plates of the membrane filtration technique or the positive tubes of the multiple-tube fermentation technique. A positive result is the production of gas in the Durham tube. If the Brilliant Green tube is positive for any of the lauryl tryptose tubes, then those lauryl tryptose tubes can be considered in determining the MPN value.

Completed Test

It is useful to spot check the results of the confirmed test with the methods of the completed test. Testing about 10% of the samples is sufficient. A sample from each positive Brilliant Green tube is streaked on a plate of MacConkey agar or LES Endo agar and incubated at 35°C for 24 hours. Isolated colonies are chosen and typical or atypical coliforms are streaked onto lauryl tryptose tubes and nutrient agar slants. The goal is to obtain purified isolates. The lauryl tryptose tubes show gas production in the Durham tube. The nutrient slant is used for Gram staining of the microorganisms. They should show Gram-negative bacilli.

E. coli-Specific Tests

EC-MUG Test

One of the enzymes in *E. coli* is ß-glucuronidase (known as GUS and encoded by *uidA*) (Arul et al., 2008) is found in 97% of isolates. The enzyme is a glycosyl hydrolase, which hydrolyses ß-glucouronic acid residues from the nonreducing termini of gly-

FIGURE 3-13 **An m-ENDO LES plate with *E. coli*.**

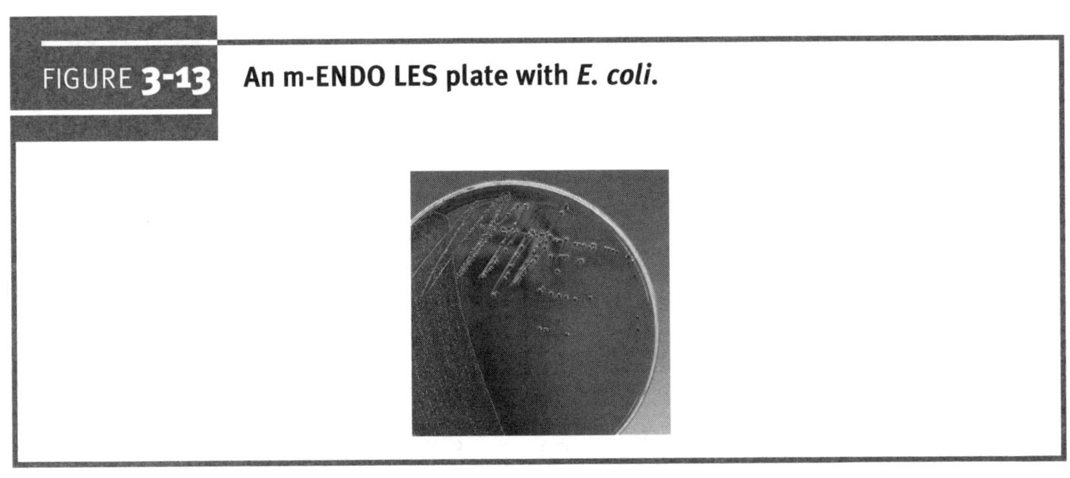

SOURCE: Photo courtesy of ThermoFisher.

cosaminoglycans. An artificial substrate has been created, 4-methylumbelliferyl-ß-D-glucuronide (known as MUG). If the enzyme is present it will cleave the MUG substrate and produce a fluorescent compound (Rice et al., 1990) that is evident as bright blue under long-wave UV light (**Figure 3-14**).

EC-MUG medium is designed to take advantage of this artificial substrate. Incubating a positive water sample at 44.5°C demonstrates a blue fluorescence within 24 h. An MPN calculation can be obtained if a series of tubes is used, as outlined previously.

GAD TEST

E. coli also has the enzyme glutamate decarboxylase (GAD). It is an intracellular enzyme and must be released by lysis in order to be effective against its substrate. The GAD reagent contains a lytic agent (Triton-X-100), glutamic acid, and bromcresol green as a pH indicator.

The reagent is added to cells from the presumptive test. The test temperature is only 35°C, and results are ready in as little as an hour. If the sample is positive, the color changes from yellow to blue. Again, an MPN calculation is possible here.

FECAL COLIFORMS

Fecal coliforms are identical to coliforms with the exception that they will grow at 44.5°C. After performing the presumptive test, each positive tube is used to inoculate a tube (with Durham tube) of EC broth. If the tube shows gas production in 24 h at 44.5°C, it contains fecal coliform. This test can be performed on agar medium by using M-FC medium and incubating at 44.5°C; positive colonies appear as blue colonies.

Other Tests

HETEROTROPHIC PLATE COUNT (HPC)

The total number of aerobic heterotrophic bacteria in a water sample is useful because it gives an indication of overall biomass in a water sample. These are often considered in lake or river water samples. A rich, nonselective medium such as R2A medium is used to grow as many bacteria as possible. A longer incubation time (as long as 7 days), and a lower temperature (22–28°C) are used because these are not solely coliform bacteria so 37°C is not needed. Ten-fold dilutions are used because many water samples have high numbers of total aerobic heterotrophs. Results are calculated based on the serial dilutions, the plating dilution, and the plate count. Multiple plates are used at each dilution to obtain statistically meaningful results.

An alternative method uses a direct count of bacterial cells. Samples are fixed to microscope slides and stained with a fluorescent dye such as acridine orange. Cells are then counted microscopically, and the sample volume is used to determine the

FIGURE 3-14 **MacConkey agar with MUG substrate and *E. coli* growth.**

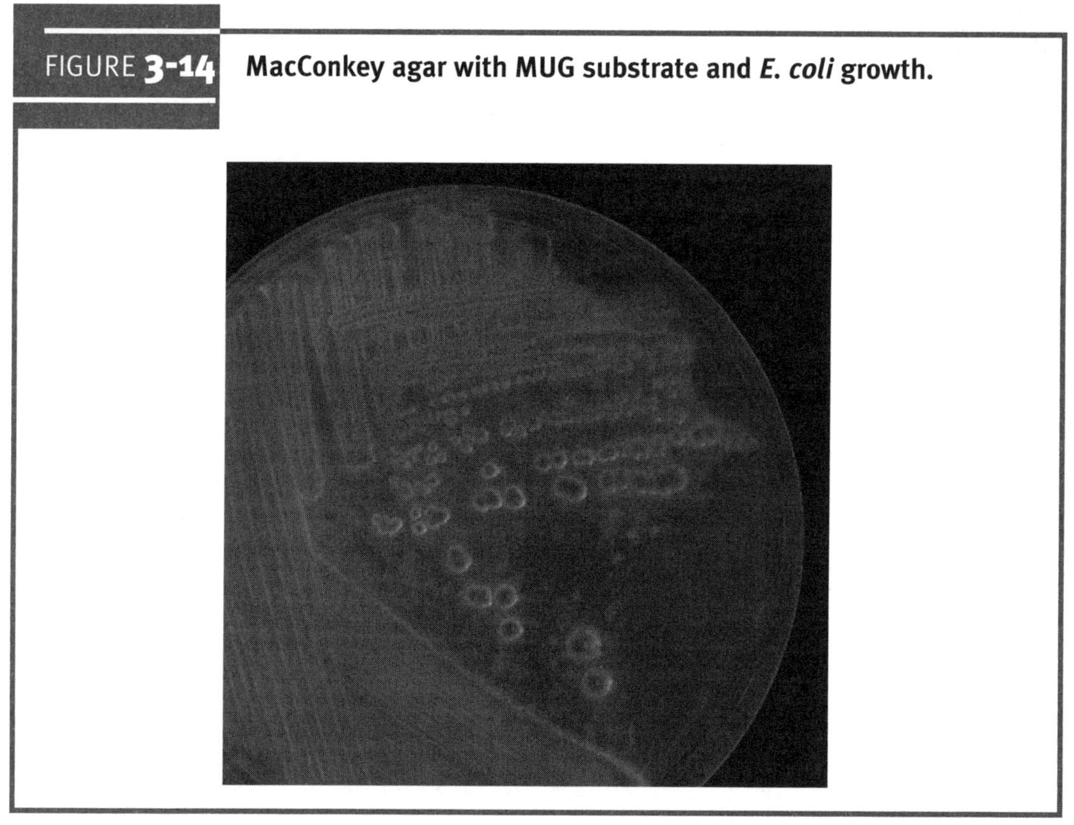

SOURCE: Photo courtesy of ThermoFisher.

concentration. However, this method is tedious, and it counts both live and dead cells.

Molecular Tests

Improvements in the monitoring of water supplies for pathogens are needed. Clearly, a test that yields definitive results in 24 h is slower than the water industry would prefer, and the sporadic nature of sampling means that some spikes in pathogen concentrations can slip through undetected. Detecting viruses and parasites that cannot be cultivated in easily performed agar-based assays is also a problem. As noted elsewhere, some pathogens, such as the parasites, do not correlate with the coliform bioindicator, and have only a weak correlation with turbidity. The ability to monitor every known pathogen in a real-time, online process would be a tremendous benefit to the water processing industry. As can easily be imagined, this is a daunting task when hundreds of millions of gallons of water must be purified each day.

However, incremental improvements can be made to move technology toward the ultimate goal. Improvements in certain parameters can be identified. For example, the **sensitivity** of the test measures the lowest number of the pathogen target that can be detected. A microorganism cannot be divided into less than 1 unit, but the volume measurement then becomes relevant. If the sensitivity of a test is 10 bacteria per liter, this is equivalent to 0.01 bacteria per ml, and might be expressed in this manner. However, the former term is more easily understood. The **specificity** of the test describes the ability of the test reagents to *not* react with targets other than the target of interest. That is, if the test is designed for *Cryptosporidium hominis*, it should not react with other parasites, such as *C. parvum* or *C. canis*, or any other species. This should not be confused with **precision**, which measures the reproducibility of a test. In any test some variability occurs because of minor variations in reagent delivery, incubation, inhibitory substances, or other confounding conditions. The question is whether a set of tests run against the same sample give results that are grouped closely, or whether the results vary. The more closely the results are grouped, the greater the precision. This does not mean that the results are better than those of another test; a different test may be more specific but may produce more variability when repeated. **Accuracy** refers to the degree to which the test comes close to the true answer.

To a great extent, these attributes are found in molecular-based techniques. To increase sensitivity and/or specificity, antibody tests are employed. These make use of the highly specific reaction between the binding site of the antibody and the antigen. More specifically, the antibody recognizes the antigenic determinant, or epitope, which is a three-dimensional structure on the antigen. Because the targets of interest are microorganisms and are usually pathogens, an antibody response to these species (specifically, to the surface proteins on the pathogens) is the typical outcome. If antibodies are produced by injecting a pathogen into an animal and then harvesting the sera and the antibody fraction, a polyclonal antiserum is obtained. If, however, a specific B cell, producing only one type of antibody, is harvested and propagated, then a monoclonal antibody is produced which is very specific to the epitope. Even in these cases the antibody might cross react with another molecule, but these occurrences can usually be eliminated by screening potential monoclonal antibodies under realistic conditions. Nonspecific binding of the antibody, which is a protein, to other substances in the test material may also occur solely on the basis of its nature as a protein.

Under the right conditions the antibody binds to the antigen on the pathogen. Excess antibody (unbound to antigen) is washed away. Without some sort of marker, this interaction is not seen. Antibodies can be conjugated with an enzyme. After addition of the proper substrate, the enzyme produces a reaction, which often results in a characteristic color change. Other enzyme conjugations to radioisotopes (which are easily detected but have fallen out of favor because of the radioactive waste problem) or fluorescent compounds may also be used. These latter conjugations are very useful when microscopy is performed because specific filter sets are available with which to view certain fluorescent molecules.

Nucleic acid tests are also highly sensitive and specific. In these tests nucleic acid (either DNA or RNA) is detected. If the correct nucleic acid sequence is used, the test can be very specific, often to the species level and sometimes even to the strain level. Each of these tests requires the harvest of the nucleic acid and then its manipulation in a test, followed by a detection step. One of the older techniques is to use a known piece of DNA as a probe. The probe is labeled, usually with a fluorescent or chromogenic molecule, and is then used to hybridize with the DNA from a sample. If conditions are produced correctly, the probe binds only to the nucleic acid to which it is complementary. The excess probe is then washed away, and the probe can be detected.

Even small numbers of microorganisms can be detected using Polymerase Chain Reaction (PCR). In this technique a small piece of the DNA molecule is selected that is distinct for that species. Short, single-stranded primers are designed to flank this piece of DNA. The PCR reaction proceeds using an enzyme that recognizes the primers and copies the DNA between the primers. Each round of copying doubles the number of copies of the DNA, thereby amplifying it greatly. If a marker is incorporated into the nucleotides as they are used to make the new DNA pieces, the amplified DNA can be easily detected. This is the basis for Real Time PCR, which has become a fast and reliable means of detecting pathogens in many types of samples.

Notably, many human viruses do not have DNA but instead utilize RNA. Some are double-stranded RNA, but many are single-stranded and may either be the so-called plus strand or the minus strand. This further complicates PCR amplification because DNA is not present. In these cases a Reverse Transcriptase PCR (RT-PCR) is used to convert the RNA strand to a DNA double strand in the initial step, and then conventional PCR amplification is performed. The Reverse Transcriptase enzyme performs this initial task very well, and protocols for its use are common.

Procedures for Swimming Pools

The heterotrophic plate count is the best indicator of disinfection efficiency, and it is performed exactly as described previously. If tests for coliforms are needed after a fecal release in the water, then the membrane filtration assay or multiple-tube fermentation assay are used, as described above. Details of tests for specific pathogens are presented in Chapter 4.

Beaches

To test coastal waters, water is collected from about 1 ft below the surface of the water and at an overall depth of approximately 3 ft, where the bathers are likely to congregate. Shallow areas and non-swimming areas where wave action does not have a good chance to mix the water thoroughly are avoided. Samples are taken at several points along the shoreline, and possible influences on water quality are noted, such as the presence of large flocks of birds or storm drain outfalls and proximity to commercial establishments (e.g., shops, marinas, and boardwalks).

In fresh water, samples are tested for both *E. coli* and for *Enterococci* (see Chapter 4 for regulatory limits). For *E. coli*, a 100-ml sample is concentrated on a filter. The water may have significant particulate matter that can clog the filter, so the use of several filters is acceptable because only the total colony count per 100 ml is needed. The filter is placed on mTEC medium. Incubation at 44.5°C is allowed overnight because it is a fecal coliform. To verify the presence of *E. coli* among the colonies that grow, the filter is immersed in a urea broth solution and is examined for yellow or yellow-brown colonies under UV light.

Enterococci can be enumerated using a multiple-tube MPN test or by counting colonies on a membrane filtration assay. The MPN test uses a series of azide dextrose broth with at least three decades represented (i.e., 10 ml, 1 ml, 0.1 ml). The samples are incubated at 35°C for 24 hours and 48 hours and any turbid tube is positive. These tubes are used for the confirmatory test. The confirmatory test uses Pfizer selective enterococcus agar (PSE). The plates are streaked with the samples from the positive presumptive test tubes and are incubated at 35°C overnight. A positive result is a brown-black colony, indicating either fecal *Streptococcus* or *Enterococcus*. The *Enterococci* grow at high temperatures (45°C) and in a high salt concentration (6.5%). A separate medium is used to determine whether the colonies are *Enterococcus*.

The membrane filter assay uses mE agar with incubation at 41°C for 48 h. Colonies that appear are transferred to EIA substrate and are incubated for an additional 20 minutes. A positive result on this indicator substrate is the appearance of pink or red colonies with a black or reddish-brown precipitate on the filter. Isolates should be catalase negative (*Staphylococcus* is catalase positive) and should stain as Gram-positive cocci. The *Enterococci* are distinguished from the fecal *Streptococci* by their growth in 6.5% salt media at 45°C. Saltwater samples are tested for *Enterococci* only. The method used is the same as for fresh water.

Pseudomonas species are typically environmental strains that carry little risk of infection (**Figure 3-15**). They prefer colder temperatures (20°C) and have a wide metabolic diversity, allowing them to subsist on a large number of carbon sources. However, *P. aeruginosa*, which is known for antibiotic resistance, grows at higher (physiological) temperatures and can be detected using either a membrane filtration or the multiple-tube assay.

P. aeruginosa is likely to be in low concentration, and a filtered sample of more than 100 ml is probably necessary to get an accurate concentration. The amount of particulate material that might clog the filter becomes a key factor. Use M-PA agar (Eaton et al., 2005) and incubate at 44.5°C for 72 h. Colonies should have a dark green to brown appearance. For confirmation, the colonies are streaked again on Milk Agar and are incubated overnight at 35°C. *P. aeruginosa* display a greenish-yellow diffusible pigment.

The multiple-tube assay uses Asparagine Broth. Sample sizes are dependent on site history, but sample sizes of 10 ml, 1 ml, and 0.1 ml are probably effective. These tubes do not have the Durham tube used in the coliform tests. Instead, the tubes are incubated at 37°C for 24 h and 48 h and are then examined under long-wave UV light for a characteristic green fluorescence. The confirmatory test procedure is to inoculate positive presumptive tubes into acetamide broth and then incubate them for 24 hours to 36 hours at 37°C. Positive samples

FIGURE 3-15 **Pseudomonas streaked on m-ENDO agar.**

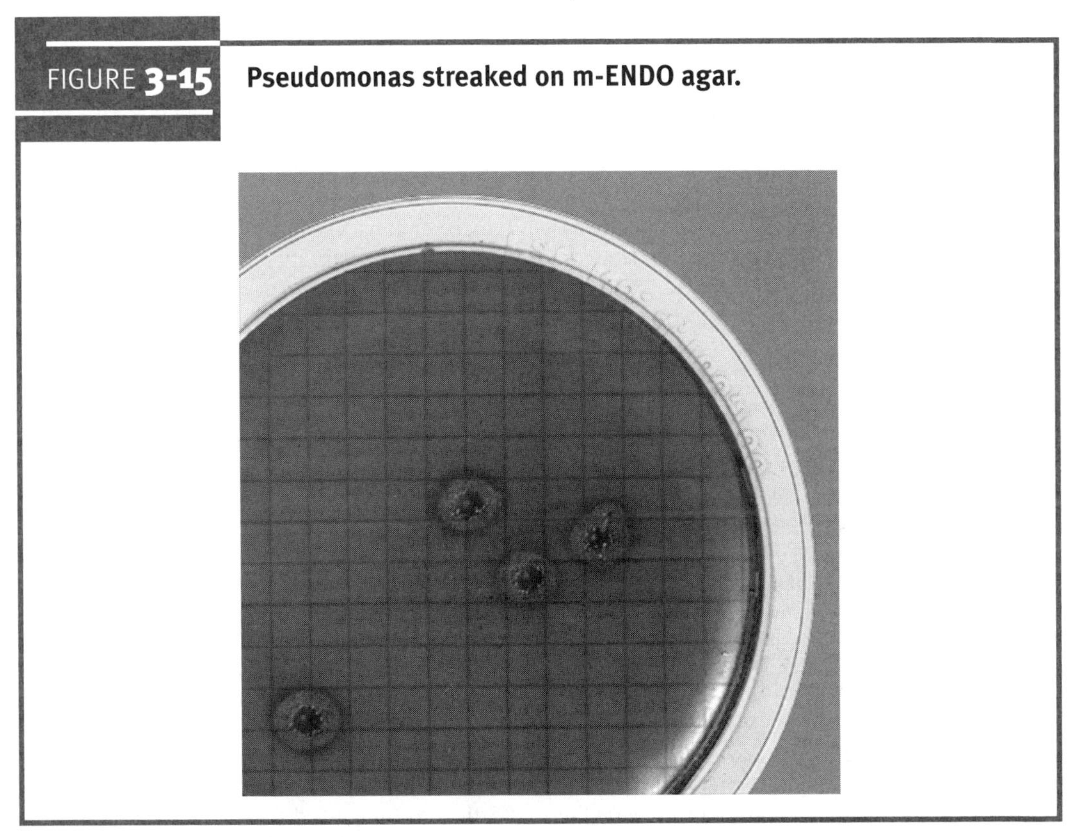

SOURCE: Photo courtesy of ThermoFisher.

turn purple, indicating an alkaline reaction with the phenol red indicator.

WATER LAWS

As might be expected of a resource that is needed by everyone and that has, if mistreated, the potential to sicken many people, the rules surrounding water treatment and monitoring are extensive and reflect the best science currently available. In the past few decades a number of federal laws have been enacted that were designed to improve the quality of drinking water and to limit disease. Legislation always maintains a balance between what is desirable and what is actually doable.

The golden age of microbiology is generally considered to have been the latter half of the 1800s, when scientists like Louis Pasteur and Robert Koch established the germ theory of disease and identified many of the microorganisms responsible for specific illnesses. At the end of 1800s and the beginning of the 1900s, it became evident that many of these known pathogens could be transmitted through drinking water (the fecal–oral route of transmission), and that some methods of purifying the water (e.g., filtration through sand and chlorination) could yield vast improvements in public health. During this time, cities began to introduce public water supplies as an alternative to household wells or water delivery services. Eventually, water quality standards were established and treatment facilities were built. But how much treatment was effective in preventing disease? Which standards would be used to define effective treatment?

A species of indicator bacteria was needed that could be easily grown and counted. Many species of bacteria had been described at this time, and many were known to be normal flora of the intestinal tract. The standard chosen was the concentration of coliform bacteria in a sample of treated water. The choice of coliforms and fecal coliforms was very sensible. Coliform bacteria could be easily grown with and identified by even the relatively unsophisticated technology of the time. The presence of coliforms was highly suggestive of fecal contamination, and in treated water supplies their presence indicated that water treatment was not wholly effective. This was very important at the outset because many waterborne pathogens were yet to be identified, and methods for their cultivation were yet to be developed. For instance, the science of virology could arguably be said to have begun in 1892 with experiments on Tobacco Mosaic Virus and waterborne diseases like Hepatitis A virus and Rotavirus, which were previously known only by the symptoms they produced. The first virus shown to infect humans, Poliovirus, was not so described until 1908.

Therefore, monitoring for fecal coliforms is a surrogate for other pathogens associated with fecal contamination. Technology has improved greatly over the years; detection methods now include the use of antibodies or molecular methods like PCR. These newer techniques have many advantages, including faster results and the ability to distinguish specific pathogens. However, these techniques are typically much more expensive than those using agar or liquid media. Growth of coliforms is uncomplicated and does not require expensive equipment or highly skilled workers. In addition, an immense amount of data about coliform monitoring by the simpler methods is available that confirms the value of these methods.

Bacteria are not evenly (or randomly) distributed throughout the water system. Instead, sampling occasionally yields a high concentration in a single sample, while the vast majority of samples show no coliforms at all (Christian and Pipes, 1983). This fact complicates monitoring because uncertainty about the hygiene of a system remains, even when most (or all) samples are negative for coliforms, as does uncertainty about what one positive sample (even one with a high concentration of coliforms) actually means for the whole system. Increasing the number of samples would give more confidence in the pattern of contamination, but sampling is expensive, and most of the time the water distribution system is very safe.

With any water distribution system, a finite chance remains that a single glass of water will have a high concentration of microorganisms that may cause disease. However, effective monitoring and treatment keeps the likelihood of this type of event very low.

The EPA follows a pattern for rule making, beginning by gathering information on the presence of contaminants in water. The EPA then sets a Maximum Contaminant Level Goal (MCLG) for each contaminant. This contaminant concentration is set at the level below which there is no predicted health risk. In addition, a Maximum Contaminant Level (MCL) allowed in the drinking water supply is established. While the MCLG is not a legally enforceable limit, the MCL is, and ideally the MCL and the MCLG would be the same. But setting the MCL as close as possible to the MCLG is more practical and takes into account the best available technology as well as testing and treatment costs. In many instances the contaminant cannot easily or cost-effectively be quantified, and in these cases a Treatment Technique (TT) is allowed. That is, a specific technique to reduce the contaminant concentration in a typical water supply to a level below the MCL is allowed.

For most of U.S. history the quality of drinking water was a local matter, unregulated by the federal government. Water was regarded as a consumable commodity that did not typically cross state borders, and therefore was only applicable for state regulation. In 1914 the Public Health Service (then a branch of the Department of the Treasury, now part of the Department of Health and Human Services) promulgated the first federal standards for drinking water. They set the allowable coliform level at 2 per 100 ml. These federal rules were only applicable to interstate travel, so they affected water served on ships and trains. It was not until late in the 1900s that federal rules became preeminent. The major federal water laws are summarized below.

The Clean Water Act of 1972

The Clean Water Act was instituted to address the problem of sewage that was contaminating public water supplies. Regulations were implemented that established the allowable concentrations of contaminants that could be discharged into a water supply. As part of this system, the EPA created the National Pollutant Discharge Elimination System (NPDES). This system issues waste discharge permits to municipalities that define their allowed sewage discharges to surface water supplies. (Groundwater supplies were not addressed.)

The Safe Drinking Water Act of 1974

The Safe Drinking Water Act empowered the Environmental Protection Agency to regulate PWSs. The SDWA established the first monitoring standards for PWSs. These included inorganic compounds, microorganisms (known pathogens), turbidity, and radionuclides. The aim was to produce clean water at the tap, and therefore did not address water source issues.

The Total Coliform Rule

The **Total Coliform Rule** (TCR) was approved by the U.S. Congress in 1989. It authorizes the EPA to establish rules for all PWSs. Specifically, it requires water suppliers to monitor their systems for the presence of total coliform bacteria using specified tests (**Table 3-4**). The concept behind this rule is that an estimate of coliform density is not as useful as frequent monitoring for the presence or absence of any coliforms. The sampling frequency is dependent on the number of customers in that system. If a sample is positive for total coliforms, it must be examined for either fecal coliforms or *E. coli*. Additional samples must also be taken from sites near any site that has produced a positive result for total coliforms.

The Maximum Contaminant Level (MCL) has been fixed at <5% positive samples for coliforms in a month when the number of samples is >40. Smaller water utilities may sample <40 times per month; in this case the MCL is no more than one positive sample per month. If the MCL is exceeded, the water utility must follow a plan of increased monitoring near the affected site and must report their findings to the state authority. The TCR also specifies when small water utilities must conduct a sanitary survey of their system. A sanitary survey is an onsite review of the public water utility, including the water source, purification facilities, equipment (including the distribution system), and the operation and maintenance procedures.

The TCR acknowledges that coliforms are occasionally present in water samples. The violation of the MCL occurs when samples from the area are repeatedly positive for coliforms.

Surface Water Treatment Rule (SWTR) 1989

The SWTR was created as a response to threats from *Giardia*, *Legionella*, and viruses. MCLGs were set at a decrease of three orders of magnitude for *Giardia* and four orders of magnitude for viruses. The SWTR applies to all PWSs using surface-water sources or groundwater sources under the direct influence of surface water. These microorganisms are difficult to enumerate, and therefore the SWTR established approved treatment techniques (e.g., filtration). Turbidity was regulated with a top limit of 5.0 nephelometric turbidity units (NTU) and a requirement that 95% of turbidity measurements in a month measure 0.5 NTU or less.

Disinfectants and Disinfection Byproducts Rule (DBPR) 1998

The DBPR addresses the problems of residual disinfectants (mainly derived from chlorine) in the water supply. Chlorination of water can result in DBPs that, over time, are associated with an increased risk of cancer (e.g., bladder cancer). Specifically, it sets Maximum Residual Disinfectant Levels (MRDL) for chlorine, chloramines, and chlorine dioxide. It also sets MRDLs for the following DBPs: trihalomethanes (e.g., chloroform, bromodichloromethane, dibromochloromethane, and bromoform), haloacetic acids (e.g., dichloroacetic acid, trichloroacetic acid), chlorite (ClO_2), and bromate (BrO_3).

Interim Enhanced Surface Water Treatment Rule (IESWTR) 1998

The IESWTR was created in response to information about parasites in water supplies, such as *Cryptosporidium* and *Giardia*. The rule affects PWSs that

TABLE 3-4 **Total Coliform Rule Monitoring Frequencies**

PWS ROUTINE Monitoring Frequencies

Population	Minimum Samples/Month
25–1,000*	1
1,001–2,500	2
2,501–3,300	3
3,301–4,100	4
4,101–4,900	5
4,901–5,800	6
5,801–6,700	7
6,701–7,600	8
7,601–8,500	9
8,501–12,900	10
12,901–17,200	15
17,201–21,500	20
21,501–25,000	25
25,001–33,000	30
33,001–41,000	40
41,001–50,000	50
50,001–59,000	60
59,001–70,000	70
70,001–83,000	80
83,001–96,000	90
96,001–130,000	100
130,001–220,000	120
220,001–320,000	150
320,001–450,000	180
450,001–600,000	210
600,001–780,000	240
780,001–970,000	270
970,001–1,230,000	300
1,230,001–1,520,000	330
1,520,001–1,850,000	360
1,850,001–2,270,000	390
2,270,001–3,020,000	420
3,020,001–3,960,000	450
≥ 3,960,001	480

NOTE: *Includes PWSs that have at least 15 service connections but serve <25 people.

supply more than 10,000 people and that use either surface water or groundwater that is under the direct influence of surface water (GWUDI). Groundwater is classified as GWUDI if it is within 500 ft of a surface-water source and either has demonstrated coliform contamination from that surface water or has demonstrated that an inadequate aquifer barrier exists. This rule acknowledges that groundwater and surface water are not mutually exclusive, and that mixing of the two can occur in some locations.

Water treatment must be designed to enable a reduction of *Cryptosporidium* oocysts by two orders of magnitude. Given the usual low concentration of oocysts in water supplies, this reduction should be sufficient. Further, because oocysts are often associated with increased turbidity, the rule established a maximum turbidity of 1.0 NTU and a further requirement that 95% of measurements in a month be less than 0.3 NTU. These required levels are substantially more stringent than those established by the 1989 SWTR. In addition, guidelines were created for reservoirs and watersheds.

Long Term 1 Enhanced Surface Water Treatment Rule (LT1ESWTR) 2002

The LT1ESWTR extends the provision of the IESWTR to include water systems that serve less than 10,000 people. It also coordinates compliance with the DBPR by requiring supporting data for changes in the disinfection processes in water treatment systems. In this way, compliance with parasite removal will not lead to higher concentrations of DBPs.

Stage 2 Disinfectant and Disinfection Byproduct Rule (Stage 2 DBPR) 2005

The Stage 2 DBPR is a more sophisticated version of the previous DBPR, in which DBPs are required to be monitored at discrete locations and in which a running annual average of total trihalomethanes and of the five haloacetic acids is calculated. This differs from the previous rule in which the running annual average was calculated for the system as a whole. An emphasis is placed on monitoring and evaluation of DBPs in the water system in order to avoid occasional "spikes" in the concentration of harmful compounds. The Stage 2 DBPR is scheduled to be phased in after several years of study, with full compliance slated for some time between 2012 and 2016.

Long Term 2 Enhanced Surface Water Treatment Rule (LT2ESWTR) 2005

The LT2ESWTR, sometimes abbreviated as LT2, was adopted to account for high levels of *Cryptosporidium* in certain high-risk water systems. The IESWTR is sufficient for most communities, but it is insufficient for high-risk areas and for water systems that do not filter their water. The rule stipulates a monitoring time period, during which coliforms and *Cryptosporidium* are enumerated. Based on the results, augmentation of the water purification methods may be required to ensure removal of oocysts. The rule is extensive and complicated, but it generally targets water systems that have proven problems with *Cryptosporidium* outbreaks. Most water systems are unaffected by this ruling, although programs must monitor the oocyst concentration in each water supply.

The water systems with *Cryptosporidium* are evaluated for oocyst removal according to their current water treatment processes. That is, they are assigned a numerical value of oocyst removal efficiency in terms of an order of magnitude ("log") of removal. If needed, they must make certain changes to increase this efficiency.

As shown in the **Table 3-5**, water systems are denoted with a "bin" number depending on their proven oocyst concentration. For bin 2 systems, the total oocyst removal target is 4.0 log, or four orders of magnitude. To cite an example, conventional treatment provides 3.0 log of that number; the other 1.0 log must be provided in another manner. For bin 3 the total is 5.0 log. For bin 4 the total is 5.5 log. The list of other treatment credits includes slow sand or diatomaceous earth filtration (3.0 log) and direct filtration (2.5 log). Treatment credit for alternative technologies is assigned by individual states.

Public water systems that serve fewer than 10,000 people and are not required to monitor *Cryptosporidium* are characterized in bin 1. Unfiltered water systems must demonstrate a 2.0–3.0 log removal efficiency, depending on the prior monitoring results.

TABLE 3-5 Oocyst Concentration and Treatment

The LT2ESWTR provides a convenient table that informs the PWS of the various methods that can be followed to lower their potential oocyst problem by orders of magnitude ("log units").

		Additional treatment requirement if your PWS uses. . . .		
Oocyst concentration	Bin number	Conventional filtration, diatomaceous earth, or slow sand filtration	Direct filtration	Alternative filtration technology
<0.075 oocysts/liter	1	None	None	None
0.075–1.0 oocysts/liter	2	1 log	1.5 log	State-determined
1.0–3.0 oocysts/liter	3	2 log	2.5 log	State-determined
>3.0 oocysts/liter	4	2.5 log	3.0 log	State-determined

Conventional treatment is defined as having separate, sequential process units for coagulation and flocculation, clarification (in which the solids are removed), and granular media filtration (such as a sand filter).

How are the additional credits obtained? PWSs can obtain treatment credit by instituting a number of changes. A "microbial toolbox" contains a variety of treatments. They include the following:

- Watershed control program (0.5 log)—following a state-approved plan, including annual surveys
- Presedimentation basin with coagulation (0.5 log)—demonstrated turbidity reduction using this method
- Two-stage lime softening (0.5 log)—a precipitation method for dissolved minerals
- Bank filtration (05–1.0 log)—a pumping well is used to recover surface water that has been filtered through soil from a river bed or a river bank. Credit 0.5 if the pump is 25 feet away from the surface water, credit 3.0 if it is 50 feet away.
- Combined filter performance (0.5 log)—as shown by decrease in turbidity
- Individual filter performance (0.5 log)—as shown by decrease in turbidity
- Bag and cartridge filters (up to 2.0 log)—as measured in tests
- Membrane filtration (depends on demonstrated efficiency)—as measured in tests
- Second stage filtration (0.5 log)—a second granular media filtration, used if the first step in the series was coagulation
- Slow sand filters (2.5–3.0 log)—credit for 2.5 if used as secondary filtration, credit 3.0 if used as primary filtration
- Chlorine dioxide treatment—(depends on performance) as measured in tests
- Ozone treatment—(depends on performance) as measured in tests
- UV light treatment—(depends on performance) as measured in tests

The rule stipulates certain other changes, such as covering reservoirs that contain finished water. This provides another layer of protection for water supplies. Water systems that change their processes must conduct a thorough review before implementing them.

QUESTIONS FOR DISCUSSION

1. Three levels of testing are used to determine whether coliforms are present in water. At each stage, how might mistaken identification occur?
2. What is a coliform? How is it different from a fecal coliform?
3. Which is the better disinfectant, chlorine or ozone? Why?
4. For a water system the k value is 10^4 l·mol^{-1} min^{-1}; the goal is to reduce the bacterial concentration from 10^8 to 10^2 bacteria per liter; and the time factor must be 3 h. What is the concentration of disinfectant that must be used?
5. In question 4, if the k value is increased to 10^5, what will happen to the D value?
6. Summarize the critical steps of water purification.
7. If a slow sand filter is used to purify water, what will happen when a toxic material enters the input water supply?
8. The difference between effective dose and the MCL is very slight. If fluoridation were a new application proposed today, would it be allowed?
9. The action of fluoridation is still not known precisely. Should we promote a method when we are not sure how it works?
10. What do you perceive as the trend in water legislation? Which pathogens do the regulators seem to care about, and how are the rules being adapted to meet these challenges?
11. The first major water legislation was produced in 1972. What changes to society have occurred since then that are relevant to the water supply?

References

Arnold, F.A., H.T. Dean, P. Jay, and J.W. Knutson. 1956. Effect of fluoridated public water supplies on dental caries prevalence. Public Health Reports 71: 652–658.

Arnold, F.A., H.T. Dean, and J.W. Knutson. 1953. Effect of fluoridated public water supplies on dental caries prevalence. Seventh year of the Grand Rapids-Muskegon Study. Public Health Reports 68: 141–148.

Arul, L., G. Benita, and P. Balasubramnaian. 2008. Functional insight for ß-glucuronidase in Escherichia coli and Staphylococcus sp. RLH1. Bioinformation 2: 339–343.

Breed, R.S. and J.F. Norton. 1937. Nomenclature for the colon group. Am. J. Public Health 27: 560–563.

Chick, H. 1908. An investigation of the laws of disinfection. J. Hygiene 8: 92–157.

Chaueret, C.P., C.Z. Radziminski, M. Lepuil, R. Creason, and R.C. Andrews. 2001. Chlorine dioxide inactivation of *Cryptosporidium parvum* oocysts and bacterial spore indicators. Appl. Environ. Microbiol. 67: 2993–3001.

Christian, R.R. and W.O. Pipes. 1983. Frequency distribution of coliforms in water distribution systems. Appl. Environ. Microbiol. 45: 603–609.

Cromeans, T.L., A.M. Kahler, and V.R. Hill. 2010. Inactivation of Adenoviruses, Enteroviruses, and Murine Norovirus in water by free chlorine and monochloramine. Appl. Environ. Microbiol. 76: 1028–1033.

Dean, H.T., R.M. Dixon, and C. Cohen. 1935. Mottled enamel in Texas. Public Health Reports 50: 424–442.

Eaton, A.D., L.S. Clesceri, E.W. Rice, and A.E. Greenberg (eds.). "Standard Methods for the Examination of Water and Wastewater," 21st Edition. 2005. American Public Health Association, American Water Works Association, Water Environment Federation.

Fawell, J., K. Bailey, J. Chilton, E. Dahi, L. Fewtrell, and Y. Magara. Fluoride in drinking-water. IWA Publishing, London, 2006.

Freeze, R.A. and J.H. Lehr. The Fluoride Wars. John Wiley & Sons, Inc. Hoboken, NJ, 2009.

Glaze, W.H. 1987. Drinking-water treatment with ozone. Environ. Sci. Technol. 21: 224–230.

Hammer, M.J. and M.J. Hammer. "Water and Wastewater Technology", 3rd ed. Prentice Hall, Englewood Cliffs, NJ. 1996.

Hapeman, C.J., C.P. Dionigi, P.V. Zimba, and L.L. McConnell. 2002. Agrochemical and nutrient impacts on estuaries and other aquatic systems. J. Agric. Food Chem. 50: 4382–4384.

Jones, S., B.A. Burt, P.E. Petersen, and M.A. Lennon. 2005. The effective use of fluorides in public health. Bull. WHO 83: 670–674.

Kuramitsu, H.K. 2003. Molecular genetic analysis of the virulence of oral bacterial pathogens: an historical perspective. Crit. Rev. Oral Biol. Med. 14: 331–344.

Johnson, C.H., E.W. Rice, and D.J. Reasoner. 1997. Inactivation of *Helicobacter pylori* by chlorination. Appl. Environ. Microbiol. 63: 4969–4970.

Leclerc, H., D.A.A. Mossel, S.C. Edberg, and C.B. Struijk. 2001. Advances in the bacteriology of the coliform group: their suitability as markers of microbial water safety. Annu. Rev. Microbiol. 55: 201–234.

Leone, N.C., M.B. Shimkin, F.A. Arnold, C.A. Stevenson, E.R. Zimmerman, and P.A. Geiser. 1954. Medical aspects of excessive fluoride in water supply. Public Health Rep. 69: 925–936.

Moore, W.E. and L.V. Moore. 1994. The bacteria of periodontal diseases. Periodontology 2000 5: 66–77.

Mullen, J. 2005. History of water fluoridation. British Dental Journal 199: 1–4.

O'Connell, H.A., L.J. Rose, A. Shams, M. Bradley, M.J. Arduino, and E.W. Rice. 2009. Variability of Burkholderia pseudomallei strain sensitivities to chlorine disinfection. Appl. Environ. Microbial. 75: 5405–5409.

Rice, E.W., M.J. Allen, and S.C. Edberg. 1990. Efficacy of ß-glucuronidase assay for identification of Escherichia coli by the defined-substrate technology. Appl. Environ. Microbiol. 56: 1203–1205.

Russell, R.R.B. 2008. How has genomics altered our view of caries microbiology? Caries Res. 42: 319–327.

Shin, G. and M.D. Sobsey. 2008. Inactivation of norovirus by chlorine disinfection of water. Water Res. 42: 4562–4568.

Thurston-Enriquez, J.A., C.N. Haas, J. Jacangelo, and C.P. Gerba. 2005. Inactivation of enteric Adenovirus and feline Calicivirus by chlorine dioxide. Appl. Environ. Microbiol. 71: 3100–3105.

Wang, L.K. Water chlorination and chloramination, in: Handbook of Environmental Engineering, Volume 4: Advanced Physicochemical Treatment Processes. L.K. Wang, Y.T. Hung, and N.K. Shammas (eds). Humana Press, Totowa, NJ. 2006. (Chap. 11) pp. 367–401.

Watson, H.E. 1908. A note on the variation of the rate of disinfection with change in the concentration of the disinfectant. J. Hygiene 8: 536–542.

CHAPTER 4

Waterborne Disease

LEARNING OBJECTIVES

- Distinguish between MCLG and MCL.
- Recognize the chemical and biological contaminants that are routinely examined.
- Describe how contaminants enter the water supply.
- Define Infectious Dose and explain its limitations.
- Describe what a case-control study is and how it contributes to our understanding of infectious disease.
- Identify the major pathogens that are associated with drinking water.
- Evaluate the likelihood of acquiring specific pathogens in a country with modern treatment systems.
- Identify the major pathogens that are associated with recreational use of water, as well as their mode of infection.
- Describe where harmful algal blooms come from and what their health consequences are.

KEY TERMS

- Carcinogenicity
- Case-control study
- Clinical cases
- Maximum Contaminant Level Goal (MCLG)
- Maximum Contaminant Level (MCL)
- Mutagenicity
- Subclinical
- Teratogenicity
- Toxicity

WHAT CAN GO WRONG WITH WATER?

The people that provide our drinking water have a tremendous responsibility. We take for granted that water from the tap will be potable anytime we use it for drinking, washing food, or brushing our teeth. In many parts of the world this is not the case; the water supply may be completely untreated. In general, the water utilities of a modern industrial country do a good job. In a modern water supply system, it is highly unlikely that a person will get sick from drinking any single glass of water. This does not mean that there are no harmful agents in the water, or that no one ever gets sick from the water supply. Rather, ingesting a concentration of any contaminant that is high enough to cause illness is highly unlikely. Generally speaking, water supplies in the United States are of high quality.

The task of the public water system is to produce water that is fit to drink. There are many substances found in water that make it unfit for consumption (i.e., chemical contaminants and pathogens), and the pH of the water can be too acidic or too alkaline. But these conditions are

TABLE 4-1 Off Odors in Water Supplies

Algal Compound	Reported Odor
Dimethyl trisulfide	Septic, garlic, putrid, swampy
n-Heptanal	Fishy, oily
n-Hexanal	Grassy, fatty
3-Methylbutyrate	Rotten, rancid
Trans-2-Nonenal	Cucumber
2-Methylpent-2-enal	Rum, marzipan
1,3 Octadiene	Earthy/mushroom
Methylamine	Ammonia
beta-Ionone	Violets
Geosmin	Earthy/musty
Limonene	Citrus
Linalool	Grassy, floral
beta-Cyclocitral	Tobacco, smoky, moldy
2-Isobutyl-3-methoxy pyrazine	Earthy/potato bin

SOURCE: From Watson, S.B. (2004).

difficult for the consumer to detect. If water is rejected by the consumer, it is likely to be because of bad taste or smell of the water, or because it has an off-color or is noticeably turbid. If surface water is the source of drinking water, it is possible that algae in the water have imparted some of these unpleasant qualities to the water. They can also add deleterious substances to the water. **Table 4-1** shows some of the odors that algae can impart to water supplies. Most are obviously bad odors, with some being highly objectionable; some (curiously) are actually pleasant but still are unwanted in the water supply.

Contaminants can enter the water supply at a number of points. Surface water can be contaminated by run-off from many different sources, such as farm fields or feedlots (i.e., carrying fertilizers, fecal wastes, or pesticides), industrial sites (i.e., carrying inorganic and organic chemicals, waste from food processing, and man-made chemicals), and urban wastes that are washed from streets after rainstorms (e.g., litter, gasoline, motor oil, anti-freeze, etc.) into the local water supply. Power plants use water for cooling purposes and may discharge heated water back into the environment. All types of organisms in the water supply are unaccustomed to heated water, and significant changes in temperature may be lethal or have other deleterious effects. Mining sites may be responsible for acid mine drainage, which lowers the pH so much that it kills all forms of life in a stream. Even recreational uses on surface water leads to significant contamination because gas and oil can easily spill into the water. People often add their septic waste or household waste to the water supply.

CHEMICAL CONTAMINANTS

Examination of water samples reveals that chemical contaminants are more likely to exceed regulatory limits than are biological contaminants. The biological agents—bacteria and parasites—can usually be removed or killed using a variety of treatment techniques, elevating water quality to acceptable levels. Chemical contaminants can occasionally and unexpectedly show high concentrations that are only found later using chemical analysis. Water utilities technicians analyze water for a wide variety of possible contaminants.

Chemical contaminants can cause one or more health effects: toxicity, carcinogenicity, mutagenicity, and teratogenicity. **Toxicity** refers to an adverse effect on the body or on a specific tissue, such that it is damaged and/or destroyed. In high doses these toxins can cause death, although this is a highly unusual occurrence. The damage may be acute (with effects observed a short time after exposure) or chronic (effects are only seen after exposure to a sustained low amount of the compound). **Carcinogenicity** is the result of chronic exposure to compounds that cause or promote cancerous growths or tumors. **Mutagenicity** occurs when the compound causes heritable changes in the DNA of living cells. Some of these mutations may be benign, while others are malignant. **Teratogenicity** causes abnormalities in a developing fetus that causes a child to be born with some type of deformity or disability.

As mentioned above, chemical concentrations in water can be high enough to give a toxic dose and an immediate illness, but this is extremely unlikely. Far more likely is the occurrence of water containing subacute concentrations of harmful chemicals that will, over time, have a cumulative effect on humans. The biggest concern is that continual exposure to small amounts of contaminants will eventually lead to cancers, such as liver or bladder cancer. Therefore, acceptable levels of chemicals are set for lifetime exposures that significantly reduce cancer risks. For example, benzene, a known carcinogen, must be held to concentrations such that it will not cause cancer, even if people drink that concentration over a lifetime. Fuel spills contaminate water with benzene, toluene, ethylbenzene, and xylene, the so-called BTEX group.

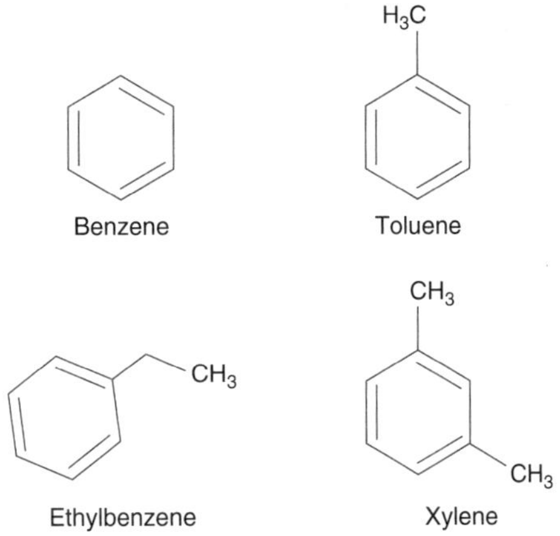

In setting these levels, two concentrations are used. One is the **Maximum Contaminant Level Goal (MCLG)**. This level may be an achievable ideal, but not a legal requirement. In contrast, the **Maximum Contaminant Level (MCL)** is a fixed regulatory figure. Concentrations of a particular chemical must be below this concentration to comply with the law (see Chapter 6). The law is based on a 70-kg person consuming 2 liters of water per day. **Tables 4-2** through **4-6** list the chemical contaminants and their current regulatory limits, as defined by the Environmental Protection Agency (EPA).

The BTEX group

Benzene

Toluene

Ethylbenzene

Xylene

Chemical contaminants enter the water supply through a number of mechanisms. As shown in Table 4-2, the organic compounds include a number of pesticides and insecticides, which may originate from household use or farm fields and which enter the water supply through runoff. The organics also include a number of solvents and industrial chemical intermediates, which probably originate from factories. The inorganics, which are mostly the heavy metals (**Table 4-3**), may also originate from an industrial factory, but substantial concentrations of these metals in the soil and rock gradually dissolve in the groundwater. The disinfectants (**Table 4-4**) result from their use in

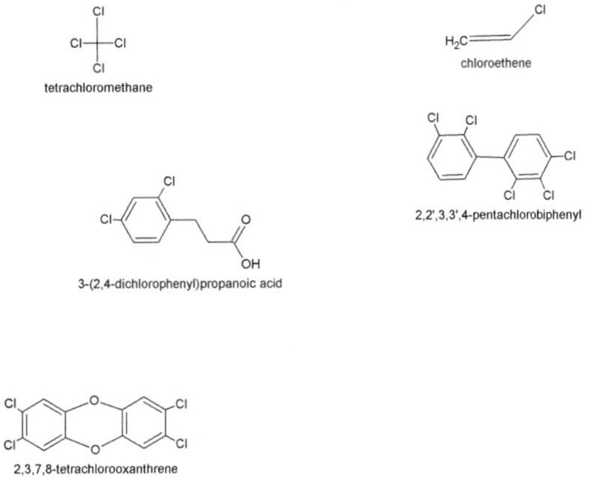

tetrachloromethane

chloroethene

2,2',3,3',4-pentachlorobiphenyl

3-(2,4-dichlorophenyl)propanoic acid

2,3,7,8-tetrachlorooxanthrene

TABLE 4-2 Organic Compounds Include Common Contaminants such as Pesticides and Petroleum Derivatives

Contaminant	MCLG (mg/l)	MCL (mg/l)	Notes
Acrylamide	zero	See note below	A vinyl polymer used industrially in the paper industry and as a flocculent agent in water treatment. A neurotoxin.
Alachlor	zero	0.002	An herbicide widely used in farming. Toxic to the liver and a probable human carcinogen.
Atrazine	0.003	0.003	An herbicide that is still used in the United States, but banned in Europe.
Benzene	zero	0.005	The basic aromatic ring compound. Found in many industrial applications, in gasoline, and in cigarette smoke. Damages the bone marrow; implicated in leukemia.
Benzo(a)pyrene (PAHs)	zero	0.0002	A 5-ringed compound that is the product of incomplete combustion of hydrocarbons. Both mutagenic and carcinogenic.
Carbofuran	0.04	0.04	A pesticide that is very effective, but so destructive of other wildlife that it is being phased out.
Carbon tetrachloride	zero	0.005	Previously used as a solvent and a cleaning agent. Toxicity is acute in liver and kidneys.
Chlordane	zero	0.002	A pesticide that was used until 1988. Toxic to many organs, including heart tissue.
Chlorobenzene	0.1	0.1	A solvent and a precursor molecule for other organic compounds. Toxic to liver, kidneys, central nervous system.
2,4-D	0.07	0.07	2,4-Dichlorophenoxyacetic acid. Widely used as an herbicide. Data are inconclusive about cancer risk.
Dalapon	0.2	0.2	2,2-Dichloropropionic acid. An herbicide that can irritate the respiratory tract if inhaled.
1,2-Dibromo-3-chloropropane (DBCP)	zero	0.0002	A pesticide that is no longer used. Probable carcinogen.
o-Dichlorobenzene	0.6	0.6	1,2 dichlorobenzene. Used to make pesticides.

Contaminant	MCLG (mg/l)	MCL (mg/l)	Notes
p-Dichlorobenzene	0.075	0.075	1,4 dichlorobenzene. The active ingredient in mothballs. Suspected of being a carcinogen, but probably only at high concentrations.
1,2-Dichloroethane	zero	0.005	Ethylene dichloride. Used to make polyvinyl chloride (PVC) and a variety of other consumer goods. Probably only dangerous (carcinogenic?) at high doses.
1,1-Dichloroethylene	0.007	0.007	1,1 dichloroethene. Used industrially in the production of vinyl chloride. Toxic to the liver at all concentrations.
cis-1,2-Dichloroethylene	0.07	0.07	1,2 dichloroethene or 1,2 DCE. Has been used as an anesthetic. Affects the central nervous system. The cis version has been shown to decrease red blood cell counts.
trans-1,2-Dichloroethylene	0.1	0.1	Very similar to the cis version.
Dichloromethane	zero	0.005	Methylene chloride. A volatile chemical used in many applications. Considered a carcinogen, but only at very high concentrations.
1,2-Dichloropropane	zero	0.005	Formerly used as a solvent and as a fungicide. Damages the liver, kidneys, and lungs.
Di(2-ethylhexyl) adipate	0.4	0.4	DEHA. A plasticizer and component of lubricants. May be responsible for liver carcinomas.
Di(2-ethylhexyl) phthalate	zero	0.006	DEHP. Very similar to DEHA.
Dinoseb	0.007	0.007	An herbicide banned by the EPA in 1986. Interferes with ATP formation.
Dioxin (2,3,7,8-TCDD)	zero	0.00000003	Refers to a general class of compound, of which this one is the archetype. By–products and contaminants of various industrial processes and from combustion. No known uses. Causes many toxic outcomes. Very toxic, LD50 = 0.001 mg/kg body weight.
Diquat	0.02	0.02	An herbicide used in controlling aquatic weeds. Similar to paraquat. Affects the gastrointestinal system, liver, and kidneys.
Endothall	0.1	0.1	An herbicide. Toxic to some fish species, but only dangerous to humans at high doses.
Endrin	0.002	0.002	A cyclodiene insecticide. Also kills rodents and birds. Only dangerous to humans in acute poisonings.

(continued)

TABLE 4-2 *continued*

Contaminant	MCLG (mg/l)	MCL (mg/l)	Notes
Epichlorohydrin	zero	See note below	A chemical intermediate used to make epoxy resins. Mutagenic and a probable carcinogen.
Ethylbenzene	0.7	0.7	Found in gasoline and as a solvent in paints. Used in styrene manufacture. Can damage either the kidney or the ear (hearing). Probable carcinogen.
Ethylene dibromide	zero	0.00005	Dibromoethane. A chemical intermediate and an insecticide for high value uses, such as termites. Probable carcinogen.
Glyphosate	0.7	0.7	N-phosphonomethyl glycine. A popular and effective broad-spectrum herbicide. Usually safe to use, although it should never be ingested.
Heptachlor	zero	0.0004	A cyclodiene insecticide. Now used only for critical needs such as fire ant eradication.
Heptachlor epoxide	zero	0.0002	A derivative of heptachlor.
Hexachlorobenzene	zero	0.001	A fungicide that was used on dry grains, and thus a risk for ingestion. Suspected teratogen. No longer used in the United States
Hexachlorocyclopentadiene	0.05	0.05	A chemical intermediate for many products.
Lindane	0.0002	0.0002	Gamma-Hexachlorocyclohexane. An insecticide that can be used topically for eradication of head lice. Can damage the liver and kidneys, but dangerous only in high amounts.
Methoxychlor	0.04	0.04	An insecticide of the dichlrodiphenyl-ethane class. Attacks the nervous system, but only dangerous at high doses.
Oxamyl	0.2	0.2	Vydate. An insecticide that can cause weakness, blurred vision, and headaches at high doses.
Polychlorinated biphenyls (PCBs)	zero	0.0005	General title for 209 congeners that have the biphenyl structure and chlorine residues. Used extensively as dielectric fluids in transformers.

Contaminant	MCLG (mg/l)	MCL (mg/l)	Notes
Pentachlorophenol	zero	0.001	General use as a biocide in the paper and wood industries. No longer used because of the other contaminants found in it. Easily absorbed through the skin.
Picloram	0.5	0.5	An herbicide used on woody plants. Only dangerous at high doses.
Simazine	0.004	0.004	An aquatic herbicide (algicide). Similar to atrazine.
Styrene	0.1	0.1	The building block of polystyrene, which is used for many consumer products. Can cause hearing loss and liver damage, and is a possible carcinogen.
Tetrachloroethylene	zero	0.005	PERC or perchloroethylene. A degreaser and a common component of dry cleaning. Probable carcinogen.
Toluene	1	1	A component of gasoline and paints, and a chemical intermediate for many products. Affects the nervous system.
Toxaphene	zero	0.003	Complex insecticide that is no longer used in the United States. Possible carcinogen. Can affect many different organs, including the immune system.
2,4,5-TP (Silvex)	0.05	0.05	2-(2,4,5-Trichlorophenoxy)propionic acid. An herbicide. Only dangerous at high concentrations.
1,2,4-Trichlorobenzene	0.07	0.07	Intermediate in herbicides and wood preservatives. At high concentrations it will affect the adrenal glands.
1,1,1-Trichloroethane	0.2	0.2	Methylchloroform. A degreaser and component of glues. No longer manufactured in the United States. Hazardous only in high amounts.
1,1,2-Trichloroethane	0.003	0.005	A solvent. Can affect many organs, including the stomach.
Trichloroethylene	zero	0.005	TCE. Solvent and degreaser. Found in limited amounts in many household products. Suspected carcinogen.
Vinyl chloride	zero	0.002	The building block of polyvinyl chloride (PVC). Known carcinogen, affecting many organs.
Xylenes (total)	10	10	Three isomers. Very similar to toluene.

NOTE: Acrylamide can be used in wastewater treatment. Epichlorohydrin is a byproduct of its use. In these cases, a separate regulatory limit is established for these chemicals. LD50 = the dose that kills half (50%) of the animals tested. Based on information from the U.S. Environmental Protection Agency Klaasen, C.D. (ed). Casarett and Doull's *Toxicology: The Basic Science of Poisons*. 6th edition. McGraw-Hill, New York. 2001.

TABLE 4-3 **Inorganic Chemicals and Metals, Frequent Contaminants in Both Groundwater and Surface Water**

Chemical name	MCLG (mg/l)	MCL (mg/l)	Notes
Antimony	0.006	0.006	Sb. Released during industrial processes, it makes its way to water supplies. Most cases of toxicity are related to occupational hazard.
Arsenic	zero	0.01	As. Chronic exposure can affect both peripheral and central nervous systems. Easily absorbed through the gastrointestinal tract. Known carcinogen.
Asbestos (fiber (fiber >10 micrometers)	*	*	For both MCLG and MCL, the fibers per liter of water. Any of several minerals containing silicon. Ingestion may lead to intestinal polyps.
Barium	2	2	Ba. A common element; becomes associated with bone. Not very dangerous.
Beryllium	0.004	0.004	Be. Released into the air through coal combustion. Carcinogenic, and causes contact dermatitis.
Cadmium	0.005	0.005	Cd. Implicated in lung and prostate cancers. Toxic to kidney tissue, and adversely affects calcium metabolism.
Chromium	0.1	0.1	Cr. Used in chrome plating and stainless steel production. Human exposure is mostly through food.
Copper	1.3	See note below Action Level = 1.3	Cu. Essential in the diet, but high levels will cause gastrointestinal upset. Chronic exposure affects the liver and kidneys.
Cyanide (as CN-)	0.2	0.2	CN. Includes HCN, NaCN, KCN. In moderate concentrations it is extremely dangerous. Chronic exposure leads to damage to thyroid or nervous tissue.
Fluoride	4	4	F. A common element that is used to prevent tooth decay. Excess causes discoloration of teeth.
Lead	zero	See note below Action Level = 0.015	Pb. Formerly found in paint, and still a problem in old houses. Human exposure is mostly through food. Affects most organ sources, and is well known for its effects on neural development.

Chemical name	MCLG (mg/l)	MCL (mg/l)	Notes
Mercury (inorganic)	0.002	0.002	Hg. A distinction is made from methyl mercury, the organic form that accumulates in many animals. Levels in water are usually very low. In excess, can cause a variety of illnesses.
Nitrate (measured as Nitrogen)	10	10	NO_3^-. Component of fertilizers. Causes Blue Baby Syndrome in infants.
Nitrite (measured as Nitrogen)	1	1	NO_2^-. Component of fertilizers. Causes Blue Baby Syndrome in infants.
Selenium	0.05	0.05	Se. An essential element in the diet, but too much gives a variety of systemic effects. Excess is easily excreted.
Thallium	0.0005	0.002	Tl. A very toxic element, but also very rare. Easily absorbed by the gastrointestinal tract, where it can damage intestines, liver, kidneys.

NOTE: A Treatment Technique (TT) is used for copper and lead. In these cases, the regulation requires the water system to monitor the water supply, but a violation only occurs if more than 10% of the samples are above the stated action levels. Based on information from the EPA, Klaasen, C.D. (ed). Casarett and Doull's *Toxicology: The Basic Science of Poisons*. 6th edition. McGraw-Hill, New York. 2001.

TABLE 4-4 Approved Residual Disinfectant Quantities

Residual disinfectant	MRDLG (mg/L)	MRDL (mg/L)	Notes
Chloramines (as Cl2)	4	4.0	Excess causes gastrointestinal distress.
Chlorine (as Cl2)	4	4.0	Excess causes gastrointestinal distress.
Chlorine dioxide (as ClO2)	0.8	0.8	Highly irritating to gastrointestinal tract.

NOTE: MRDLG = Maximum Residual Disinfectant Level Goal; MRDL = Maximum Residual Disinfectant Level. Based on information from the EPA, Klaasen, C.D. (ed). Casarett and Doull's *Toxicology: The Basic Science of Poisons*. 6th edition. McGraw-Hill, New York. 2001.

TABLE 4-5 Disinfection Byproducts

Contaminant	MCLG	MCL	Notes
Bromate	Zero	0.01	BrO_3^-. Probable carcinogen.
Chlorite	0.8	1	ClO_2^-. Irritant to gastrointestinal tract. Causes respiratory distress.
Haloacetic acids	See note below	0.06	HAA5. MCLG depends on the chemical species. Probable carcinogens.
Total Trihalomethanes	See note below	0.080	TTHM. Bromoform and chloroform are examples. Probable carcinogens, damage to several organs.

NOTE: Some MCLG values: Bromoform, zero; dibromochloromethane, 0.06; trichloracetic acid, 0.3. Based on information from the EPA, Klaasen, C.D. (ed). Casarett and Doull's *Toxicology: The Basic Science of Poisons*. 6th edition. McGraw-Hill, New York. 2001.

TABLE 4-6 Radioactivity

Contaminant	MCLG	MCL	Notes
Alpha particles	none	15 picocuries per Liter (pCi/L)	Relatively low energy particles. Overexposure leads to increased risk of cancer.
Beta particles and photon emitters	none	4 millirems per year	Results from radioactive decay of isotopes. Overexposure leads to increased risk of cancer.
Radium 226 and Radium 228 (combined)	none	5 pCi/L	Rare natural element, often found as a gas. Overexposure leads to increased risk of cancer.
Uranium	zero	30 μg/L	Rare natural element. Overexposure leads to increased risk of cancer.

NOTE: pCi = picoCuries. Based on information from the EPA.

treating drinking water. The disinfection byproducts (**Table 4-5**) result from disinfectants that produce unwanted products during their use. The radioactive compounds (**Table 4-6**) usually enter the water supply through dissolved mineral compounds. All contaminants can be removed to satisfactory levels, but because they are not a major concern for this text the procedures are not given in detail.

BIOLOGICAL CONTAMINANTS

In contrast to chemical contaminants, biological contaminants (pathogens) in the water supply cause an immediate illness, that is, within the incubation time of the pathogen (Craun, 1988). Certain pathogens will cause chronic conditions that will, over time, cause additional injury. For example, Hepatitis B will cause a chronic condition that can lead to hepatocellular carcinoma. However, the waterborne pathogens typically cause an immediate illness, usually concerning the gastrointestinal system.

Predictably, microorganisms are easily passed to humans through water. Water is essential to life, and microorganisms thrive in it. However, we need water for drinking, food preparation, washing, and many other purposes. This gives the microorganisms living in water an opportunity to invade and cause an infection. In general, they can be ingested (food or water), they can be inhaled (causing an infection of the respiratory system), or they can contact skin or mucous membranes (causing superficial infections, irritations, dermatitis, etc.). By far the most prevalent of these routes of exposure is ingestion, and results can be devastating. The World Health Organization estimates that 1.5 million children around the world die each year from diarrhea. That is equivalent to 20% of childhood deaths from all causes. This is a tragic statistic and emphasizes the importance of safe, potable water as well as the need for better and quicker treatment, monitoring, and prevention.

As mentioned in Chapter 3, individual pathogenic species are too numerous to routinely examine in water samples; techniques for identification can be costly and time-consuming. Instead, either Treatment Techniques (TTs) are used to ensure reduction of pathogen numbers (such as for *Giardia* and viruses) or bioindicator organisms (e.g., coliform bacteria) are cultured as a surrogate for fecal contamination of water. Both of these examples can be found in **Table 4-7**. A bioindicator is a microorganism that is easily found and that has a significant coincidence with pathogens of interest. The coliform group represents many pathogens of fecal origin. It is likely that there are many more pathogens that have yet to be characterized. In one study of tap water versus filtered water, it was estimated that 35% of the gastrointestinal upset experienced by the test group was due to the water supply (Payment et al., 1991). Although this was likely due to viruses causing a mild infection, our understanding and methods of water treatment can still be improved.

THE CONCEPT OF INFECTIOUS DOSE

On a superficial level an "infectious dose" is easy to define; it is the number of microorganisms needed to cause disease in an individual. However, many factors complicate this definition in actual practice. The elements of risk from a pathogen include the type and number of the pathogen, the number of different pathogens the person is actually exposed to, the health of the pathogens, the methods used to treat the pathogens prior to human exposure, and the health of the individuals and their possible prior exposure to the pathogen(s).

The effective dosage for one person may be quite different from another person. For instance, groups with higher susceptibility (i.e., the young, the elderly, and the ill) will typically have a lower infectious dosage compared with a healthy adult (**Box 4-1**). The very young—from birth through perhaps two years of age—have an immature immune system. In addition, their small size makes them very susceptible to dehydration following diarrhea and vomiting. Ordinary diarrhea is a big killer worldwide, usually of small children. In the elderly—possibly defined as 70 years old and above—have generally diminished health and a weakened immune system. The very ill already have an underlying medical condition and thus are weakened. Many chronically ill people fit into this category. Those with compromised immune systems are also especially vulnerable, including people who are HIV positive and people undergoing cancer therapy

TABLE 4-7 Microbial Contaminants of Water

Contaminant	MCLG	MCL	Potential Health Effects from Ingestion of Water	Sources of Contaminant in Drinking Water
Cryptosporidium *sp.*	zero	Note 1	Gastrointestinal illness	Human and animal feces
Giardia lamblia	zero	Note 2	Gastrointestinal illness	Human and animal feces
Legionella	zero	Note 3	Legionnaire's Disease, a type of pneumonia	Found naturally in water; multiplies under specific nutrient conditions
Heterotrophic Plate Count	Not applicable	Note 4	No specific disease	Overgrowth of bacteria is suggestive of other problems.
Total coliforms (including fecal coliform and E. coli)	zero	Note 5	Bioindicators that suggest when fecal contamination is present. Some are also pathogenic.	Coliforms are naturally present in the environment as well as feces fecal coliforms and *E. coli* only come from human and animal fecal waste.
Viruses (enteric)	zero	Note 6	Gastrointestinal illness	Human and animal feces
Turbidity	Not applicable	Note 7	Indicator of possible contamination	Natural phenomenon

NOTE: Microorganisms differ from chemicals in that they cannot be expressed as mg/l values. Instead, they might be expressed as colony forming units (CFU) for bacteria or plaque forming units (PFU) for viruses. Many species cannot be cultivated on artificial media. In these cases, Specific TTs are used that are sufficient to remove log values of the microbes.

Note 1: A TT should be used to remove 99% of *Cryptosporidium* oocysts.

Note 2: A TT should be used to remove 99.9% of *Giardia* cysts.

Note 3: No stated limit for *Legionella* bacteria, but procedures that are effective for *Giardia* also remove *Legionella*.

Note 4: No more than 500 CFU/ml. Water sources typically have substantial microbial counts, but high counts are not advised.

Note 5: No more than 5% of samples in a month's time may be positive for coliforms. For detailed information, see Total Coliform Rule (TCR).

Note 6: A TT should be used to remove or inactivate 99.99% of viruses.

Note 7: Nephelometric turbidity units (NTUs) must be below 1 at all times, and must be below 0.3 NTU in at least 95% of samples.

BOX 4-1 At-Risk Populations

The very young (newborns to about 2 years of age). Immature immune system is unable to fight off major pathogens effectively.

The very old ("the elderly"). Immune system no longer works as efficiently.

The very sick (age varies). Immune system is compromised by medical treatment (e.g., chemotherapy or organ transplant) or by disease (e.g., HIV).

because their immune systems are destroyed by a pathogen in the former case and by the treatment in the latter case. It also includes those with organ transplants, because they must take immunosuppressive drugs to stop their immune systems from rejecting the organ, as well as those with long-term illnesses (e.g., diabetes or chronic obstructive pulmonary disease).

Even otherwise healthy people vary greatly in susceptibility to infection. People who have been vaccinated or who have acquired immunity through previous infection with the same organism will have more resistance to infection. Subtle differences in nutrition may affect the immune system's efficacy. Other small changes can have major effects. People taking antacids will reduce stomach acidity, but the acidity kills many bacteria that would otherwise invade the bowel. These people thus have a higher susceptibility of infection via this route.

Aspects of different microorganisms also vary, affecting their infectiousness. Bacteria expressing pathogenicity factors have an advantage once inside the body, but the environmental signals that promote gene expression are not entirely known. Some of these pathogenicity factors include pili for cell attachment, toxins, and secretory systems for invasive proteins. In dosage studies the bacteria are often grown in standard laboratory media where their numbers can be estimated with some accuracy, but this is not how they exist *in situ*. These factors require much more research before a complete understanding can be obtained. In contrast, many of the microorganisms in environmental conditions may be significantly weakened by chemical and physical (e.g., heat and ultraviolet light) sources. Even a high concentration of these microbes may be unable to cause infection.

Therefore it is impossible in any practical sense to determine an infectious dose; there are too many variables. Yet some benchmark is needed. The benchmark is typically referred to as the ID50, or the number of pathogens needed to make 50% of individuals in a population sick. This is a more realistic assessment than a "one size fits all" approach.

Another animal species (typically a mammal) can be used as a model. Mice, rats, rabbits, and guinea pigs have all been used to test infectious dose. Typically, a group of animals is fed doses at different concentrations, and the animals are monitored to determine the rate of illness. A similar experiment can be used to test for the lethal dose, leading to a LD50 value, the number of pathogens needed to make 50% of individuals in a test population die. This is often done for chemical exposures. The major drawbacks here include cost and the likelihood that the animals are fairly homogeneous and thus are not good predictors of exposure to a diverse human population. However, such tests can establish rough benchmarks.

It is unethical to deliberately infect people who might suffer severe consequences due to the test, even if informed consent is obtained (and for infants, informed consent is problematic). However, circumstances can occasionally create a situation in which data on infectious dose becomes available. The **case-control study** is a retrospective analysis of an outbreak of disease (**Box 4-2**). The study attempts to determine what caused the outbreak, who was affected (and which groups were disproportionately affected), and how effective the response to the outbreak was. Valuable information can be collected in this manner, both on demographic information (e.g., sex, race, and age) and on infectious dose. The greatest difficulty is obtaining an estimate of how much of the contaminated material any particular person came into contact with. Once the affected product (typically food or water) is identified, the researcher depends on the accuracy of the ill person's memory to determine

BOX 4-2 The Case-Control Study

Case: A person with an illness that we are interested in, usually because of an outbreak of disease. The researcher must define the disease as well as possible and must produce a set of conditions that the case meets so that it is accurately counted.

Control: A person who has not been ill with the disease of interest. Matching cases and controls as closely as possible is usually desired. This may mean geographical proximity but may also include factors like age, sex, race, general health, physical measurements (i.e., height and weight), social factors (i.e., rich, middle-class, or poor), or lifestyle conditions (e.g., smoking or drinking).

how much (and thus the approximate dose) was consumed. Having samples of the original affected material is very helpful, although the product has often changed substantially (i.e., water) or has been discarded (i.e., food).

When the extent of illness is calculated, it can be reported in several ways (**Box 4-3**). If the **clinical cases** are reported, the illness is so severe that the person must seek help from a medical professional (e.g., a doctor). Anyone who contacts a doctor, hospital, health authority, or who dies from the illness becomes a clinical case. Far more common are the **subclinical** cases. These are people who become sick but do not seek medical help. This definition includes people who go back to work while ill, those who simply take time off and recover on their own, those who self-medicate to get well, and anyone else who experiences symptoms of illness but who recovers without formal medical intervention. In most cases of gastrointestinal illness, the subclinical cases may outnumber the clinical cases 100-fold. Subclinical cases should not be ignored; these people are also ill and may promote secondary spread of illness.

BOX 4-3 Infections and Illness

An **infection** occurs when a microorganism enters the body and begins to multiply in a tissue where it is not wanted. For instance, when *Staphylococcus aureus* enters through a wound and begins to multiply. This may or may not lead to an illness.

An **illness** occurs when the microorganism has caused a decline in health, as evidenced by signs and symptoms. **Symptoms** are complaints that a patient reports to the doctor (e.g., achiness, tiredness, upset stomach, etc.). **Signs** are data collected by the physician that demonstrate a change from a normal condition (e.g., temperature, blood count, heart rate, etc.). An example of an illness is the presence of *Salmonella* (as determined from a stool culture) in the gastrointestinal tract of a patient who reports the symptom of severe diarrhea.

Most waterborne disease outbreaks are never identified. They are usually self-limited and do not cause fatalities. When identified, *Shigella* and *Campylobacter* dominate as causes. Here, all of the potential pathogens are described. The waterborne pathogens listed in this chapter can infect people in a number of ways (i.e., foodborne, aerosols, contact with wounds), but they are included here because they enter the host through contaminated water. The illness they cause, some of their notable pathogenicity factors, and a description of past outbreaks are included.

POTABLE WATER PATHOGENS: BACTERIA

Campylobacter

These are slender, spiral-shaped cells that are very common in the environment. The two major species of concern are *Campylobacter jejuni* and *C. coli*. *Campylobacter* species are unusual pathogens because they are microaerophilic and have an optimal temperature of 42°C. They are associated with many animal species, especially birds. While they are spread through the fecal–oral route of transmission, the influence of bird feces on water supplies is significant. (Note: The body temperature of a bird is typically a little higher than the normal human temperature.) Large flocks of birds that utilize the water source, such as a reservoir or river, or the presence of local poultry farms and their associated waste, may be a source for dissemination of *Campylobacter*. However, *Campylobacter* are difficult to cultivate on artificial media, and therefore may be difficult to find during outbreaks (**Figures 4-1** and **4-2**).

Campylobacter species cause gastroenteritis in humans, typically manifesting as flu-like symptoms and acute diarrhea. Severe abdominal pain is often the major symptom, and an initial diagnosis of appendicitis is not unusual. The disease is self-limiting but may last a week. The infectious dose probably ranges in the hundreds of cells, although definitive testing has not been published. Contamination may come from food products (e.g., meat and poultry), or perhaps from food that has been washed with contaminated water. The U.S. Centers for Disease Control (CDC) estimates the annual U.S. case rate at 13 per 100,000 making it the most common cause of bacterial gastroenteritis. Case rates in developing countries can be far higher; these bacteria have been estimated to account for 5–14% of all diarrhea cases. *Campylobacter* are easily killed

FIGURE 4-1 **Scanning electron microscope image of *Campylobacter jejuni*. Note the corkscrew shape.**

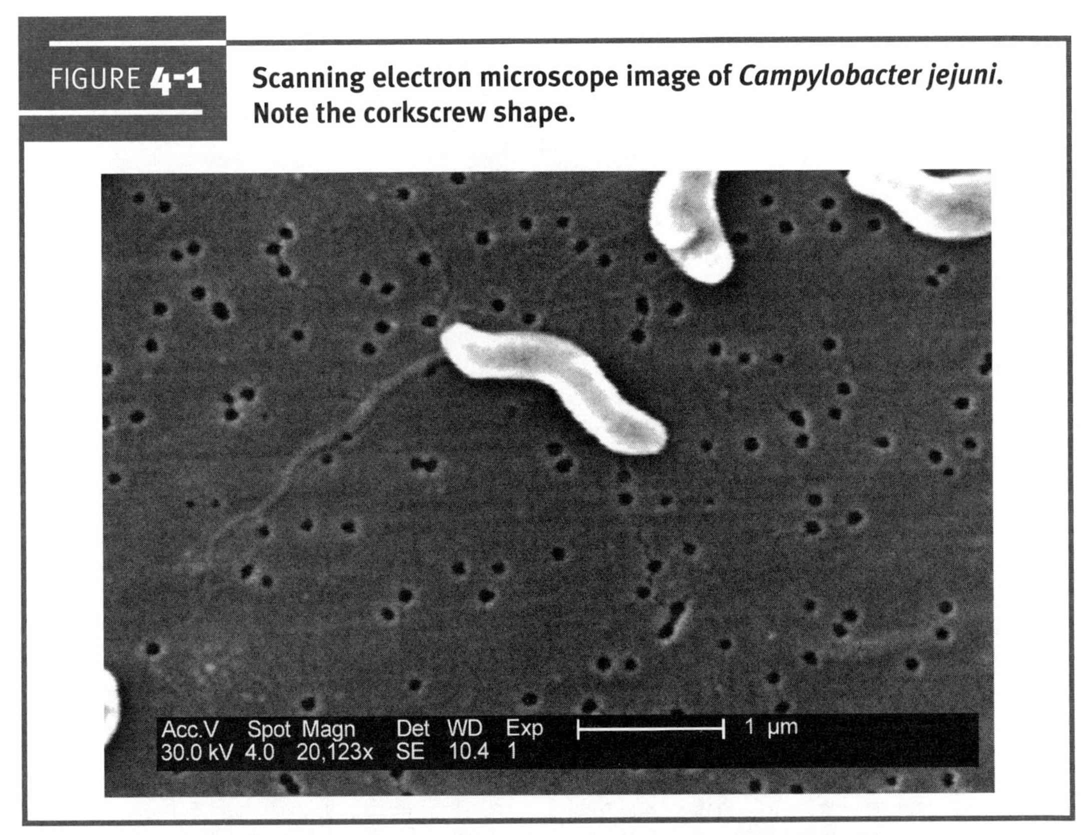

SOURCE: Courtesy of Dr. Patricia Fields and Dr. Collette Fitzgerald, Centers for Disease Control and Prevention (http://phil.cdc.gov/phil/quicksearch.asp) #5780.

FIGURE 4-2 Campylobacter colonies on a blood agar plate.

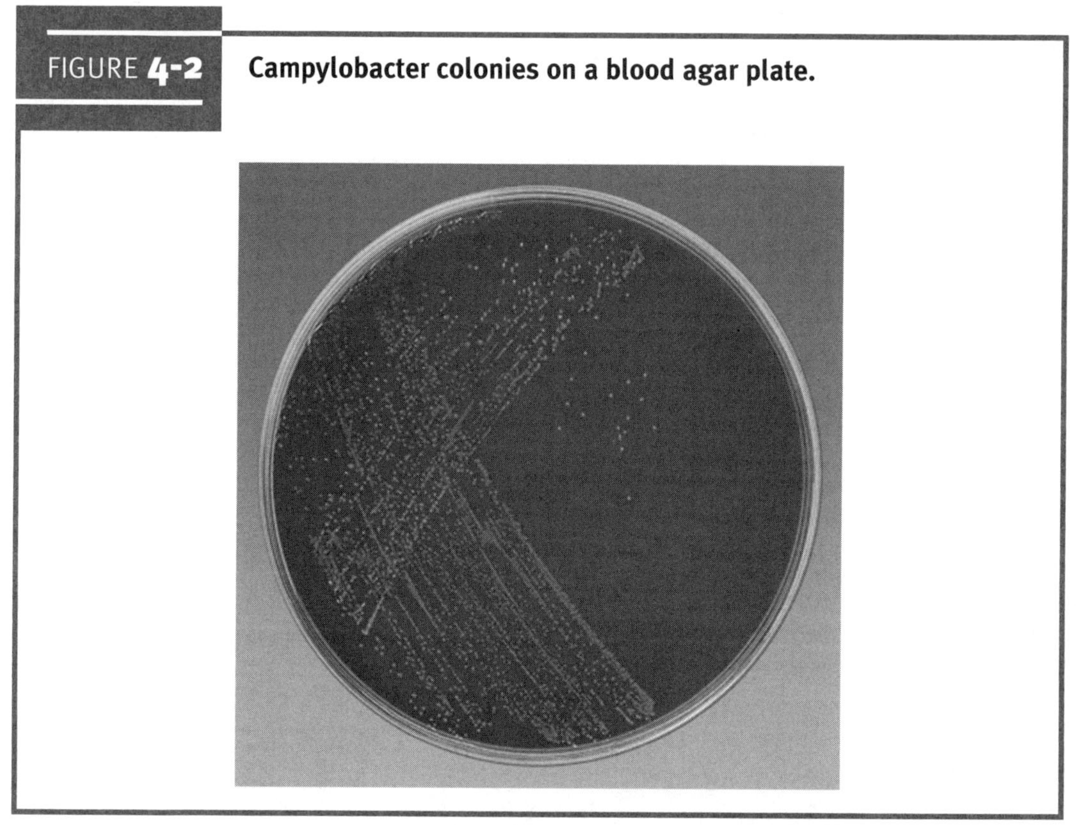

SOURCE: Courtesy of ThermoFisher.

by chlorination, so treated water is usually free of contamination. However, cross contamination or infiltration of a water source by sewage can cause illness.

Escherichia coli

E. coli (a fecal coliform bacterium) is a normal constituent of the intestinal tract of humans, other mammals, and birds used extensively as a bioindicator bacterium in the detection of fecal contamination of water. These indicator strains of *E. coli* are not hazardous; however, many other strains have acquired pathogenicity genes and have become dangerous contaminants. These strains are denoted enterohemorrhagic *E. coli* (EHEC), and they produce a bloody diarrhea. The enterotoxins that they produce damage the intestinal lining and cause bleeding. Enterohemorrhagic *E. coli* is typically self-limiting in adults with an illness of about 7 days, but in small children it can lead to Hemolytic Uremic Syndrome (HUS) and can be fatal. Even in nonfatal cases, the kidneys can sustain permanent damage. Monitoring has shown that EHEC are present in low numbers in cattle feces. Given the right conditions, these pathogens multiply to much higher levels.

Several types of pathogenic strains have been described. Enteropathogenic *E. coli* (EPEC) is commonly found in infants. It presents a danger because diarrhea can quickly dehydrate an infant and become life threatening. Enterotoxigenic *E. coli* (ETEC) is commonly referred to as Traveler's Diarrhea. The infectious dose has been estimated at between 10^8 and 10^{10} bacteria; this seems high, but this amount could easily be found in a glass of water. Enteroinvasive *E. coli* (EIEC) produces a watery diarrhea. The enteroaggregative *E. coli* (EAEC) and the Diffuse Adherent *E. coli* (DAEC) are other known types. The most well-known strain of this pathogen is *E. coli* O157:H7, an EHEC (**Figures 4-3** and **4-4**). Here the O refers to the core antigen, part of the lipopolysaccharide; the H refers to the flagellar antigen.

E. coli was chosen as the archetypal bioindicator partly because of its ubiquity in the environment. The pathogenic strains are also persistent. They can survive environmental extremes, although chlorination is very effective in killing them. *E. coli* in the water supply is always an indication of fecal contamination. The infectious dose appears to be high (10^8 organisms), but infections have been reported from swimming in contaminated pools and in natural water sources.

E. coli O157:H7 can be detected on special sorbitol MacConkey agar plates. In contrast to other *E. coli* strains, O157 does not utilize sorbitol and therefore will remain colorless on these special plates. A description of an *E. coli* outbreak is presented in Chapter 6.

FIGURE **4-3** **A transmission electron micrograph of *E. coli* O157.**

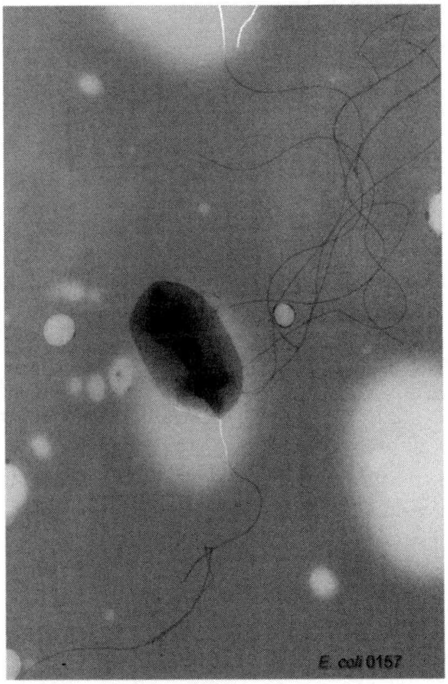

SOURCE: Courtesy of Elizabeth H. White, M.S./CDC.

Flavobacterium

Flavobacterium are aerobic, Gram-negative bacilli that are considered environmental isolates. As such, they are persistent in water systems, probably through biofilm growth. They require very little carbon, and water systems that are not flushed regularly may have enough organic build up to support *Flavobacterium* growth.

These bacteria can cause a mild gastroenteritis, although in susceptible populations they can even cause septicemia and meningitis. *Flavobacterium* are known for resistance to chlorine, which also helps them persist in water pipes.

Helicobacter pylori

Helicobacter pylori is a very slender Gram-negative bacillus. It forms a tight spiral, almost like a corkscrew (**Figure 4-5**). This bacterium is now well known as the causative agent of gastric and duodenal ulcers (Marshall and Warren, 1984). It is a very common infectious agent, even in industrialized countries, where as much as 45% of the population may be infected (although most are asymptomatic). In developing countries this rate can be as high as 90%.

Beyond the suffering due to the ulcer itself, the *H. pylori* infection is also associated with a greater risk of developing a gastric cancer. It has also been linked to gastric adenocarcinoma, one of the most common cancers worldwide (Herrera and Parsonnet, 2009; Nomura et al., 1991), and to gastric non-Hodgkins lymphoma (Zucca et al., 1998). *H. pylori* is found in water supplies, and currently the best evidence indicates that the source of *Helicobacter* is drinking water. Because humans are the only known reservoir, the carrier rate in the general population may be quite high.

FIGURE 4-4 ***E. coli*** **growing on Violet Red Bile Agar plates. The plate in the foreground has the MUG substrate, which allows easy identification.**

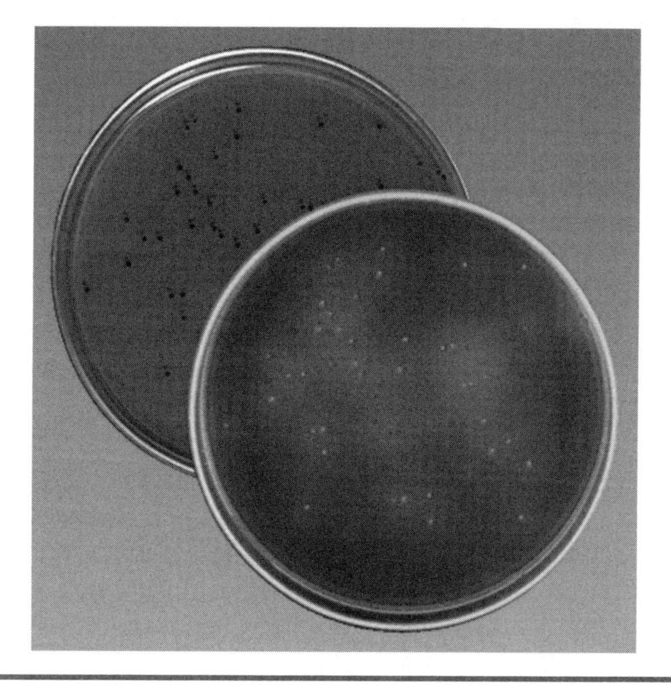

SOURCE: Courtesy of ThermoFisher.

Helicobacter pylori have remarkable means for survival in the hostile environment of the stomach (Dunn et al., 1997). If the ulcer becomes a clinically significant enough to diagnose, the microbes can be killed with antibiotic therapy. In a water system, *H. pylori* are very susceptible to chlorination. However, a high percentage (by some accounts, 50%) of the population is infected with *Helicobacter*, although human infection by another species, *H. flexispira* (Romero et al., 1988) has also been reported. Much about how this microbe is passed through populations remains unknown.

Klebsiella

Klebsiella is another typical coliform bacterium that is common in the intestinal tract (**Figure 4-6**). Some species are found in other mammals, and some are associated with plants. These are not typically dangerous but can confound coliform counts if the plant material gets into the water supply. This microorganism is known for creating a number of infections (e.g., urinary tract infections and pneumonia), but in relation to water exposure *Klebsiella* is probably most important for those swimming in contaminated waters, especially if water vapor is inhaled or if contaminated water enters an open wound.

Plesiomonas shigelloides

Originally *Aeromonas shigelloides*, these microorganisms are like the *Aeromonas* genus in that they are Gram-negative, oxidase-positive, and facultative

FIGURE 4-5 **Electron micrograph of *Helicobacter pylori*, the causative agent of gastric ulcers. Note the curved shape of the cells. They will burrow into the lining of the stomach to protect themselves from the highly acidic conditions.**

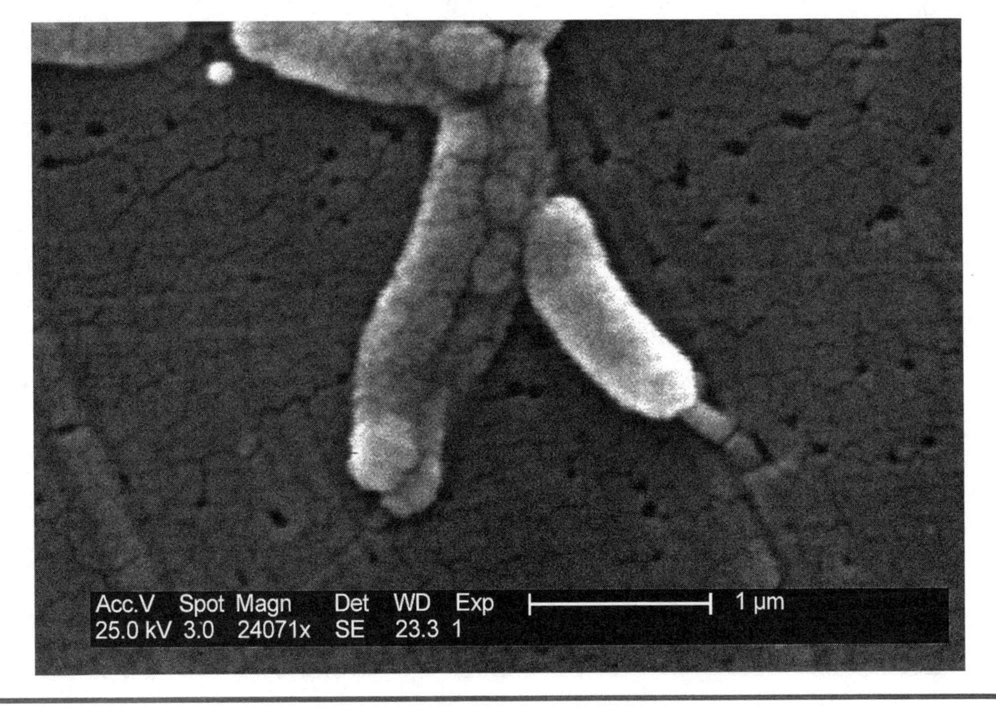

SOURCE: Courtesy of Dr. Patricia Fields and Dr. Collette Fitzgerald, Centers for Disease Control and Prevention. (http://phil.cdc.gov/phil/details.asp). #5775.

anaerobes. They are distinguished from *Aeromonas* by inositol fermentation and the DNase test.

Plesiomonas are found in water supplies and in uncooked shellfish. In healthy people they cause severe gastroenteritis characterized by diarrhea (often accompanied by blood or mucus), vomiting, and fever. *Plesiomonas are* usually self-limiting. In susceptible populations *Plesiomonas* can cause septicemia or meningitis.

In a case study from New York (Van Houton et al., 1998), at least 60 people became ill after attending a party in which several catered food items (i.e., macaroni salad, potato salad, and baked ziti) were contaminated by *Plesiomonas* and *Salmonella* serotype Hartford. An examination of the catering business revealed that it was based at a local convenience store that received its water supply from a private well. The well was very shallow (only 10 feet deep) and was downstream from a nearby chicken farm (1/4 mile away). It was assumed that the groundwater supply was affected by recent rainfall that may have carried waste from the farm to the well. In addition, the automatic chlorinator at the well was not properly maintained, and there was no chlorine in the water supply. Under these circumstances the water probably contaminated

FIGURE 4-6 **A scanning electron micrograph (SEM) of Klebsiella. These bacteria have no flagella and are not motile.**

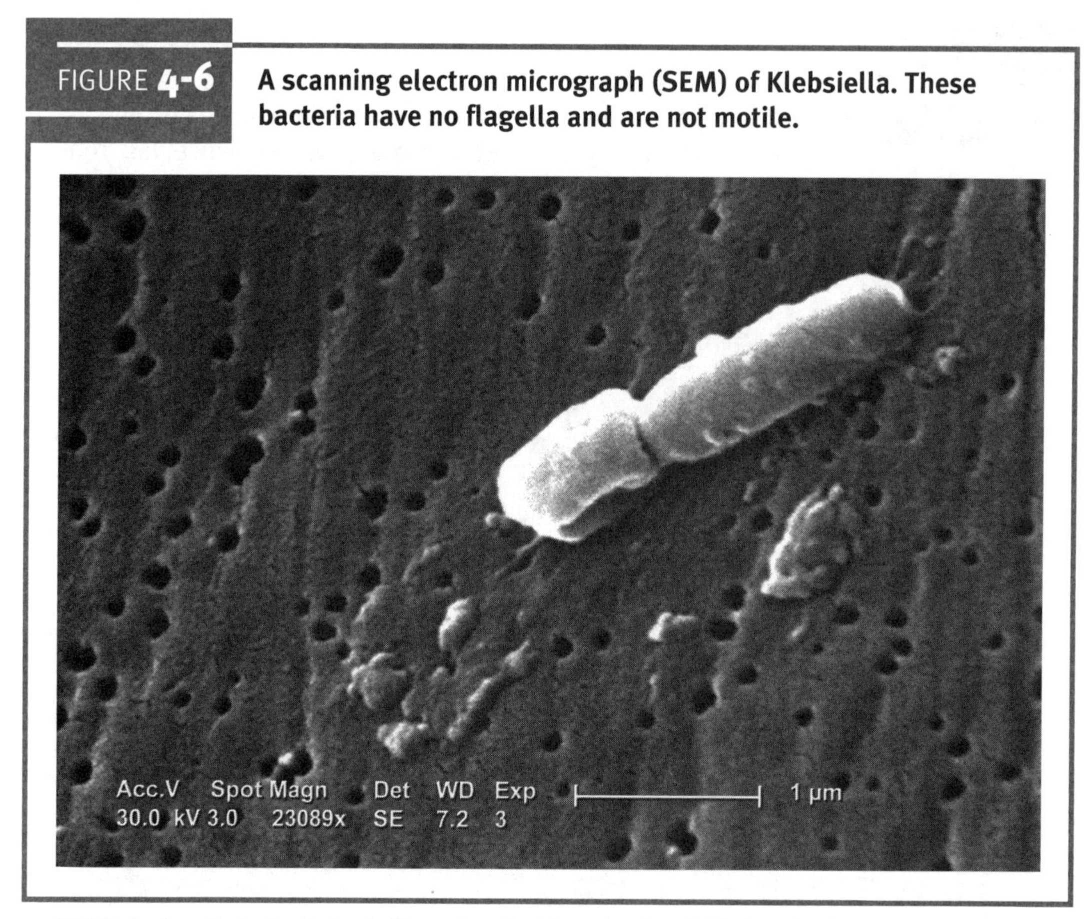

SOURCE: Courtesy of Janice Carr, Centers for Disease Control and Prevention. (http://phil.cdc.gov/phil/details.asp). #6834.

the food during a rinsing step. This case demonstrates that *Plesiomonas* (and other) infections can occur when a series of mistakes lead to a contaminated water supply.

Salmonella

Salmonella bacteria are facultative, anaerobic, Gram-negative bacilli, but they are not coliforms (**Figures 4-7** and **4-8**). The infectious dose is high, at least 10^8 bacteria. There are a large number of serotypes that have been described, but they fall into two species, *S. enterica* and *S. bongori*. *Salmonella enterica* includes a great many subspecies; all pathogens are found in this group. For convenience, the subspecies are typically referred to as species names.

Salmonella enterica is a prevalent source of gastroenteritis around the world. It causes fever, abdominal cramps, and diarrhea. This species is a frequent contaminant of food products, such as chicken and eggs (see Chapter 8). Chickens are not the only bird to carry the pathogen; many other avian species, as well as many mammals, reptiles, and amphibians can carry it. For example, pet turtles may spread the bacteria to a small child.

Salmonella typhi is the causative agent of typhoid fever, which is rare in the United States today but which caused significant outbreaks during the 1700s and 1800s with high fatality rates. Fatality rates of 10–20% were not uncommon, although the

FIGURE 4-7 **Scanning electron micrograph of *Salmonella* sp.**

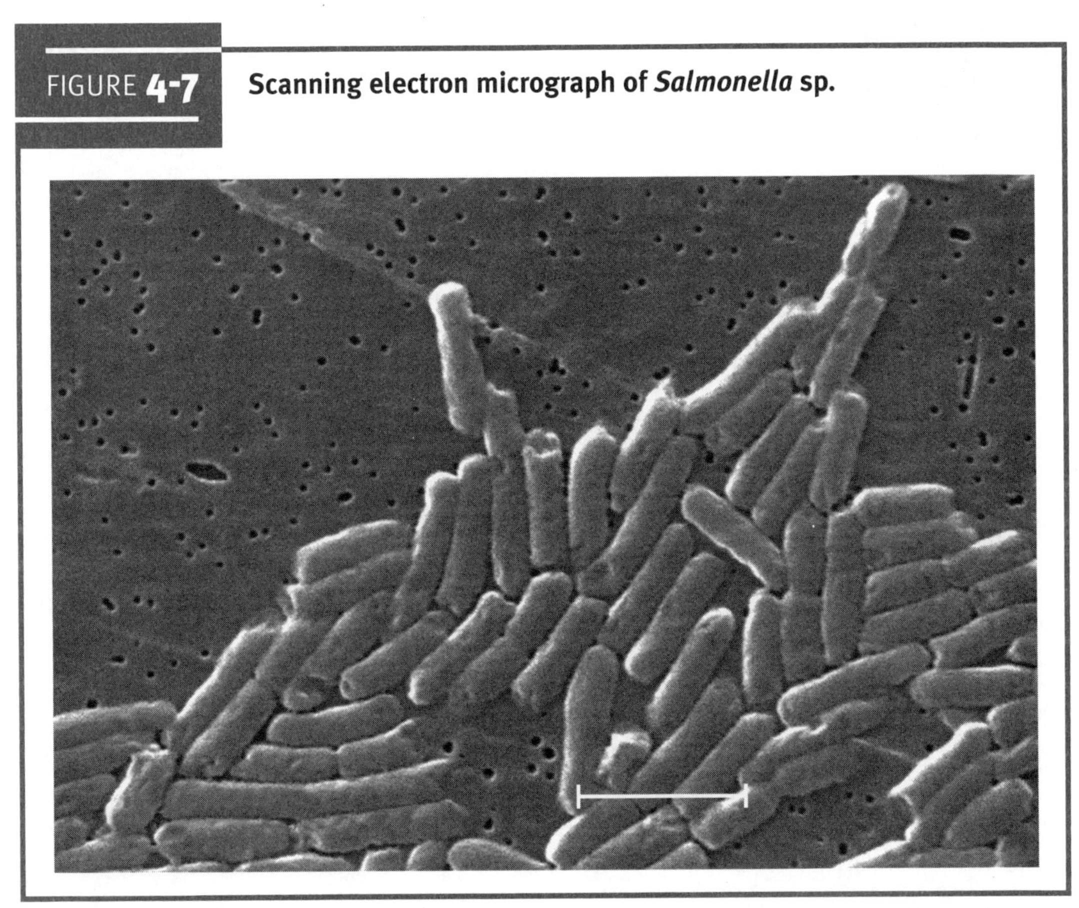

SOURCE: Courtesy of Janice Haney Carr, Centers for Disease Control and Prevention. (http://www.cdc.gov/media/subtopic/library/diseases.htm).

use of antibiotics has decreased this rate to about 1%. In contrast to other *Salmonella* species, *S. typhi* has a relatively low infectious dose, and it is still a significant pathogen worldwide. Many people who recover from the illness become carriers and shed the bacteria in their stool. A milder form is caused by *S. paratyphi*.

Serratia

Serratia are coliform bacteria, although they ferment lactose weakly (**Figure 4-9**). They are ubiquitous environmental bacteria and colonize many surfaces through biofilm formation, which may account for their reported resistance to chlorination. Although they are not usually a major contaminant, a number of reports are available on contamination of finished medical products.

Shigella

Shigella are Gram-negative, facultative anaerobic bacilli that are closely related to *E. coli*, but are not coliform (**Figure 4-10**). They cause severe gastroenteritis and dysentery. *Shigella* contain a set of genes for Shiga toxins, which are extremely destructive. Four species are responsible for most

FIGURE 4-8 **Salmonella on Hektoen Enteric Agar plate.**

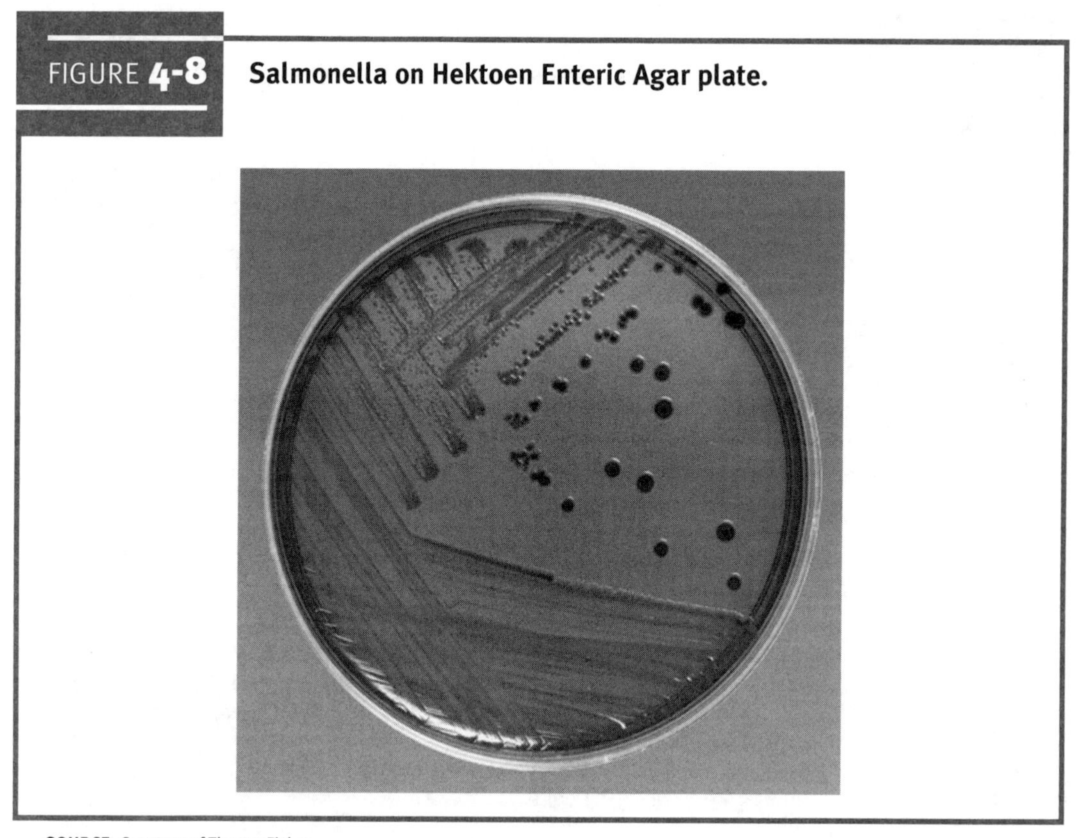

SOURCE: Courtesy of ThermoFisher.

illness: *S. dysenteriae*, *S. sonnei*, *S. boydii*, and *S. flexneri*. *Shigella sonnei* is found most often in the United States. In contrast to *Salmonella*, which it resembles superficially, the infectious dose is very low, probably less than 100 microorganisms.

Breaking the fecal–oral route of transmission is the key to removing the threat of *Shigella*. Notable examples of outbreaks demonstrate that breakdowns in the sewage treatment system are all that is necessary for large numbers of people to become infected. *Shigella sonnei* was found in an outbreak in Idaho in 1995 (Arnell et al., 1996). The cause appeared to be contaminated well water, although the source of the well's contamination was not identified. Coliform bacteria were found in the water, but *Shigella* were not. This is not unusual in outbreaks of *Shigella*. The pathogen was finally isolated from stool samples of patients. The low infectious dose of *Shigella* means that many cases may be caused by person-to-person (secondary) spread of bacteria.

Vibrio Cholerae

Vibrio cholerae is the causative agent of cholera, which has been, and continues to be, one of the most lethal infections worldwide (**Box 4-4**). Of a number of known serotypes, O1 and O139 are the most prominent. In the United States cholera is very rare and is confined to a region along the Gulf coast. Chlorination effectively kills these bacteria, and no large outbreaks have occurred in the United States in the past 100 years.

FIGURE 4-9 ***Serratia* colonies on an agar plate.**

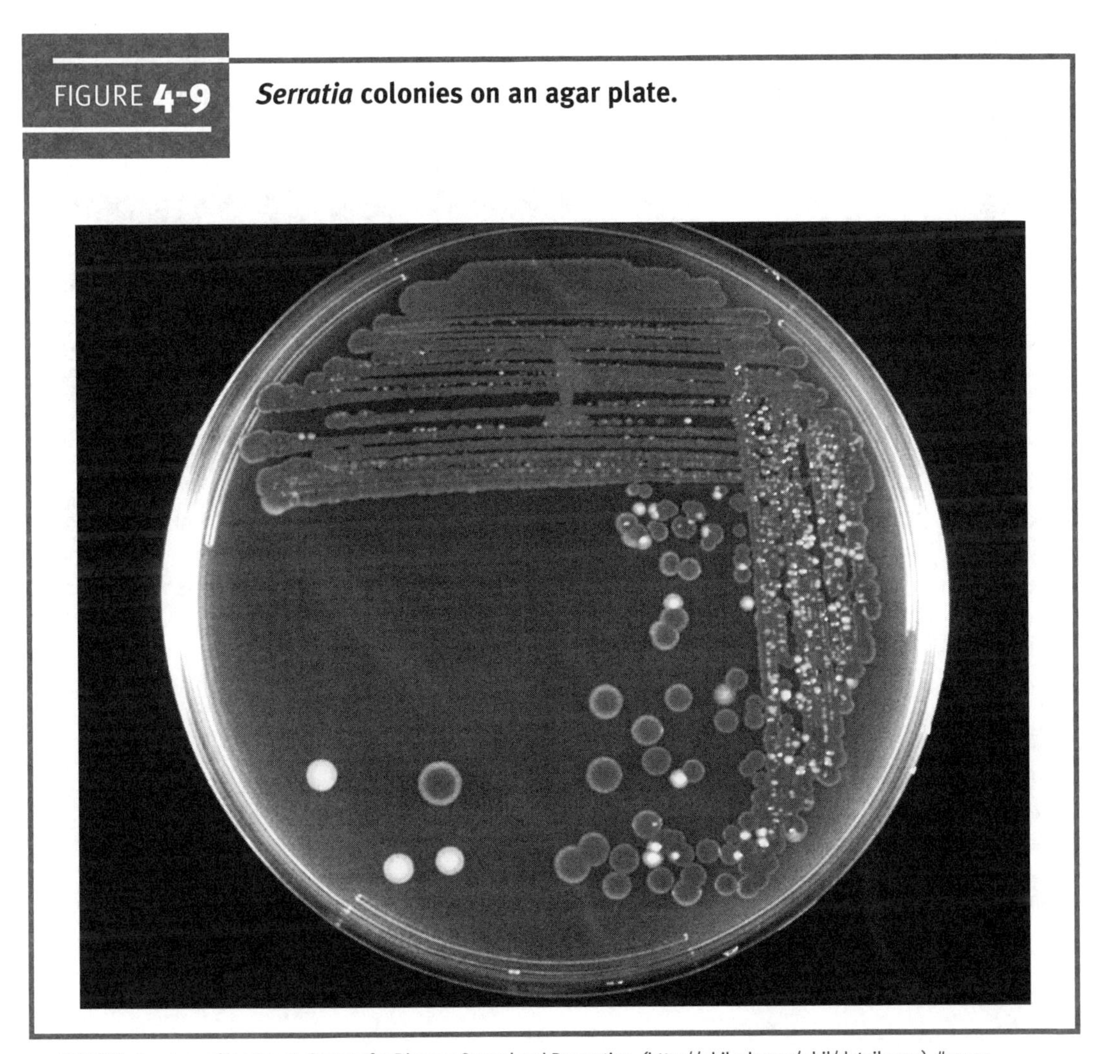

SOURCE: Courtesy of Dr. Negut, Centers for Disease Control and Prevention. (http://phil.cdc.gov/phil/details.asp). #10544.

Vibrio are Gram-negative curved bacilli (**Figure 4-11**) that prefer brackish water (such as the water found in the Mississippi River delta). Drinking water with some contamination by brackish water is a plausible source of infection, although people are unlikely to drink brackish water directly. Food that has been washed with contaminated water is a more likely route of contamination. The infectious dose is high, but if an infection occurs it leads to a very severe diarrhea. A high percentage of cases are fatal, with deaths due to massive loss of fluids and electrolytes; therefore the treatment is usually fluid-replacement therapy. Under these circumstances the disease should be self-limiting (**Figure 4-12**). A description of a curious outbreak of cholera is presented in Chapter 6.

FIGURE 4-10 **Shigella infection on intestine of Rhesus monkey. Note petechial hemorrhages.**

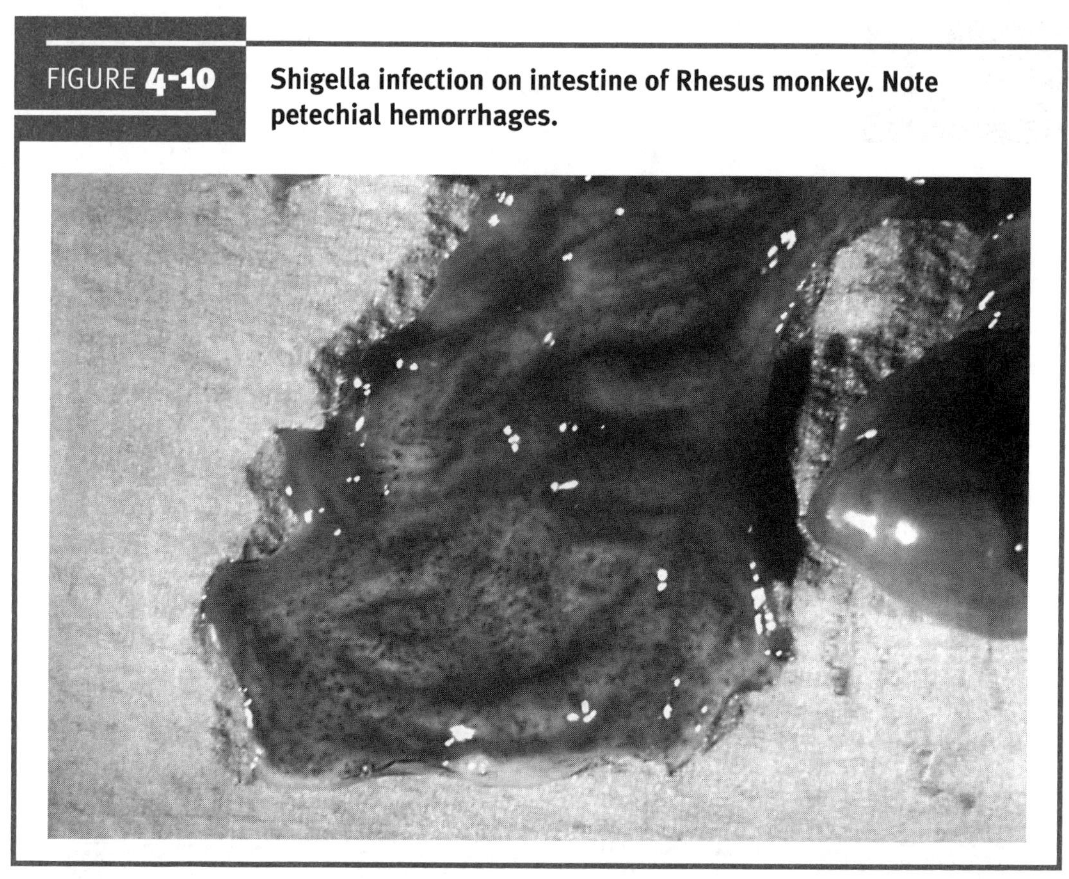

SOURCE: Courtesy of the Centers for Disease Control and Prevention. (http://phil.cdc.gov/phil/details.asp). #5164.

POTABLE WATER PATHOGENS: VIRUSES

Adenoviruses

Adenoviruses are usually associated with person-to-person transmission (e.g., causing a number of respiratory disease syndromes), although they are also spread by the fecal–oral route, and they also cause gastroenteritis. They are medium-sized double-stranded DNA viruses that do not have an envelope (**Figure 4-13**). These infections are usually seen in small children, and they rarely cause major illness. Occasionally a small outbreak occurs if a swimming pool is poorly chlorinated; in these cases adenovirus conjunctivitis occurs. Epidemics of adenovirus spread through respiratory droplets, in which an acute respiratory disease occurs, is more common.

Astrovirus

Astroviruses are found in the family Astroviridae; they are small, nonenveloped, positive-stranded RNA viruses. They are an important cause of gastroenteritis around the world, and they are probably also responsible for sporadic outbreaks and for a large percentage of early childhood diarrhea. Astroviruses can be found in contaminated food and water, and they are generally passed on via the fecal–oral route. Both birds and mammals act as carriers. As RNA viruses, they are detected using RT-PCR. Serological tests are also available (Guix et al., 2005).

Caliciviruses

This family includes four genera: *Norovirus, Sapovirus, Vesivirus,* and *Lagovirus.* Only the first two are known to cause disease in humans. They are small, nonenveloped, positive, single-stranded RNA viruses. Caliciviruses are extremely environmentally resistant, which gives them the opportunity to spread quickly.

Norovirus

This virus was formerly named Norwalk Virus and Norwalk Agent because of an outbreak at a school in Norwalk, Ohio. It has only recently been renamed *Norovirus* (**Figure 4-14**). As few as ten viral particles may constitute an infectious dose, and epidemics quickly occur in confined spaces, such as cruise ships and care facilities. Aerosolized viral particles may be sufficient to spread the infection. Passage probably occurs as people who clean up the feces and vomit of sick people become physically close to the concentrated pathogen (Lawrence, 2004). *Noroviruses* are relatively chlorine-resistant, which also contributes to their infectivity; nevertheless, large waterborne outbreaks have not yet occurred.

The number of cases of gastroenteritis (severe diarrhea and vomiting) that *Noroviruses* cause is difficult to estimate because it is difficult to monitor. However, *Noroviruses* are probably responsible for millions of cases yearly and might be as common as the *Rotaviruses.* The CDC estimates that *Noroviruses* are responsible for 23 million cases of gastroenteritis each year. The illness is usually self-limiting, with only supportive therapy needed (i.e., fluid replacement).

In the aftermath of Hurricane Katrina (2005) a large number of people were evacuated from New Orleans to neighboring states. Some 24,000 people were sheltered in Houston at Reliant Park (i.e., the Astrodome). Over 1,100 people became ill with diarrhea and vomiting, many became ill after being in the facility for less than a week (Palacio et al., 2005). Laboratory tests determined that the cause of the

BOX 4-4 The Scourge of Cholera

Cholera is a waterborne disease that has caused several recorded epidemics, often with high mortality. It is caused by *Vibrio cholerae*, a Gram-negative curved rod that prefers brackish water. In 1854 it caused an epidemic in the Soho area of London. A local physician, Dr. John Snow, patiently tracked each case of cholera to determine the factors they had in common. His research led him to believe that the common thread was a water supply, the Broad Street pump. In those days the water for an area of the city was supplied by well water, and the inhabitants would pump the well to draw up a bucket of water. The water was then hauled back to their homes.

A person suffering from cholera had taken up residence in the area and had used one of the local privies (sanitation was rather crude at the time). The fecal matter had escaped from this septic system and had infiltrated the groundwater supply, which then infected many people in the neighborhood. Snow successfully argued his findings before the local authorities, who removed the pump handle. This forced the local people to obtain water from another source, and gradually the epidemic ceased.

Because of his excellent work in describing the outbreak and the means to halt it, Snow is regarded as the father of epidemiology. He did not know the cause of cholera; the organism would only be described by Robert Koch about two decades later. But he demonstrated that water can be a carrier of disease, and that an epidemiological approach can be an effective tool to fight disease.

FIGURE **4-11** **Scanning electron micrograph of *Vibrio cholerae* O1 bacteria.**

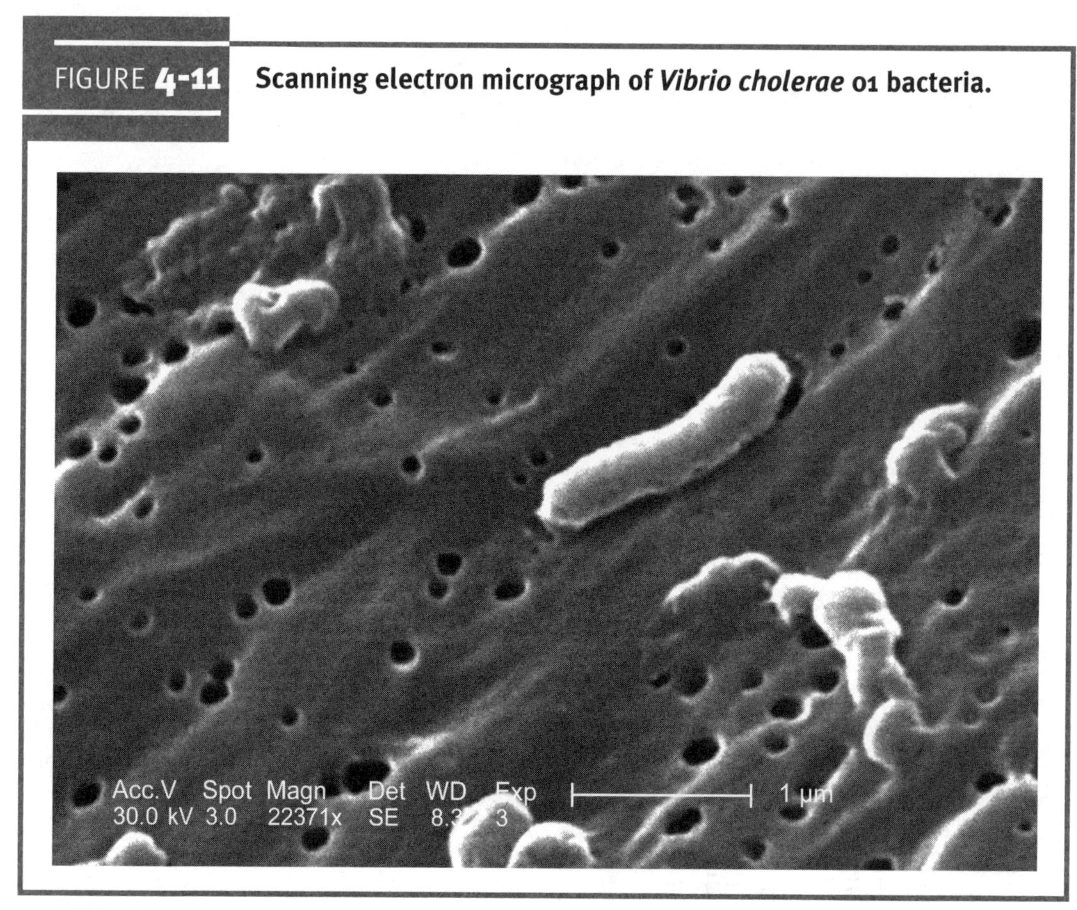

SOURCE: Courtesy of Janice Carr, Centers for Disease Control and Prevention. (http://phil.cdc.gov/phil/details.asp) #7818.

illness was *Norovirus.* In a number of cases the virus was acquired through secondary spread (either person-to-person or through fomites). This report highlights the rapid spread of *Norovirus* in densely packed communities, such as cruise ships or camps, and the low infectious dose (possibly <100 virus particles). In this case, even aggressive hygiene procedures promoted by the public health authorities were insufficient to stop the outbreak completely.

A demonstration of the power of *Norovirus* to infect many otherwise healthy people quickly was made by an outbreak at the U.S. Air Force Academy in Colorado (Warner et al., 1991). Hundreds of cadets at the academy suddenly became ill after eating chicken salad. The cause was determined to be celery that had been rinsed in contaminated water.

Sapovirus

The type species is named Sapporo Virus after the city in Japan where it caused an outbreak in an orphanage in 1977 (Chiba et al., 1990; Jiang et al., 1997). *Sapovirus* typically causes a mild gastroenteritis, usually in children, although adults can also be affected.

Enteroviruses

This group includes polioviruses, coxsackieviruses (A and B), echoviruses, and other enteroviruses. While polio has been nearly eradicated worldwide,

FIGURE 4-12 **A cholera ward in Peru. During times of epidemic, emergency measures are needed to accommodate the large numbers of infected people. Providing enough trained medical personnel can be another obstacle to effective care.**

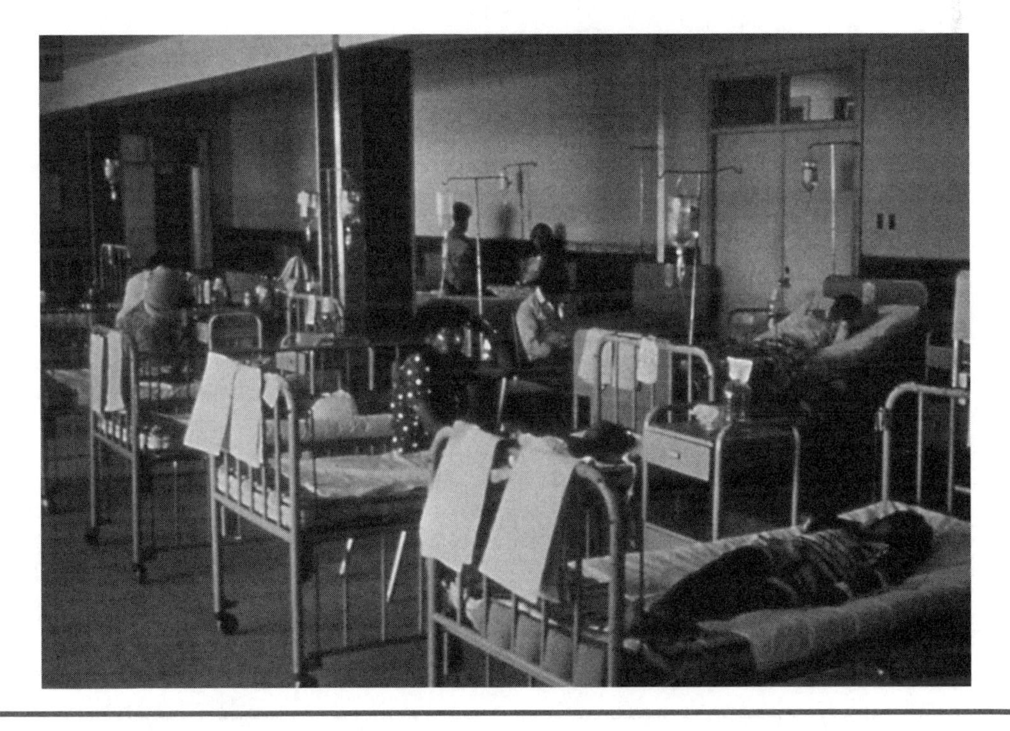

SOURCE: Courtesy of the Centers for Disease Control and Prevention. (http://phil.cdc.gov/phil/details.asp). #5322.

the other enteroviruses are quite common and probably cause a large fraction of waterborne disease. A recent study of enterovirus in sewage and in clinical samples found that viruses were quite common in sewage and that the incidence of serotypes in sewage were approximately equivalent to the incidence in clinical samples (Sedmak, 2003). These virus groups are quite distinct.

Poliovirus

Poliovirus is the causative agent of polio, once a dreaded disease, which has now been almost entirely eradicated. There are three serotypes of poliovirus, each of which is represented in the polio vaccines (i.e., the inactivated Salk vaccine and the attenuated Sabin vaccine). The vaccine was one of the greatest accomplishments of the 20th century; it removed one of the most feared pathogens from the list of epidemic diseases. Because of the international campaign against polio, the virus has been eradicated from the western hemisphere and nearly so from the rest of the world. Sporadic cases occur, but vigilance on the part of the World Health Organization may soon lead to the announcement that it has been completed eliminated.

As an environmentally resistant, nonenveloped virus, poliovirus caused a great many illnesses. In fact, it was quite common to get a poliovirus

FIGURE 4-13 **A transmission electron micrograph image of Adenovirus particles.**

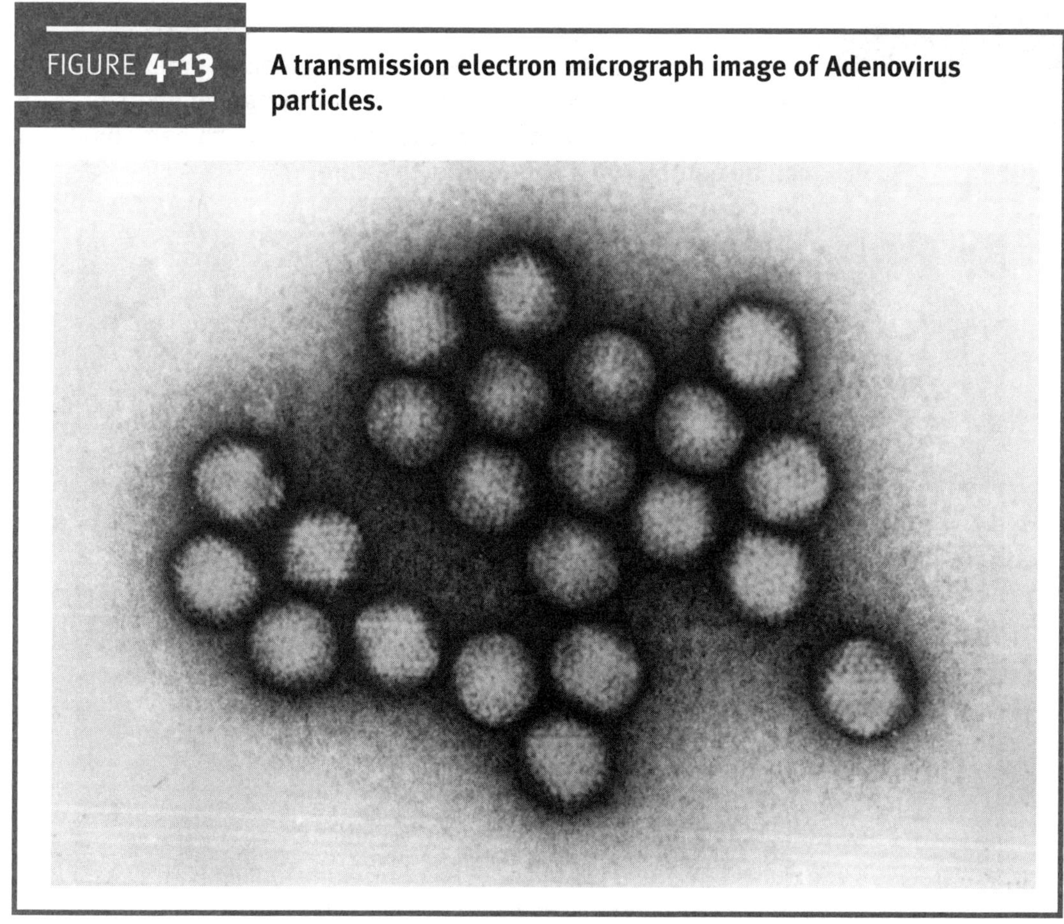

SOURCE: Courtesy of D. G. William Gary, Jr., Centers for Disease Control and Prevention. (http://phil.cdc.gov/phil/details.asp) #10010.

infection, although in the vast majority of cases the afflicted person got a "summer cold" or similar mild illness. In some small fraction of cases there was noted a transient muscle paralysis, and in a smaller fraction this paralysis became permanent. Most of the time the paralysis was debilitating but not life threatening. However, in a very small percentage of cases the paralysis became severe, leading to major loss of movement or function, and in some cases death. For instance, paralysis of the muscles responsible for breathing resulted in asphyxiation.

Coxsackie A Virus

There are 23 known Coxsackie A serotypes, the most studied of which is A9. Another well-known disease, hand, foot, and mouth disease (HFMD), is caused by serotype A16. However, it is transmitted by contact with other infected persons and through fomites.

Coxsackie B Virus

There are six known Coxsackie B virus types. Types B2 and B5 are the most prevalent. These viruses

FIGURE 4-14 **A transmission electron micrograph of *Norovirus* particles.**

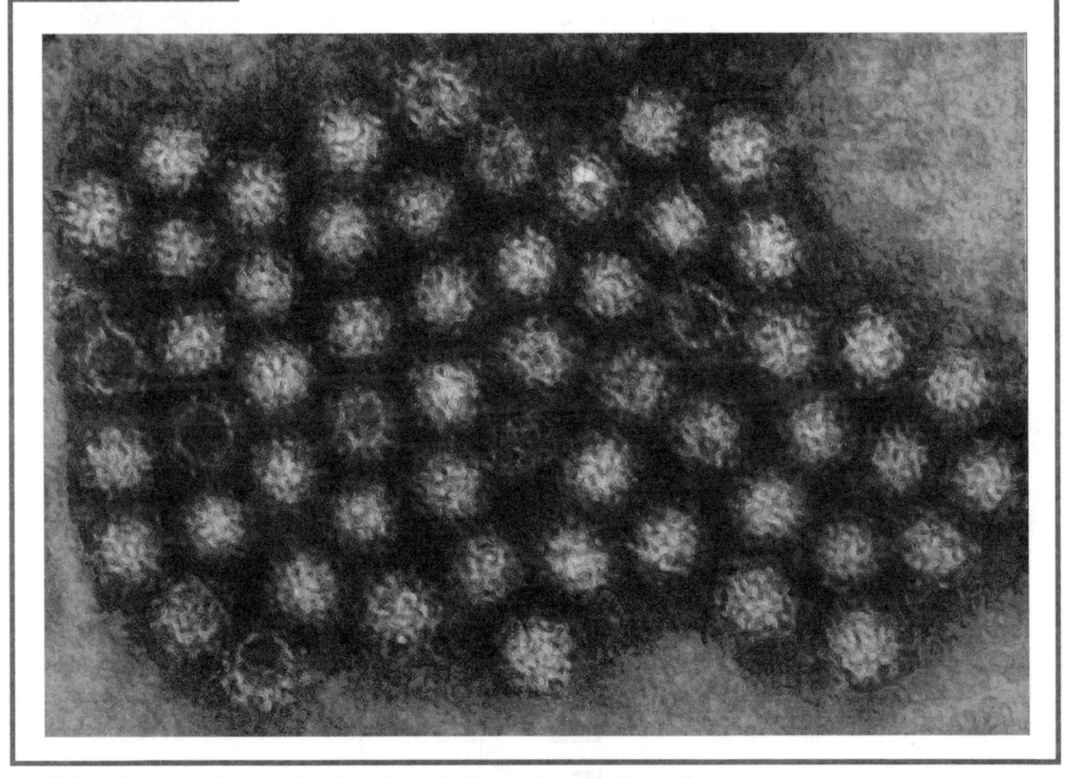

SOURCE: Courtesy of Charles D. Humphrey, Centers for Disease Control and Prevention. (http://www.cdc.gov/media/subtopic/library/diseases.htm).

target cardiac tissue and can cause myocarditis, although more common symptoms are gastroenteritis and a general feeling of malaise. Asymptomatic cases are probably very numerous. Little is known about its transmission.

Echovirus

The *Echoviruses* cause very common infections in children, and some cases are very severe, resulting in myocarditis or meningitis. They may also cause rashes when contacted during swimming. They are easily transmitted through water or on fomites.

Parechovirus

There are currently 14 known types of *Parechoviruses.* They are probably very common pathogens, although rarely do they cause clinical symptoms. They are probably most often mistaken for a common cold because symptoms are a mild gastritis and respiratory secretions.

Other Enteroviruses

It is unlikely that all of the enteroviruses have been discovered. Therefore, some known, but poorly characterized, examples remain.

Hepatitis A Virus (HAV)

Hepatitis A virus is a nonenveloped, single-stranded, plus-stranded RNA virus. It is very environmentally stable, although current water purification methods are sufficient to reduce viral levels by four orders of magnitude, which is usually sufficient to avoid any outbreaks (Cuthbert, 2001). Untreated, the viral particles can remain viable in water for months. Airborne droplets may carry the virus and infect people within about 100 yards of a source. This is a relevant concern to people who work in sewage treatment systems. An effective vaccine is available and is recommended for these workers. Notably, infected people become infectious before they experience symptoms, and this may be a factor in the spread of the virus.

Hepatitis A virus infections are acquired through contact with fecal matter, either in the water supply or directly (**Figure 4-15**). This includes a risk of infection during certain sexual practices, especially among Men who have Sex with Men (MSM). Many small outbreaks occur because food handlers with unhygienic personal habits contaminate food during preparation. Other cases are due to shellfish contaminated with HAV that are eaten raw.

The CDC estimates that there are about 25,000 HAV cases per year, although figures in the past few decades have been far higher (**Figure 4-16**). The disease is self-limiting, displays typical symptoms of hepatitis, and is not chronic. The fatality rate is very low. Also, a good vaccine is now available. Studies of serological conversion demonstrate that

FIGURE 4-15 **Hepatitis A Virus routes of exposure.**

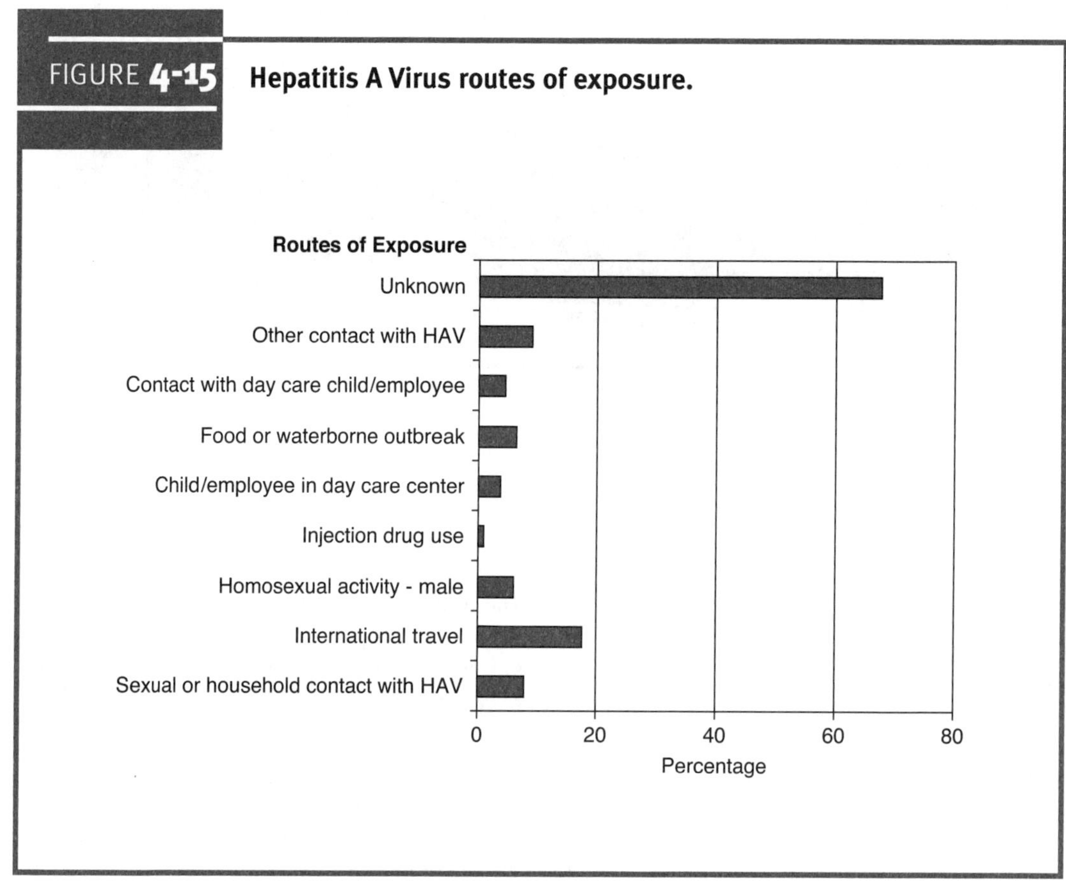

SOURCE: Based on data from the Centers for Disease Control and Prevention.

the pathogen infects far more people than actually become symptomatic; it may be that only 10% of infected people become clinical cases. The disease may be very mild in children, but as people age the symptoms of the disease will become more severe. Unlike other hepatitis viruses, HAV does not result in a chronic infection.

Hepatitis E Virus (HEV)

Relatively little is known about HEV, although it does not appear to be a threat in developed countries. It is spread through the fecal–oral route of transmission, and therefore has the potential to cause epidemics in areas with poor sanitation. Related viruses have been discovered in other mammal and avian species, and transmission has been demonstrated in some cases (Chandra et al., 2008).

HEV is a nonenveloped, single-stranded, positive-strand RNA virus. The hepatitis that it produces is very similar to HAV. Little is known about its infectivity and its pattern of epidemic spread. Seropositivity can be high (i.e., 30%) even in the United States. Immunity to HEV is assumed to be relatively short lived, allowing successive epidemics in the same populations. HEV may be the most widespread hepatitis virus worldwide. A particularly severe epidemic in New Delhi, India in 1955–1956

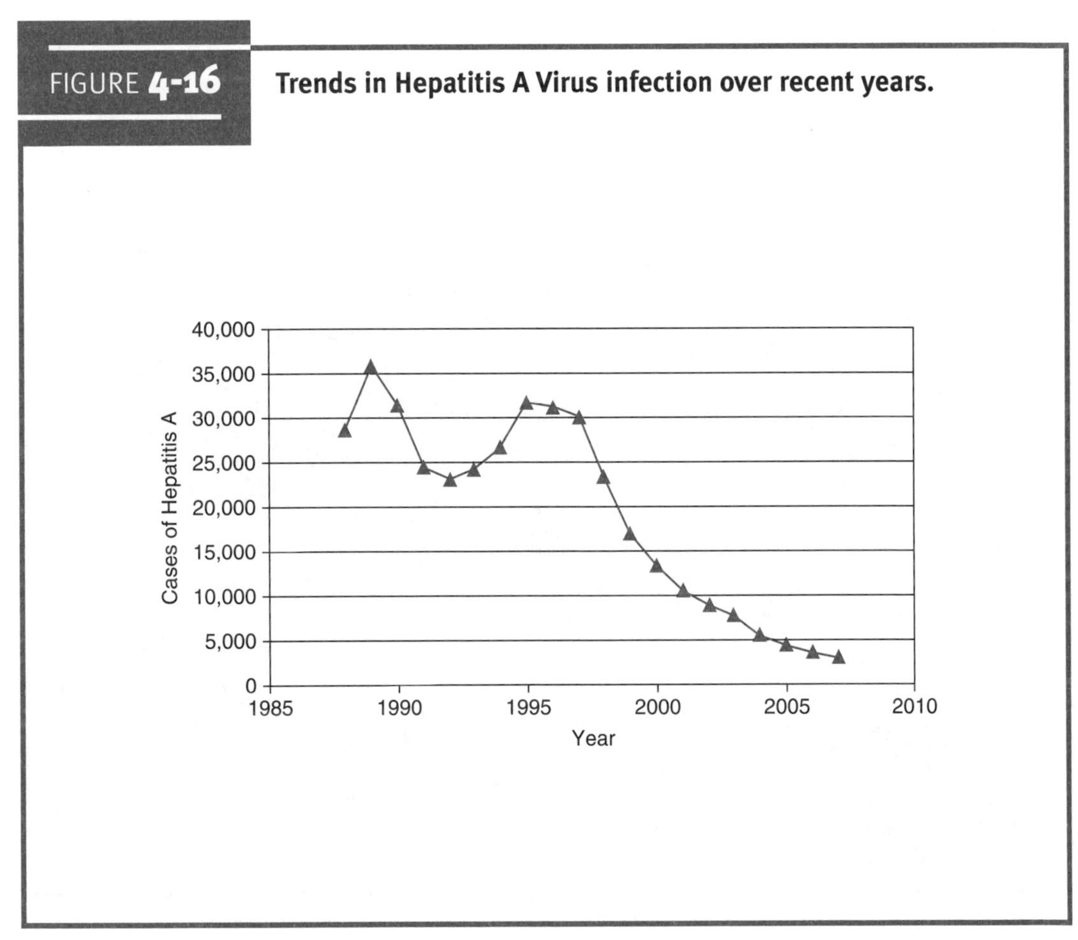

FIGURE 4-16 **Trends in Hepatitis A Virus infection over recent years.**

SOURCE: Based on data from the Centers for Disease Control and Prevention.

has now been attributed to HEV; as many as 30,000 people were affected.

Most infections appear to be asymptomatic, but the virus can cause an acute infection. The infection is usually self-limiting. Most people recover with only supportive therapy, although in pregnant women the disease can be more severe. HEV typically has a low mortality (<1%), although in pregnant women it can be as high as 15%. A vaccine for HEV is not available at this time, although promising results in clinical trials have been reported. A serological test for the diagnosis of HEV has not been approved by the Food and Drug Administration (FDA), although tests are available for use in research (Fitzsimons et al., 2010).

Rotavirus

A member of the Reovirus family and a non-enveloped, double-stranded RNA virus (**Figure 4-17**). The virus is very stable and is easily passed between people or through a contaminated water supply. Outbreaks are fairly common among very young children,

FIGURE 4-17 **Electron micrograph of Rotavirus virions.**

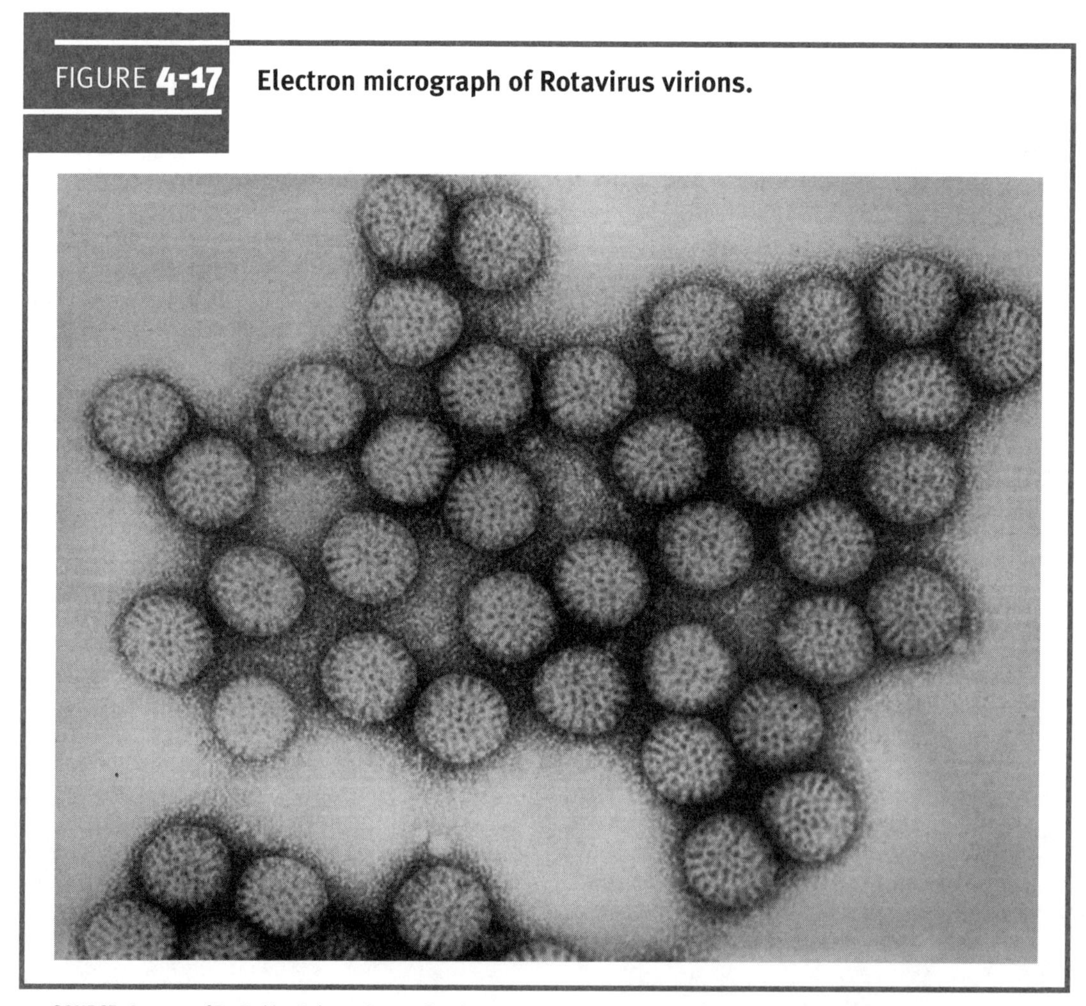

SOURCE: Courtesy of Dr. Erskine Palmer, Centers for Disease Control and Prevention. (http://phil.cdc.gov/phil/details.asp). #273.

especially in day care facilities. Although most are relatively mild, there are probably hundreds of thousands of cases per year. While humans are the primary carrier, birds and other mammals appear capable of carrying the virus as well.

Ordinarily the disease is self-limiting, although in very young children there is a threat of dehydration. Hospitalization may be required to closely monitor these young patients. An effective vaccine is now available. Current water treatment techniques, especially chlorination, are sufficient to kill rotavirus. Worldwide, rotavirus is a significant problem, with the diarrhea it causes responsible for many deaths in children. Two vaccines are currently available for rotavirus. Merck has developed Rotateq, while GlaxoSmithKline has developed Rotarix.

POTABLE WATER PATHOGENS: PROTOZOA

Acanthamoeba

Acanthamoeba is a free-living microorganism that is not uncommon in water supplies, although it is usually held to low concentrations (**Figure 4-18**). Its life cycle is shown in **Figure 4-19**. Most of the time it causes no problems, but occasionally *Acanthamoeba* will cause a potentially severe eye infection (i.e., keratitis, an inflammation of the cornea) among wearers of soft contact lenses. If untreated it can infect deeper tissues and can destroy part of the patient's vision. While some cases result from contact with freshwater sources, other cases are due to poor hygiene with contact lens care. Occasionally a person will try to produce his/her own saline solution with a home brew, not appreciating that the solution must be sterile. This can result in the direct deposition of the pathogen in the eye.

In immunocompromised people, *Acanthamoeba* has the potential to cause a disseminated infection that can affect many organs, but it is particularly severe in the central nervous system. This disease is called Granulomatous Amoebic Encephalitis, an illness similar to *Naegleria* infections. Fatality rates are quite high for this illness. Another pathogenic protozoan, *Balamuthia mandrillaris*, acts in the same manner.

Balantidium coli

This pathogen is passed via an environmentally tough cyst that is found in fecal matter from infected humans and pigs. It causes a severe diarrhea and period of dysentery, which may last weeks or months. The parasite burrows into the intestinal lining, causing infections that produce pus in the feces. Significant weight loss is the likely result. While an active case of diarrhea is ongoing, only the trophozoite is expelled. However, after the diarrhea abates the cyst is expelled in the feces. The disease is very rare in the United States.

Cryptosporidium

Cryptosporidium is responsible for the largest waterborne outbreak of disease in 20th-century America in 1993, when over 400,000 people in Milwaukee became ill, and more than 100 people died (see Chapter 6). Because of this outbreak, a number of new rules came into practice for the treatment of water. There is still a very real threat from fecal releases in community swimming pools, where people may inadvertently ingest some water. The infectious dose may be as low as ten oocysts (**Figure 4-20**). Published research reports many different estimates of the infectious dose, yet the number remains uncertain because of the great difficulty in ascertaining whether oocysts are actually alive and infectious. Because of this low number, the risk is present for secondary transmission to other people.

The environmentally resistant form of the parasite is the oocyst, which is very resistant to chlorination. Ozone and ultraviolet light are more effective at killing oocysts. Ozone is effective at reducing oocyst concentrations by two orders of magnitude if a concentration of 1 mg/l for 4.5 minutes is used. No vaccine is available, and treatments have variable results, with supportive treatment still the major option. Under these circumstances, it is best to protect the public by eliminating the threat before it can infect humans. Its life cycle, which includes both sexual and asexual cycles, is shown in **Figure 4-21**.

A number of species of this parasite can cause disease in humans, although *C. hominis* (associated with humans) and *C. parvum* (associated with cattle) appear to be the most significant (Rose, 1988). Many

FIGURE 4-18 **Electron micrograph of *Acanthamoeba*.**

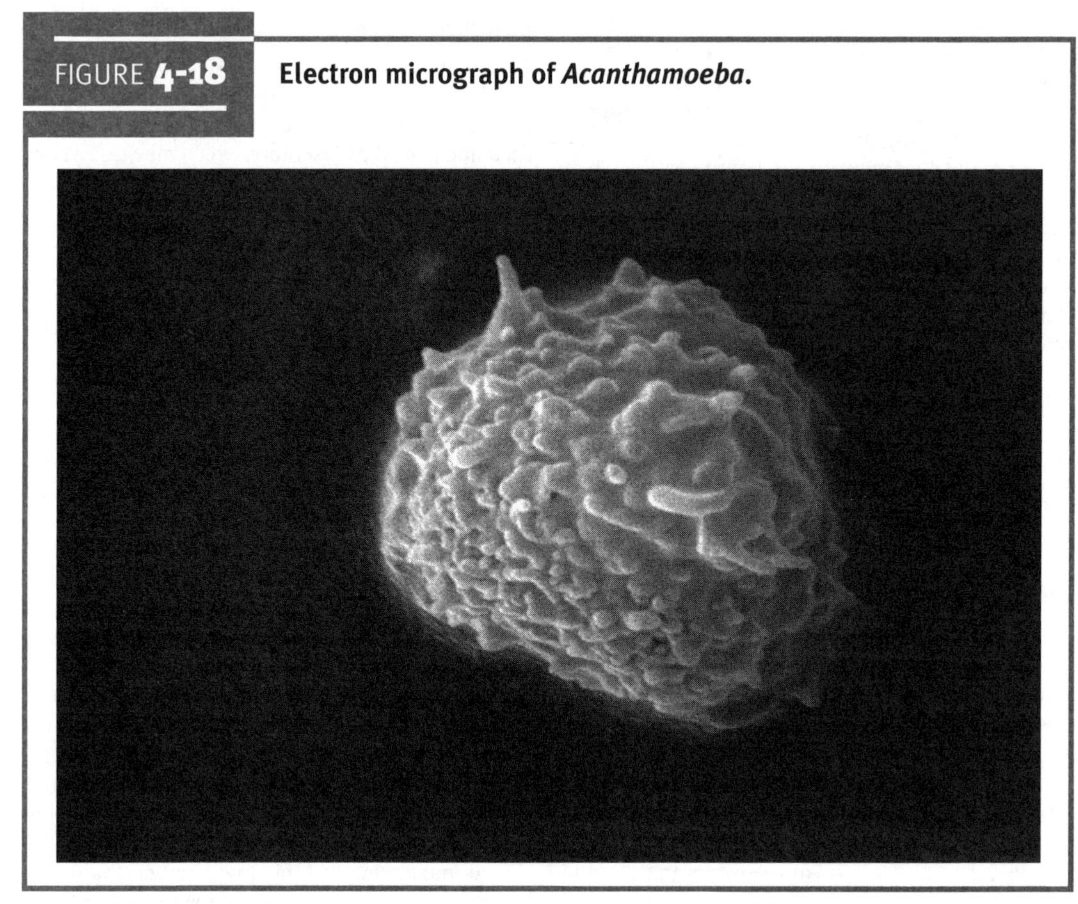

SOURCE: Courtesy of Catherine Armbruster and Margaret Williams, Centers for Disease Control and Prevention. (http://phil.cdc.gov/phil/details.asp). #11899.

other mammal species have their own specific species (e.g., cats have *C. felis*), and some of these are known to cross infect to humans. Birds do not seem to harbor the parasite, however, which is distinct from their ability to carry other human pathogens. On the other hand, gulls feed on material that may be contaminated with *Cryptosporidium*, and therefore may pass the oocysts in their feces, even though they do not acquire the disease (Rose, 1997).

Cryptosporidium causes a watery diarrhea that may last from 7 to 10 days. It is usually self-limiting, although in immunocompromised persons the illness can be fatal (Clark, 1999). Oocysts may be shed in the feces for several weeks after symptoms abate. The fate of the oocysts in the environment remains a mystery. They can persist in water sources for several months, although their exact location is unknown.

Low numbers of both *Cryptosporidium* and *Giardia* in surface water supplies is not unusual, as evidenced by an extensive study (LeChavalier et al., 1991). These researchers noted the correlation of increased water turbidity with higher numbers of the parasites. A seasonal effect has been noted for the magnitude of *Cryptosporidium* oocysts in sur-

FIGURE 4-19 **Schematic of the *Acanthamoeba* life cycle.**

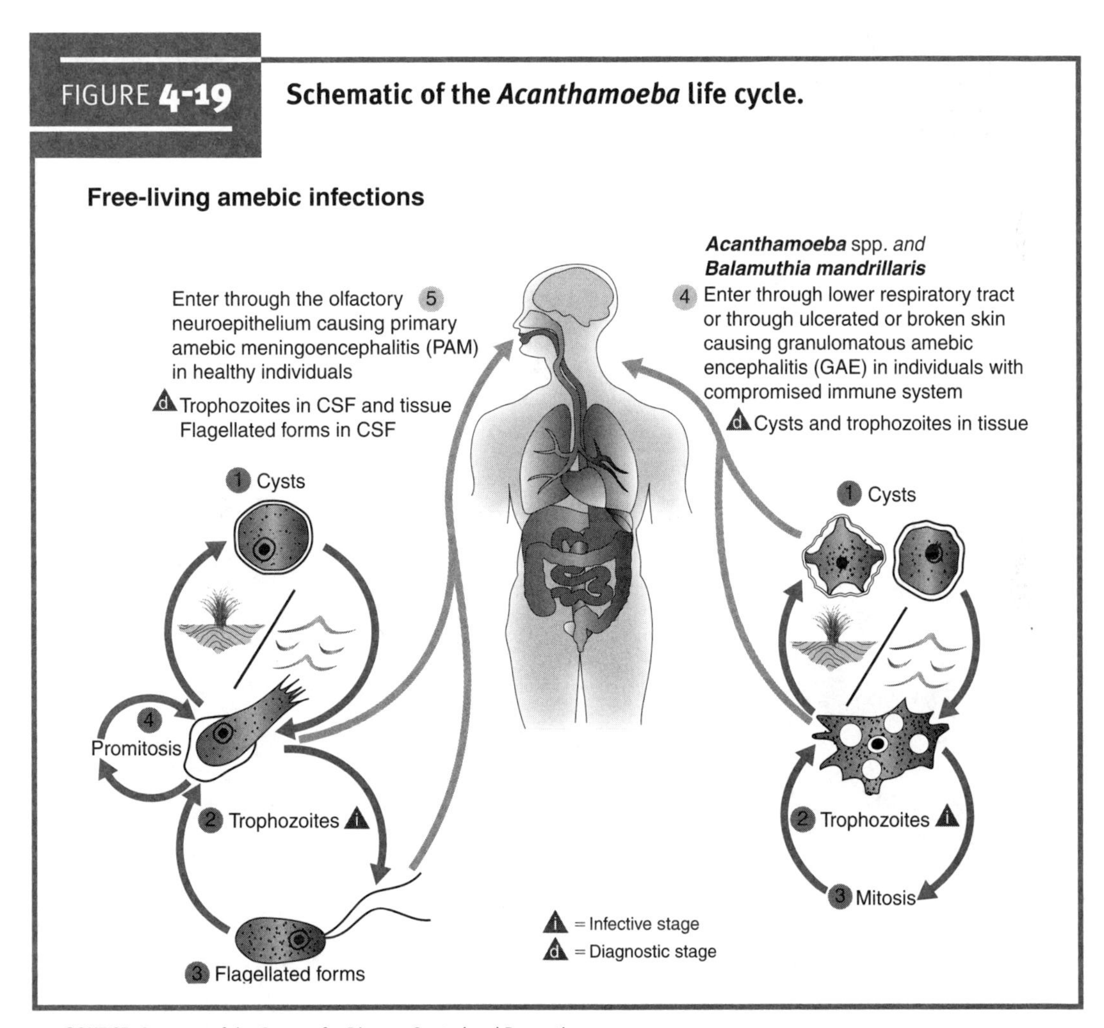

SOURCE: Courtesy of the Centers for Disease Control and Prevention.

face waters, as well as a correlation with areas of dairy farming (Hansen and Ongerth, 1991). The ubiquity of *Cryptosporidium* oocysts was demonstrated by Jiang et al. (2005), who examined water samples from watersheds and found them in the great majority of cases. This included *C. hominis*.

Cyclospora

The *Cyclospora* parasite resembles *Cryptosporidium* in many respects, such as its oocyst formation, environmental stability, and the watery diarrhea that it produces (**Figure 4-22**). However, it is slightly larger than *Cryptosporidium,* and humans appear to be the only host. It causes an illness that may last as long as 7 weeks. *Cyclospora* are often seen not in water but in fresh produce. Recent outbreaks have been traced to contaminated berries. The oocysts were supposedly deposited on the produce via a contaminated water supply, but proof is often lacking.

FIGURE 4-20 ***Cryptosporidium parvum* oocysts.**

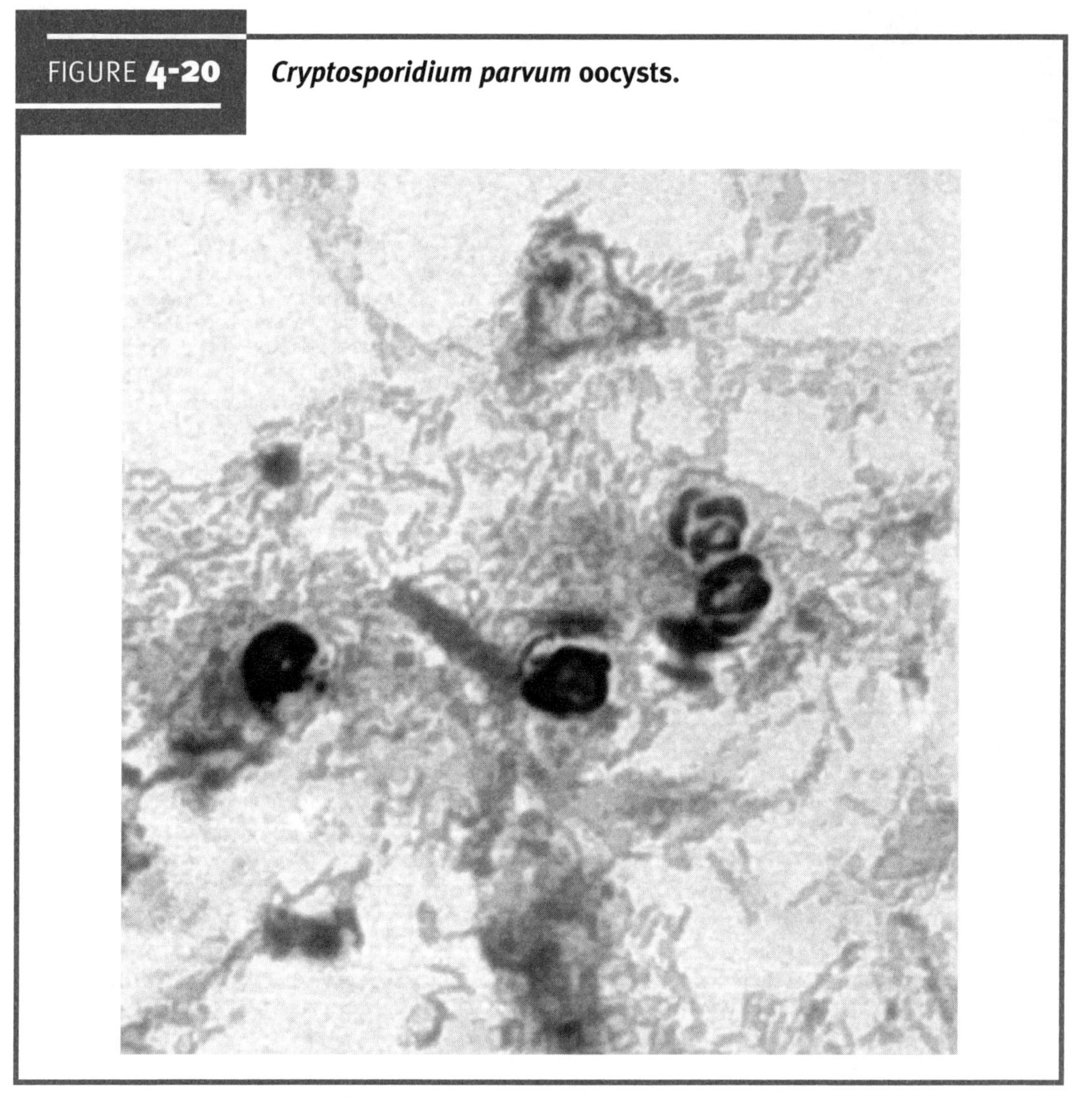

SOURCE: Courtesy of DPDx/the Centers for Disease Control and Prevention.

Cystoisospora belli (Isospora belli)

The only member of this genus that causes disease in humans is *Cystoisospora belli*, although other species infect animals such as dogs and cats. This parasite makes an oocyst that enables the organism to survive for long periods of time in the environment. Standard water purification methods appear to be sufficient for removing the oocysts, which are somewhat larger than the similar structures of *Cryptosporidium* and *Cyclospora*.

Although it is a relatively rare disease in the United States, it has been seen in immunocompromised individuals; it is more common in tropical

FIGURE 4-21 ***Cryptosporidium* life cycle, including the complex stages in the intestine. Note the asexual and sexual stages.**

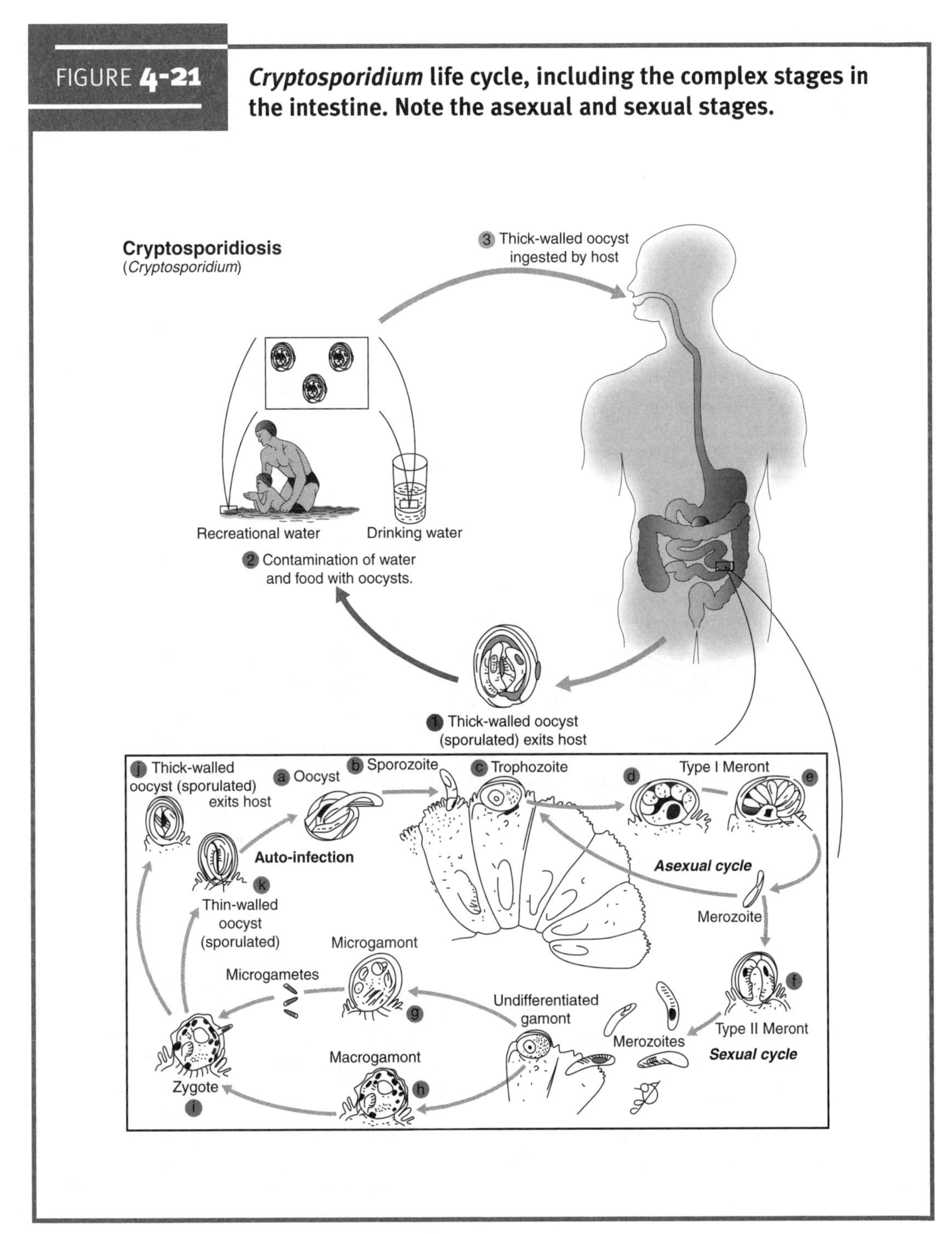

SOURCE: Courtesy of the Centers for Disease Control and Prevention.

FIGURE 4-22 A *Cyclospora* cyst.

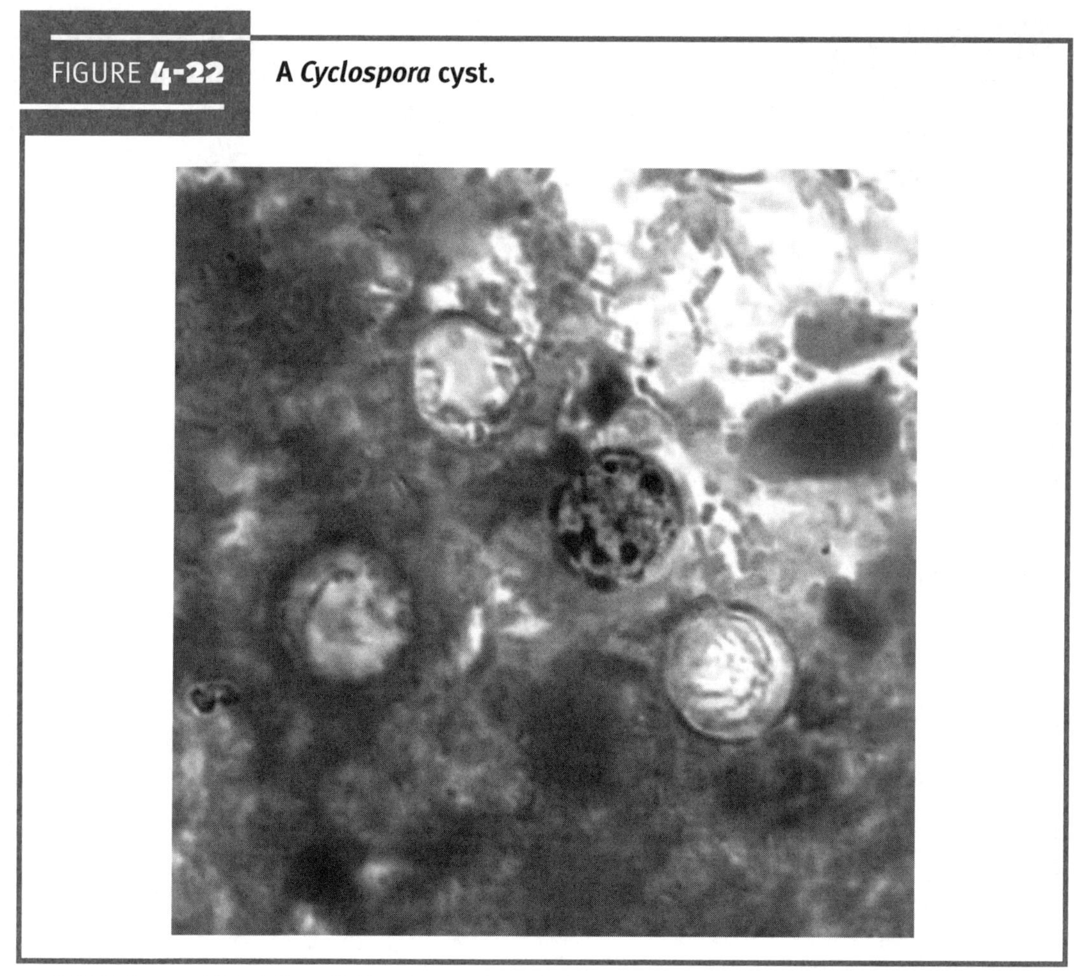

SOURCE: Courtesy of Melanie Moser, Division of Parasitic Disease Information, Centers for Disease Control and Prevention. (http://phil.cdc.gov/phil/details.asp). #7827.

areas of the world. *Cystoisospora belli* is unusual in that it causes a diarrheal disease that may last for a year. This can be very debilitating in a person who is already ill with another disease.

Entamoeba histolytica

The *Entamoeba histolytica* parasite causes amebiasis, which, in its severe form, is referred to as amebic dysentery and which can spread systemically. Most people who ingest the parasite will not become ill, and most who do become ill will have a relatively mild bout of diarrhea. People who are currently experiencing symptoms shed trophozoites in their feces, but these are not infectious. During the asymptomatic period they will shed cysts, which are infectious. This may continue for three months. Because of the many asymptomatic carriers who are still able to shed the organism, the recurrence of *E. histolytica* is always a possibility.

Although this disease is rare in the United States and is confined mostly to travelers returning from developing countries, in some countries amebiasis

FIGURE 4-23 ***Giardia intestinalis*** **cyst under differential interference contrast microscopy.**

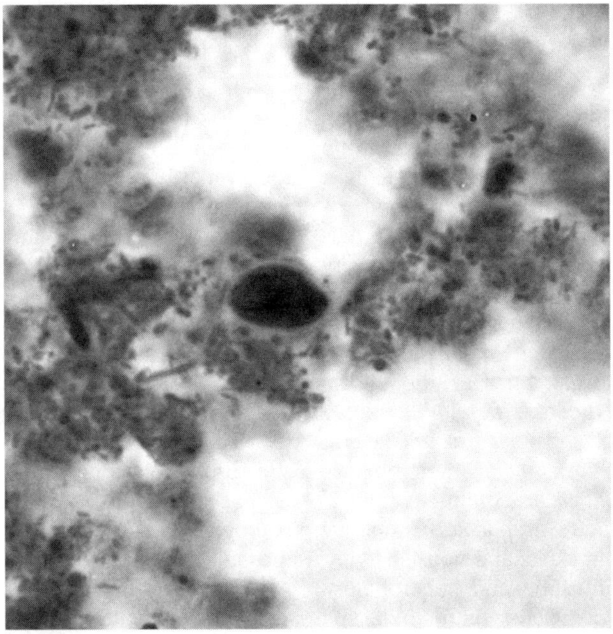

SOURCE: Courtesy of Melanie Moser/CDC.

remains a significant cause of death. It is often mistaken for the relatively harmless *E. dispar* or other species. The environmental particle is the cyst, which is not as resilient as those of many other parasites.

Giardia lamblia

Giardia lamblia is the most commonly isolated parasite in the United States (**Figure 4-23**). It causes a range of gastrointestinal symptoms, from asymptomatic carriers to severe diarrhea with weight loss. Although the diarrhea is not usually life threatening, it can last anywhere from 2 to 6 weeks. Many reports have been made of "nonstop symptoms" with *Giardia* infection, which is currently unexplained. *Giardia* destroys the intestinal villus layer, which both promotes diarrhea and decreases nutrient uptake. Because of the asymptomatic carriers, and the potential reservoir in many other mammalian species, this parasite will undoubtedly continue to be a problem. Secondary spread is very likely. In addition, the cyst survives for months under cool aqueous conditions.

Infections arise from exposures in swimming pools. Fortunately, standard water purification methods are sufficient to remove the *Giardia* cysts, which are larger than most parasite cysts. Chlorination is more effective with *Giardia* than with *Cryptosporidium*, while ozone and ultraviolet light are also very effective. An outbreak of *Giardia* in Bergen, Norway was studied extensively (Robertson et al., 2006). Over 1,500 people became ill, and the most likely explanation was a contamination of the groundwater supply through sewage infiltration from nearby septic systems. This outbreak prompted a study of the incidence of *Giardia* in the community, and a subsequent report (Robertson et al., 2008) found that the same genotypes were present in area sewage sources a year later. A low but steady rate of infection was likely occurring, and another large outbreak might potentially occur. This pattern most likely occurs in other communities as well.

Microsporidia

Microsporidia is a term that refers to a large group (at least 1,200 species) of opportunistic parasites in the phylum Microsporidia that form environmentally resistant spores that are very small, approximately the size of bacteria. They are probably very common, but they are difficult to track and enumerate. Many carrier species, including birds and mammals, have been suspected. Human pathogenic species include (with targeted tissues) the following: *Brachiola algerae* (skeletal muscle), *Brachiola connori* (intestine), *Brachiola vesicularum* (skeletal muscle), *Encephalitozoon cuniculi* (hepatobiliary tract), *Encephalitozoon hellem* (conjunctiva, lung, sinuses, and urinary tract), *Encephalitozoon intestinalis* (urinary tract, conjunctiva, and sinuses), *Enterocytozoon bieneusi* (intestine and hepatobiliary tract), *Microsporidium ceylonensis* (cornea), *Microsporidium africanum* (cornea), *Nosema ocularum* (cornea), *Pleistophora ronneafiei* (skeletal muscle), *Trachipleistophora hominis* (skeletal muscle), *Trachipleistophora anthropophthera* (heart, brain, kidneys), and *Vittaforma corneae* (cornea).

As seen in this list, the Microsporidia comprise a large group with a number of different targets. They share a common focus of infection through water, either by drinking or washing (hence the potential for eye infections). Most of the illnesses are mild, however, and the number of infections is probably greatly underreported. Only occasionally will an infection become serious, resulting, for example, in a case of hepatitis or blindness. Current water treatment techniques are variable for removal of these spores.

WATER CONTACT HAZARDS

The ingestion of pathogens in drinking water is an obvious health risk, but there are additional health risks from contact with water that contains pathogenic microorganisms. This happens daily as people bathe, although usually the water that comes into our homes is suitable for both drinking and bathing. Water contact occurs in other ways, such as swimming in swimming pools (both home and municipal) and in natural water sources. Maintenance of the pool water and monitoring of the natural water are important components of public health.

Some pathogens of great importance are covered here and may be described elsewhere as well. Pathogens are found in natural water systems but are primarily related to food products harvested from those water sources (see Chapter 8). For example, *Vibrio vulnificus* is known to inhabit brackish waters and to infect various species of shellfish. Consumption of raw clams or oysters may result in a severe gastrointestinal illness. Infection by *Legionella* is also not covered here, although the source is typically fresh water. The route of infection by *Legionella* is typically through aerosols (see Chapter 13 on airborne diseases). *Plasmodium*, the parasite responsible for malaria, depends heavily on water sources but is covered under the insect-borne pathogens (see Chapter 14) because it requires the mosquito for transmission.

Water is used in many other facets of our society, including water used for agricultural purposes and for industrial purposes. While the full extent of these uses cannot be adequately covered here, principles for effective use and certain relevant examples are offered.

Recreational Uses

People love the water; it is the site of many different activities: swimming, diving, scuba diving, water polo, rowing and canoeing, sailing and power boating, waterskiing, white water rafting, fishing, etc. Both fresh water (e.g., lakes, rivers) and marine waters (i.e., oceans) are utilized by millions of people every year. Most of these activities take place in the warmer months, although in parts of the world the temperature is agreeable year round. Some water sources are used in the winter months, although this use tends toward ice activities (e.g., ice skating, hockey, ice fishing), and contact with the water in these cases is a concern more for hypothermia than infectious disease.

Swimmers may inadvertently swallow some water, even if they are swimming in saltwater. If pathogen concentrations are high enough, even this small amount is sufficient to cause illness. In these cases the problem is essentially the same as if the person ingested the microorganisms in a glass of drinking water. Another major problem with recreational waters is that the swimmer or bather is typically immersed in the water; if the swimmer has any open wounds, contaminated water can infiltrate the wound and cause an infection. Water may also intrude into susceptible mucous membranes: the eye, the ears, the nose, the oropharynx, and the urogenital system. Ear infections are sometimes referred to as "swimmer's ear," which is typically caused by *Pseudomonas aeruginosa* or *Staphylococcus aureus*. They can be difficult to eradicate because of reinfection from the same source and because of antibiotic resistance. In addition, contact dermatitis (an immune reaction) may occur when specific compounds contact the swimmer's skin.

People enjoy swimming so much that they have constructed swimming pools that can be placed anywhere. Individual homes sometimes have pools, and neighborhoods may have community pools. Water parks have become a huge industry. These large pool complexes are usually attached to a hotel; conventional swimming pools are augmented with splash fountains, water slides, wave pools, indoor surfing, and other clever adaptations. Visitors move from pool to pool, and they can relax in a whirlpool or a hot tub.

The relatively low water volume is the main concern with swimming pools. The water is very confined, meaning that no dilution effect is available. The water is treated and recycled back into the pool. This highlights the major problem with water contamination in pools: fecal contamination. Swimmers certainly shed many substances in water: body oils, hair, dead skin cells, and various personal products like suntan lotion, cologne, or cosmetics. They also shed bacteria from the skin such as *Staphylococcus* (usually *S. epidermidis*, although *S. aureus* is possible). Material may blow in occasionally, such as leaves or insects, and, if left unattended, the pool will eventually produce algae.

Again, the major problem with pools is the introduction of fecal material into the water. Some fecal contamination is unavoidable because human bodies are immersed in the water; birds might also defecate in the water. A person may have a bowel movement in the pool, releasing an enormous number of bacteria (as much as 10^{11} per gram of feces) into the water. While adults typically have bodily control, infants frequently do not. If an adult loses bowel control, it is often due to a gastrointestinal illness, which means that the pathogen will also be expelled into the water. Anyone who comes into contact with these bacteria is at great risk of contracting an infectious dose.

Remediation

Standard household pool filters are not designed to produce potable water. They are designed to remove particles from the water and thus to inhibit the growth of microorganisms. The materials of interest include oils from human bodies, hair, skin cells, debris that blows into the pool, and similarly sized particles. Sand filters utilize relatively large sand grains that trap particles. Pool water, as anyone who has ever received a mouthful of it can attest, has a high chlorine content. However, while pool water is highly chlorinated, there is a still a contact time requirement. People in the pool might contact the pathogen before the chlorine has had a chance to work. This is especially true for chlorine-resistant microorganisms like *Cryptosporidium*. The best option is to close the pool to swimmers, drain it, and thoroughly rinse the surface before refilling. Most people probably just overchlorinate, let the filter run, and wait until the water appears clear.

The same is true for community pools or water parks: the major problem is introduction of fecal material (and the associated pathogens) into the water. Such events have been given various prosaic names by water park staff: unplanned fecal release (UFR) is sufficiently clinical, and "Code Brown" is sardonic but the meaning is clear. Public or corporate entities need a specific plan to respond to water contamination to limit litigation.

The small and shallow pools that are reserved for very young children ("kiddie pools") present a special hazard. The pools characteristically have very little volume, and chlorine can dissipate quickly. In addition, there is very little dilution effect. Children are so young that they cannot always accurately predict when they must have a bowel movement or when they are feeling gastrointestinal illness. Also, the children are slightly more likely to swallow pool water or at least to hold some in their mouths. The combination of risk factors makes it likely that other children in the pool will be quickly exposed if one child defecates.

Other features of water parks make them prone to the spread of infectious disease after a fecal release. In addition to having the pool areas for swimming, they have many "splash" features. That is, some structures are designed to pour water or spray water over the patrons. This is certainly a lot of fun, but this action results in more aerosols. After a fecal release, the aerosols may be inhaled, leading to respiratory diseases as well as gastrointestinal diseases.

The safest treatment for a fecal release in a large pool is the same as for a small one: remove the people from the pool quickly, drain the water, rinse the interior thoroughly, and then refill. However, the usual compromise is to remove people from the pool and then hyperchlorinate. This treatment depends on whether the fecal material is in a formed stool or is loose, like a case of diarrhea. The former may simply be a case of poor bowel control, such as with a young child. The latter suggests a current illness, with the likelihood that pathogens have been released. For formed stools, the first action is to remove the fecal material and dispose of it appropriately. The chlorine content of the pool should be raised to 2 ppm (assuming that the water pH is no higher than 7.5 and the temperature is at least 25°C). Contact time should be at least 30 minutes. For loose stools it is more difficult to remove the fecal material, but at least some should be removed. The presence of fecal material has been shown to increase the concentration × time (Ct) needed to kill *Cryptosporidium* oocysts (Carpenter et al., 1999). The chlorine content of the pool water should be raised to 20 ppm (assuming the water pH is no higher than 7.5 and the temperature is at least 25°C). Contact time is much longer: 12.75 hours (Shields et al., 2008). This gives a Ct value that should be sufficient to kill *Cryptosporidium* oocysts. During this time the pool filter should run, removing particulates. Oocysts may be caught, depending on the efficiency of the filter. In any case, the filter should be backwashed at the end of the treatment, and the backwash water should be disposed of in a sanitary sewer line. If possible, the filter material should be changed at this time. Note that the remaining chlorine content is very high; the chlorine level should be reduced before swimmers are allowed back in the pool.

Natural Waters

Swimming in natural water sources is also very popular, and here the category can be further broken down to freshwater areas and saltwater areas. (Some water parks have saltwater attractions, but most parks use fresh water.) Freshwater sources include lakes, ponds, and rivers. Saltwater sources include shorelines along the oceans. Both of these are fundamentally different from those areas mentioned above. As natural sources they are not as clean as a swimming pool, which is so pure that you can easily see the bottom of the pool. That means that, even if there was a fecal release, it would escape notice. The second point is just as important: natural water sources are typically much larger, creating a greater dilution effect. In fact, the combination of vast volume and wave action means that any fecal release will be spread around and diluted rather quickly. An occasional fecal release is not generally considered a problem.

Natural water sources have another great disadvantage—they are unprotected from run-off of all sorts and are susceptible to sewage infiltration. Ponds and lakes may receive run-off from farm fields, including manure used as fertilizer and from feed lots. Shore birds are probably a significant contributor of fecal material to beaches because they are attracted to food waste left behind by beachgoers and tend to defecate locally. Their excrement contains coliform bacteria, which may raise bioindicator counts to unacceptable levels (McLellan, 2004). Rivers acquire pollutants from every source along their path, especially after rainfall. Large lakes may receive treated wastewater, which should be free of pathogens but may have many other microorganisms. Oceans acquire contaminants from countries around the world, and the waste materials there include a shocking diversity of hazardous materials, household waste, and wastewater. Once contaminated, these large water bodies cannot be cleaned with any degree of reliability. We depend on the natural bioremediation of waste to decrease the organic contaminants.

Under these circumstances, monitoring becomes important, but the challenges are daunting. Unlike swimming pools with their constant volumes of water, natural water volumes change constantly. Large lakes may have currents, and wind action is always a factor. Oceans have wave action and tides. These physical forces move contaminants around. Another factor that makes monitoring problematic is the lack of real-time results. Bacterial indicators, like those used for monitoring drinking water, are tested in tubes of media or on agar plates. They require 24 hours to produce a presumptive result; therefore, the results that are posted for a particular beach are the results for yesterday, not today. The quality of the water today may differ significantly.

Monitoring

Routine monitoring of pools or natural recreational waters is usually comprised of a search for coliform bacteria and for total aerobic heterotrophic bacteria. These tests are identical to the bioindicator models used in monitoring potable water. Coliforms are important because they indicate fecal contamination; total heterotrophs are important because they relate to the total biomass in the system. If too much biomass is present, the risk of allergic reactions is greater. When chlorinated bodies of water are examined, sodium thiosulfate must be used to neutralize the chlorine, otherwise the time between sampling and analysis will contribute to the overall *Ct* value, and the bacterial count may be artificially low.

Additional bioindicators have been useful for recreational water sources. The presence of *Enterococcus* species in the water has been shown to be an effective bioindicator of fecal contamination (Ferley et al., 1989). This is especially true for saltwater monitoring, and the *Enterococcus* bioindicator is often used in conjunction with fecal coliforms (Pruss, 1998). Searching for specific pathogens, such as *Staphylococcus aureus* and *Pseudomonas*, which are implicated in many wound infections, may also be useful.

Enterococcus is a genus that is very similar to *Streptococcus*. In fact, it was formerly classified with the *Streptococci* as Lancefield group D. The Lancefield designations are based on response to specific immune sera, and the *Enterococci* have the same Group D antigen (glycerol teichoic acid antigen) on the cell wall as do the true *Streptococci*. However, they are distinct enough to merit their own genus. Many of the original *Streptococcus* Group D species were therefore converted to *Enterococcus* species, including *faecalis*, *faecium*, (the two most common species), *durans*, *gallinarum*, and *avium*. Some Group D *Streptococcus*, such as *equinus* and *bovis* remain. The *Enterococci* are commonly listed as fecal streptococci (FS) and are used in a ratio with fecal coliforms (FC).

Monitoring a natural water source is more difficult than monitoring a water treatment system. In water treatment there is a directional flow of the product, and samples can be taken from the same point at the same time every day. In natural water systems the water changes direction; the range of contaminants is greater; significant differences in water quality can occur when sample points are only a meter apart, and the depth of the water often varies. Creating a predictive matrix for all of these factors is quite difficult (Fleisher et al., 1993).

The limits shown in **Table 4-8** are not based on the avoidance of all illness, but rather on a realization that some small, acceptable level of gastroenteritis will occur because of the uncontrollable nature of natural water supplies. A 30-day geometric mean is used for these calculations. Notably, the illness (gastroenteritis) rate of swimmers in freshwater is 8 per 1000. These limits were created according to the following formulas:

The geometric mean for *E. coli* is defined as

$$\text{antilog}_{10}\ [(\text{illness rate} + 11.74)/9.40]$$

For *Enterococci* in freshwater the geometric mean is

$$\text{antilog}_{10}\ [(\text{illness rate} + 6.28)\ /\ 9.40]$$

For *Enterococci* in saltwater the geometric mean is

$$\text{antilog}_{10}\ [(\text{illness rate} - 0.20)\ /\ 12.17]$$

The single-sample maximum in each case (that is, for each type of beach) is found using a separate equation:

Single-sample limit = antilog_{10} [($\log_{10}$ geometric mean) + (confidence level factor × $\log_{10}$ standard deviation)]

where confidence-level factor = 0.675 (C1), 0.935 (C2), 1.28 (C3), and 1.65 (C4), and $\log_{10}$

TABLE 4-8 **Standards for Fresh Water (per 100 ml)**

Bioindicator	Illnesses per 1000	Geometric mean	Single-sample maximum (beach designation) C1	C2	C3	C4
E. coli	8	126	235	298	409	575
Enterococci	8	33	61	78	107	151
Standards for saltwater (per 100 ml)						
Enterococci	19	35	104	158	276	501

standard deviation = 0.4 for freshwater and 0.7 for saltwater.

For example,

$$235 = \text{antilog}_{10}[\log_{10} 126 + (0.675 \times 0.4)] = \text{antilog}_{10}(2.10 + 0.27)$$

The Clean Water Act specified acceptable levels of bioindicators in coastal waters, which are defined as the coasts of the Great Lakes (freshwater) and the marine coastal waters in which water contact is likely. A further subcategory denotes the extent of use. Designated bathing beach waters (called C1) are typically the beaches that people frequent for the express purpose of swimming, and they are often commercially well developed. Moderately used coastal recreation waters (C2) are typically not bathing beaches, but they may attract swimmers anyway. They are used less than half as much as C1 beaches. Lightly used coastal recreation waters (C3) have even fewer swimmers than C2 waters. Infrequently used coastal recreation waters (C4) are, as the name says, used infrequently for swimming.

States that do not have water quality standards of their own are subject to Title 40 Code of Federal Regulations Part 131, which specifies acceptable levels of *E. coli* and *Enterococci* for freshwater areas and of *Enterococci* for saltwater areas. *Enterococci* survive much longer than *E. coli* in saltwater, and therefore they are better indicators in the marine environment.

SPECIFIC PATHOGENS

In an analysis of outbreaks of gastrointestinal disease in recreational settings due to viruses, Sinclair et al. (2009) found that most (46%) were caused by noroviruses, followed by adenoviruses (24%) and echoviruses (18%). Not surprisingly, most outbreaks were associated with swimming pools (48%), followed by lakes and ponds (40%). Because people inadvertently swallow water during recreational activities, all of the pathogens listed as water consumption hazards apply here as well. However, to delineate the particular hazards due to simple water contact, a more focused list is presented.

Adenovirus

If swallowed, adenoviruses can cause gastroenteritis (discussed previously). By contact, adenoviruses can cause conjunctivitis and skin rashes. If aerosolized, they may be inhaled and may cause both pharyngitis and lower respiratory tract illness. Adenoviruses can replicate in adenoids, tonsils, and in the intestinal tract. Infections typically involve children. A good example of the potential for an adenovirus outbreak is demonstrated by a report from an Australian school in 2000. While the origin of the pathogen remains unknown, the illness that it caused among children using the school's swimming pool was significant, with about 40% affected at the peak of the outbreak. Typical symptoms included sore throat, fever, headache, vomiting and/or diarrhea, and conjunctivitis. The school's saltwater pool was implicated because the residual chlorine level was only 1 ppm, half the recommended effective level (Harley et al., 2001).

Aeromonas hydrophila

This bacterial species is found in freshwater and in brackish water. *Aeromonas hydrophila* can cause a severe gastroenteritis if ingested (although the infectious dose appears to be very high), but if it comes into contact with skin, it causes infections in open wounds.

Leptospira

Leptospirosis is an uncommon disease in the United States, but it can readily be found in many tropical and subtropical countries. It is characterized by high fever, chills, headache, and muscle aches. If *Leptospira* gets into the eyes, it can cause conjunctivitis. Fortunately, it responds well to antibiotic therapy, but if untreated, major organs can be affected, and in rare cases the patient may die. *Leptospira* cells are tightly twisted coils (**Figure 4-24**).

Leptospirosis is a zoonotic disease, meaning that it is found in animals. However, the usual exposure to humans occurs through water that has been contaminated with urine from an infected animal. Because *Leptospira* are found in many domesticated animals (e.g., cattle, horses, and pigs), an infected herd could deposit large numbers of *Leptospira* into the local water supply. Animal handlers are also at great risk of infection, which can occur through ingestion of contaminated food or water but also through contact with contaminated water that infiltrates a wound or contacts the mucous membranes.

An outbreak of leptospirosis among U.S. citizens was noted in 2000 after a large group of outdoor enthusiasts journeyed to Malaysian Borneo to compete in a multisport endurance race. This race featured hiking through the jungle, biking, swimming and kayaking in fresh water and in marine water, cave exploration, and climbing. Obviously the participants were in good physical shape. After returning to the United States many became ill, and a retrospective analysis (Sejvar et al., 2003) found that as many as 80 people were infected and that 29 needed hospitalization. Fortunately, no one died. The *Leptospira* microorganism was probably acquired through swimming in freshwater, although it was impossible to pinpoint the source of infection.

Mycobacterium

These bacterial species can be found in any type of water. *Mycobacterium marinum* causes skin ulcerations when it invades small wounds, especially on the extremities. While it can be contracted while swimming in a natural water source (either fresh or salt water, despite its name), it is more commonly associated with fish tanks. When a person cleans a fish tank, his or her arms are exposed to the pathogen, giving it a chance to invade. The resulting infection has been called "swimming pool granuloma" because of the peculiar symptoms it causes—typically small, red nodules. Other *Mycobacterium* species cause similar soft-tissue infections. *M. ulcerans* causes Buruli ulcers on the skin, probably after infiltration of a small break in the skin. However, it might also be transmitted through insect bites. This disease is confined to tropical and subtropical countries.

Pseudomonas

These large, Gram-negative bacilli are common in the environment (e.g., soil, water), but they are not coliforms (**Figure 4-25**). *Pseudomonas aeruginosa* is a feared opportunistic pathogen because it acquires antibiotic resistance genes readily, making therapy very difficult. It is especially dangerous for immunocompromised patients. These bacteria are noted for causing persistent pneumonia in cystic fibrosis patients, in wound infections of all types, and especially in burn patients. They survive very well in water, creating a major cause of concern. In water it may invade open wounds and start a suppurative infection. This is one of the pathogens implicated in persistent "swimmer's ear" infections and also in "hot tub rash" syndrome, in which red bumps appear in clusters on the skin, often becoming pus-filled. *Pseudomonas aeruginosa* survives well at higher temperatures, and if the hot tub or whirlpool is not adequately maintained, *P. aeruginosa* can spread from person to person through the shared water.

If *P. aeruginosa* contaminates a food or medical product, it may persist for an extended period. It is

FIGURE 4-24 ***Leptospira* cells. Note the tightly twisted cells, looking like braided hair.**

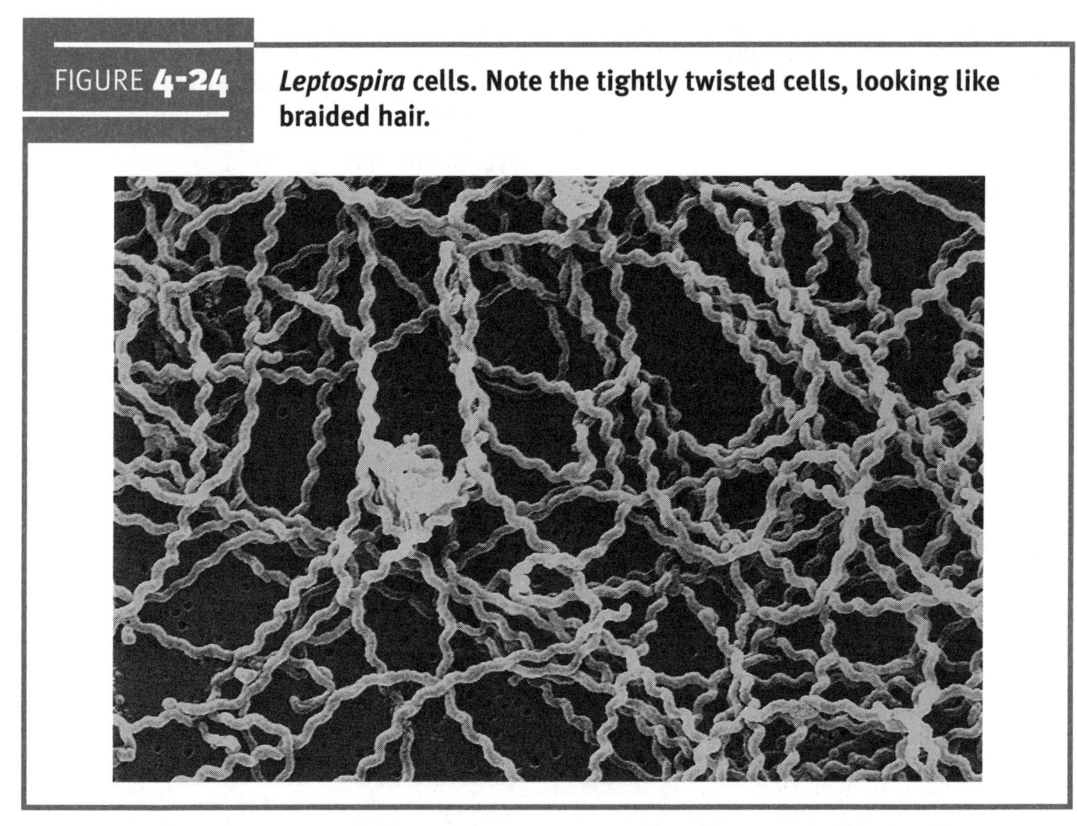

SOURCE: Courtesy of Rob Weyant, Centers for Disease Control and Prevention. (http://phil.cdc.gov/phil/details.asp). #138.

a constituent of biofilms and so survives on pipes and in water treatment plants. *Pseudomonas aeruginosa* can be killed using shock treatments (suddenly high concentrations) of chlorine.

Vibrio (V. vulni cus, V. parahaemolyticus, V. alginolyticus, and V. mimicus)

Vibrio cholerae is, of course, the causative agent of cholera and is certainly a waterborne pathogen; however, it must be ingested. Other species are associated with wound infections and occasionally with ear infections. If untreated, *Vibrio* species may cause septicemia, which is potentially fatal. As with all *Vibrio* species, these strains prefer brackish water. The Gulf Coast of the United States sees the highest incidence of *Vibrio* infections. They are also associated with gastroenteritis in people who eat contaminated shellfish (filter feeders), especially when they are consumed raw.

Cyanobacteria

Cyanobacteria are blue-green algae; they are autotrophic bacteria. They produce toxins that can cause diarrhea if ingested, although these events are rare. They can multiply to very high numbers, causing harmful algal blooms (described in the next section). The algal blooms may release toxins into the water. Swimmers that contact these toxins may suffer from dermatitis (e.g., skin rash or hives) with mild to severe symptoms.

Naegleria fowleri

Naegleria fowleri is a free-living amoeba found in soil or natural freshwater sources, particularly those

FIGURE 4-25 ***Pseudomonas aeruginosa* cells. Note their large width.**

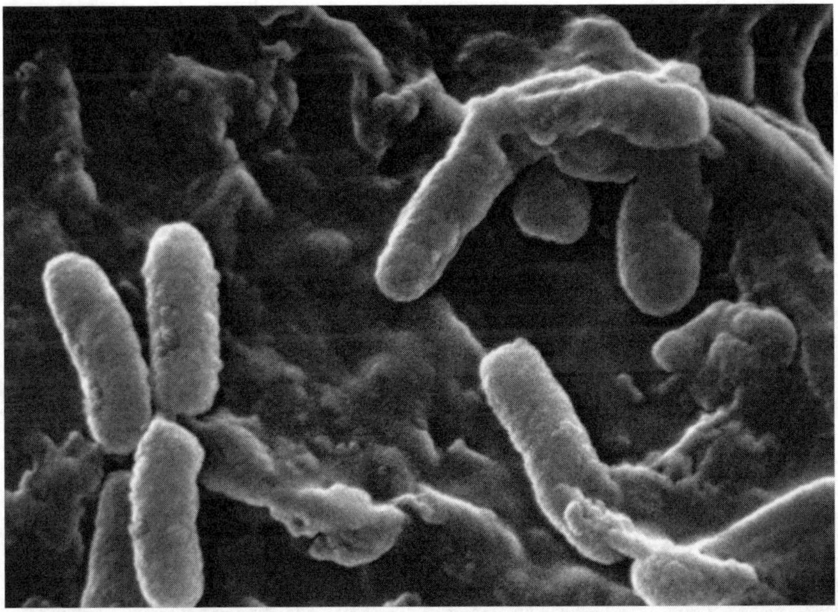

SOURCE: Courtesy of Janice Haney Carr, Centers for Disease Control and Prevention. (http://phil.cdc.gov/phil/details.asp). #10043.

with relatively high temperatures (**Figure 4-26**). These organisms are not found in properly maintained swimming pools or other chlorinated water sources. In neglected pools there is a possibility of *Naegleria* blowing in via soil particles. *Naegleria fowleri* is most often associated with "ole swimming hole" ponds, particularly in the southwestern United States. These organisms are, fortunately, very rare; perhaps two or three infections per year are recorded.

Infection by *N. fowleri* can be devastating, apparently infecting the body through water that enters the nose. It crosses into the brain and spinal cord by a method that is still unknown and creates an infection known as Primary Amebic Meningoencephalitis (PAM). Brain tissue is destroyed quickly, and death is almost assured. The fatality rate of PAM has improved somewhat in recent years as physicians begin to spot the infection earlier; early treatment is essential for survival. The oocysts can stay alive for several months, but the trophozoite stage actually infects. A sample report form used for *Naegleria* outbreaks is shown in **Figure 4-27**.

Schistosoma

Three species of these parasitic worms can cause disease in humans, known as schistosomiasis or bilharzia (**Figure 4-28**). All are extremely rare in the United States (i.e., they are largely confined to people returning from endemic areas), although they are common in tropical areas of the world. Some 200 million cases are estimated worldwide. These parasites have a complex life cycle that includes an egg stage and a larval stage carried by snails. When people enter water containing these

FIGURE 4-26 *Naegleria fowleri.*

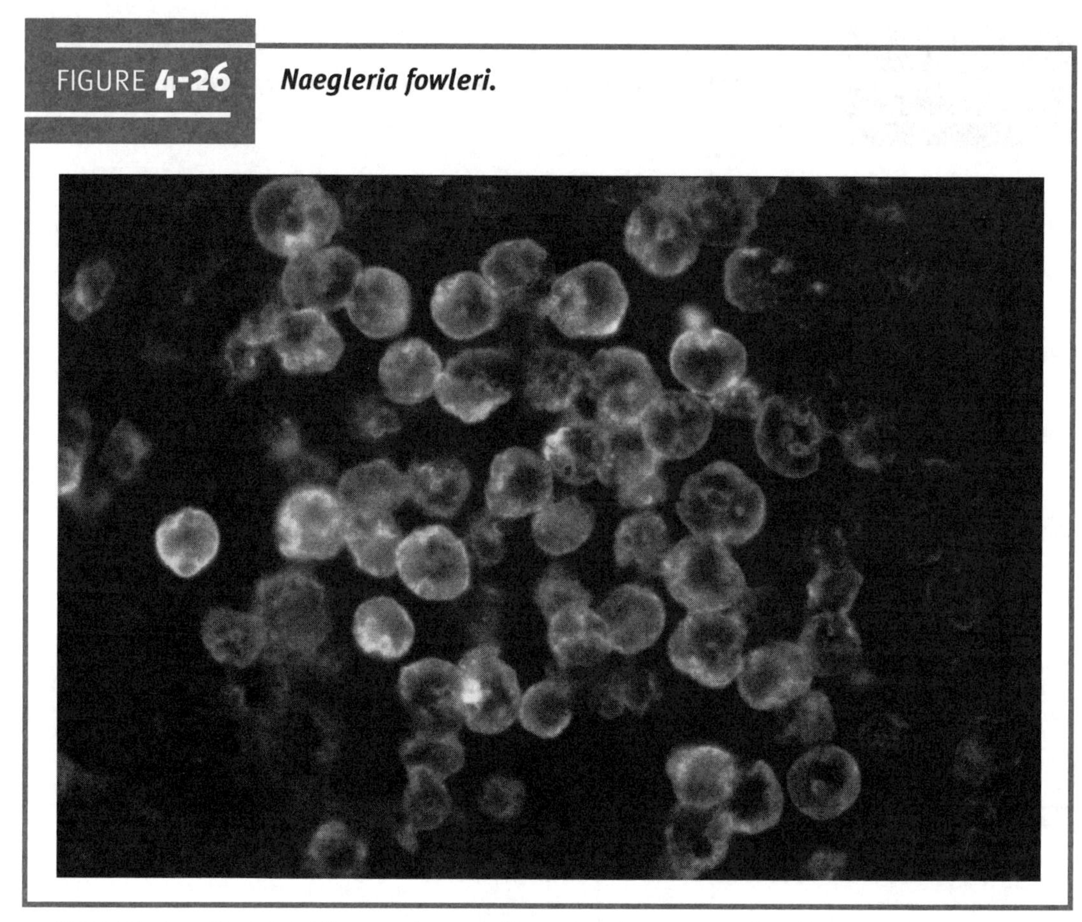

SOURCE: Courtesy of Dr. Govinda S. Visvesvara, Centers for Disease Control and Prevention. (http://phil.cdc.gov/phil/details.asp). #408.

infected snails, there is a risk of infection. The symptoms result from a delayed-type hypersensitivity (DTH) reaction to the parasite.

The pathogen is spread by a fecal–oral route of transmission, which suggests that it might be eradicated in areas where it is endemic. As a free-living worm, it can only survive in the water for about 2 days.

OTHER VENUES TO CONSIDER

Industrial

A great many types of industries are at work today, and more will undoubtedly emerge over time; each type has its own water requirements. Many, like power generation, use water simply to cool energy-intensive processes. Food manufacture and processing requires water that is at least as good as household water. Produce must be rinsed to remove soil particles. Food is typically cooked in water (which is generally sufficient to kill any microorganisms). The requirements of food processing are addressed in greater detail in Chapter 10.

Medical industries frequently require a higher quality of water. This is especially true for the manufacture of injectable medicines, such as intravenous fluids and antibiotics. The requirement for fewer chemical contaminants and no biological contaminants results in a higher quality standard.

FIGURE 4-27 Reporting form for Primary Amebic Meningoencephalitis.

Primary Amebic Meningoencephalitis Case Report Form

Report Source

Date Reported to CHD:_____ Reporting Source (Check all that apply): ☐ Lab ☐ Hospital ☐ Physician ☐ Public Health Agency ☐ Other____

Reporter name:________________ Reporter phone:________________

Demographic Information

Patient's Last Name First M.I.	Date of Birth / /	Age	Gender ☐ Male ☐ Female ☐ Unknown
Address City	State	Zip	Phone Number:

Occupation/Grade Level	Place of Employment/School/Daycare	Ethnicity: ☐ Hispanic ☐ Non-Hispanic	Race: ☐ White ☐ Black ☐ Asian/PI ☐ Am. Indian ☐ Unknown ☐ Other_____	Marital Status: ☐ Single ☐ Married ☐ Widowed ☐ Divorced

Clinical Information

Date Diagnosed: / / M.R. #:	Data/Time of Onset: / / ______ a.m./p.m.	Date/Time of Recovery: / / ______ a.m./p.m.	Treatment: ☐ Yes ☐ No Dates:______to______ Drug(s):________________
Hospitalized? ☐ Yes ☐ No ☐ ER only	Admission Date: / /	Discharge Date: / /	Dates:______ to ______ Drug(s):________________
Hospital Info: ________________ ( )________________		Outcome: ☐ Recovered ☐ Died / /	Physician Info: ________________ ( )________________

Symptoms: ☐ Headache ☐ Fever ☐ Nausea ☐ Vomiting ☐ Stiff neck	☐ Confusion ☐ Loss of balance ☐ Loss of bodily control ☐ Lack of attention to people and surroundings ☐ Seizures ☐ Hallucinations	Laboratory Information: ________________		Specimen Type: ☐ Cerebral Spinal Fluid ☐ Brain Tissue ☐ Other________	Results: ________
		Date Collected: / /	Lab Report Date: / /	Confirmed by State Lab: ☐ Yes ☐ No ☐ Unknown ☐ NA	Confirmed by CDC Lab: ☐ Yes ☐ No ☐ Unknown ☐ NA

Risk Factors

Yes No Don't Know NA

☐ ☐ ☐ ☐ Recreational water exposure

☐ Man-made Lake ☐ Freshwater Lake ☐ River ☐ Canal
☐ Swimming Pool ☐ Wading pool ☐ Spa ☐ Pond
☐ Water park ☐ Fountain ☐ Hot springs
☐ Other:______

Exposure Date: / /

Source:________________

Location of water activities:________________

Water Temp: ______F/C Ambient Air Temp: ______F/C

Estimated Depth:______ Turbidity:______

Water level:______ If treated water, chlorine level:______

☐ ☐ ☐ ☐ Did person dive into the water?

☐ ☐ ☐ ☐ Did person jump into the water?

☐ ☐ ☐ ☐ Did person swim in the water?

Yes No Don't Know NA

☐ ☐ ☐ ☐ Did person participate in other water sports? Which ________________

☐ ☐ ☐ ☐ Did person splash in the water?

☐ ☐ ☐ ☐ Did person inhale any water up the nose?

☐ ☐ ☐ ☐ Did person swallow any water?

☐ ☐ ☐ ☐ Did person wear a nose clip or hold your nose shut when jumping or diving in the water?

☐ ☐ ☐ ☐ Did you see any signs posted regarding "No Swimming"?

☐ ☐ ☐ ☐ Do any power plants dump water into the water? or Are any power plants located near the water?

☐ ☐ ☐ ☐ Did person have any contact with soil/sand?

☐ ☐ ☐ ☐ Were you aware of the health risks associated with swimming in warm man-made/fresh water lakes? If yes, how:
☐ Healthcare Provider ☐ Newspaper ☐ TV ☐ Radio
☐ Health Department ☐ Friend/Relative ☐ Sign
☐ Prior to Illness ☐ After Illness

☐ ☐ ☐ ☐ Did you know of press releases warning about the risks?

Most likely exposure/site:________________ Site name/address:________________

Where did exposure probably occur? State and County:__________ ☐ US but not______ ☐ Outside US ☐ Unknown

Investigator:________________ **Phone:**________________ **Date:**________________

SOURCE: Courtesy of the Milwaukee Health Department.

Chemical contaminants can be removed through filtration and ion exchange chromatography, and biological contaminants can be removed through filtration and then sterilization by autoclaving or irradiation.

Agricultural

The biggest problem with agricultural water in recent years has been the mixing of wastewater with freshwater, promoting the fecal–oral route of transmission. However, this agricultural contamination is more indirect because the water itself is not consumed but instead is used to irrigate crops, where the pathogens become associated with the produce. In some cases it appears that contaminated wash water was used to rinse the produce, yielding the same contamination. Farms are responsible for using uncontaminated water, or water that comes from a monitored and treated water supply (Corkal et al., 2004).

HARMFUL ALGAL BLOOMS

The term "algae" refers to a complex group of aquatic (fresh and marine) organisms that form an important part of the biosphere. Some, the phytoplankton, are photosynthetic eukaryotes that account for a large fraction of recycled oxygen, by some estimates, 50%. Some, like the cyanobacteria (blue-green algae), are prokaryotes. Others are

FIGURE **4-28** **Schistosome parasites.**

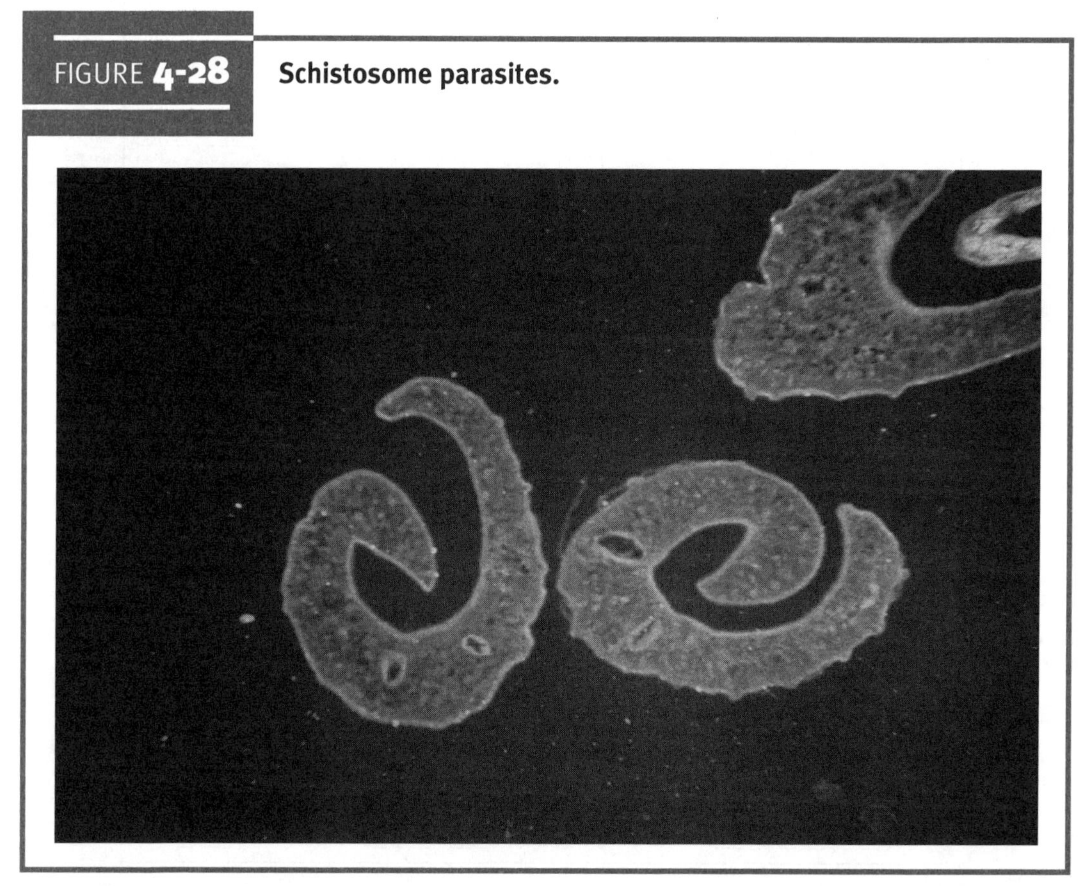

SOURCE: Courtesy of Marianna Wilson, Centers for Disease Control and Prevention. (http://phil.cdc.gov/phil/details.asp). #11205.

BOX 4-5 *Pfiesteria Piscicida*

Pfiesteria piscicida is an algal species that has been implicated in fish kills (note the species name) and in deleterious effects in humans. It was also the focus of intense media coverage following reports in 1997 detailing its role in outbreaks along the eastern coast of the United States, notably in Chesapeake Bay. An examination of this case reveals some of the difficulties in making public health policy.

Unusual fish kills were noticed along the eastern shore in 1996, and these continued into 1997. The fish had lesions that were described as "punched out" (i.e., deep, penetrating ulcers). Also in 1997, several fishermen reported illness associated with fishing in areas that corresponded to the fish kills. At this time, the illnesses were associated with a possible outbreak of estuarine-associated syndrome (PEAS), because the algal blooms and fish kills were all in estuary areas.

The principal complaint was memory loss, and it was assumed that the same toxin that was causing the lesions was aerosolized and inhaled by the fishermen. A retrospective analysis of these affected people and occupationally matched controls found that people who were exposed to the algae were significantly more likely to experience new or increased forgetfulness, to experience headaches, to have skin lesions, and to have a burning sensation on the skin when in contact with water (Gratten et al., 1998). A separate report from recreational swimmers in the same areas further supported this assumption. Case studies of individuals who were exposed to algal blooms revealed an even larger set of symptoms, some of them quite unusual and long-lasting (Shoemaker and Hudnell, 2001). No deaths were attributed to this intoxication, and most people respond well to treatment with cholestyramine. This drug was originally intended as a treatment for high cholesterol, but has now been shown to be effective in toxin elimination. The popular press seized on these outbreaks, creating a great deal of public awareness and a significant amount of public panic. Based on these complaints and on observations, the CDC created a surveillance system for *Pfiesteria*.

Subsequent research on this algal species failed to reveal the toxin responsible for the observed health effects (Rubin et al., 2001). The association of the illness with *Pfiesteria species* (e.g., *P. piscidia*, *P. shumwaye*, and perhaps others) appears firm, although the conditions that express the toxin, or the genes responsible for toxin production, have not been described. This has led some workers to suggest that *Pfiesteria* is not the proximal cause of the illnesses and fish kills; there are other algal species and bacterial species present in every affected fish that might contribute to the illness (Samet et al., 2001). However, this does point out the difficulty in attributing some illnesses to specific causes, and the role that the media can play in affecting public opinion.

macroscopic, such as the well-known seaweed. Finally, some are nonpigmented eukaryotic cells, like the zooplankton. This grouping includes several phyla with so many structural forms and names that it is sometimes difficult to remember that all algae are closely related. Sizes range from single cells to the large brown algae beds or seaweed that may be several meters in length.

In aquatic systems it is not unusual to find low concentrations of algal species in the water,

TABLE 4-9 Toxins Data

	Anabaena	Aphanizomenon	Apanocapsa	Cylindrospermopsis	Hapalosiphon	Microcystis
Anatoxin -a	√	√				
Anatoxin -a(s)	√					
Saxitoxin	√	√		√		
Cylindro spermopsin		√		√		
Microcystins	√		√		√	√
Nodularins						
Lyngbyatoxin						
Aplysiatoxin						

SOURCE: Adapted from Carmichael, 2001.

sometimes confined to a specific niche, such as a layer of water below the surface. To a certain extent the algae are able to control their buoyancy and remain at this optimum survival depth. Occasionally the concentration of algae increases dramatically in a water supply because the environmental conditions change to favor their growth; these exact conditions depend on the particular algal species. At this point these "algal blooms" are more visible. The species grows until the conditions change.

Algal blooms are often associated with pollution; that is, by excessive nutrient loading from human sources, but this is not entirely accurate. Periodic algal blooms are a normal part of many ecosystems. For example, springtime algal blooms of local diatom species occur cyclically when the mixing of the water results in nutrient availability. This may be caused by winds, temperature shifts, or increased mixing caused by melting ice and snow. For diatoms, which have a silicate cell wall (or frustrule), the concentration of available silica in the water is a determining factor for blooms. When the available silica are depleted, the bloom spontaneously fades. These types of blooms are normal, if not always desired.

Some algal blooms, however, are caused by or supported by anthropogenic pollution. This is especially true for algae that are dependent on nitrogen compounds (N) and phosphate (PO_4^-), which are typically introduced into aquatic systems as wastewater discharge or as run-off from farm fields contaminated with excessive fertilizer. In other instances human engineering of the environment plays a key role in algal blooms. Creation of still water where formerly it ran freely, such as when dams or ponds are built, can create the conditions for algal growth. If a dam prevents nutrient sources from continuing down river, then an additional factor is present to promote algal growth. Regardless of the source, algal blooms are usually unwelcome. Blooms that recede into the water source may be of little concern, but blooms that wash up on shore will begin to rot, producing a foul odor. Some algal blooms are merely unsightly, such as those that cause "scum" formation on the ocean surface or "foam" on the shore.

Nodularia	Lyngbya	Oscillatoria	Umezakia	Schizothrix	Nostoc	Affected organ
		√				Nerve
		√				Nerve
	√					Nerve
			√			Liver
		√			√	Liver
√						Liver
	√					Skin
				√		Skin

Other algal blooms can have more serious effects. Large blooms may deplete dissolved oxygen in the area and may kill animal species. This occurs when high levels of respiration and oxygen utilization caused by biomass decomposition occurs. The biomass can possibly inhabit man-made structures, such as water pipes and filters, and clog them.

Some algal blooms are clearly dangerous; they are referred to as harmful algal blooms (HABs). Occasionally they are reported by their colors, such as the frequently cited "red tide." The danger from these blooms results from specific toxins that the algae produce, and which are released in significant concentrations in the water. These toxins may be dangerous to other plant species, to fish or other animals, or to humans (**Box 4-5**). The toxins bioaccumulate in shellfish, for example, making them unsuitable for human consumption.

Like mold toxins, toxin production by algae is not well understood. However, under certain conditions algae will form toxins that are released into the water and affect other species (**Table 4-9**). The algae may be ingested by predators, leading to an intoxication, or the toxins can bioaccumulate in certain species and then affect organisms farther up the food chain. With the power of wave action, individual cells may break open and release their toxins into the air. This can also lead to aerosolization of the toxin.

Paralytic Shellfish Poisons

This designation refers to certain dinoflagellates, such as *Karenia brevis*, that have a reddish-brown color. When a large bloom occurs, they take on a distinctive red appearance and have been referred to as "red tide." These events have been observed in the Atlantic and Pacific Oceans, and in the Gulf of Mexico, as well as in estuaries. Swimming through these algae is dangerous, although the more common danger is eating shellfish that have accumulated the toxin from ingesting algae. These toxins kill many types of shellfish, such as oysters, scallops, and mussels, devastating the fishing industries that depend on the quality and availability of viable seafood sources.

Somewhere between 30 minutes and 2 hours after ingestion, numbness or tingling of the face and lips begins and may spread to the arms and legs. Other symptoms include headache, dizziness, nausea, and muscular paralysis. Oddly, some people report a floating sensation. Death is possible if enough toxin is ingested. In 1996, the deaths of several manatees in Florida were attributed to these toxins.

Neurotoxic Shellfish Poisons

Karenia brevis or related species also cause neurotoxic shellfish poisons. These toxins have been found in oysters, clams, and mussels. About 1 to 3 hours after ingestion general numbness and tingling begin, along with diarrhea, nausea, and vomiting. An odd symptom that has been reported is the reversal of hot and cold sensations. This poisoning does not usually result in death. In aerosol form the toxin causes irritation in the respiratory tract and mucous membranes.

Amnesic Shellfish Poisons

A relatively rare intoxication is caused by another red algae, *Pseudo-nitzschia* or *Nitzchia pungens*. The active agent is domoic acid, which affects neurotransmission. It results in nausea, vomiting, and diarrhea but also includes neurological symptoms such as short-term memory loss (thus the name of the poison), disorientation, paralysis, and confusion. In rare cases it results in death (Teitelbaum et al., 1990).

Diarrhetic Shellfish Poisons

Other diarrhetic shellfish illnesses are caused by okadaic acid, one of the toxins produced by *Dinophysis*. As the name implies, *Dinophysis* causes diarrhea and other gastrointestinal distress symptoms, but it does not cause neurological symptoms and it is not regarded as a risk for death. This type of poisoning has been seen in many parts of the world.

Azaspiracid Shellfish Poisons

The azaspiracids are a group of related toxins. They may derive from algae of the genus *Protoperidinium*. These toxins are unusually stable and can remain active for 6 months. The symptoms of this intoxication are typically confined to severe gastrointestinal upset.

Ciguatera Food Poisoning

Ciguatera is the bioaccumulation of ciguatoxins from various algal species in the tissues of reef fish of warm waters (e.g., mackerel, grouper, and snapper). In tropical areas it is well known that certain fish should not be eaten during specific times during the year because of the possibility of ingesting a toxic fish. In the United States, Florida and Hawaii have been sites of ciguatera, but tropical waters worldwide are at risk.

Ciguatera leads to severe gastrointestinal upset, but it includes some unusual symptoms: the sensation that your teeth are loose, the confusion of hot and cold sensations, and a metallic taste in the mouth. Symptoms may persist from months to years. Deaths are known to result from ciguatera, but they are rare. Annual case rates in the United States range from 50 per 100,000 in Florida to 700 per 100,000 in the U.S. Virgin Islands (Lehane and Lewis, 2000).

Even outside of ciguatera-endemic areas, predatory fish may still have high levels of ciguatoxins. For example, in 2004 a barracuda was caught off the coast of South Carolina, and a husband and wife ate fillets of the barracuda and experienced severe symptoms of ciguatera (Villareal et al., 2006). Although both recovered, the wife required an extensive treatment and reported some tingling sensations a full 18 months after the episode. Barracuda is one of the worst fish species for ciguatera, and consuming them is probably best avoided.

Cyanobacteria Toxin Poisoning

Cyanobacterial blooms occur in either fresh water or brackish water. In large algal blooms of these bacterial species, the water may be tinged a distinctive blue, brown, red, or green. The toxins that these microbes produce are quite toxic, and ingestion can be fatal. However, bioaccumulation is not usually a problem. Instead, overall toxicity is related to algal biomass. Even immersion in water can be dangerous, and a number of reports state the first indication as either a dead animal that consumed the water or a sick dog that was playing in the water. Even aerosols can contain toxic amounts of these compounds.

Cyanobacteria that contribute to this HAB number at least 40 species, and include *Microcystis aeruginosa, Anabaena, Aphanizomenon,* and *Cylindrospermopsis* (Carmichael, 2001). They produce a variety of toxins, some of which have neurological effects (e.g., muscle weakness, difficulty in breathing, and convulsions), and some of which damage organ tissues (usually liver and kidney). (See **Table 4-9**.) Human deaths are very rare; a set of errors in a dialysis unit has been fatal in one notorious case (Carmichael et al., 2001; Jochimsen et al., 1998).

QUESTIONS FOR DISCUSSION

1. What is the fecal–oral route of transmission? Give examples.
2. What is the difference between the MCLG and the MCL?
3. Why aren't specific pathogens enumerated in water samples?
4. What problems arise when attempting to determine an infectious dose?
5. Which are the most prevalent pathogens associated with drinking water? With recreational water exposure?
6. What is the difference between a clinical and a subclinical case?
7. Pick any major pathogen and describe any factors that increase or decrease its infectivity.
8. Why is Naegleria such a feared pathogen?
9. Which products of algae can be deleterious to humans?
10. Why do algae bloom only occasionally?

References

Arnell, B., J. Bennett, R. Chehey, and J. Greenblatt.1996. *Shigella sonnei* outbreak associated with contaminated drinking water—Island Park, Idaho, August 1995. Morbid. Mortal. Weekly Rep. 45: 229–231.

Carmichael, W.W. 2001. Health effects of toxin-producing cyanobacteria: "The CyanoHABs." *Human Ecol. Risk Assess.* 7: 1393–1407.http://www.cdc.gov/hab/cyanobacteria/facts.htm—top

Carmichael, W.W., M.F.O. Azevedo, J.S. An, et al. 2001. Human fatalities from cyanobacteria: chemical and biological evidence for cyanotoxins. Environ. Health Perspect. 107: 663–668.

Carpenter, C., R. Fayer, J. Trout, and M.J. Beach. 1999. Chlorine disinfection of recreational water for Cryptosporidium parvum. Emerg. Infect. Dis. 5: 579–584.

Chandra, V., S. Taneja, M. Kalia, and S. Jameel. 2008. Molecular biology and pathogenesis of Hepatitis E virus. J. Biosci. 33: 451–464.

Chiba, S., S. Nakata, K. Numata-Kinoshita, and S. Honma. 2000. Sapporo virus: history and recent findings. J. Infect. Dis. 181 (Sup 2): S303–S308.

Clark, D.P. 1999. New insights into human cryptosporidiosis. Clin. Microbiol. Rev. 12: 554–563.

Corkal, D., W.C. Schutzman, and C.R. Hilliard. 2004. Rural water safety from the source to the on-farm tap. J. Toxicol. Environ. Health A 67: 1619–1642.

Craun, G.F. 1988. Surface water supplies and health. J. Am. Water Works Assoc. 80: 40–52.

Cuthbert, J.A. 2001. Hepatitis A: old and new. Clin. Microbiol. Rev. 14: 38–58.

Dunn, B.E., H. Cohen, and M.J. Blaser. 1997. *Helicobacter pylori.* Clin. Microbiol Rev. 10: 720–741.

Ferley, J.P., D. Zmirou, F. Balducci, B. Baleux, P. Fera, G. Larbiagt, E. Jacq, B. Moissonnier, A. Blineau, and J. Boudot. 1989. Epidemiological significance of microbiological pollution criteria for river recreational waters. Intl. J. Epidem. 18: 198–205.

FitzSimons D., G. Hendrickx, A.Vorsters and P. Van Damme. 2010. Hepatitis A and E: update on prevention and epidemiology. Vaccine 28: 583–588.

Fleischer, J.M., F. Jones, D. Kay, R. Stanwell-Smith, M. Wyer, and R. Morand. 1993. Water and non-water-related risk factors for gastroenteritis among bathers exposed to sewage-contaminated marine waters. Internation. J. Epidem. 22: 6998–6708.

Grattan, L.M., D. Oldach, T.M. Perl, M.H. Lowitt, D.L. Matuszak, C. Dickson, C. Parrott, R.C. Shoemaker, C.L. Kauffman, M.P. Wasserman, J.R. Hebel, P. Charache, and J.G. Morris. 1998. Learning and memory difficulties after environmental exposure to waterways containing toxin-producing *Pfiesteria* or *Pfiesteria*-like dinoflagellates. Lancet 352: 532–539.

Guix, S., A. Bosch, and R.M. Pinto. 2005 Human astrovirus diagnosis and typing: current and future prospects. Lett. Appl. Microbiol. 41: 103–105.

Hansen, J.S. and J.E. Ongerth. 1991. Effects of time and watershed characteristics on the concentration of *Cryptosporidium* oocysts in river water. Appl. Environ. Microbiol. 57: 2790–2795.

Harley, D., B. Harrower, M. Lyon, and A. Dick. 2001. A primary school outbreak of pharyngoconjunctival fever caused by adenovirus type 3. Commun. Dis. Intell. 25: 9–12.

Herrera, V. and J. Parsonnet. 2009. Helicobacter pylori and gastric adenocarcinoma. Clin. Microbiol. Infect. 15: 971–976.

Jiang, J., K.A. Alderisio, and L. Xiao. 2005. Distribution of *Cryptosporidium* genotypes in storm event water samples from three watersheds in New York. Appl. Environ. Microbiol. 71: 4446–4454.

Jiang, X., W.D. Cubitt, T. Berke, W. Zhong, X. Dai, S. Nakata, L.K. Pickering, and D.O. Matson. 1997. Sapporo-like human caliciviruses are genetically and antigenically diverse. Arch. Virol. 142: 1813–1827.

Jochimsen, E.M., W.W. Carmichael, J. An, et al. 1998. Liver failure and death after exposure to microcystins at a hemodialysis center in Brazil. New England J. Med. 338: 873–878.

Lawrence, D.N. 2004. Outbreaks of gastrointestinal diseases on cruise ships: lessons from three decades of progress. Curr. Infect. Dis. Reports 6: 115–123.

LeChevallier, M.W., W.D. Norton, and R.G. Lee. 1991. Occurrence of *Giardia* and *Cryptosporidium* spp. in surface water supplies. Appl. Environ. Microbiol. 57: 2610–2616.

Lehane, L. and R.J. Lewis. 2000. Ciguatera: recent advances but the risk remains. Int. J. Food Microbiol. 61: 91–125.

Marshall, B.J. and J.R. Warren. 1984. Unidentified curved bacilli in the stomach of patients with gastritis and peptic ulceration. Lancet 1 (8390): 1311–1315.

McLellan, S.L. 2004. Genetic diversity of *Escherichia coli* isolated from urban rivers and beach water. Appl. Environ. Microbiol. 70: 4658–4665.

Nomura, A., G.N. Stemmermann, P.H. Chyou, I. Kato, G.I. Perez-Perez, and M.J. Blaser. 1991. Helicobacter pylori infection and gastric carcinoma among Japanese Americans in Hawaii. New Engl. J. Med. 325: 1132–1136.

Palacio, H., U. Shah, C. Kilborn, et al. 2005. Norovirus outbreak among evacuees from Hurricane Katrina—Houston, Texas, September 2005. Morbid. Mortal. Weekly Rep. 54: 1016–1018.

Payment, P., L. Richardson, J. Siemiatycki, R. Dewar, M. Edwardes, and E. Franco. 1991. A randomized trial to evaluate the risk of gastrointestinal disease due to consumption of drinking water meeting current microbiological standards. Am. J. Public Health 81: 703–708.

Pruss, A. 1998. Review of epidemiological studies on health effects from exposure to recreational water. Intl. J. Epidem. 27: 1–9.

Robertson, L.J., L. Hermansen, B.K. Gjerde, E. Strand, J.O. Alvsvag, and N. Langeland. 2006. Application of genotyping during an extensive outbreak of waterborne giardiasis in Bergen, Norway, during autumn and winter 2004. Appl. Environ. Microbiol. 72: 2212–2217.

Robertson, L.J., T. Forberg, and B.K. Gjerde. 2008. *Giardia* cysts in sewage influent in Bergen, Norway, 15–23 months after an extensive waterborne outbreak of giardiasis. J. Appl. Microbiol. 104: 1147–1152.

Romero, S., J.R. Archer, M.E. Hamacher, S.M. Bologna, and R.F. Schell. 1988. Case report of an unclassified microaerophilic bacterium associated with gastroenteritis. J. Clin. Microbiol. 26:142–143.

Rose, J.B. 1988. Occurrence and significance of *Cryptosporidium* in water. J. Am. Water Works Assoc. 80: 53– 58.

Rose, J.B. 1997. Environmental ecology of Cryptosporidium and public health implications. Annu. Rev. Public Health 18: 135–161.

Rubin, C., M.A. McGeehin, A.K. Holmes, L. Backer, G. Burreson, M.C. Earley, D. Griffith, R. Levine, W. Litaker, J. Mei, L. Naeher, L. Needham, E. Noga, M. Poli, and H.S. Rogers. 2001. Emerging areas of research reported during the CDC National Conference on *Pfiesteria*: From Biology to Public Health. Environ. Health Perspec. 109: 633–637.

Samet, J., G.S. Bignami, R. Feldman, W. Hawkins, J. Neff, and T. Smayda. 2001. *Pfiesteria*: review of the science and identification of research gaps. Report for the National Center for Environmental Health, Centers for Disease Control and Prevention. Environ. Health Perspec. 109: 639–659. http://ehp.niehs.nih.gov/members/2001/suppl-5/639-659samet/EHP109s5p639PDF.pdf

Sedmak, G., D. Bina, and J. MacDonald. 2003. Assessment of an Enterovirus sewage surveillance system by comparison of clinical isolates with sewage isolates from Milwaukee, Wisconsin, collected August 1994 to December 2002. Appl. Environ. Microbiol. 69: 7181–7187.

Sejvar, J., E. Bancroft, K. Winthrop, et al. 2003. Leptospirosis in "Eco-Challenge" athletes, Malaysian Borneo, 2000. Emerg. Infect. Dis. 9: 702–707.

Sellner, K.G., G.J. Doucette, and G.J. Kirkpatrick. 2003 Harmful algal blooms: causes, impacts and detection. J. Ind. Microbiol. Biotechnol. 30: 383–406.

Shields, J.M., V.R. Hill, M.J. Arrowood, and M.J. Beach. 2008. Inactivation of Cryptosporidium parvum under chlorinated recreational water conditions. J. Water Health 6: 513–520.

Shoemaker, R.C. and H.K. Hudnell. 2001. Possible estuary-associated syndrome: symptoms, vision, and treatment. Environ. Health Perspec. 109: 539–545.

Sinclair, R.G., E.L. Jones, and C.P. Gerba. 2009. Viruses in recreational water-borne disease outbreaks: a review. J. Appl. Microbiol. 107: 1769–1780.

Sobel, J. and J. Painter. 2005. Illnesses caused by marine toxins. Clin. Inf. Dis. 41: 1290–1296.

Szewzyk, U., R. Szewzyk, W. Manz, and K.H. Schleifer. 2000. Microbiological safety of drinking water. Annu. Rev. Microbiol. 54: 81–127.

Teitelbaum, J.S., R.J. Zatorre, S. Carpenter, D. Gendron, A.C. Evans, A. Gjedde, and N.R. Cashman. 1990. Neurologic sequelae of domoic acid intoxication due to the ingestion of contaminated mussels. New Engl. J. Med. 322: 1781–1787.

Van Houten, R., D. Farberman, J. Norton, J. Ellison, J. Kiehlbauch, T. Morris, and P. Smith. 1998. Plesiomonas shigelloides and Salmonella serotype Hartford infections associated with a contaminated water supply—Livingston County, New York. Morbid. Mortal. Weekly Rep. 47: 394–396.

Villareal, T.A., C. Moore, P. Stribling, et al. 2006. Ciguatera fish poisoning—Texas, 1998, and South Carolina, 2004. Morbid. Mortal. Weekly Rep. 55: 935–937.

Warner, R.D., R.W. Carr, F.K. McCleskey, P.C. Johnson, L.M. Elmer, and V.E. Davison. 1991. A large nontypical outbreak of Norwalk virus. Gastroenteritis associated with exposing celery to nonpotable water and with Citrobacter freundii. Arch. Intern. Med. 151: 2419–2424.

Watson, S.B. 2004. Aquatic taste and odor: a primary signal of drinking water integrity. J. Toxicol. Environ. Health A 67: 1779–1795.

Zucca, E., F. Bertoni, E. Roggero, G. Bosshard, G. Cazzaniga, E. Pedrinis, A. Biondi, and F. Cavalli. 1998. Molecular analysis of the progression from *Helicobacter pylori*-associated chronic gastritis to mucosa-associated lymphoid-tissue lymphoma of the stomach. New Engl. J. Med. 338: 804–810.

CHAPTER 5

Wastewater Treatment

LEARNING OBJECTIVES

- Describe the major issues involved in wastewater treatment, including magnitude.
- Define BOD and COD, and know how they are applied to waste management.
- List the components of wastewater treatment, their order of use, and what they achieve.
- Describe how the various microbial groups, electron acceptors, and elemental cycles are used to handle waste.
- Evaluate the most common problems with wastewater treatment and treatment facilities and the likelihood of their occurrence.
- Explain how dry waste (sludge) can be disposed of, and which methods are preferred.

KEY TERMS

- 10:10:10 rule
- 30:20 rule
- Activated Sludge
- Assimilatory Nitrate Reduction
- Dissimilatory Nitrate Reduction
- Nitrogen Fixation
- Primary Sedimentation
- Secondary Treatment
- Sludge

CONSIDER WASTEWATER

Nobody wants to think about wastewater; they just want it to go away. The U.S. Environmental Protection Agency (EPA) estimates that 30 billion gallons of wastewater are treated in the United States every day. Wastewater includes more than the waste that gets flushed down the toilet; it also includes every bit of water that enters a drain. For households, that includes the drains in kitchen and bathroom sinks, the washing machine, and showers and bathtubs. This creates a surprising mixture of materials in the sewage pipes: fecal material, urine, soaps, shampoos, detergents of all sorts, toothpaste, body oils and skin cells, and hair. In homes with garbage disposals built into the sink, the waste stream includes macerated food as well. And, of course, a variety of materials do not belong in sewage at all. Every parent has a story about how their child flushed something down the toilet that should not have been flushed. Estimates report that only 0.06% of all wastewater is dissolved or suspended solids; the vast majority (99.94%) of wastewater is simply water.

It is good that wastewater is mainly water. This helps it flow through the sewage lines to the sewage treatment plants. If the material were denser, it would simply clog the pipes and begin to putrefy. Wastewater is a mix of all sorts of materials and must be treated before it is sent back to the environment. It is not acceptable to simply dump untreated waste; it has the potential to carry numerous pathogens. The goals of wastewater treatment are simple: destruction of all pathogens and reduction of the carbon and nitrogen load that will be returned to the environment. Killing the pathogens breaks the fecal–oral route of transmission, which is essential for the health of society. In fact, it is difficult to overemphasize how important this is. If fecal materials were simply dumped into the street, as occurred in European cities in the Middle Ages, the populace would suffer from all manner of illness, and the stench would be horrible. Eventually the wastewater would find its way to a river or lake, and the waste would then contaminate the drinking water supply. To say that modern public health depends largely on the availability of potable water and the effective treatment of wastewater is not an exaggeration.

CHARACTERIZATION

Wastewater is nutrient rich. If it is dumped directly into the receiving water, it will cause the growth of many types of organisms, such as algae. This would lead to the loss of dissolved oxygen in the water and the death of any animal (e.g., fish, snails, crustaceans, and protozoa).

Therefore, wastewater must be treated before it can be returned to the environment. It is possible

FIGURE **5-1** **A testing laboratory for a sewage treatment plant looks much like a university research laboratory.**

to clean up wastewater to drinking water levels, but this is unnecessary and the expense would be great. The wastewater must be cleaned to an acceptable level, but not to the quality of drinking water or natural water. How, then, should wastewater be characterized, given that it is such a mix of materials? There really is no practical way to define wastewater. Therefore, surrogate measures have been created to characterize wastewater and the efficiency with which treatment is able to reduce the amount of waste (**Figure 5-1**). These are suspended solids, chemical oxygen demand (COD), and biochemical oxygen demand (BOD).

Suspended Solids

The first measure is an analysis of suspended solids (**Figure 5-2**). As the name implies, these are solid materials in the wastewater. They are easily caught on a filter, dried, and weighed. The difference between the weight of the filter before and after is the suspended solids, reported in mg per liter of water.

Chemical Oxygen Demand

The second test determines COD (**Box 5-1**). It is simple to understand but complicated to perform. The question that this test answers is this: What

FIGURE 5-2 **Suspended solids settle out from samples taken throughout the sewage treatment plant. The water becomes progressively (left to right) clearer, demonstrating that BOD has been removed.**

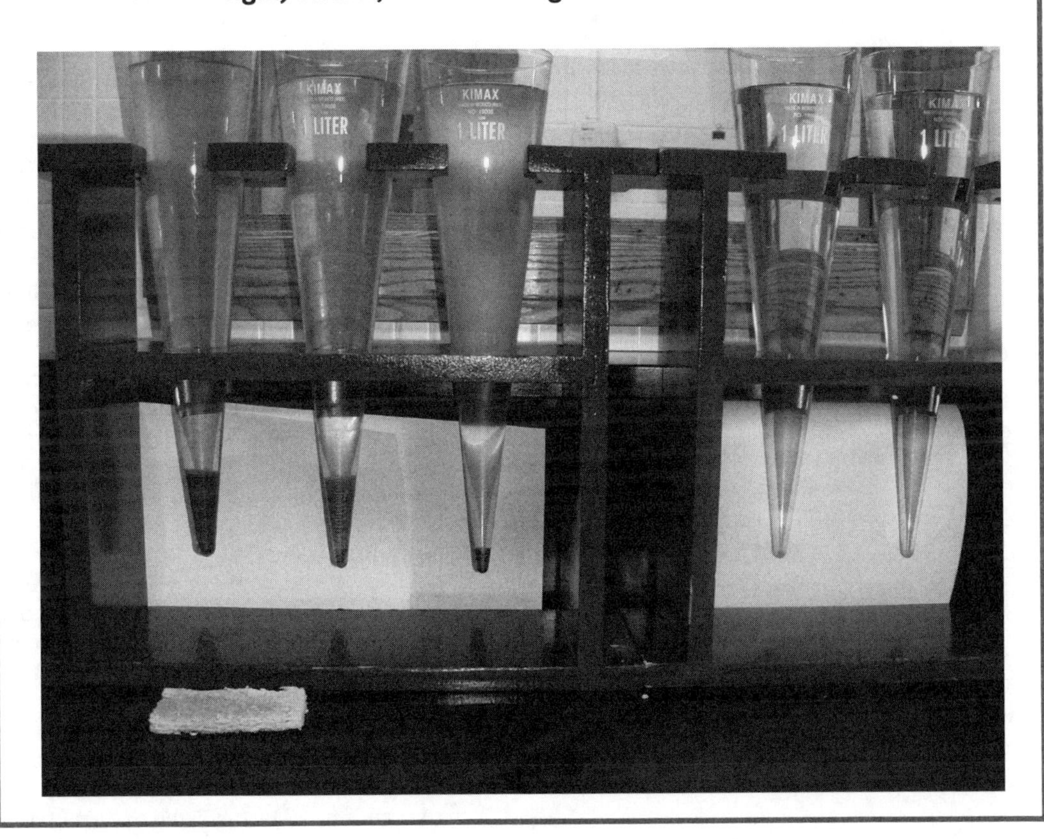

BOX 5-1 COD Explained

The oxidant in the COD process is potassium dichromate ($K_2Cr_2O_7$). The reflux reaction is so aggressive (150°C for 2 hours) that both organic and inorganic components are oxidized. In practice, the organic fraction is far greater than the inorganic fraction, and no distinction is made between the two in ordinary practice. The oxidant is added in excess to the reflux vessel, which contains a wastewater sample. Sulfuric acid is also added. A control solution must be run at the same time in which no wastewater is added. This will be the 100% unused dichromate figure. The initial equation during the reflux reaction is

$$[\text{organics}] + Cr_2O_7^{2-} + H^+ \rightarrow CO_2 + H_2O + 2Cr^{3+}$$

To determine how much dichromate (oxidant) was used in this reaction, it is necessary to ask how much is left. This is done by titrating the dichromate with ferrous ammonium sulfate (FAS) and a ferroin indicator. The color change should be a noticeable switch from blue-green to reddish-brown. COD is calculated as

$$COD(mg/l) = \frac{(ml_{blank} - ml_{sample}) \times [FAS] \times 8000}{ml_{sample}}$$

where 8,000 is used because of the conversion factor in which 1 gram = 1,000 mg and a factor of 8 because the equivalent weight of oxygen is 8. Note that the negative control is used for the 100% value.

The colorimetric method for calculating COD is based on the appearance of the reduced Cr^{3+} species. Both methods are good at estimating COD within the range of 100–400 mg/l. Higher COD values can be found after appropriate dilutions.

amount of oxygen would be needed to fully oxidize all of the carbon in a wastewater sample? That is, how much oxygen is needed to take every organic molecule and turn the carbon into carbon dioxide? This is done using a vigorous chemical reaction, and the measurement of oxygen used is complex.

Biochemical Oxygen Demand

The third major test determines BOD, sometimes referred to as biological oxygen demand (**Figure 5-3**). This test is more challenging from a conceptual point of view, but it is rather easy to perform as long as a few rules are followed. Like COD, the test measures how much oxygen is used in treating a wastewater sample, although in this case the oxygen is used by microorganisms in the wastewater. This oxygen is the dissolved oxygen (DO) in the water or, more typically, the DO in a diluent that is mixed with a small sample of wastewater.

While the atmosphere is about 20% oxygen, the concentration of DO in water is quite low, perhaps 10 mg/l. This concentration depends on the temperature of the water and the barometric pressure at the surface of the water. Generally, the higher the temperature, the more the gases are driven off from water, thus reducing the DO concentration. Also, the higher the barometric pressure, the higher the concentration of DO in water. With these two factors in mind, it is typical to see concentrations between 3 and 10 mg/l.

If a sample of wastewater was mixed with an equal volume of oxygenated water, the oxygen would be quickly used up. This is one way to measure BOD, but it is not especially accurate. The classical method of measuring BOD actually uses a period of 5 days, or the 5-day BOD. Therc are two reasons that this time period is used; the first is historical. The test was invented in Great Britain in response to the very question we are asking: What happens to wastewater that ends up in the rivers? The scientists of the time estimated that waste dumped into the river would take a maximum of 5 days to transit that river and be dumped into the ocean, where it was no longer their concern—hence, 5 days.

But the test would not have survived in that form if it had not also been practical. In fact, it is practical to measure at that time because of the rate at which oxygen is used. By day 5 the oxygen utiliza-

FIGURE 5-3 **Biochemical oxygen demand (BOD) for influent (diamonds) and effluent (squares) for the month of March at a sewage treatment facility. Note the variability of the influent BOD, as well as the uniform low value for the effluent. Treatment is highly effective at removing BOD to regulatory standards.**

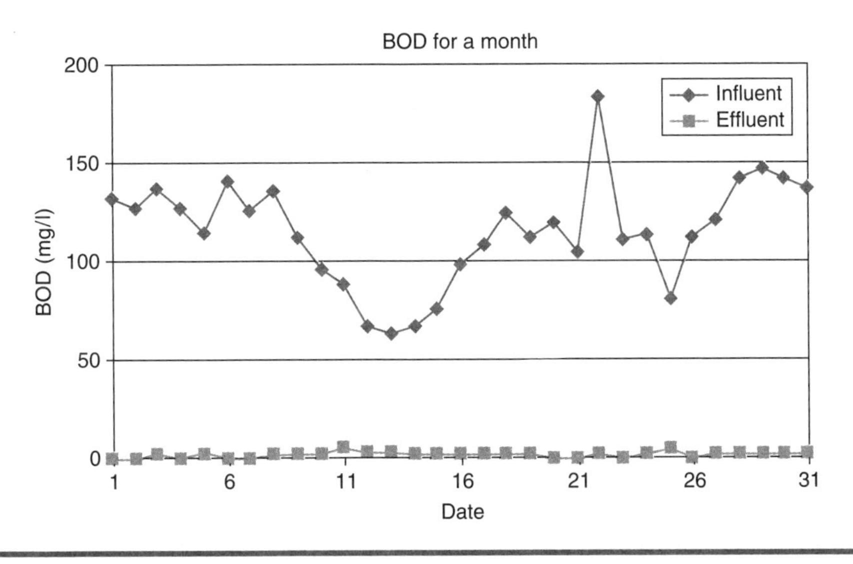

SOURCE: Based on data from the Fox River Water Pollution Control Center.

BOX 5-2 BOD Theoretical Calculations

BOD is easy to calculate for a given sample, but the figure is an approximation at day 5, and not what the microbial community is ultimately able to digest. Because the BOD curve approaches a maximum asymptotically, it is difficult to estimate what the final BOD would be. The culture can achieve a theoretical maximum, usually written as "ultimate BOD."

$$BOD_t = \text{ult.BOD}(1 - 10^{-kt})$$

That is, the BOD at some day t is equal to the ultimate BOD times a correction factor. The exponential t is the same unit of days, while the k value is a constant for the system that must be found empirically, but which is usually 0.1–0.2. Therefore, by day 5 and $k = 0.1$, and with a BOD of 225 mg/l, the ultimate BOD = 331 mg/l.

The k factor can be found by plotting the cubed root of (time/BOD) versus time, where all time factors are in days (**Figure 5-4**). The straight line of the graph will have a y-intercept. The k factor = 2.61 (slope/y-intercept). Fortunately, this calculation does not have to be performed often.

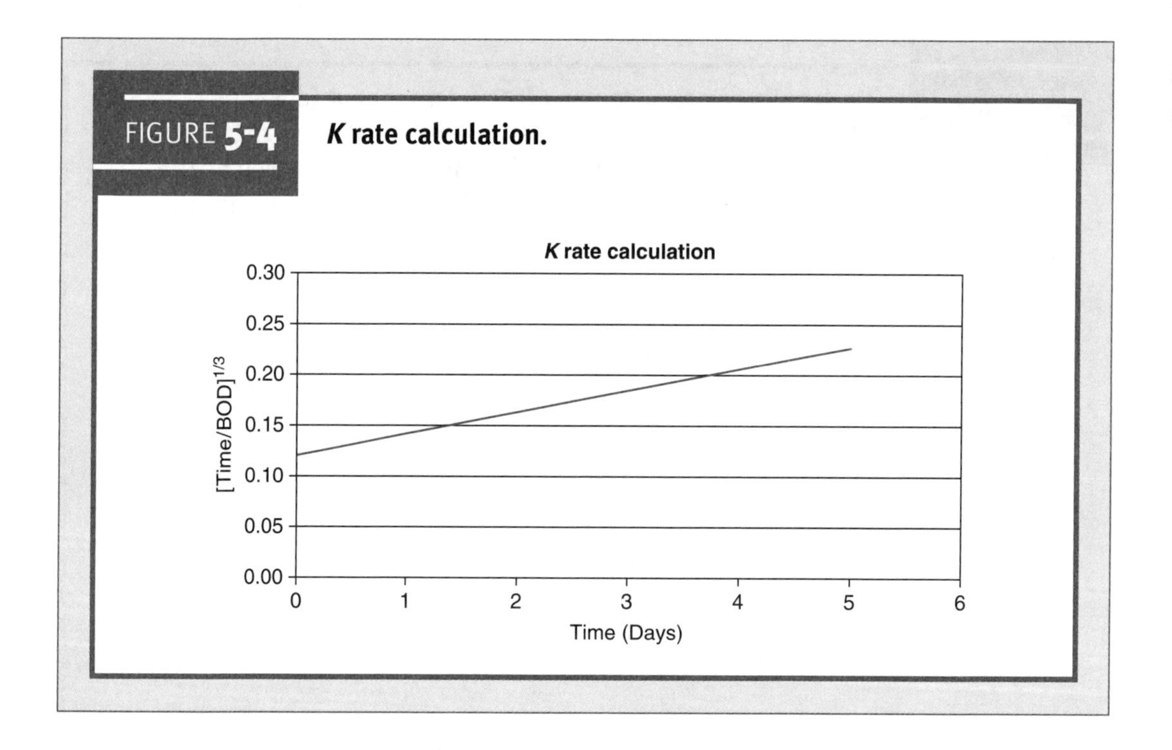

FIGURE 5-4 *K* rate calculation.

tion has passed the rapid reaction phase, and the curve is starting to flatten out (**Figure 5-5**). Estimations of BOD are thus more accurate at this point (**Box 5-2**). An ultimate BOD can then be calculated (**Box 5-3**).

Sometime after day 5, oxygen utilization from nitrification begins. Nitrifying bacteria use ammonia, and other nitrogen-containing compounds use more oxygen. This additional BOD can be calculated as well if the need arises, but the test is usually completed on day 5, and the additional effect is ignored.

A few concepts are important to the performance of the BOD test. The DO concentration should change by at least 2 mg/l to be accurate. If it does not change by at least this much, then either the sample was overly diluted or toxins were in the sample. In the former case, sufficient oxygen was available, but the bacteria had difficulty contacting the substrates because of the great dilution. In the latter case, the toxins killed some or all of the microorganisms, leaving no way to determine the real BOD. This is one reason that authorities constantly remind the public to avoid flushing certain substances (e.g., motor oil or antifreeze) down the drain: chemicals can have a toxic effect on the very microorganisms that are helping to treat our wastewater.

Another figure to keep in mind is the final DO concentration: at least 2 mg/l. Below this mark, too much substrate is present in the water, and the oxygen is quickly used up. A minimum number for BOD under these conditions can be calculated, but it may not provide an accurate estimation of the true figure. Even very low amounts of DO are suspect; at a certain level, plenty of substrate may be available for consumption by aerobic microbes, but finding the low levels of oxygen is very difficult. Again, accuracy is not ideal; the ideal situation is to utilize all the available carbon in the sample without using up all the oxygen. Oxygen concentration is easily and quickly measured with a DO meter. Standard BOD bottles often have an opening that fits the oxygen probe tightly.

Incubation should be carried out at 20°C for 5 days in the dark. Autotrophic microorganisms in the sample might use sunlight for photosynthesis,

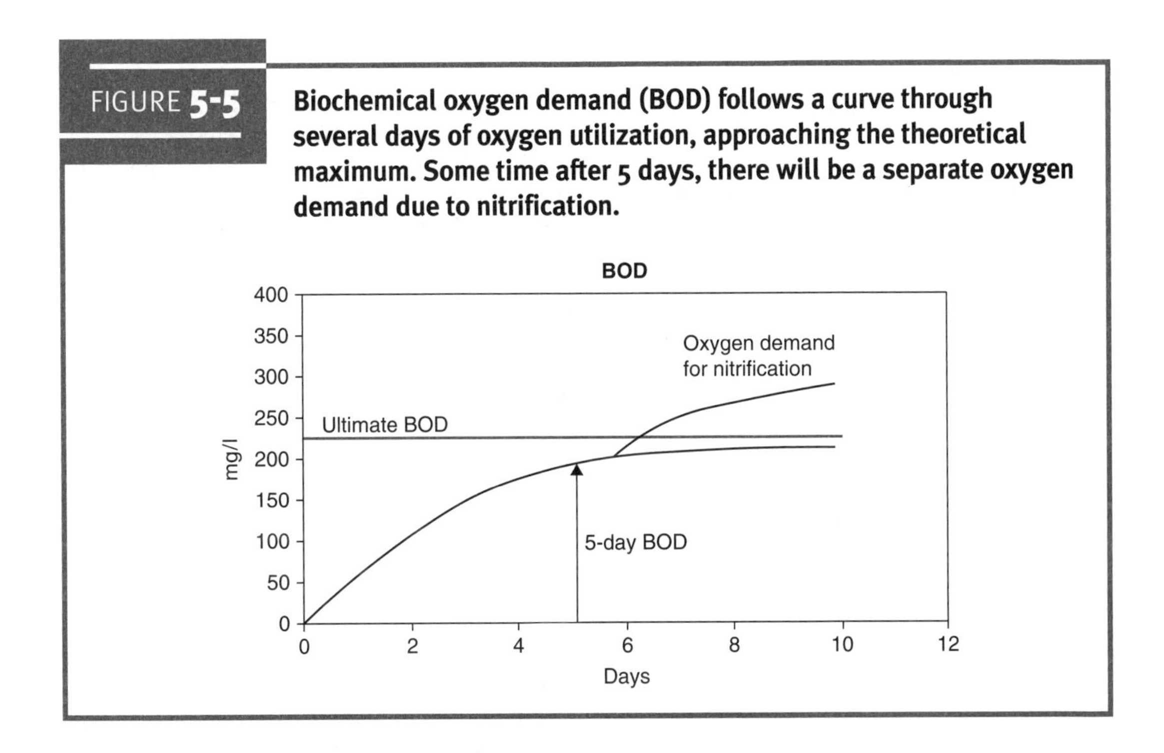

FIGURE 5-5 **Biochemical oxygen demand (BOD) follows a curve through several days of oxygen utilization, approaching the theoretical maximum. Some time after 5 days, there will be a separate oxygen demand due to nitrification.**

replacing some of the oxygen and giving erroneous results. Although the range of acceptable values seems very narrow, in practice the true BOD can be estimated based on experience with the system. Typical wastewater BOD values in the United States are between 100 and 350 mg/l. In Europe, the values are higher, reflecting less water dilution in their wastewater stream. Notably, BOD values can be calculated for waste streams other than sewage (**Table 5-1**). For example, industries such as food processors, which create large amounts of carbon-containing wastes, can calculate the BOD for treatment of their waste stream to determine whether the waste stream has sufficient microbial content to oxidize the carbon compounds in the effluent. In these cases, a seed culture can be added, such as wastewater from a sewage treatment plant. The seed culture also contributes to the total BOD, but

TABLE 5-1 **Percentage of Sample to Use for Different Water Sources**

Water source	Percentage of sample in BOD bottle
Raw sewage	0.5–3%
Settled sewage	1–7%
Secondary sewage effluent	1–10%
River water	20–100%

BOX 5-3 Calculating BOD

BOD is calculated based on the oxygen consumed in 5 days and the percentage of wastewater used in the BOD bottle. The following equation is used for this calculation.

$$BOD = \frac{[O_2]_{init} - [O_2]_{final}}{\frac{\text{waste sample volume}}{\text{total volume}}}$$

For example, if the initial oxygen concentration is 8.9 mg/l, the final concentration is 3.4 mg/l, and a 3-ml sample of wastewater is mixed with diluent in a standard 300-ml BOD bottle, the BOD is calculated as 8.9 − 3.4 = 5.5, divided by 3/300, or 0.01, resulting in BOD = 550 mg/l.

this can be estimated using a separate BOD bottle with only diluent and the seed culture.

Local laws may require industrial corporations to treat their waste at the factory to remove some BOD before emptying it into the municipal sewage system. The waste that a new factory will contribute to the waste stream must be considered by the municipal host; requirements should be negotiated before (rather than after) the factory starts production.

BOD is always less than COD, even though they both start with the same amount of carbon. This is because COD is a vigorous reaction that achieves complete oxidization. BOD is dependent on microbial reactions with the carbon compounds, and many compounds are difficult (recalcitrant) to use, especially long polymers of aromatic compounds. These compounds break down more slowly in the environment, but they eventually do break down and are recycled. Of course, contamination with synthetic carbon compounds is not broken down by microbes at all, although it is broken down by the COD process. The strength of the BOD test is that it measures the waste material (through the oxygen demand) that is readily broken down by microbes, and therefore it mimics a natural system. Because the wastewater is ultimately destined for a natural environment, this is a relevant scenario.

The targets of suspended solids at 30 mg/l and BOD at 20 mg/l are summarized in a goal called the **30:20 rule**. In many municipalities, this rule is applied by National Pollutant Discharge Elimination System (NPDES) permit. Permits are authorized through the Environmental Protection Agency (EPA) for specific wastewater operators according to the typical wastewater they handle, the receiving water for the treated waste, and the distance between communities. This last factor is important, because the next (e.g., downstream) community may be using the same body of water as its drinking water source. While it is not unreasonable to expect the surrounding communities to purify their water according to the principles described in Chapter 3, the polluting community has a major responsibility to clean its wastewater as well as possible. In some densely populated areas, the NPDES is more stringent and may require additional parameters. The **10:10:10 rule** (all in mg/l) refers to suspended solids / BOD / ammonia. The inclusion of ammonia acknowledges the important role of fixed nitrogen in eutrophication of water sources. Techniques for removing excess nitrogen are shown below. Other parameters that might be included are pH, grease content, and phosphate concentration.

Typical ranges for input wastewater at the treatment plant are BOD at 100–350 mg/l, ammonia at 1–50 mg/l, and phosphorus at 6–20 mg/l. These figures must be reduced before the wastewater can be returned to the environment.

WASTEWATER TREATMENT PROCESSES

Like water purification, wastewater treatment is not an especially difficult process (**Figure 5-6**). Regulations require the "best available technology" to be used in the treatment process. This gives the operators some latitude in choosing the methods that are actually available to a community; as long as the parameters are met, it does not matter which of the approved techniques are used. When wastewater enters the facility, it brings with it not only household waste but also infiltration from groundwater and degraded pieces of infrastructure. Pipes in the ground have a certain lifetime, and over time cracks and breaks inevitably appear and joints between pipes corrode. Thus, groundwater enters the system. In small amounts this is not a problem, but if it becomes a significant portion of the water stream, the pipes must be extracted and replaced. This is an expensive but essential duty of government. In addition, pipes, especially those made of concrete, chip and pit over time, allowing gritty material to enter the wastewater stream.

Occasionally, unacceptably high levels of hydrogen sulfide (H_2S) occur in the wastewater stream. Aeration can remove H_2S by channeling an air stream through a biotower that contains a microbial biofilm community on an inert substrate (**Figure 5-7**). The biofilm community destroys the hydrogen sulfide and eliminates its characteristic rotten-egg odor.

Some materials can be easily removed prior to wastewater treatment (**Figure 5-8**). This includes the removal of solid material that is more dense than water and material that is less dense than water (i.e., "sinkers" and "floaters"). Floating materials are easily skimmed off the surface and are disposed of in a sanitary waste dump. Dense particles like grit, stones, and chunks of concrete from pipes quickly settle out in a grit chamber and are disposed of in a landfill (**Figure 5-9**). Large particles that survive this step are ground thoroughly in a comminuter, which is designed to shred large items before they can clog or jam the equipment in the treatment facility.

Primary sedimentation is the first major component of wastewater treatment. The wastewater is pumped into a sedimentation tank (or clarifier) and is held there for several hours (**Figure 5-10**). Greases and oils float to the surface, where they are skimmed off. Solid materials gradually settle out during this time, forming **sludge** (or primary

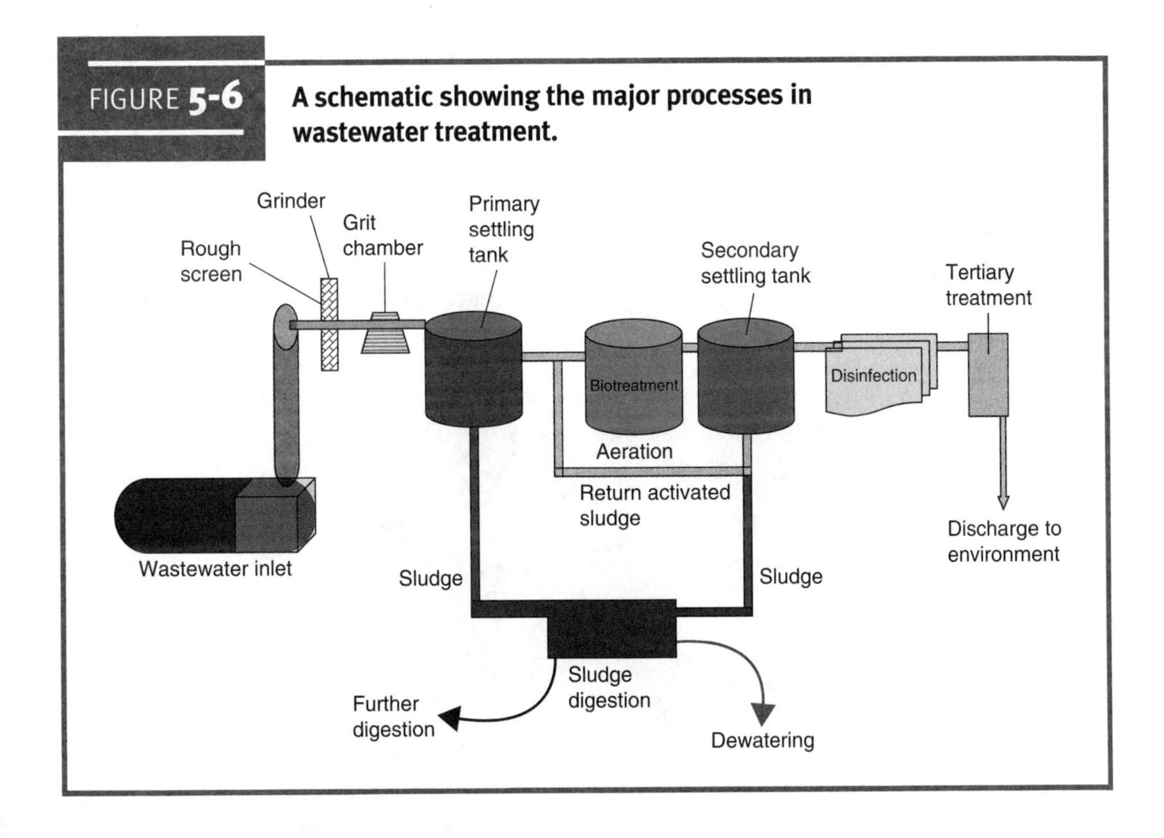

FIGURE 5-6 A schematic showing the major processes in wastewater treatment.

FIGURE **5-7**

Odors in a sewage treatment plant can be controlled through either a biotower that scrubs the air of volatile compounds or a chemical tower (shown here) that achieves the same result.

FIGURE **5-8**

The first step in wastewater treatment is removal of solid items that can be physically removed from the wastewater stream by raking over a rough filter. These materials are shredded and sent to a landfill.

FIGURE 5-9 **The grit chamber removes particles the size of sand grains from the wastewater stream.**

sludge). This material collects at the bottom of the settling tanks and is periodically scooped out, dehydrated, and then disposed of as a solid. This simple system is surprisingly effective, resulting in the removal of 55% of suspended solids and 35% of total BOD. Disposal of the solids is described below. The remaining water is sent to the next stage of treatment, which uses a biological process.

Secondary treatment is dependent on a biological system. The strength of these systems is that, as biological systems, they are inexpensive and do an excellent job of removing suspended solids and BOD. They utilize both dissolved and suspended organic carbon. Their weakness comes from the fact that as biological systems they can be easily damaged and replacement may require a long period of time.

The first method depends on a liquid suspension of microorganisms. **Activated sludge** starts out as sludge, but it has been improved. Sludge and wastewater in general have very high concentrations of microorganisms (**Figure 5-11**). These microorganisms are fed by the ample nutrients in the wastewater, and they multiply and create flocculent material ("flocs"), which is denser than water and thus settles out of the water easily. At some point in the development of wastewater technology, it became evident that recycled sludge utilized nutrients faster than original sludge and that the more times sludge was recycled, the more efficient it became at degrading the nutrients in wastewater, (i.e., it was "activated"). By recycling sludge repetitively, the treatment time of the water was shortened, and the whole process became more efficient.

The recycling process, then, finds the balance between the right population and proportion of microorganisms to handle the wastewater stream. The great strength of this method is that it depends on the biological component to break down sewage waste. Its weak point is that biological systems are

FIGURE **5-10** The primary settling tank shown empty (a) and filled (b). Wastewater will reside here for a few hours while the denser materials settle to the bottom of the tank, where the boom moves the material toward a central collection site in the center of the apparatus

(a)

(b)

FIGURE 5-11 **The mixing point for activated sludge and the incoming wastewater is shown here. This mixture is referred to as mixed liquor. One of the key determinants for processing is the amount of suspended solids in this sample, or the mixed liquor suspended solids (MLSS).**

easily damaged. If, for instance, the wastewater treatment plant is flooded (which is a likelihood because they are located at the low point of the collection system), then the floodwaters may wash away the activated sludge. An activated sludge system can always be restarted, for instance, using a seed culture from a different facility, but the efficiency will not return quickly. In addition, the specific mixture of microbes may be highly adapted to a particular waste stream.

Activated sludge may be composed of as much as 10^{10} microbes per gram. This complex community consists of bacteria, some fungi, and protozoa that graze on bacteria on the surface of the sludge. One study found that 80% of the bacteria in activated sludge came from the proteobacteria of the alpha, beta, or gamma subclasses (Wagner et al., 1993). Sludge does not ordinarily contain pathogens; however, the mixture is still far too complex for any degree of precise analysis or for any real understanding of the forces that favor one microbe over another. Furthermore, the mixture also contains dead cells as well as a tremendous amount of extracellular polymeric substance (EPS) that binds the microbes in large flocs.

Solid wastewater systems are used as well. Trickling filters are comprised of any type of packing material (often rather loosely packed) that allows wastewater to flow through, such as sand filters. Over time, biofilm builds up on the inert filter material, and the filter catches suspended solids (**Figure 5-12**). The biofilms are efficient at utilizing the dissolved organics. Over time, the biofilm builds up and the filter clogs; backwashing the filter

FIGURE 5-12 **Analysis of total suspended solids (TSS) in a wastewater treatment plant during a typical month. Note that the lines representing TSS after second treatment and TSS for effluent largely overlap.**

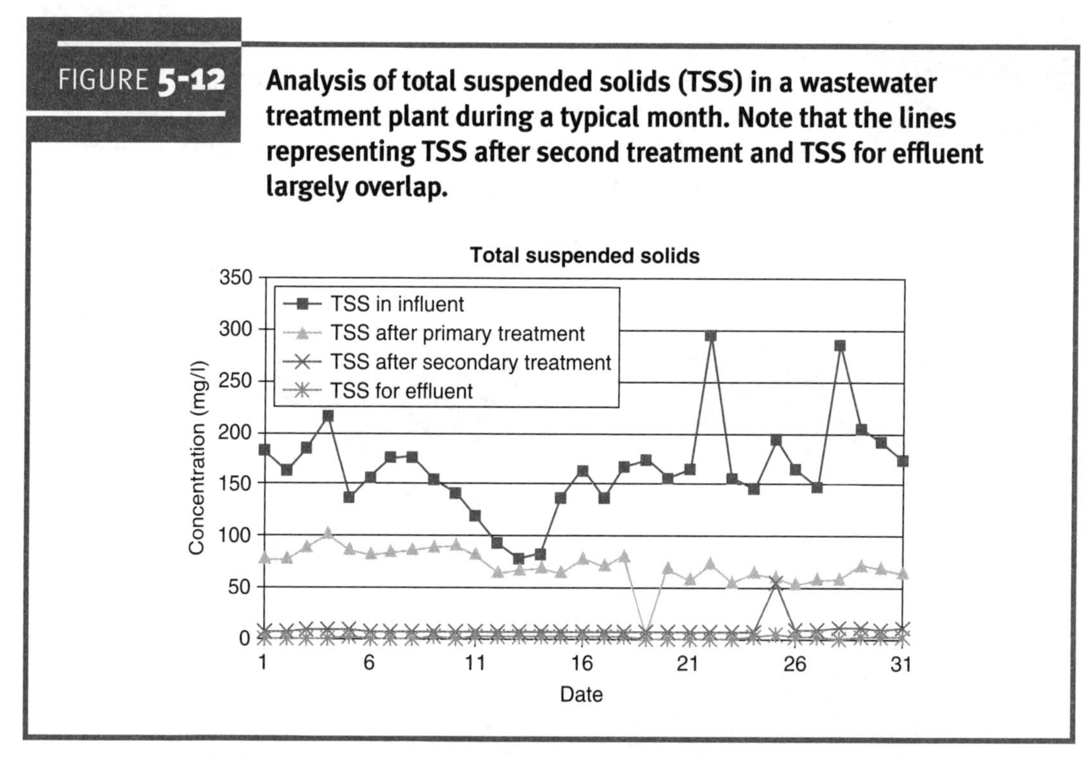

SOURCE: Based on data from the Fox River Water Pollution Control Center.

is sufficient to remove most of the excess. This concentrated backwash must be disposed as solid waste to remove it from the system.

The critical factor in these trickling filters is surface area, which then influences contact time with the biofilm. If the filter has too little surface area, the wastewater passes through untouched. Too much, and the filter clogs repeatedly. One way to standardize the surface area uses rotating biological contactors (RBCs), which resemble a stack of old-fashioned records placed side by side and connected through a central axis. Biofilm builds up on the plates; the plates turn continuously and the wastewater stream moves across them. The biofilm on the plates picks up nutrients and then oxygen from the air. Excess biofilm can be physically scraped from the plates.

If volatile substances remain, they can be stripped out using activated carbon filters. If pH is still outside the normal range of 6.0–9.0, a neutralization step is required. A set of chemical standards must be met as well. Generally speaking, the chemical concentration goals are the same for wastewater as for drinking water. When the suspended solids and BOD have been reduced to meet or exceed permitted standards, the water can be returned to the environment (**Figure 5-13**). A final shock treatment with chlorine is typical to kill any residual pathogens in the water, and a coliform concentration of 200 per 100 ml is usually sufficient for release of the water (**Figure 5-14**).

Wastewater will, of course, contain a high number of coliform bacteria, the bioindicators of greatest interest. This is not unusual and is not a concern as long as the subsequent treatment reduces this number to reasonable levels. Tests for coliform bacteria are essentially the same as for freshwater monitoring (described in Chapter 4), although the sample volumes and dilution schemes are modified to account for the higher expected concentrations. **Table 5-2** lists typical sample volumes and treatments.

FIGURE **5-13** Discharge of treated wastewater directly into an area stream. Note the otherwise pristine nature of the surroundings. The treated water is quite low in nutrients at this point.

FIGURE **5-14** The chlorinator nozzle for treated wastewater. A shock treatment is sometimes used to reduce the number of bacteria before discharging the wastewater to a receiving water. The mixing of chlorine takes place as the wastewater follows a "racetrack" course to increase turbulence.

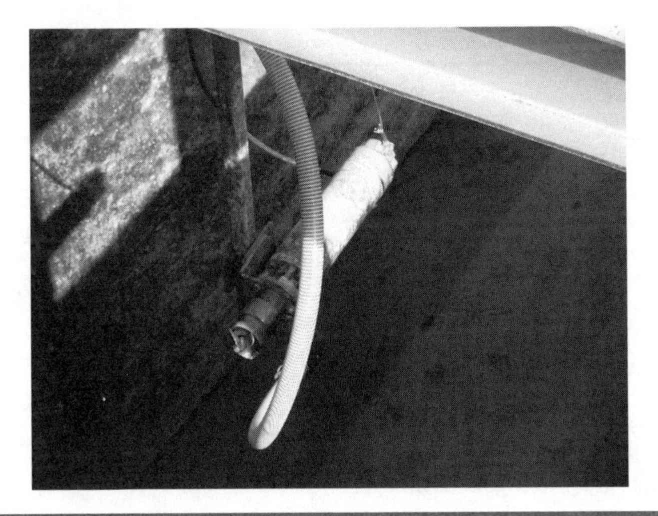

TABLE 5-2 Sample Volume to Use for Different Water Sources

Water source	Lowest volume	Plating scheme	Highest volume	Plating scheme
Clearwell water	100 ml	Concentrate on filter	1000 ml	Concentrate on filter
Tap water, household water	100 ml	Concentrate on filter	1000 ml	Concentrate on filter
Surface waters (lakes)	50 ml	Concentrate on filter	100 ml	Concentrate on filter
Groundwater (wells)	50 ml	Concentrate on filter	100 ml	Concentrate on filter
GWUDI	50 ml	Concentrate on filter	100 ml	Concentrate on filter
Drinking water pretreatment	1 ml	Concentrate on filter	50 ml	Concentrate on filter
Bathing beaches	1 ml	Concentrate on filter	50 ml	Concentrate on filter
Rivers	10 µl	Dilute 10-fold, plate 100 µl	1 ml	Concentrate on filter
Surface water in agricultural area	10 µl	Dilute 10-fold, plate 100 µl	1 ml	Concentrate on filter
Bathing areas subject to runoff	10 µl	Dilute 10-fold, plate 100 µl	1 ml	Concentrate on filter
Groundwater suspected of contamination	1 µl	Dilute 100-fold, plate 100 µl	100 µl	Plate directly
Surface water affected by treated sewage	100 µl	Plate directly	10 ml	Concentrate on filter
Treated sewage	100 µl	Plate directly	10 ml	Concentrate on filter
Surface water affected by untreated sewage	1 µl	Dilute 100-fold, plate 100 µl	100 µl	Plate directly
Raw sewage	0.1 µl	Dilute 1000-fold, plate 100 µl	10 µl	Dilute 100-fold, plate 100 µl
River water		20–100%		

The Nitrogen Cycle

Other activated sludge systems are used to remove things other than BOD. Excess nitrogen can be removed by taking advantage of the nitrogen cycle. Wastewater typically has an abundance of fixed nitrogen, considering that wastewater is protein rich (composed of amino acids) and receives urea ($H_2N-CO-NH_2$) in urine. Fixed nitrogen (such as ammonium compounds) is the limiting nutrient in most environmental systems. Ammonium is relatively difficult to produce and is metabolically expensive to create. Therefore, organisms scavenge fixed nitrogen efficiently. If a large source of fixed nitrogen were to be introduced suddenly into the environment, the local microbial community, including the cyanobacteria, would grow quickly. This results in eutrophication of the receiving water.

O
H_2N NH_2
urea

Clearly, excess nitrogen disposal occurs in two ways: as sludge or as nitrogen gas. Nitrogen is common in the environment, although the vast majority occurs as nitrogen gas in the air, where it makes up about 80% of the atmosphere (**Figure 5-15**). This is only sparingly bioavailable. **Nitrogen fixation** is the process by which some species of bacteria (e.g., *Rhizobium*, *Azotobacter*, and *Azospirillum*) break the triple bond of molecular nitrogen and form ammonium ions. This is a vital part of the ecosystem. Once formed, it can be incorporated into other biomolecules and can then be passed through the food chain and used by other organisms. This mechanism is an important aspect of nitrogen removal from wastewater (**Figure 5-16**).

The other aspect that is important for removal is **dissimilatory nitrate reduction**. This is distinct from **assimilatory nitrate reduction**, in which nitrate is incorporated into cells and is reduced to form ammonium ions. This is a common process that functions because this nitrogen does not have the triple bond of the molecular nitrogen. The reaction follows the well-described glutamate synthetase pathway and is followed by transamination, which moves the amino group to another molecule. The process is useful for creating enough ammonium ions for the synthesis of amino acids, and therefore, it is sensitive to feedback inhibition. However, this process is insensitive to molecular oxygen.

In contrast, dissimilatory nitrate reduction uses nitrate as a terminal electron acceptor. Therefore, it is not sensitive to feedback inhibition but is an anaerobic process, and at least some components are sensitive to molecular oxygen. In this process, nitrate is successively reduced to nitrite, nitric oxide, nitrous oxide, and finally back to molecular nitrogen, which as a gas rejoins the atmosphere. The wastewater is passed sequentially through tanks that are aerobic and anaerobic. In the aerobic tanks, nitrification reactions take place, and the excess nitrate is incorporated into flocs, which then are removed in a settling tank. This is an activated sludge step, so it is efficient at the removal of nitrate. BOD is greatly reduced in this step. Any of a large number of aerobic heterotrophs are used

FIGURE 5-15 **The nitrogen cycle.**

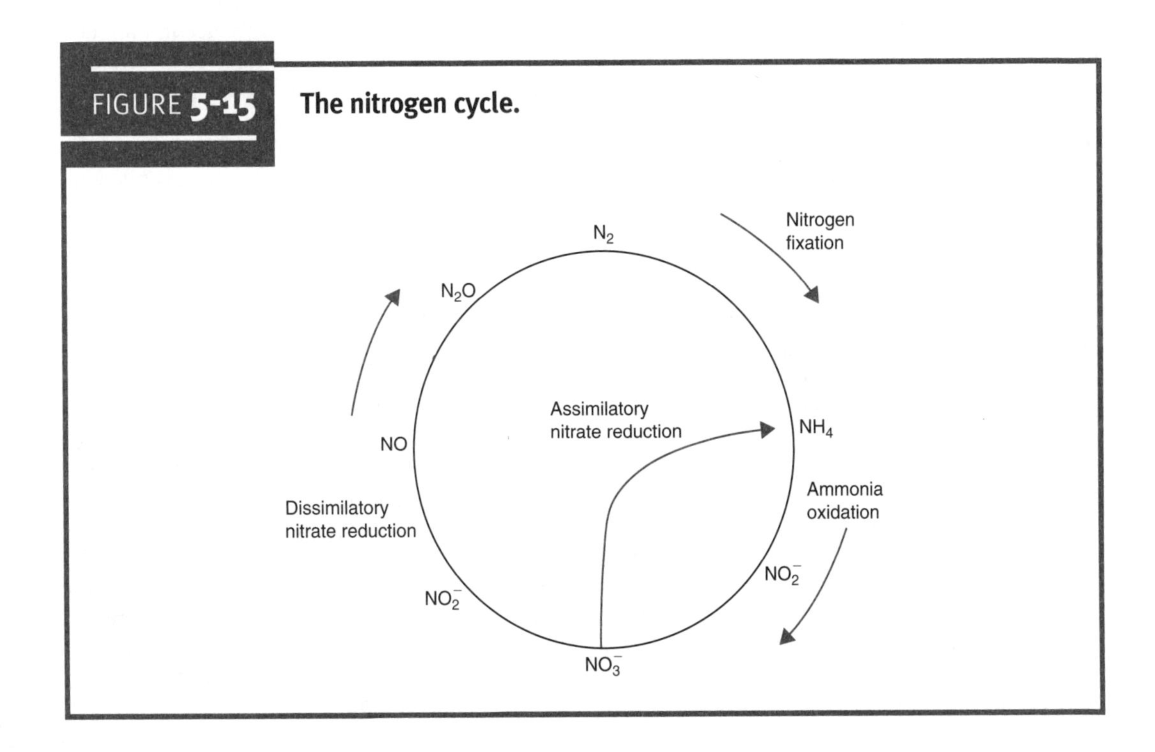

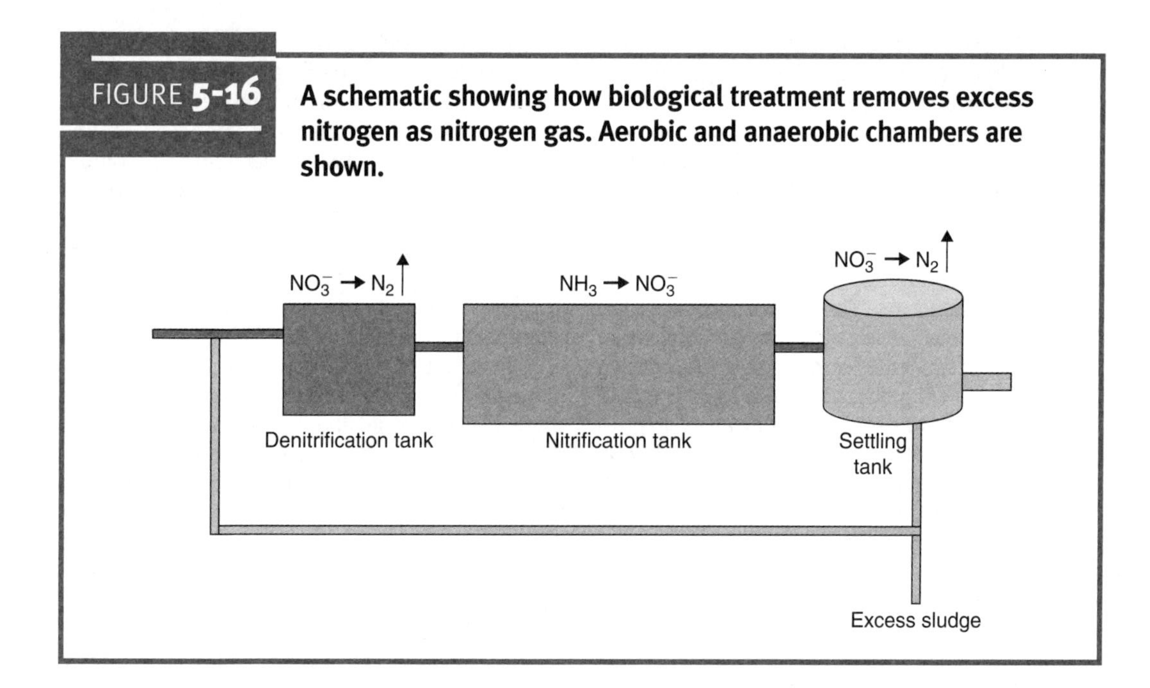

FIGURE 5-16 A schematic showing how biological treatment removes excess nitrogen as nitrogen gas. Aerobic and anaerobic chambers are shown.

here. The aerobic tank is typically sparged with air to provide sufficient oxygen (**Figure 5-17**).

In the anaerobic tank, the nitrate is used as a terminal electron acceptor during fermentation of the organic compounds of the wastewater. Complete reduction produces nitrogen gas, which then returns to the atmosphere and reduces the overall amount of nitrogen in the wastewater. A variety of different microbial species may be active in this step. Many are ordinary aerobic heterotrophs that have enzymes along part of the denitrification pathway. Some have enzymes for the entire pathway. Others utilize molecular hydrogen and fix carbon dioxide (*Hydrogenomonas*); some are able to fix nitrogen (*Azospirillum*); and some are photosynthetic (*Rhodopseudomonas*). *Thiobacillus denitrificans* is unusual in its ability to denitrify and to oxidize reduced sulfur compounds, such as hydrogen sulfide (Knowles, 1982).

Pseudomonas species are often active in the denitrification tank even though they are typically considered obligate aerobes. They might not have the nitrate reductase enzyme, but are able to utilize nitrite. They might also only reduce nitrite to nitrous oxide, which is released as a gas. Some concern has been expressed over this fact because nitrous oxide is a greenhouse gas. But the amounts emitted are probably very small, and other species in the denitrification tank further reduce it. Pure cultures, or even reasonably definable cultures, never occur in these tanks. They are consortial communities so complex that they have been impossible to model. *Alcaligenes*, *Bacillus*, *Chromobacterium*, and *Neisseria* are some of the genera that carry out denitrification. The anaerobes that participate in these reactions are often very sensitive to oxygen. The tank typically has a greatly reduced oxygen concentration, but molecular oxygen may be found in niches of the flocs. This may slow down the overall process, but it cannot easily be avoided.

The process can remove 60–80% of the nitrogen content, making this a valuable approach. If the amount of available carbon is insufficient to drive the denitirification reaction, then additional carbon sources can be added to the tank. Methanol has been used for this purpose; it is utilized by *Hyphomicrobium* strains. If other, cheaper carbon sources are available, the microbial community might adapt to it over time, improving efficiency. A one-stage reactor that is efficient at converting ammonium directly to nitrogen gas has been studied (Kuai and Verstraete, 1998).

FIGURE 5-17 **Secondary treatment for wastewater. This is the aerobic side, in which sparging of air appears as bubbling on the surface. This water will then be cycled to the anaerobic side for reduction of BOD and nitrogen.**

Each tank can be arranged to perform as an activated sludge tank or as a solid medium, such as sand and gravel. In the latter case, a significant layer of biofilm forms on the particles and the water passes over it. Wastewater can be pumped over the top of a solid matrix column and allowed to percolate through, or it can be pumped from the bottom and allowed to spill over the top. In each case, care is taken to maintain the flow rate so that the active bacteria are not flushed from the system and low oxygen conditions are continuous.

Experience with the individual system dictates the operating parameters that yield the best result. This includes the amount of air sparged into the aerobic tanks, the need for additional carbon sources, and the retention times of the wastewater in the aerobic and anaerobic tanks.

Excess Phosphorous

Because phosphorous may occur in excess (6–20 mg/l), a mechanism must be used to reduce the levels to a manageable amount (**Figure 5-18**). Phosphorous (or typically phosphate, the biologically useful form) is not usually the limiting nutrient in biological systems, but reducing the amount is generally a good idea. In the past, chemically based removal methods, such as the addition of iron sulfate, have been used, although these are becoming expensive.

Fortunately, a biological method based on an activated sludge process can remove excess phosphate, referred to as enhanced biological phosphorous removal (EBPR). Certain bacterial species accumulate phosphorous as polyphosphate molecules. Polyphosphate compounds are energy rich and can be used effectively under certain growth conditions (e.g., when activated sludge is mixed with sewage from the primary sedimentation tank). These bacteria form flocs and settle out of the liquid in the settling tanks. This reduces the overall BOD. Many of the species involved are rather poorly defined in every aspect, and they are difficult to grow in pure cultures on artificial media. In a recent work, one of the major phosphorous removal species was

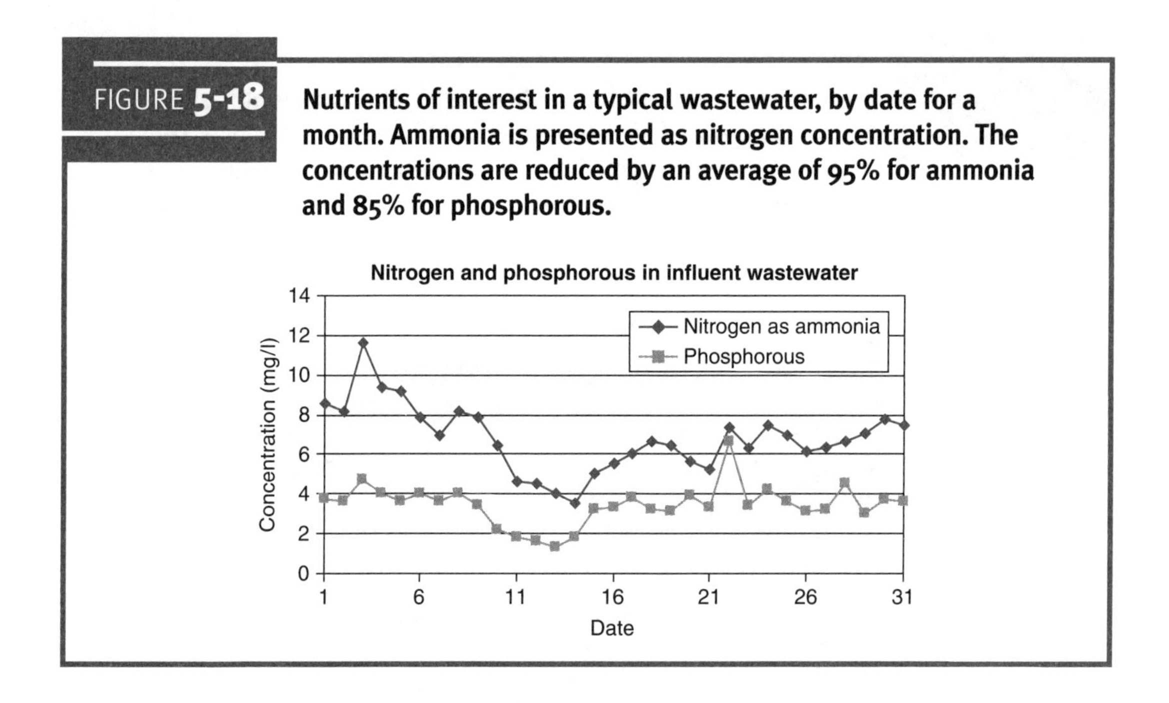

FIGURE 5-18 **Nutrients of interest in a typical wastewater, by date for a month. Ammonia is presented as nitrogen concentration. The concentrations are reduced by an average of 95% for ammonia and 85% for phosphorous.**

partially characterized and tentatively named *Accumulibacter phosphatis* (Martin et al., 2006). Whether or not this name is adopted, it is clear that this microorganism is very efficient at accumulating phosphate.

WHAT HAPPENS TO THE SLUDGE?

There are two major problems with the sludge that is separated from wastewater: (1) it is very rich in nutrients and (2) there is a lot of it. Sludge may also contain pathogens, which is a fact that must be dealt with in considering final disposal options. Therefore, the major goals of this treatment step are these: destruction of pathogens, reduction of microbial activity, reduction of noxious odors, and reduction of organic matter (so that the sludge will not putrefy during final disposal). If this is done properly, 30–40% of the solid fraction will be converted to gases (e.g., carbon dioxide and methane) and will not need further disposal.

A decrease in the mass of the sludge is important for the overall economic health of the facility. Large sludge masses dictate that it be disposed off site. Therefore, it must be transported to another site. Trucking costs are figured by weight, so removing as much weight as possible on site is cost-efficient. The easiest way to remove weight is to dehydrate the sludge. Sludge begins as a gooey mass, high in water content. Dehydration can be accomplished by physically squeezing out the water or by centrifugation. In municipalities that have relatively low populations and available land, the sludge is sometimes laid out in rows on sand beds where the sun can dry the material before it is removed from the site. Freezing weather in some areas makes this difficult to accomplish.

Diluted wastewaters are often treated exclusively with aerobic systems, which produce an effluent that is sufficiently reduced in BOD and is allowed to return to the environment. More concentrated wastes, such as sludge, can be effectively treated with an anaerobic system. The system is often referred to as a two-stage system, although a single tank is used. The result is a reduction in the solids because carbon is released in gaseous form. Maintenance of the anaerobic bioreactor requires more care, with factors such as temperature (typically 35–37°C), pH (6.0–8.0), and retention time of the sludge among the critical issues. The concen-

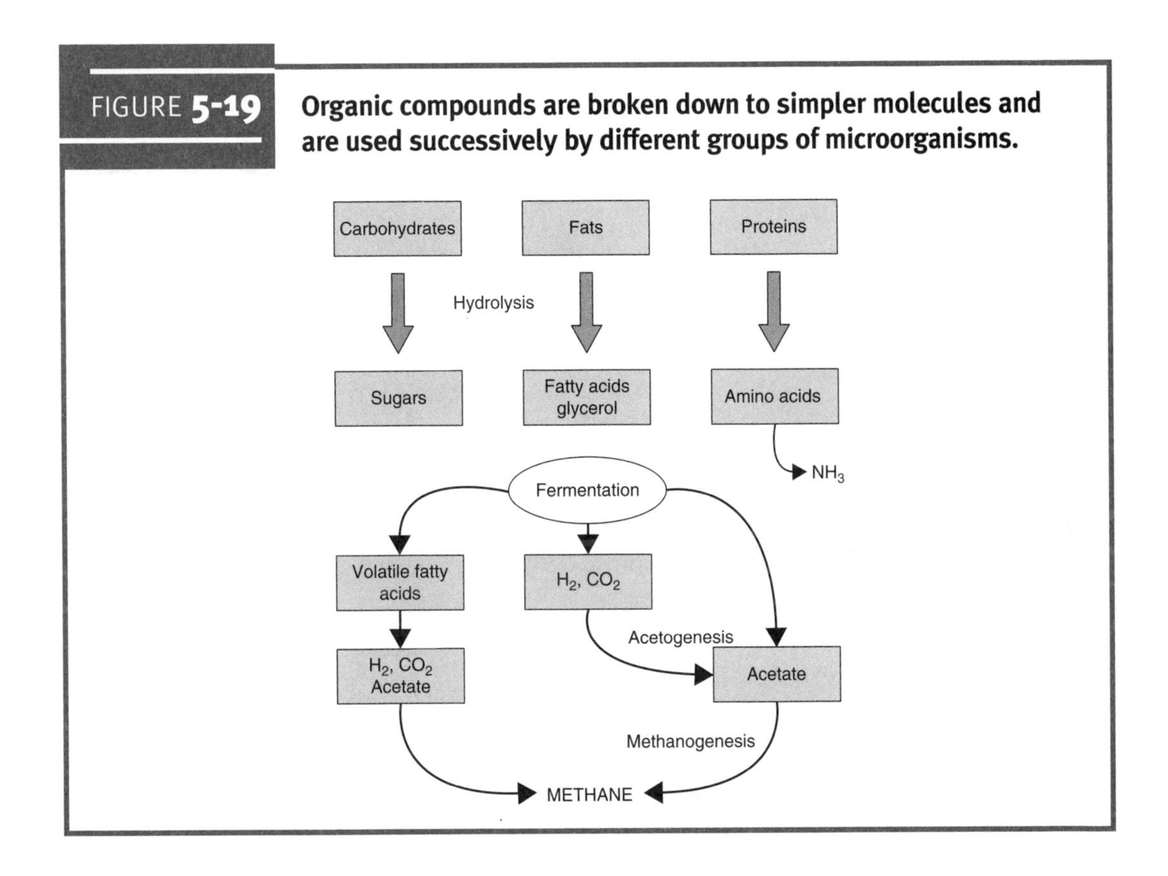

FIGURE 5-19 **Organic compounds are broken down to simpler molecules and are used successively by different groups of microorganisms.**

tration of all pathogens decreases by several orders of magnitude under these conditions, and if done properly, it enables the digested sludge to be used as a fertilizer on agricultural or horticultural lands.

The first stage of the anaerobic process is the utilization of various complex compounds in the waste material. That is, polymeric substances are broken down into monomers. Carbohydrates are hydrolyzed to simple sugars, proteins are converted to amino acids, and fats are broken into fatty acids and glycerol. These compounds are, in turn, converted to carbonic acids, alcohols, hydrogen gas, carbon dioxide, and ammonia (**Figure 5-19**). Volatile fatty acids such as acetate, butyrate, proprionate, valerate, and caprolate are generated during these fermentation reactions. The organic compounds are converted to acetic acid, hydrogen, and carbon dioxide. These reactions are carried out by a great variety of obligate anaerobic or facultative bacteria.

H_3C O OH

Acetic acid

H_3C O OH

Butanoic acid

H_3C O OH

Hexanoic acid

TABLE 5-3 **Methanogenic Reaction**

Substrate	Methanogenic reaction	G°′ (kJ/mol)	Representative genera
Hydrogen/ carbon dioxide	$4H_2 + HCO_3^- + H^+ \rightarrow CH_4 + 3H_2O$	−135	*Methanobacterium* *Methanococcus* *Methanothermus* *Methanospirillum*
Formate	$4HCO_2^- + H^+ + H_2O \rightarrow CH_4 + 3HCO_3^-$	−130	*Methanobrevibacter* *Methanogenium* *Methanoculleus*
Alcohols	$2CH_3CH_2OH + HCO_3^- \rightarrow 2CH_3COO^- + H^+ + CH_4 + H_2O$	−116	*Methanospirillum* *Methanomicrobium*
Methanol	$4CH_3OH \rightarrow 3CH_4 + HCO_3^- + H_2O + H^+$	−105	*Methanosarcina* *Methanohalobium* *Methanolobus* *Methanococcoides*
Hydrogen/methanol	$CH_3OH + H_2 \rightarrow CH_4 + H_2O$	−113	*Methanosphaera*
Methylamines	$4(CH_3)_3\text{-}NH^+ + 9H_2O \rightarrow 9CH_4 + 3HCO_3^- + 4NH_4^+ + 3H^+$	−74	*Methanohalobium* *Methanosarcina* *Methanohalophilus*
Dimethylsulfide	$2(CH_3)_2\text{-}S + 3H_2O \rightarrow 3CH_4 + HCO_3^- + 2H_2S + H^+$	−49	*Methanolobus*
Acetate	$CH_3COO^- + H_2O \rightarrow CH_4 + HCO_3^-$	−31	*Methanothrix* *Methanosarcina*

NOTE: All ΔG° are given per mole of methane produced. They are calculated based on the data in Thauer et al. (1977).

At this point, the second stage of the reaction takes place, in which the methanogenic bacteria use the carbon dioxide as a terminal electron acceptor and utilize the acetate and hydrogen gas as an energy source. Other compounds can be utilized as well (**Table 5-3**). A number of different methanogens may participate in this process, but the end result is methane gas. Methane is, of course, a valuable energy source (**Figure 5-20**). In the past, it was released to the atmosphere or flared off as it built up in the tank. However, methane is increasingly being recovered at the waste treatment plant or at waste burial sites (discussed in the next section) and used to power the treatment plant. Some can be used to heat the tank to the required temperature, while the rest of the methane can be burned to provide power for lighting and heating the plant. The result is the release of carbon dioxide into the atmosphere, and the burning of methane also releases carbon dioxide. This results in a decrease in BOD in the sludge.

In practice, the anaerobic sludge tank is operated with minimal mixing, but heating of the tank to an optimum temperature and monitoring ensure that the processes do not become unbalanced. The tank itself often has a floating lid to accommodate the gases produced. This treatment is relatively slow, sometimes requiring a month of incubation before

FIGURE 5-20 **The floating lid on a methanogenesis tank. Methane produced here is recycled by the plant to provide heat for the treatment plant. The amount of sludge inside determines whether the lid should be raised or lowered.**

digested sludge is removed and new sludge is introduced (Sterritt and Lester, 1988).

The composition of human waste reflects the substances that people ingest and ultimately flush into the sewage system. Sometimes this includes anthropogenic substances that cannot be easily degraded; sometimes this reflects medical products (pharmaceuticals) that have unusual chemical compounds. Levels of heavy metals, such as cadmium or zinc, are sometimes higher than normal. In high enough concentrations, the biological reactions necessary for wastewater processing are inhibited.

In areas where the population is low and the available space is high, there may not be a need for a costly treatment facility with multiple settling tanks because waste stabilization ponds are used instead. These ponds are essentially open pits into which the sludge is deposited on a regular basis. The solid fraction settles to the bottom of the pond, while the upper layer gradually develops a scum of algal growth. This method differs greatly from the settling tanks described above, but is useful because the algae use sunlight to regenerate oxygen in the top layer of the pond. Thus an aerobic digester is created at the top of the pond, an anaerobic layer is created at the bottom of the pond, and the sludge digests slowly over time.

While oxygen enters the system via photosynthesis and diffusion from air, gases such as carbon dioxide, nitrogen, and methane gradually exit. The length of time required for full decomposition of the sludge varies with temperature; these ponds are not artificially heated like those at the treatment plant. Eventually, however, the sludge will be suitable for disposal. The water can be drained away, and the dried sludge can be trucked away for landfill.

Sludge processing includes a limited number of outcomes. Sludge can be used via land application as a fertilizer (**Box 5-4**). However, the amount of sludge processed is vast, and the possible agricultural uses are actually rather limited. Treated sludge can be applied to farm fields only at certain times of the year because constant fertilization is not needed. Also, natural fertilizer cannot be applied near harvest time, and excess natural fertilizer simply runs off during a rainstorm. And, of course, a nonorganic item occasionally gets into the sludge and is applied to a farm field. Farmers do not like litter on their fields.

Natural fertilizer can most easily be used on inedible crops. Christmas tree farms, forests cultivated for timber, horticultural land (e.g., land used to grow ornamental plants and shrubs or commercial flowers), and the median strips of highways are all acceptable and safe areas for sludge application. However, the amount of sludge that can be used in this manner is dwarfed by the amount of sludge actually generated.

Sludge can be disposed of by ocean dumping, because oceans are extremely large ecosystems that can handle added nutrients. However, sloppy practices in the past have given ocean dumping a very bad reputation, and laws prevent this practice in many areas. As a future alternative, ocean dumping is probably a poor idea.

Sludge is still very high in energy, and if dried out thoroughly, it will form a combustible product. The early pioneers of the prairie states discovered that dried buffalo dung could be burned as a fuel, which was handy in areas with few trees. Waste incinerators produce electricity by burning sludge. This is a safe practice, although incineration is not generally favored by the public.

Landfill dumping is probably the best current option for sludge disposal. Waste sludge is not like household waste, which is a mixture of all sorts of things (e.g., plastic, organics, wood, paper, metal, etc.). Waste sludge is a largely organic cake of material, and although it may be hard to define, it is also reliably degradable, at least in large part and over long time periods. Waste sludge is dumped in landfills in special areas because the land will eventually be reclaimed as productive property. The time frame for this event may be 30 or 40 years, but the organic material gradually degrades in nutrient value and a composted material is created that is harmless to human health.

The first task in creating a landfill is to identify a large property in which drainage can be controlled. The topsoil is removed and sold as a valuable commodity. The waste sludge is then added in rows and compacted in place. Each day of application, the waste is covered with soil to keep birds and vermin away. Row after row is added until a full layer is complete, then the next layer is begun. In this manner a hill is gradually created; its overall height depends on local zoning issues. Some of these hills can be quite impressive in size.

Decomposition of the waste continues after burial. The indigenous microbes can utilize the substrates for decades. Aerobic respiration is dependent on molecular oxygen, however, and the supplies of oxygen soon run out. Other terminal electron acceptors can be used by the anaerobes, and complex communities are created as some microorganisms convert polymers to monomers and create fermentation products. Other microbes utilize these fermentation products, and gradually the character of the waste changes. A central component of this system is the available terminal electron acceptors. After oxygen, nitrate is the preferred acceptor, followed by the iron and manganese acceptors (which are relatively rare), sulfate, and finally carbon dioxide (or carbonate). This last electron acceptor is utilized by the methanogenic bacteria as they receive energy from molecular

BOX 5-4 Milorganite

Waste sludge can be converted to a useful fertilizer such as Milorganite, which is produced by the Milwaukee Metropolitan Sewage District. This high-quality product is sold for gardens, landscaping, parkland, and related uses, an outcome that is preferable to the truck-and-landfill approach. (Courtesy of Milwaukee Metropolitan Sewage District.)

hydrogen and pass off their electrons to create methane. Methane is an energy-rich gas, and pockets of methane form below the surface in the methanogenic zone. This gas can be recovered and used as a power source. In some cases this gas is merely flared off, but in some facilities the gas is collected and used, making the landfill site a net energy producer (**Figure 5-21**).

THINGS THAT CAN GO WRONG WITH WASTEWATER

The biggest risk from wastewater is that it will be returned to the environment untreated, carrying along pathogens and such high nutrient loads that the receiving water is degraded and community health is impaired. The treatment process for wastewater has been described in detail because of its importance. However, there are other potential problems with wastewater. Each community has miles of pipes connecting homes to sewers, and additional miles of sewer lines lead to the sewage treatment plant. Problems can occur all along the system.

Within the sewage treatment plant, facilities must be maintained on a regular basis by doing things like changing filters and monitoring process efficiency. Human error is always an issue, although adoption of standard procedures greatly facilitates

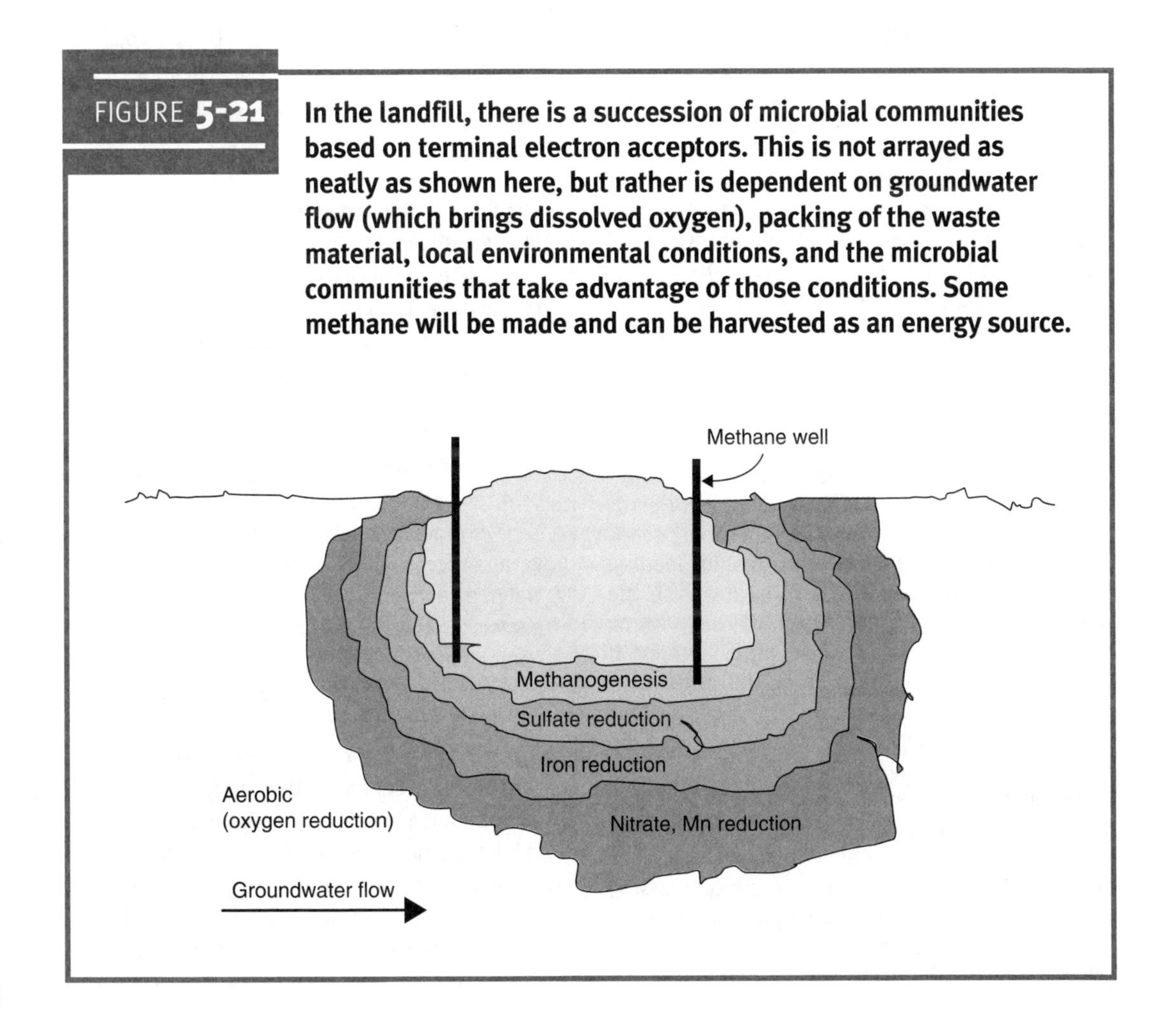

FIGURE 5-21 **In the landfill, there is a succession of microbial communities based on terminal electron acceptors. This is not arrayed as neatly as shown here, but rather is dependent on groundwater flow (which brings dissolved oxygen), packing of the waste material, local environmental conditions, and the microbial communities that take advantage of those conditions. Some methane will be made and can be harvested as an energy source.**

effective management. Training and supervision are vital for both safety of the personnel and for smooth, continuous operation. Most people have no idea where the community sewage treatment plant is located in their community. The plants are usually quiet, give off no unpleasant odors, and are so well run that accidents are very rare.

The daily operation of a wastewater treatment plant is usually smooth and efficient, but these are living systems and subject to change. Certain microbial species, filamentous bacteria, are "nuisance" bacteria that can overgrow the system and cause fouling and clogging of the treatment plant. The culprit species include *Sphaerotilus*, *Leptothrix*, *Gallionella*, *Clonothrix*, *Beggiatoa*, and *Crenothrix*, among others. In the sedimentation tanks, the flocs produced by these filamentous bacteria do not sediment very well, and they can be carried over in the effluent. That is, the floc is loose and bulky instead of compact and dense; therefore, this condition has been called "bulking." These filamentous microbes can biofoul wells also, leading to reduced flow. Their proliferation is not understood well enough to avoid these problems completely. However, the biggest problem from the sewage system itself is the pipe breakage. Nothing lasts forever, and in the very challenging environment of water-based high nutrients and microorganisms, the effects on the pipes of all types (i.e., plastic, metal, concrete) are significant. Biofilms form almost immediately. If biofilms become too prominent, the flow of water may be slowed, leading to clogged pipes. Corrosion of pipes, often involving the microbes, eventually occurs. Also, the ground can shift, leading to stresses on the pipes and separations in the seals between pipe units. Some of these allow groundwater into the sewage pipes, where it increases the volume of material that must ultimately be treated.

Cracks and breaks that allow sewage out of the sewer pipes are also a risk. This type of problem is significant because it spills wastewater into an area that is not designed to receive it. Most sewer pipes are buried underground at a sufficient depth for the region so that there is little danger of freezing, and they are certainly deep enough to avoid shallow digging on site. However, the pipes may be below the water table, causing contamination of the

BOX 5-5 Use of the Nitrogen Cycle in Wastewater Treatment

Because it is cyclic, the nitrogen cycle has no real beginning or end. For discussion, we start with nitrification, the use of ammonia as an energy source. Autotrophic bacteria (see equations) use the energy to fix carbon from carbon dioxide. Ammonia is not usually present in high enough amounts to warrant its use as an energy source, but certain microorganisms have the capacity to use it if it is available. This is an aerobic reaction that results in a pH decrease. Air can be sparged into the tanks to promote this activity.

$$2NH_4^+ + 3O_2 \leftrightarrow 2NO_2^- + 2H_2O + 4H^+ \qquad \Delta G = -66 \text{ kcal/mole}$$

$$2NO_2^- + O_2 \leftrightarrow 2NO_3^- \qquad \Delta G = -17 \text{ kcal/mole}$$

The first reaction is carried out by microbes like *Nitrosomonas* and *Nitrosococcus*. The second reaction involves *Nitrobacter*. This set of reactions reduces the overall BOD and results in the accumulation of biomass through flocculent material, which is then removed as sludge. Again, an activated sludge is produced that performs this task efficiently. The end result is the production of the completely oxidized species, nitrate.

Overproduction of nitrate also poses dangers. If excess nitrate gets into the drinking water, it can be reduced to nitrite by microorganisms in the gut and can then combine with hemoglobin (Umbreit, 2007). Excess nitrates are rarely a problem for large water supplies, but they can occur in private wells. In some cases, nitrates in the water supply are found with bacterial contamination and enough of a carbon source to cause the reduction to nitrite. In other documented cases, nitrate levels were elevated due to industrial contributions to the waste stream. Nitrosamines are associated with an increased risk of cancer; most of the members of this chemical class have been found to be carcinogenic. Because these compounds are metabolized by the liver, hepatic cancer is a risk if they are consumed over a long period of time. The use of nitrates for preserving meats is also a concern because the amine groups of amino acids are readily available in the body and may combine with nitrates to produce nitrosamines.

Some level of nitrates in drinking water is not usually a problem. Nitrates can be converted to nitrites in the mouth because the normal flora includes many species that are able to reduce nitrates. However, this is considered a normal occurrence. The bulk of ingested nitrates is absorbed in the duodenum and is excreted in the urine in normal individuals. Many microbes in the ileum can convert nitrate to nitrite, but only very small concentrations of nitrate ever get that far.

Newborn infants are, however, quite susceptible to nitrates in the drinking water, and a specific illness is associated with ingestion of nitrites. This is referred to as "Blue Baby Syndrome," although the syndrome actually has many causes and describes the observation of cyanosis of the infant due to insufficient oxygen. In this particular case, excess nitrite oxidizes Fe^{2+} to Fe^{3+}, causing a condition called methemoglobinemia (Bruning-Fann and Kaneene, 1993). In its oxidized form, the hemoglobin holds the oxygen molecule very tightly, and normal delivery of oxygen to the tissues is impaired. In fact, it is normal for a small fraction (1–3%) of blood to be in the Fe^{3+} state. At this level there is no danger, but excessive nitrite consumption can produce symptoms if the percentage goes above 20% and can produce severe effects after 50%. Newborns are especially susceptible to methemoglobinemia because a significant fraction of their blood is still fetal hemoglobin, which itself binds oxygen tightly. In addition, infants have a more difficult time converting methemoglobin to hemoglobin enzymatically (Mensinga et al., 2003).

Other populations have an increased susceptibility to nitrates. Individuals who do not make enough stomach acid (achlorhydria) or who suppress their stomach acid artificially through the use of certain drugs or heavy antacid use will have a stomach environment that is no longer as inhospitable to microbial growth, as is normally the case. This permits colonization with species that can reduce nitrates, which are then passed into the duodenum and are taken up as the nitrates are. In severe cases of diarrhea, nitrates can be transported quickly to the ileum and can complicate the illness with increased methemoglobin.

This introduces the other side of the nitrogen cycle, which is denitrification. Nitrate has little energy content, but makes an effective terminal electron acceptor if another energy source is available and if oxygen is not available. That is, denitrification is an anaerobic process. The enzymes involved have varying susceptibilities to atmospheric oxygen, and in some cases the enzymes are completely inhibited by even small concentrations of oxygen. A variety of common microbial species have some (e.g., *Bacillus*, *Klebsiella*, *Vibrio*) or all (e.g., *Paracoccus*, *Thiobacillus*) of the enzymes for the conversion of nitrate to nitrogen gas. Even some species that are regarded as obligate aerobes (e.g., *Pseudomonas stutzerith*) have enzymes along part of the pathway. The overall equation for denitrification can be represented in which a carbon source (in this case, methanol) is used and nitrate is the terminal electron acceptor. As shown, the products include two gases—nitrogen and carbon dioxide—that are released and hydroxyl radicals, indicating that the result will produce an alkaline pH.

$$5CH_3OH + 6NO_3^- \leftrightarrow 3N_2 \uparrow + 5CO_2 \uparrow + 7H_2O + 6OH$$

In practice, wastewater may have enough available carbon to support this reaction, although an additional source like methanol can also be added to drive the reaction in the anaerobic tank. The series of chemical changes to nitrate are summarized in the following equation.

Nitrate→	Nitrite→	Nitric Oxide→	Nitrous Oxide→	Nitrogen
NO_3^-	NO_2^-	NO	N_2O	N_2

Unlike assimilatory nitrate reduction, this process is not subject to feedback inhibition and is sensitive to oxygen in at least some aspects. Typical nitrogen removal rates include 30–35% that is removed as nitrogen gas and 20–25% that is removed in organic sludge. If nitrogen levels are still too high, additional rounds of nitrification and denitrification can be performed.

groundwater supply in the area. After a while, this contamination may rise to the surface, where the odor becomes evident.

There are two sewer systems in a typical community. The septic sewer takes in household wastewater (fecal material) and has received most of the attention here. The storm sewer system is, ideally, separate. This system is evident in the sewer grates in the street, where they collect rainwater and snowmelt and funnel it all away before it can cause street flooding. Before industrialization, rainwater simply hit the ground and percolated through the soil, recharging the groundwater, or it became runoff and headed quickly into the nearest creek or stream. Now, with large areas paved with roads that are impermeable to water, and with roofs that redirect the water through gutters and into the street, an ordinary rainstorm would cause a good deal of water to pool locally if not for the storm sewer system. Storm sewers simply take precipitation downhill directly to the receiving water. This water is not completely clean—it does carry whatever materials it picks up from the streets and rooftops—but it is not wastewater.

In newer communities the septic system and storm system are separate, but in many older cities the two systems are linked. The rationale was that two separate systems were expensive, and the storm sewer system could be fed into the septic sewer lines. The extra water would dilute the waste material and flush the sewer lines, keeping the waste moving along. This works well as long as rain events fall within expected normal ranges. Unfortunately, large rainstorms or sudden thaws occur occasionally, creating large volumes of runoff. Water from the overwhelmed combined system (including wastewater) may then back up into homes through the toilets—a very unpleasant experience. At such times, the option of bypassing the sewage treatment plant and allowing the excess water to flow directly into the receiving water is often adopted temporarily. This method avoids the sewer backup, but it dumps untreated waste into the environment. There is no easy solution to this problem.

Fortunately, wastewater in sewer pipes keeps moving. Continuous flushing of the pipes means that the waste material will not decompose in place. However, occasionally the volume in the pipes can fall dramatically. For instance, in areas where the population changes dramatically seasonally, such as retirement communities in which many people leave for warmer or cooler climates during the year, water usage is not constant. This has certain dangers. Sudden increases in water volume can clog the pipes as well as create dangerous gases.

Hydrogen sulfide (H_2S) is a toxic gas that is produced by sulfate-reducing bacteria (SRB). At levels above 1400 mg/m^3, the effects can be sudden and fatal. Methane is a powerfully explosive gas that is produced by methanogenic bacteria. Before entering a sewer line, workers typically test for these gases. If they are present, a fan can sometimes reduce gas concentrations to safe levels. A self-contained breathing apparatus can also be used, although the danger of methane explosion should never be ignored. Methane is odorless, but hydrogen sulfide has a powerful rotten-egg odor.

The Sulfur Cycle

Hydrogen sulfide, as a reduced compound, is quite energy rich. It can be oxidized to release this energy.

Even elemental sulfur (S^0) can be oxidized as an energy source. This is carried out by facultative chemolithotrophs such as *Thioplaca*, *Thiothrix*, *Thiobacillus*, and *Beggiotoa*. In hot sulfur springs, an abundance of sulfur is available, and the unusual *Sulfolobus* bacteria grow here under extreme conditions. The photosynthetic sulfur bacterium *Chlorobium* can also utilize H_2S, applying the energy to the fixation of carbon from atmospheric carbon dioxide, and again, sulfate is the final product (**Figure 5-22**).

$$H_2S + 0.5O_2 \rightarrow S^0 + H_2O \qquad \Delta G = -50.1 \text{ kcal/mole}$$

$$S^0 + 1.5O_2 + H_2O \rightarrow H_2SO_4 \qquad \Delta G = -149.8 \text{ kcal/mole}$$

As the second equation shows, full oxidation leads to the production of sulfuric acid (H_2SO_4), a strong mineral acid. This is a significant problem in old mine sites, where waste rock ("slag") is piled up outside the mine. This rock has significant concentrations of pyrite (FeS_2), which is often referred to as "Fool's Gold" because of its distinct color. Locked underground, these minerals oxidize slowly, if at all. However, in slag piles, with rainwater percolating through them and with the introduction of sulfur oxidizers, a much more rapid biochemical reaction occurs, typically involving *Thiobacillus*.

$$2FeS_2(s) + 7O_2(g) + 2H_2O(l) \rightarrow 2Fe^{2+}(aq) + 4SO_4^{2-}(aq) + 4H^+(aq)$$

This is called acid mine drainage or acid rock drainage because it can occur wherever mined rock is piled up, and it can yield significant amounts of sulfuric acid. The iron (Fe^{2+}) will oxidize to Fe^{3+} as well. The acid is produced by the microbial oxidation of the iron pyrite, which then washes out the bottom of the slag pile and continues downhill until it enters a water supply (e.g., a nearby stream). Here it lowers the pH significantly, killing off all indigenous forms of life. This is another component of water pollution (covered in Chapter 3).

In pipes, the buildup of sulfuric acid can be destructive. If the waste is not regularly flushed, it builds up in one location, and the rest of the sulfur cycle is set in motion (**Box 5-6**). Sulfur is plentiful and is not usually a limiting nutrient. Some sulfur is

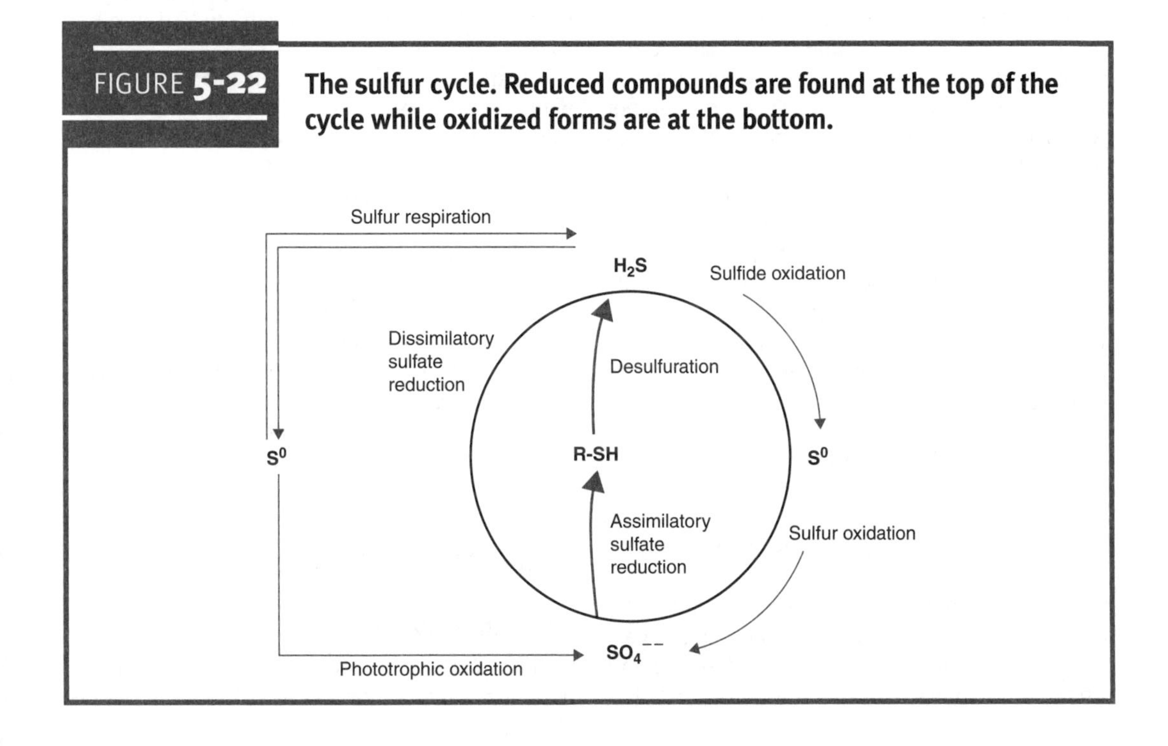

FIGURE 5-22 **The sulfur cycle. Reduced compounds are found at the top of the cycle while oxidized forms are at the bottom.**

BOX 5-6 Sulfur Species

S^{2-} = sulfides, mercaptans
S^0 = elemental sulfur
S_2O_4 = hyposulfite
SO_3^{2-} = sulfite
SO_4^{2-} = sulfate

required by cells, such as the minor amounts in the amino acids cysteine and methionine. Sulfur is acquired through assimilatory sulfate reduction, in which sulfate combines with the high-energy product adenosine triphosphate (ATP) to create adenosine 5′-phosphosulfate (APS) using the enzyme ATP sulfurylase. In this form, the sulfur atom can be easily passed to other biomolecules. Although this is a common trait among bacteria, it does not overly concern us here.

The reduction of sulfate takes place because it is a handy alternative terminal electron acceptor to oxygen. Microorganisms can utilize carbon compounds in the waste and then remove electrons in dissimilatory sulfate reduction. This is carried out by a group of microorganisms called the sulfate-reducing bacteria (SRB), which include species such as *Desulfuromonas*, *Desulfovibrio*, and *Desulfotomaculum*. These are obligate anaerobic bacteria: they use sulfur compounds for terminal electron acceptors (and not oxygen). In fact, they primarily use acetate or lactate, which are not typical in wastewater but are waste products of the fermentation of waste materials by other microbial groups (Wang, 2006).

$$\underset{\text{Acetate}}{CH_3COOH} + 2H_2O + 4S^0 \rightarrow 2CO_2 + 4H_2S$$

$$H_2 + SO_4^{2-} \rightarrow H_2S + 2H_2O + 2OH^-$$

As seen in the first equation, the end result is the production of the energy-rich hydrogen sulfide gas. The gas leaves the anaerobic waste material and comes into contact with the aerobic *Thiobacillus* species, which oxidizes it again to sulfuric acid, which drops down into the waste pile and is reused as an electron acceptor.

Of course, before these reactions occur, the powerful sulfuric acid may also damage the pipes, especially concrete pipes. Concrete is an aggregate of gravel and sand that is held together with cement. The acid degrades the cement, effectively producing free sand and gravel particles. This sort of material has no strength whatsoever, and collapse of the pipe becomes a real possibility.

If iron or steel pipes are used, the sulfide reacts with and causes pitting, producing the black compound ferrous sulfide. This is also referred to as microbially influenced corrosion (MIC). It is a complex process and difficult to study because complex biofilm communities are involved that are difficult to replicate with any degree of certainty. In addition, many factors (e.g., the types of ions involved and their concentrations at various depths within the biofilm, pH values, redox potentials, and nutrient availability) contribute to the process (**Figure 5-23**). Dissimilatory sulfate reduction processes and iron dissimilatory processes have been implicated. For example, the effects of the iron dissimilatory microbe *Shewanella* have been studied in detail (Beech et al., 2005).

Clearly, the process is an electrochemical one with specific anodic and cathodic sites. Corrosion also proceeds in the absence of microbial influence, albeit with qualitative differences. For example, when sulfate-reducing bacteria attack steel, the major mineral produced is FeS, called mackinawite. When the bacteria are removed, the major mineral is FeS_2, or pyrite (Videla and Herrera, 2005). Even when the extracellular polymeric substance alone is used, the biocorrosion appears biotic, and the exoenzymes that are important in corrosion are fairly stable in the EPS material. Until better methods are created to examine the changes that occur at the biofilm–metal interface, the process will remain difficult to model.

SIMPLE TREATMENT SYSTEMS

The Septic System

Septic systems are useful in areas that are sparsely populated, such as farmhouses, vacation cabins,

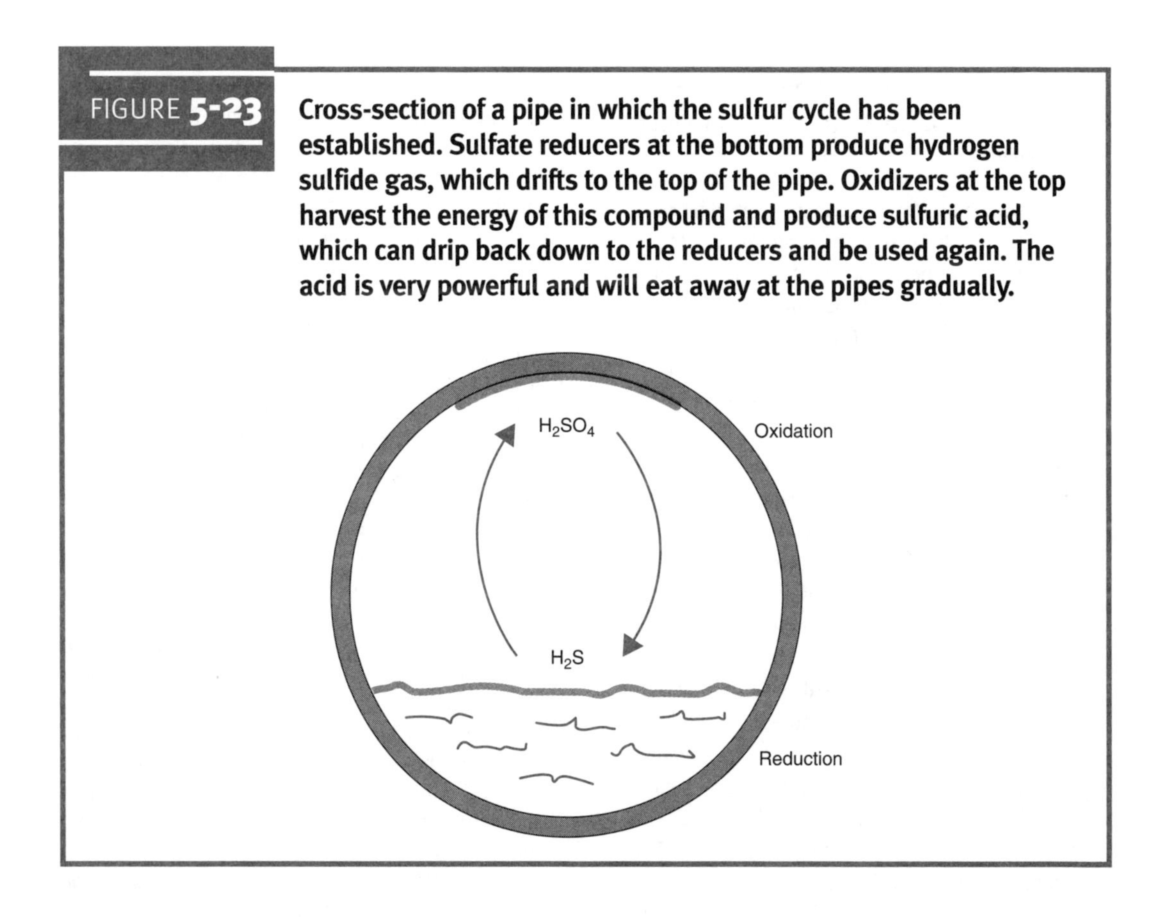

FIGURE 5-23 **Cross-section of a pipe in which the sulfur cycle has been established. Sulfate reducers at the bottom produce hydrogen sulfide gas, which drifts to the top of the pipe. Oxidizers at the top harvest the energy of this compound and produce sulfuric acid, which can drip back down to the reducers and be used again. The acid is very powerful and will eat away at the pipes gradually.**

homes built on islands, and the like. Typically these homes also have a well for drinking water, and of course, care must be taken to ensure that the two systems do not mix. The centerpiece of this system is the septic tank, which is buried in the yard outside the home and to which all the wastewater runs. Septic tanks are designed to handle total household water, but certain additions should not be routed through it. For example, swimming pool water should be drained directly over a field rather than through the septic system. A water softener that uses rinse water to recharge the system may be too toxic for the septic system; often the rinse line is rerouted by a plumber, if that is permitted by law.

The septic tank is simply a large container underground that collects wastewater. The waste separates here into sludge that sinks to the bottom and the water above it, which exits the tank through pipes that lead into the drainage field or leach field. The microorganisms in the waste material decompose the BOD of the wastewater, producing sludge. If the microbes are not doing the job effectively, a freeze-dried mixture of microbes is commercially available that can be flushed down the toilet and into the tank. This should help break down the waste. If left untended, the tank is eventually filled with sludge, sending untreated wastewater out through the leach field. So the tank must be pumped out and its contents hauled away by a professional service every 4–5 years.

The leach field is a set of ventilated pipes buried at a shallow depth that receive the water that exits the septic tank. The water is still high in nutrients, but distributing it over a grassy area with a high

BOX 5-7 What Exactly Is Compost?

Many kinds of waste can be treated by composting. Most household gardeners are familiar with the concept of composting, in which organic matter is piled together and occasionally turned or mixed. Microorganisms inside the compost pile grow on the organic material and produce heat. It is not unusual to turn over a compost pile on a cold day and see water vapor rise from the pile. The reason for this is that thermophilic bacteria are active, taking advantage of conditions that no longer favor the typical anaerobes in the pile. Thermotolerant bacteria grow at elevated temperatures (45–60°C). This community can be succeeded by the thermophiles, which grow at 60–70°C.

Composting either yard waste or other types of waste, such as wastewater sludge, has a number of advantages. After 3 days at a temperature of at least 55°C, the pathogens in the pile are destroyed. Much of the mass is reduced as carbon dioxide is produced. The final product is greatly reduced in carbon content. However, this product is not a fertilizer, as many people think; it actually has few nutrients that plants need. Nonetheless, it is an excellent soil conditioner, helping to hold water in the soil and preventing excessive erosion.

Composting of sludge can be performed if sufficient land is available to hold the sludge. The piles can be turned with a bulldozer, and the final product can be sold (or given away) to landscapers and horticultural companies. An alternative is dumping the compost in the landfill as described previously.

level of microbial activity allows the nutrients to be utilized and the pathogens to die before they contact another host. The leach field is usually grassy; trees grow roots that can interfere with the pipes. Using the area for parking risks compacting the soil or damaging the pipes. The site should be situated so that the indigenous microbes have plenty of time to work on the water before the water percolates through the soil to the groundwater.

Problems can occur if the septic tank develops a leak or crack that allows waste to enter the soil system directly. Flooding of the leach field by precipitation can wash wastewater into the groundwater system quickly. An overfilled tank can cause all of the flushed water to quickly exit; in this case, the leach field may become swampy with waste and start to stink. If the leach field is located in an area where the water table changes seasonally and no allowance is made for it, then the wastewater may contact the groundwater during certain times of the year.

A variation of this concept is found on the community level. Lagooning uses a cesspool as a collection and storage area for wastewater, and lagoons are sufficient in areas with small populations and ample land. The wastewater fills the pool or pond and remains there for period of time, perhaps several months. During this time, the microorganisms in the waste reduce the BOD and the suspended solids. After this time, the water can be drained away and the solids can be removed for disposal, involving either land application (as fertilizer) or landfill. In effect, time is not essential because the volume is so small, so a long, slow process is possible. This is a low-maintenance method, and unless the pool ruptures, little can go wrong.

QUESTIONS FOR DISCUSSION

1. Why is it so difficult to compare one wastewater stream with another?
2. For a BOD measurement, the initial oxygen concentration was 10.6 mg/l, the final was 3.4 mg/l, and the sample was 1.5 ml in a 300-ml bottle. What is the BOD?
3. Using the other conditions in question 2, the final oxygen concentration was zero. What was the BOD?
4. What are the main components of a good municipal wastewater treatment system?
5. What is the 30:20 rule?
6. What is sludge? Activated sludge?

7. What are the differences between dissimilatory and assimilatory nitrate reduction?
8. How does a septic tank work? How is it different from a municipal system?
9. Under what conditions is the accumulation of sulphuric acid a problem?
10. What are biofilms? Why are they important?

References

Beech, I.B., J.A. Sunner, and K. Hiraoka. 2005. Microbe-surface interactions in biofouling and biocorrosion processes. Internat. Microbiol. 8: 157–168.

Bruning-Fann, C.S., and J.S. Kaneene. 1993. The effects of nitrate, nitrite and N-nitroso compounds on human health: a review. Vet. Human Toxicol. 35: 521–538.

Knowles, R. 1982. Denitrification. Microbiol. Rev. 46: 43–70.

Kuai, L., and W. Verstraete. 1998. Ammonium removal by the oxygen-limited autotrophic nitrification-denitrification system. Appl. Environ. Microbiol. 64: 4500–4506.

Lovley, D.R., F.H. Chapelle, and J.C. Woodward. 1994. Use of dissolved H_2 concentrations to determine distribution of microbially catalyzed redox reactions in anoxic groundwater. Environ. Sci. Technol. 28: 1205–1210.

Martin, H.G., N. Ivanova, V. Kunin, F. Warnecke, et al. 2006. Metagenomic analysis of two enhanced biological phosphorous removal (EBPR) sludge communities. Nature Biotechnology 24: 1263–1269.

Mensinga, T.T., G.J.A. Speijers, and J. Meulenbelt. 2003. Health implications of exposure to environmental nitrogenous compounds. Toxicol. Rev. 22: 41–51.

Sterritt, R.M., and J.N. Lester. 1988. Microbiology for environmental and public health engineers. E. & F.N. Spon, London. p. 186.

Thauer, R.K., K. Jungermann, and K. Decker. 1977. Energy conservation in chemotrophic anaerobic bacteria. Bacteriol. Rev. 41: 100–180.

Umbreit, J. 2007. Methemoglobin—it's not just blue: a concise review. Am. J. Hematol. 82: 134–144.

Videla, H.A., and L.K. Herrera. 2005. Microbiologically influenced corrosion: looking to the future. Internat. Microbiol. 8: 169–180.

Wagner, M., R. Amann, H. Lemmer, and K.H. Schleifer. 1993. Probing activated sludge with oligonucleotides specific for proteobacteria: inadequacy of culture-dependent methods for describing microbial community structure. Appl. Environ. Microbiol. 59: 1520–1525.

Wang, L.K. 2006. Waste chlorination and stabilization, ch. 12. *In* L.K. Wang, Y.T. Hung, and N.K. Shammas (eds.), Handbook of environmental engineering, vol. 4: advanced physicochemical treatment processes. Humana Press, Totowa, NJ.

CHAPTER 6

Case Studies on Water and Wastewater

LEARNING OBJECTIVES

- Recognize the salient features of the case-control study.
- Critically evaluate anecdotal reports from the literature.
- Identify some of the engineering systems that are susceptible to failure.
- Construct a retrospective study to effectively gather data.
- Analyze real-world data and compare situations with current conditions.

THE *CRYPTOSPORIDIUM* OUTBREAK IN MILWAUKEE, 1993

In late March and early April 1993, the city of Milwaukee, Wisconsin, suffered through the largest outbreak of waterborne disease in America in the 1900s. In all, some 400,000 people were affected, with 44,000 visiting doctors or emergency rooms and about 4400 persons hospitalized (Corso et al., 2003). More than 100 people died. The cost was enormous, estimated at $96 million; about a third of this was due to medical treatment, and the other two thirds was in lost productivity (Corso et al., 2003). The pathogen was *Cryptosporidium*, and an analysis of the outbreak gives a good overview of how events like this one can happen and why vigilance is so important.

The outbreak was not sudden; the first sign that something was wrong came from local physicians who called the Milwaukee Health Department to report a sharp increase in severe diarrhea. Diarrhea is not a reportable illness, and the doctors did not have to contact the Health Department at all, but this is part of the informal information network that makes society function.

The Health Department started to investigate and found that the outbreak was actually widespread throughout the city. At first the authorities had no idea what was causing the illness. Diarrhea can be caused by a number of pathogens, and there were no obvious pathogens in fecal cultures. *Cryptosporidium* was not unknown at the time—small but notable outbreaks had occurred in other parts of the country—but it was not a common pathogen, certainly not for the magnitude of the illness that was occurring.

It was apparent that the cause was something in the water supply; a gastrointestinal illness is almost always caused by an ingested pathogen (food or water). The outbreak was so great that only a contaminated source that was common to all people could be to blame, and the water supply was the most obvious source. On April 5, the Health Department had enough anecdotal evidence to suggest that something was wrong. On April 7, it announced a "boil water" alert. On April 9, it had identified the water plant that was the likely source of the problem and closed it down. Over the next several weeks, the epidemic gradually faded. In the meantime a medical technologist at a local hospital discovered *Cryptosporidium* oocysts in a fecal sample and the cause was, at last, known (**Figure 6-1**).

Identifying *Cryptosporidium* was not all that helpful; there are no completely effective treatments for this illness. In 1993, the treatment was entirely palliative and/or supportive. Most people recover on their own, although the diarrhea can last 10–14 days. However, in immunocompromised people, the illness is very serious. During the Mil-

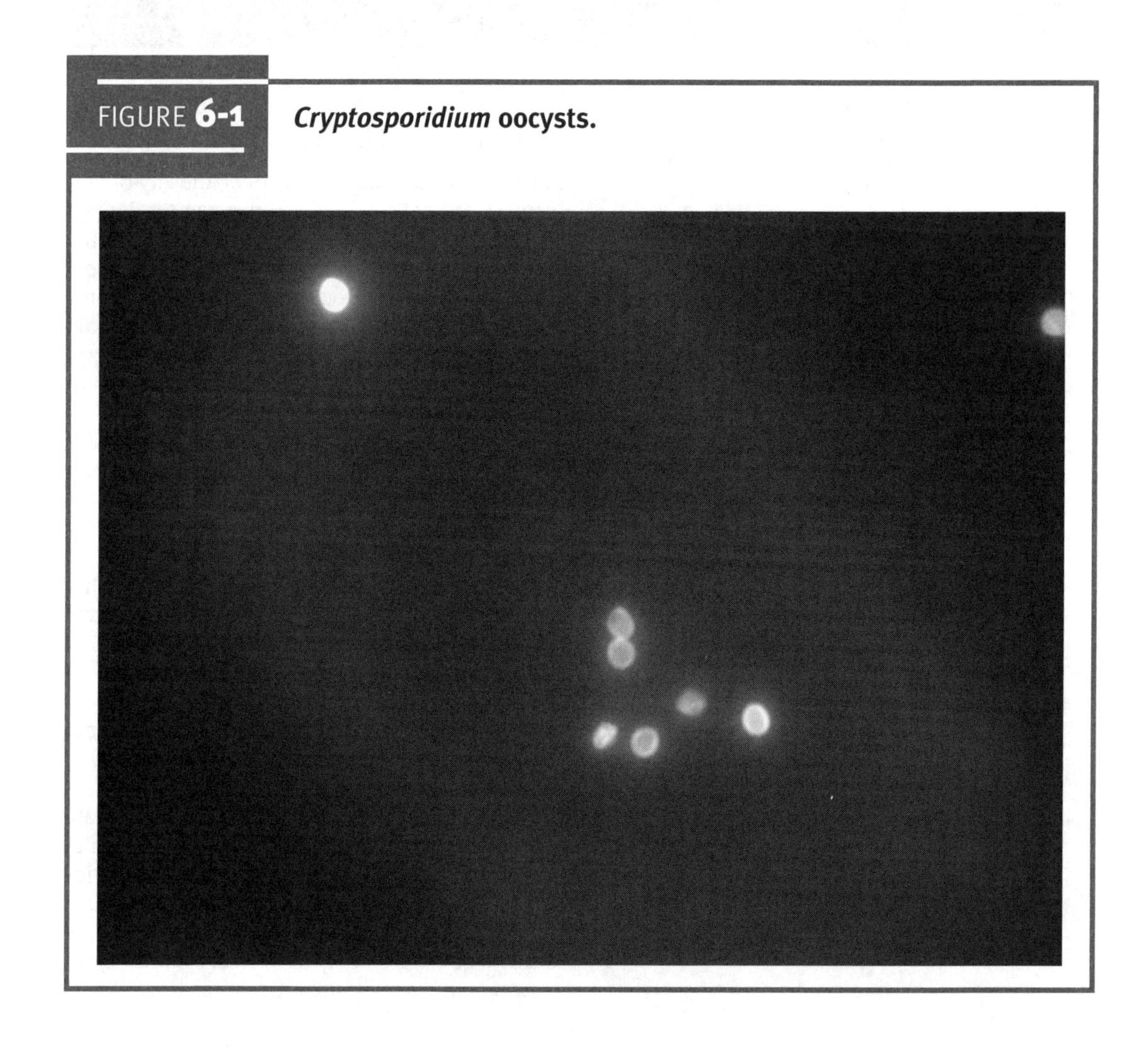

FIGURE **6-1** ***Cryptosporidium* oocysts.**

waukee outbreak, 104 people died as a result of the infection. Most of these were immunocompromised, including a large number who had HIV. In other outbreaks, the mortality rate among the HIV-positive population was approximately 50%. In the Milwaukee outbreak, about 14% of the hospitalized cases were AIDS patients (Corso et al., 2003), and AIDS patients were heavily represented in the fatality figures.

The obvious question is how a pathogen could invade the water supply suddenly and affect so many people. A typical drink of water would have to contain an infectious dose of *Cryptosporidium*. This suggested that a high concentration of oocysts was present in the water supply. Where had they come from?

There are many species of *Cryptosporidium*, and most can infect humans. The major species are *Cryptosporidium hominis*, which primarily infects humans, and *C. parvum*, which primarily infects cattle but can certainly infect humans. The species that was involved in the Milwaukee outbreak was *C. parvum*. The city of Milwaukee is on Lake Michigan, which receives water from three rivers at the port of Milwaukee: the Milwaukee River, the Menomonee River, and the Kinnickinnic River. The first two rivers run north to south and drain a large watershed that includes substantial farmland. Specifically, the watershed includes a large expanse of dairy farmland, which means the presence of a lot of cattle. Both wild animals and domesticated animals have been reported to contribute *Cryptosporidium* to streams (Jellison et al., 2002).

Cryptosporidium parvum infections are not uncommon in cattle. On some farms, the calves are more likely than not to become infected. When a calf becomes ill, it is isolated in a pen, where it is cared for during bouts of diarrhea. Calves may produce more than 10 million oocysts every day in their feces. The straw bedding is usually simply discarded. Only rarely does a calf die from this illness. Because the oocysts remain viable for an extended period, a strong rainstorm or sudden snowmelt might have flushed the oocysts into a local stream of the Milwaukee watershed. The oocysts would then reach Lake Michigan and be pulled into the water supply. It is also not unusual to find oocysts in surface waters; studies have found them in about 10% of samples, although typically in very low numbers. *Cryptosporidium* oocysts are rare in groundwater, but they might be found when surface water has infiltrated a well or when a large concentration of oocysts gradually contaminates the groundwater. It is possible for oocysts to be flushed below the ground surface and to travel laterally for many miles in the groundwater.

A degree of uncertainty about the source of the oocysts remains. Some evidence indicates that a low-level epidemic of *Cryptosporidium* was occurring for several months before the late March outbreak (Morris et al., 1998). Therefore, a small number of cases possibly resulted in a large input of oocysts into the wastewater treatment system. If the treatment was ineffective in removing oocysts (which is possible), then they would be returned to Lake Michigan along with other treated wastewater. Whether it was as a sudden high dose or a result of a magnified event over many months cannot now be determined. What is known is that a significant number of oocysts (enough to cause disease in otherwise healthy persons) got into the water supply in March.

The Milwaukee Water Works has two water purification plants: one in the northern part of the city and one in the south. Both draw water from Lake Michigan and purify it, and both feed an integrated water supply system that serves the entire city. In the event of a breakdown at one plant, the other is able to supply reasonable water requirements for the whole city. Both plants are capable of providing more than 100 million gallons of water per day. During the outbreak, the southern plant was responsible for the influx of *Cryptosporidium*.

The intake pipeline for freshwater for the southern plant was located south of the discharge pipeline for the treated wastewater, although the two were several miles apart. What role this played in the outbreak is unknown. The currents in the lake in this region typically run north to south. After the outbreak subsided, the water works responded in several ways, one of which was to extend the intake pipeline farther out into the lake. The direct contamination of the intake water with treated wastewater cannot be excluded as a factor.

The major factor appears to be the turbidity of the intake water. The turbidity of water—basically a measure of the concentrations of particles in the system—is measured in nephelometric turbidity units (NTUs) and is easily measured with an NTU meter (**Figure 6-2**). Low turbidity is desirable (ideally below 0.1 NTU, which indicates that the water is nearly clear). Whichever particles remain in the water should be removed by coagulation and sedimentation of the flocculent material. Most of the time, both the northern and southern plants had very clear water. However, during late March, the turbidity suddenly rose to a high level, at one point reaching 1.70 NTUs (Morris et al., 1996; **Figure 6-3**). Notably, this level was still far below the regulations of the time. At that time, they also routinely backwashed their sand filters and then passed the backwashed water back through the filter. Most water plants no longer do this because of the problem of reintroducing captured oocysts back into the system.

Even such a high NTU level could have been handled with normal procedures as long as adjustments were made in the coagulation step. However, a new coagulant was in use at the time, the utility technicians had little experience using it, and one of the plant flow meters was incorrectly installed. It is generally assumed that the failure to remove particulates in the water occurred coinci-

FIGURE **6-2** **Turbidimeter.**

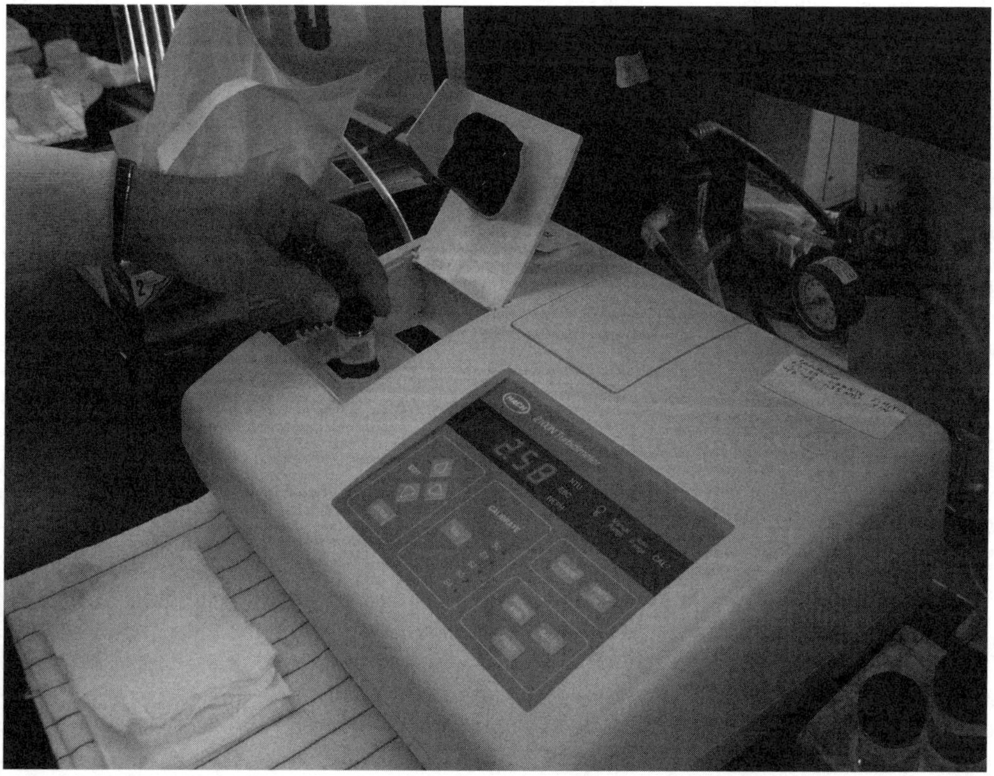

SOURCE: Courtesy of the Milwaukee Water Works. These photos may not be used to infer or imply Milwaukee Water Works or City of Milwaukee endorsement of any product, company, or position.

FIGURE 6-3 **Turbidity during the time of the outbreak, as expressed in NTU. Note the significant spike at the southern plant.**

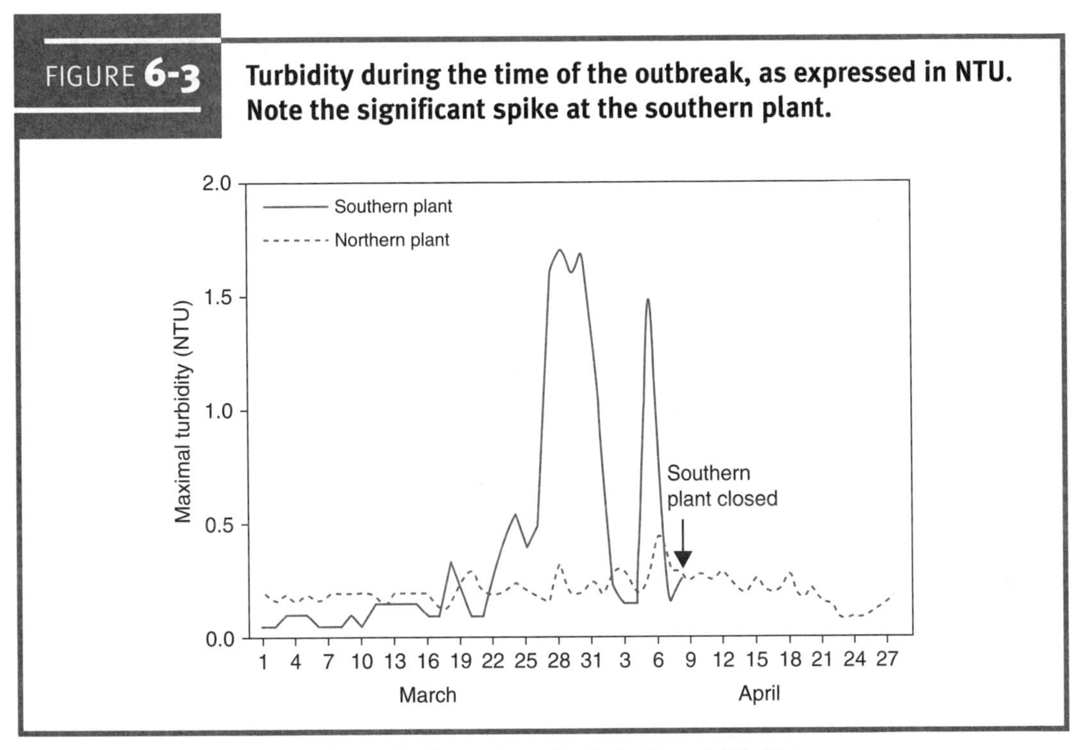

SOURCE: Figure from Mackenzie et al., 1994. Used by permission, *New England Journal of Medicine.*

dentally with the high concentration of oocysts, and that this was the proximal cause of the outbreak. Although the water was also chlorinated, oocysts are resistant to chlorine (Shields et al., 2008; **Figure 6-4**).

Once in the water distribution system, the oocysts were diluted with the rest of the water. The infectious dose of *Cryptosporidium* is probably below 10 oocysts, so it does not take much—perhaps a single drink of water—to produce an infection under these conditions. But what was the average dose? At the time, the priorities were to determine the cause of the outbreak and to fix the problems; no one saved any water samples. However, a retrospective analysis of the outbreak included a search for water that had been saved by companies and individuals. Based on saved water (e.g., ice from an ice plant that operated during the outbreak), the concentration of oocysts was between 0.7 and 13.2 per 100 liters, which is a very low concentration. However, recovery of oocysts is an inefficient process; 100 liters, or even 10 liters, is a lot of water to filter without losing a 5-μm particle. The evidence suggests that the true concentration may have been 10-fold higher. Of course, how many of these oocysts were viable and infectious is unknown. Oocysts of other species may also have been counted, some of which might not have been pathogenic at all. Much remains uncertain about this *Cryptosporidium* outbreak.

In all, about 26% of the population was affected by this outbreak. As noted previously, the vast majority never even saw a physician but treated the illness themselves. Estimates of the total number affected are based on epidemiological surveys that were conducted by telephone. To gain some perspective on the magnitude of the outbreak, in any typical time period in Milwaukee, the number of people with gastrointestinal upset is only 16,000.

FIGURE 6-4 The *Ct* values for *Cryptosporidium* are high, making effective killing of the oocysts by chlorine of questionable efficacy.

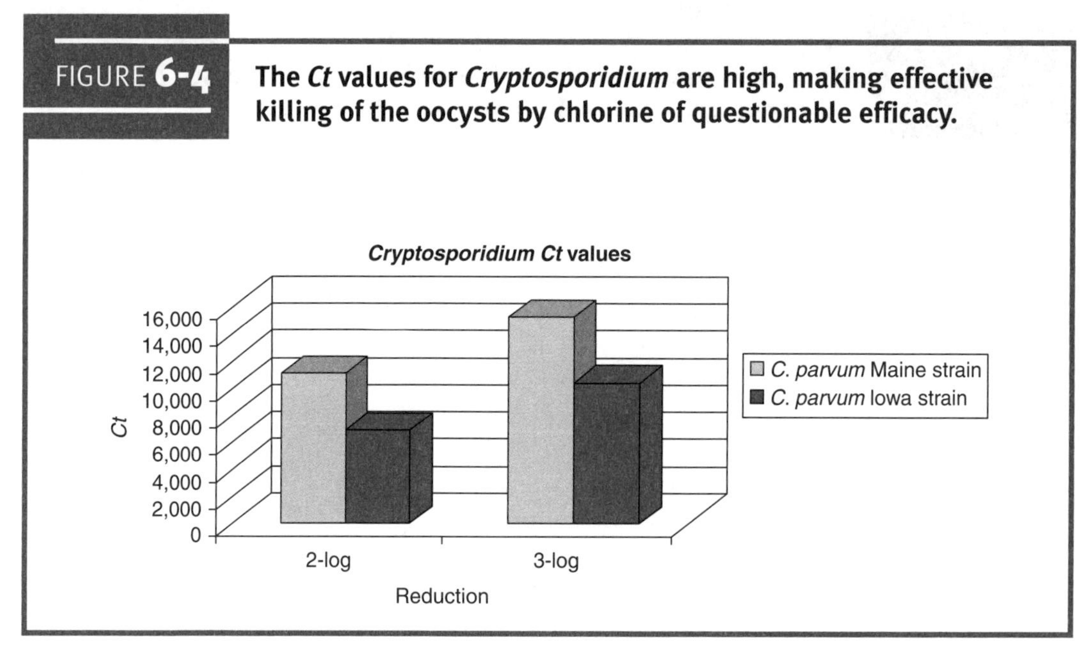

SOURCE: Based on data from Shields et al., 2008.

The water works responded to this situation by making several changes (**Figure 6-5**). As mentioned previously, they moved the intake pipeline farther offshore. Filtered backwash is no longer recycled but is instead treated as wastewater. An ozone-generation system was also installed (**Figure 6-6**). Ozone is expensive but effective at destroying oocysts. While a small number of cryptosporidiosis cases still occur every year (perhaps 10–20), a recurrence of a *Cryptosporidium* epidemic, or of any other parasite, has not been seen since this event.

Monitoring has improved over the years as well. A direct fluorescent antibody test is now used; this test is sufficient to find oocysts but does not distinguish living and dead oocysts. The major problem with this method seems to be filtering enough water (perhaps 100 liters) to get a reasonable number of oocysts, which are usually found in extremely low concentrations. As noted in Chapter 4, standard monitoring for coliform bacteria is not a good predictive index for intestinal parasites like *Cryptosporidium* and *Giardia*. In fact, both are found in about 50% of the *Cryptosporidium* outbreaks. This percentage is too low for use as a predictive index, but it is not insignificant. A preferable method of online monitoring remains to be developed.

Interestingly, the vast majority of medical costs (89%) were attributable to the 1% of infected people who were hospitalized; most home remedies for diarrhea are inexpensive. Two thirds of these hospitalizations involved people with compromised immune systems, such as AIDS patients (Corso et al., 2003). An analysis of death certificates following the outbreak found that 85% of the deaths due to *Cryptosporidium* infection were of people with AIDS (Hoxie et al., 1997). The more advanced the HIV infection was, the greater the likelihood of adverse outcomes (Vakil et al., 1996).

Defensive Measures

Survival of oocysts depends on many factors. Oocysts are susceptible to desiccation and will not last long on the ground surface (Robertson et al., 1992). Once they inhabit a water source, this is not

FIGURE 6-5 **Water treatment scheme.**

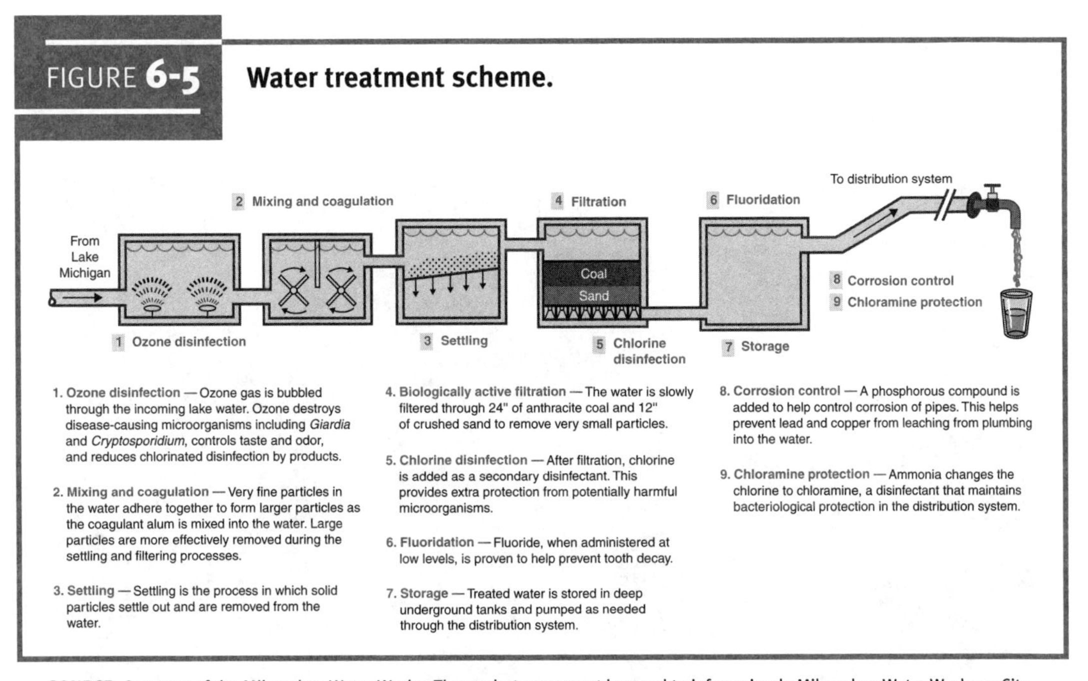

SOURCE: Courtesy of the Milwaukee Water Works. These photos may not be used to infer or imply Milwaukee Water Works or City of Milwaukee endorsement of any product, company, or position.

an issue. They are susceptible to heat, losing viability as temperatures exceed normal body temperature (37°C). When water contamination is suspected, it is advisable to boil drinking water before consuming it; this is sufficient to kill oocysts. While it is probably impractical to heat drinking water on a regular basis, it is likely that a large fraction of oocysts are destroyed by heat before they ever enter the water supply. Cattle dung starts to compost, raising the temperature above 40°C—enough to kill oocysts. Several days at such temperatures are probably sufficient to kill the oocysts (Fayer, 1994). However, if a rainstorm washes the oocysts into a stream before sufficient composting time has passed, then the oocysts are transported alive. The combination of drying and high temperature is probably effective at killing most oocysts in agricultural areas, even if complete elimination is not achieved.

Ultraviolet (UV) light is also effective at killing oocysts, and sunlight is effective as long as the oocysts are not covered by other materials that block or absorb the UV light. However, UV light does not penetrate very well, and so only oocysts near the water's surface are affected (King and Monis, 2007).

Once in the water, the oocyst is subject to additional pressures (King and Monis, 2007). Warm water tends to rise in the water column, and oocysts are then closer to UV light. Colder water tends to sink, and oocysts are then more protected from heat and UV light. The density of oocysts is slightly greater than water, so they sink, but slowly. If they are associated with other particles in the water, the settling will occur more rapidly (Brookes et al., 2004), but the oocysts are slightly negative in charge and do not bind well to soil particles as a general rule.

Oocysts may also be trapped by soil and vegetation. Atwill et al. (2006) found that a grassland buffer strip of only 1–2 meters was sufficient to stop the progress of most oocysts under runoff conditions.

FIGURE 6-6 **Ozone generator. Ozone is effective against *Cryptosporidium* oocysts.**

SOURCE: Courtesy of the Milwaukee Water Works. These photos may not be used to infer or imply Milwaukee Water Works or City of Milwaukee endorsement of any product, company, or position.

This important physical barrier can be installed easily and cost-effectively in agricultural areas, especially around feedlots.

Criticism of the Milwaukee Water Department in the wake of this outbreak is predictable, and much criticism occurred. However, a more objective view is needed. At the time of the outbreak, the Water Department authorities were following all applicable laws on water purification. Even the critical parameter of turbidity was well within limits. Of course, after this episode the limits were adjusted downward substantially, but occurred in response to the outbreak. Other problems at the Howard Avenue Water Purification Plant were relatively minor and could have occurred anywhere. Of course, no definitive explanation for how this outbreak became so widespread has been rendered. Other water authorities are mindful of this fact and are accordingly vigilant. Many more parasites might cause outbreaks in the future (Steiner et al., 1997), and the unknown factors in these outbreaks are a constant source of concern.

E. coli O157 AND *Campylobacter* DUE TO A BROKEN PUMP, 1995 (JONES AND ROWORTH, 1996)

The village of Fife is a small town in Scotland. A stream runs through the town, and the wastewater treatment plant discharges treated waste into the stream. In 1995, a major outbreak of gastrointestinal disease occurred in Fife. At least 633 people became ill, some 65% of the population at risk. The problem was traced to a vegetable farm located just outside of town. This was surprising because they had never had such a problem before, and several causes for the event were eventually uncovered.

The farm usually used well water for irrigation and washing produce, but the well pump had broken. In its place, they decided to use the river water because it was nearby and because all they had to do was drop a hose into the river and use another standard pump to bring the water to the farm.

As mentioned, the wastewater treatment plant discharged treated waste into this stream. In general, wastewater is not completely safe but is acceptable for discharge if the waste can be handled by the indigenous flora of the river before it reaches another community as its freshwater supply. In this case, the coliform count at the point where the farm was removing water was in the range of 14,000–17,000 coliform bacteria per 100 milliliter. This was obviously far greater than the drinking water standard, which is <1 per 100 milliliter.

The water was meant for washing produce, and even for this task the stream water was unsuitable. However, the farm's water lines were connected to the city water lines, and the pump used on the farm had a pressure that was 1.8 kg/cm^2 greater than the town's water pressure. Under these conditions, the stream water was pumped against the town water line and back into the homes of the townspeople. Installing a simple one-way valve at any point between the farm and the town would have easily prevented this from happening, but this necessity was overlooked.

Under these conditions, the townspeople began drinking stream water without realizing it. Illness quickly followed. The affected people reported typical gastrointestinal complaints: diarrhea, nausea, vomiting, and abdominal pain. The health authorities reacted quickly, immediately suspecting that the water supply was affected. They developed an epidemiological approach in which the town was divided into zones, and the number of sick people in each zone was determined (**Figure 6-7**).

The results were revealing. The zones nearest the farm had the highest illness rates (**Table 6-1**), more than 80% of the population. Clearly, these people received high doses of bacteria. Farther away from the farm, the illness rates were lower. Examining the water in each consecutive peripheral zone, the coliform count in zone 1 was 1000 per 100 milliliters, while in zone 4 it was zero. Therefore, the source of the contamination was easy to trace, and the mystery was solved.

Both *E. coli* O157 and *Campylobacter* were found in stool cultures from the sick people. Fortunately, everyone survived. This is a good example of a retrospective epidemiological study. The public health authorities reacted quickly, identified the source of the contamination, and moved to limit the effects. Their division of the town into zones reflected their early belief that a common water source was the likely problem. This example also shows that, in a well-designed and well-engineered system, many things must go wrong at the same time to create an outbreak of illness. In this case, the effluent from their water treatment plant was suboptimal, the farm had a critical pump breakdown and no adequate substitute, the water from the stream was used without regard for potential problems, their plumbing did not have a one-way valve to stop backflow, and the pump they were using was stronger than the water pressure pump from the town's water supply. If any one of these problems had not occurred, the effects would have been either greatly diminished or completely avoided.

Jones and Roworth (1996) included many useful items in their epidemiological analysis, such as age of the cases and controls and dates of illness. Other items could have been included in the analysis, such as sex and race, although in this particular village the racial component is probably not very diverse. An estimate of the dosage (quantity of water consumed) would also have been useful.

FIGURE 6-7 **The city of Fife was divided into these zones to facilitate its epidemiological study of the outbreak. Note the location of the farm that was the source of contamination.**

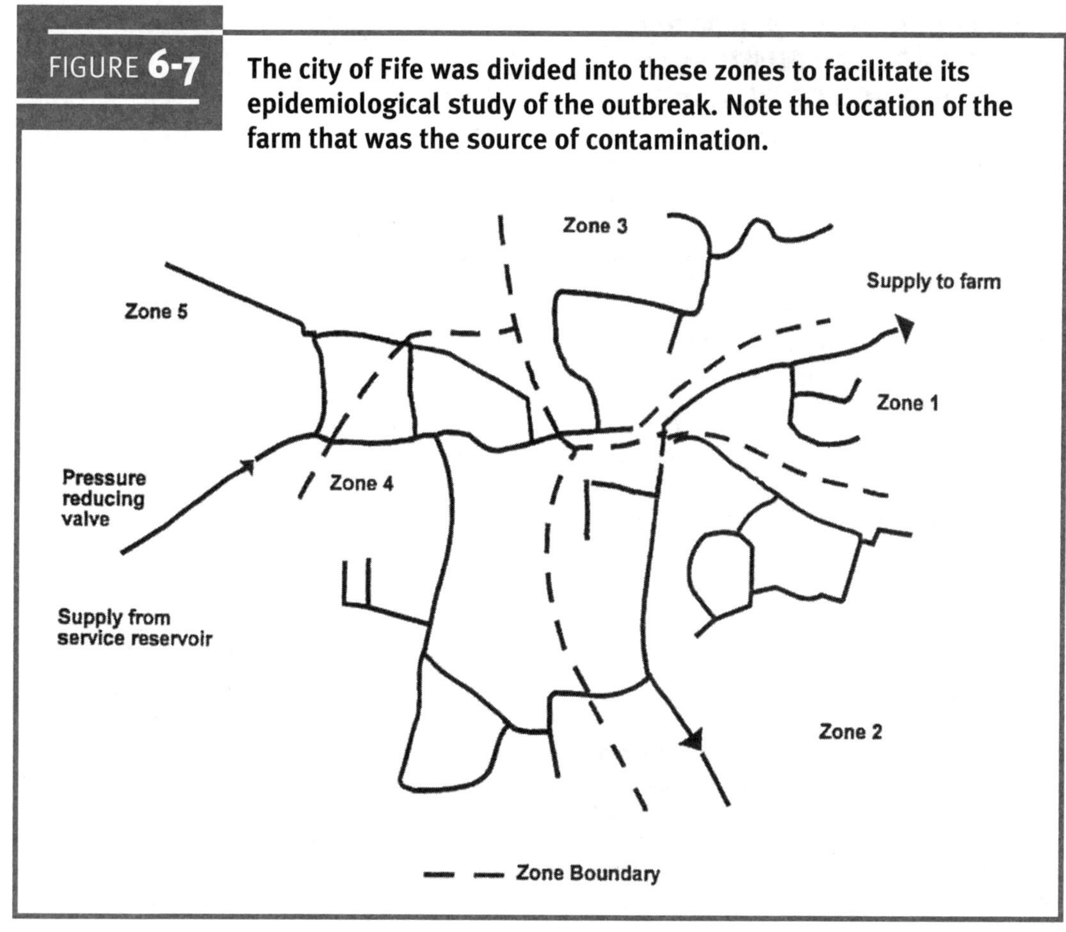

SOURCE: Data from Jones and Roworth, 1996. Used by permission.

CRYPTOSPORIDIUM ACQUIRED IN A WATER PARK, 2007 (JUE ET AL., 2009)

In 2007, an outbreak of cryptosporidiosis from *Cryptosporidium hominis* occurred among patrons of a water park in Idaho. This was a splash park, in which water is sprayed over people from a variety of fountains. Splash parks often also have a conventional pool and water slides near the fountains. The problem with these facilities is that the spray makes it easier to swallow water that is sprayed around the faces of visitors.

After several patrons reported an illness resembling cryptosporidiosis, the local health authorities examined the pool and surrounding areas. Oocysts were found in the splash fountains and in the drinking water fountains. Their epidemiological survey found 50 cases out of 154 patrons, an infection rate of 32%. All had a relatively mild case of diarrhea, considering the pathogen involved.

The relative risk (RR) was 4.7 for exposure to a splashing water feature (with a 95% confidence interval of 1.8–11.9). When both splashing water and exposure to the drinking water were examined, the RR was 8.6 (CI = 3.2–23.3). (See Chapter 16.)

In all likelihood, a fecal release in the pool went unnoticed, and the filtration system was not able to remove the oocysts. Some backflow into the munic-

TABLE 6-1 Analysis of the Outbreak in Fife

Place	Zone	Number exposed	Number ill	Number with abdominal symptoms	Number of cases	Attack rate percentage
Village	1	64	59	57	51	80
	2	323	290	279	261	81
	3	176	147	138	117	66
	4	176	122	103	87	49
	5	112	56	48	48	43
	9[a]	75	60	57	40	53
Subtotal 1–5 + 9[a]		926	734	682	604	65
Workplace		42	31	29	29	69
Total		968	765	711	633	65

[a]Those whose address within the village was not recorded.

Source: Data from Jones and Roworth, 1996. Used by permission.

ipal water system also probably occurred. The water park was closed and thoroughly cleaned, and a more aggressive treatment system was installed. This case demonstrates the high risk of acquiring pathogens in a swimming pool. The filtration systems on most pools, even large municipal pools, are probably inadequate to remove pathogens. The resistance to chlorination that *Cryptosporidium* has shown suggests that the chlorination levels in swimming pools, while much higher than in drinking water, still require a long period (total *Ct* factor) to be effective. In the meantime, everyone in the pool remains at risk.

CHOLERA ON A GULF COAST OIL RIG, 1981 (JOHNSTON ET AL., 1983)

Cholera is a rare disease in the United States, but one worth monitoring. This illness can be devastating, with uncontrollable diarrhea that leads to dehydration and even death. Interestingly, antibiotic treatment is usually not needed; however, the rapid replacement of fluids is critical to effective treatment. Cholera in the United States is largely confined to a crescent of land along the Louisiana–Texas coastline. Typically 2–3 cases occur each year, presumably from drinking contaminated water. This microbe prefers brackish water like that found in the bayou areas, and occasionally the brackish water infiltrates a local well.

Where the first case of cholera (the "index case") was acquired in this outbreak is unknown, but the spread of the pathogen makes a fascinating case study. The index case involved an oil rig worker in Louisiana. Deepwater drilling rigs are huge platforms that sit in areas like the Gulf of Mexico, and workers commit to extended shifts of work (i.e., weeks). In this case, a smaller oil rig was involved. This was a movable oil rig, one that is towed from one place to another in relatively shallow water. These types of rigs drill shallow wells in the bayou and remain for only a short period of time. However, these operations still require a time commitment of weeks from the workers, who live on the rig the entire time.

Each drilling rig has a small freshwater supply, just enough for cooking and drinking. The toilets utilize water from the bayou and discharge it back to the bayou; the facilities are required to have discharge permits for small amounts of wastewater.

The only exception to this procedure is when the rig is under tow, when the intake port for bayou water is closed and the freshwater system is used to run the toilet. During this time, only a single toilet is utilized, and a set of valves are used to shut off the brackish water and to open the freshwater line. The schematic shows the plumbing system on the oil rig (**Figure 6-8**). After the rig arrives at its new destination, the valves are reversed again.

The cause of the outbreak was actually the index case. Suffering with cholera, he would have used the toilet often, excreting a high concentration of *Vibrio cholerae* into the bayou. The brackish water provided the perfect conditions for the microbe to flourish. The water flowed from the sewage output toward the water intake. Some bacteria were probably taken into the oil rig plumbing system in this way. This should not have caused a problem

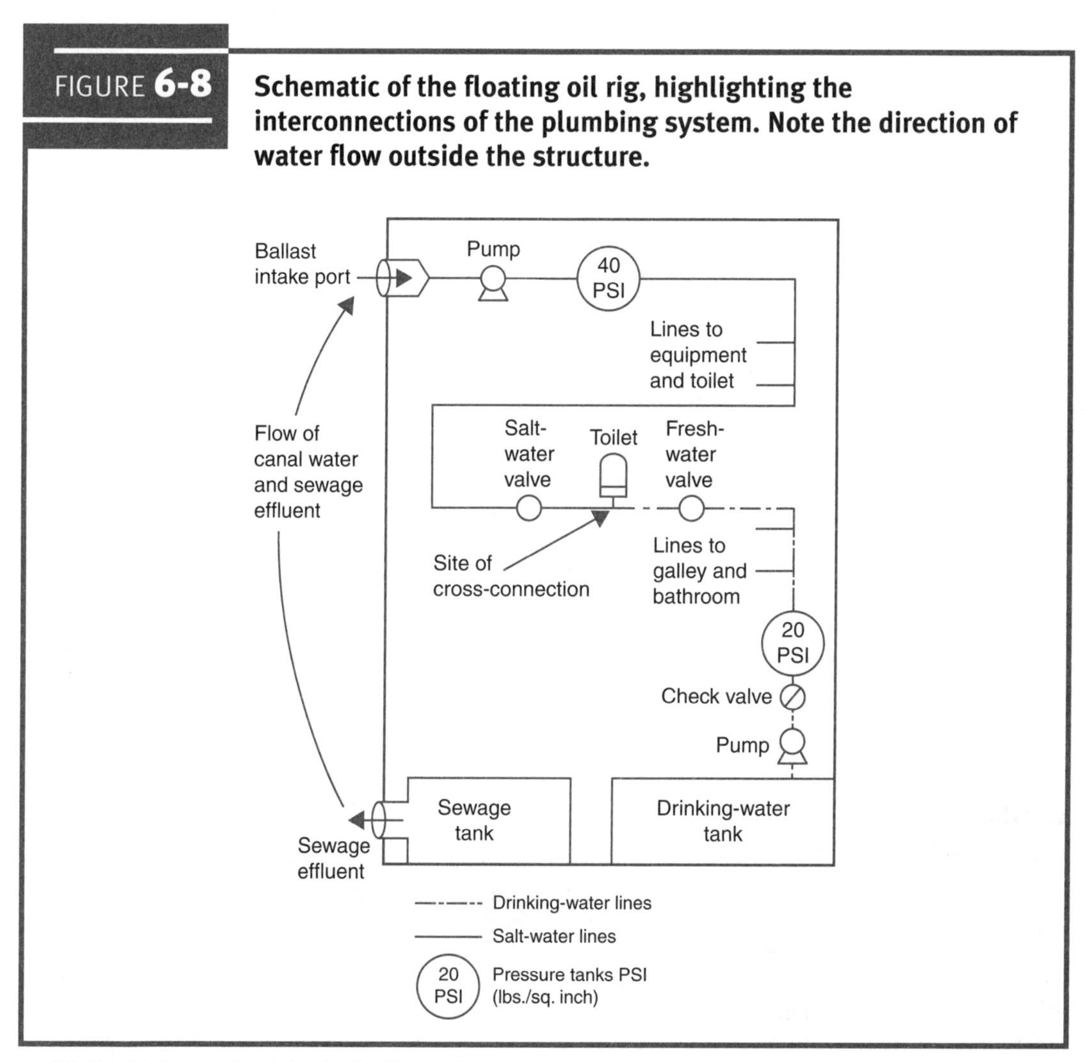

FIGURE **6-8** **Schematic of the floating oil rig, highlighting the interconnections of the plumbing system. Note the direction of water flow outside the structure.**

SOURCE: Data from Johston et al., 1983. Used by permission.

because the intake water was meant for the toilets, but during the rig's last move, the valves on the toilet were both left open. The water pressure in the sanitary line was 40 PSI, while the pressure in the drinking water line was only 20 PSI, so mixing of the sanitary water with the freshwater occurred.

As unappealing as this sounds, the concentration of *Vibrio* that was consumed in this manner was probably far too low to cause illness. This was not the proximal cause of infection. Yet another source of contamination was implicated. Because the work on an oil rig is constant and because schedules are so uncertain, the galley must have food ready at all times so that the workers can grab a quick meal at all times of day. One of the ways the cooks do this is to prepare a large pan of rice and keep it on a steamer tray. Periodically, the cook wets down the surface with water to keep it from drying out. Water contaminated with *Vibrio* was sprayed on the rice. Rice is a good substrate for bacterial growth, and it is likely that the bacteria grew quickly there to infectious dosage levels, and then the workers ate it. Fifteen people were infected with *V. cholerae* in this way.

While this is an odd and interesting case, it is also unlikely to be repeated. As mentioned above, outbreaks in an industrial setting require a confluence of errors. Systems are engineered to prevent contamination. Generally, the older the technology is, the less that goes wrong with it. Old things are not necessarily intrinsically better; older technology has been tested and corrected over time. We learn from the mistakes and their unforeseen consequences. An entire industry is dedicated to trying to determine what could possibly go wrong with a new technology and then engineering it to avoid the problem. Yet some accidents will always happen. The oil rig described here is an example of a contained system. At least, it was mostly contained. The freshwater supply should have been more closely protected because it was the only source. Another example of a contained system is a nuclear submarine, which is designed to cruise under water for months at a time (i.e., the availability of food is the major limitation). Oxygen on the sub is regenerated, water is purified from seawater, and energy is obtained from the nuclear reactor. If something goes wrong with the quality of the food or water, everyone becomes ill, and the submarine will be unable to complete its task. Fortunately, they can easily return to a port.

The International Space Station is another closed system. Its oxygen supply is regenerated, and its wastewater is purified into drinking water. If the astronauts fail to purify the water, they will become ill, and the journey home is long and difficult, taking days. Future attempts to colonize the moon or to explore other planets will require closed systems that are well engineered. A group of colonists on the moon would be entirely dependent on their own water supply, even if food was shipped to them on a regular basis. It is unlikely that they could locate sufficient water on the moon. A trip to Mars would be even more perilous. The journey would take at least a year (using current technology), and the astronauts would not have the option to simply turn around. If humans are ever to leave Earth and colonize other worlds, they will need reliable methods for recycling water in a purified form while avoiding infection from all sources.

EPIDEMIC DIARRHEA AT CRATER LAKE, 1975 (ROSENBERG ET AL., 1977)

In Oregon, Crater Lake National Park is one of the most popular parks due to its natural lake, which was formed in the caldera of a now-dormant volcano. Thousands of people visit the park every year, some of whom camp there. The park has its own water supply—a shallow spring—and its own sewer system.

In June and July 1975, an outbreak of gastroenteritis (diarrhea and vomiting) occurred among visitors and staff at the park. The cause was finally discovered in July when the last of the snow melted and a sewer was found to be overflowing. The sewer line had become obstructed, causing the sewage behind it to emerge through the sewer cover. A fluorescent-dye test was used to trace the path of contamination between the sewer and the groundwater. The fact that the snow had covered it for so long is not surprising: the park averages about 45 feet of snow per year.

The sewage traveled downhill and percolated through the soil and into the groundwater, where it

FIGURE 6-9 **Figure of *E. coli* collected on a filter and growing on selective agar.**

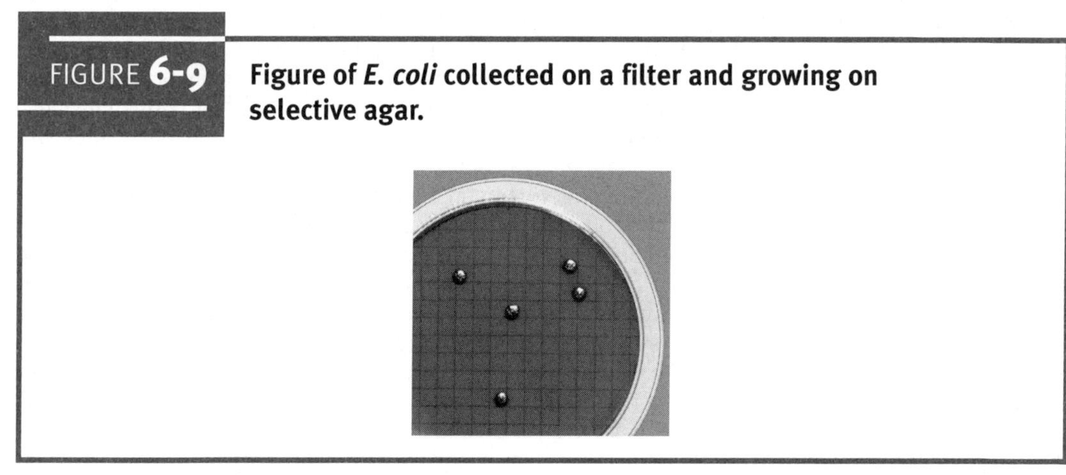

SOURCE: Courtesy of ThermoFisher.

emerged as spring water. In all likelihood, the sewage overflow had continued for some time, making successive waves of visitors ill and thus amplifying the concentration of bacteria. Case estimates were difficult to obtain because so many people had visited the park only briefly, but a good estimate is that more than 2000 people became ill.

The pathogen was an enterotoxigenic *E. coli* strain (ETEC) typed as O6:K15:H169 (**Figure 6-9**). The O antigen is from the cell's lipopolysaccharide, the K antigen is from the capsular material surrounding the cell, and the H antigen is from the flagellum. In contrast to the deadly O157 variety of EHEC, this strain is the well-known "Traveler's Diarrhea" and is only occasionally fatal. In this case, no fatalities were recorded.

When the problem was discovered, the park was closed until the sewer line problem could be fixed and the water supply could be certified as safe. The water supply had its own chlorination system, but the investigation found that it was inadequate at the time of the outbreak. Even small water systems have the potential to be the source for large outbreaks of disease.

QUESTIONS FOR DISCUSSION

1. From the examples given in this chapter, which was the most avoidable? Which one was inevitable? Why?
2. If you were a member of a jury when a victim of one of these outbreaks sued for damages, what sorts of evidence would you like to see?
3. Do you believe that outbreaks of these types are inevitable, or will we gradually eliminate them as we learn more?

References

Atwill, E.R., K.W. Tate, M.D. Pereira, J. Bartolome, and G. Nader. 2006. Efficacy of natural grassland buffers for removal of *Cryptosporidium parvum* in rangeland runoff. J. Food Protect. 69: 177–184.

Brookes, J.D., J. Antenucci, M. Hipsey, M.D. Burch, N.J. Ashbolt, and C. Ferguson. 2004. Fate and transport of pathogens in lakes and reservoirs. Environment Internat. 30: 741–759.

Corso, P.S., M.H. Kramer, K.A. Blair, D.G. Addiss, J.P. Davis, and A.C. Haddix. 2003. Cost of illness in the 1993 waterborne *Cryptosporidium* outbreak, Milwaukee, Wisconsin. Emerg. Infect. Dis. 9: 426–431.

Fayer, R. 1994. Effect of high temperature on infectivity of Cryptosporidium parvum oocysts in water. Appl. Environ. Microbiol. 60: 2732–2735.

Hoxie, N.J., J.P. Davis, J.M. Vergeront, R.D. Nashold, and K.A. Blair. 1997. Cryptosporidiosis-associated mortality following a massive waterborne outbreak in Milwaukee, Wisconsin. Amer. J. Public Health 87: 2032–2035.

Jellison, K.L., H.F. Hemond, and D.B. Schauer. 2002. Sources and species of cryptosporidium oocysts in the Wachusett reservoir watershed. Appl. Environ. Microbiol. 68: 569–575.

Johnston, J.M., D.L. Martin, J. Perdue, L.M. McFarland, C.T. Caraway, E.C. Lippy, and P.A. Blake. 1983. Cholera on a Gulf Coast oil rig. New Engl. J. Med. 309: 523–526.

Jones, I.G., and M. Roworth. 1996. An outbreak of *Escherichia coli* O157 and campylobacteriosis associated with contamination of a drinking water supply. Public Health 110: 277–282.

Jue, R., T. Schmalz, K. Carter, and R.J. Nett. 2009. Outbreak of cryptosporidiosis associated with a splash park—Idaho, 2007. MMWR 58: 615–618.

King, B.J., and P.T. Monis. 2007. Critical processes affecting Cryptosporidium oocyst survival in the environment. Parasitology 134: 309–323.

Mackenzie, W.R., N.J. Hoxie, M.E. Proctor, M.S. Gradus, K.A. Blair, D.E. Peterson, J.J. Kazmierczak, D.G. Addiss, K.R. Fox, J.B. Rose, and J.P. Davis. 1994. A massive outbreak in Milwaukee of *Cryptosporidium* infection transmitted through the public water system. New Engl. J. Med. 331: 161–167.

Morris, R.D., E.N. Naumova, and J.K. Griffiths. 1998. Did Milwaukee experience waterborne cryptosporidiosis before the large documented outbreak in 1993? Epidem. 9: 264–270.

Morris, R.D., E.N. Naumova, R. Levin, and R.L. Munasinghe. 1996. Temporal variation in drinking water turbidity and diagnosed gastroenteritis in Milwaukee. Amer. J. Public Health 86: 237–239.

Robertson, L.J., A.T. Campbell, and H.V. Smith. 1992. Survival of *Cryptosporidium parvum* oocysts under various environmental pressures. Appl. Environ. Microbiol. 58: 3494–3500.

Rosenberg, M.L., J.P. Koplan, I.K. Wachsmuth, J.G. Wells, E.J. Gangarosa, R.L. Guerrant, and D.A. Sack. 1977. Epidemic diarrhea at Crater Lake from enterotoxigenic *Escherichia coli*. An. Intern. Med. 866: 714–718.

Shields, J.M., V.R. Hill, M.J. Arrowood, and M.J. Beach. 2008. Inactivation of *Cryptosporidium parvum* under chlorinated recreational water conditions. J. Water Health 6: 513–520.

Steiner, T.S., N.M. Thielman, and R.L. Guerrant. 1997. Protozoal agents: what are the dangers for the public water supply? Annu. Rev. Med. 48: 329–340.

Vakil, N.B., S.M. Schwartz, B.P. Buggy, C.F. Brummitt, M. Kherellah, D.M. Letzer, I.H. Gilson, and P.G. Jones. 1996. Biliary cryptosporidiosis in HIV-infected people after the waterborne outbreak of cryptosporidiosis in Milwaukee. New England J. Med. 334: 19–23.

CHAPTER 7

Food

LEARNING OBJECTIVES

- Describe the difference between toxic food and contaminated food.
- Define "spoilage," and describe how it occurs.
- Identify the major pathogens associated with food and foodborne illness.
- Identify the different levels of food handling and which are associated with foodborne illness.
- Describe viable but not culturable (VBNC) bacteria, and explain why they are significant.

KEY TERMS

- Consumption
- Food processing
- Production
- Retail step
- Wholesaling

THE DANGERS OF FOOD

Just as everyone needs water, everyone needs food. And just as everyone needs water without pathogens, so also do they need food that is pathogen free. Unfortunately, this is not always possible. Even with modern food-preparation techniques, illness due to contaminated food is still possible. The CDC estimates that about 76 million cases of foodborne illness occur in a typical year. The vast majority of these cases are very mild and may not even result in a missed day of work or school, but about 325,000 hospitalizations and 5000 deaths occur each year because of foodborne illnesses (Mead et al., 1999). Obviously, there is room for improvement. Interestingly, most of these illnesses are never characterized with regard to pathogen(s) involved. While understandable for the vast majority of subclinical cases, even among fatalities the pathogen is not always identified. Clearly, more work is needed to rapidly and accurately identify pathogens. Of those that are known, *Salmonella*, *Listeria*, and *Toxoplasma* are responsible for the majority of deaths due to foodborne illness (Mead et al., 1999). It is very possible that additional pathogens are yet to be identified. The list of foodborne pathogens that are notifiable diseases is shown in **Table 7-1**. These figures appear high, but approximately 1 billion meals are consumed every day (365 billion per year) in the United States (i.e., three meals per day for the more than 300 million people, not to mention snacks). The fraction of

TABLE 7-1 **Notifiable Foodborne Pathogens**

Pathogen	Type	Notes
Bacillus anthracis	Bacteria	Anthrax; found in contaminated meat.
Brucella sp.	Bacteria	Causes brucellosis; found in milk and meat.
Clostridium botulinum	Bacteria	Botulism toxin found in canned found and under anaerobic conditions. Both foodborne and infant.
E. coli O157:H7	Bacteria	Can cause hemolytic uremic syndrome.
Listeria monocytogenes	Bacteria	Causes meningitis in some cases.
Salmonella enterica	Bacteria	Common cause of diarrhea. *S. typhi* causes typhoid fever and has been known to be passed by food handlers.
Shigella sp.	Bacteria	Can cause hemolytic uremic syndrome.
Vibrio cholerae	Bacteria	Causes cholera, a severe diarrhea. Usually found in water.
Hepatitis A virus	Virus	Causes hepatitis; can be found in shellfish.
Cyclospora	Parasite	Several cases associated with food, especially berries.
Trichinella	Parasite	Causes trichinosis.

meals that result in an illness is really quite small (**Figure 7-1**).

A trip to a modern supermarket reveals the amazing range of foods available today. Some are fresh; some are processed; some are canned for long-term storage. Some supermarkets even print maps so shoppers do not wander up and down aisles searching among the literally thousands of different items for the one they want. It has been reported that humans eat from a greater variety of living organisms than any other creature. This includes plants of many sorts, fungi (i.e., mushrooms), mammals, birds, reptiles, amphibians, fish, and a variety of invertebrates (e.g., insects, clams, and mussels).

There is a major difference between the handling of food and water. Water is a commodity that is required by all people that can be distributed through a common system to individual homes. While commercial bottled waters suit individual tastes, this is a minor component of the total water needed per household. Government authorities (water boards) are very good at providing one commodity to everyone equally. They are not able, however, to provide a diversity of products to suit individual tastes. Therefore, they cannot, with any degree of efficiency, supply food as supermarkets do. The government also cannot effectively bring onto the market new foods or food products.

This does not mean that government should play no role in promoting wholesome food. Consider the evolution of food production and distribution. When the United States was a largely agrarian society, people raised their own food. If a farmer butchered a hog and had extra meat, he might trade it to his neighbors for other commodities. Neighbors trusted each other to provide good, fresh meat and other goods in a trustworthy fashion. If they got sick from spoiled meat, they knew the exact source.

As transportation and preservation processes improved, the ability to transport food over long distances became a reality. People started using grocery stores for their greater variety of foods. Some suppliers remained local, and the grocer knew them, but others were unknown. This presented the temptation to cheat occasionally on the quality of the food. After all, if some meat was a little old, it might be sold anyway. How could it be traced?

FIGURE 7-1 **Incidence of the major foodborne pathogens, per 100,000 population.**

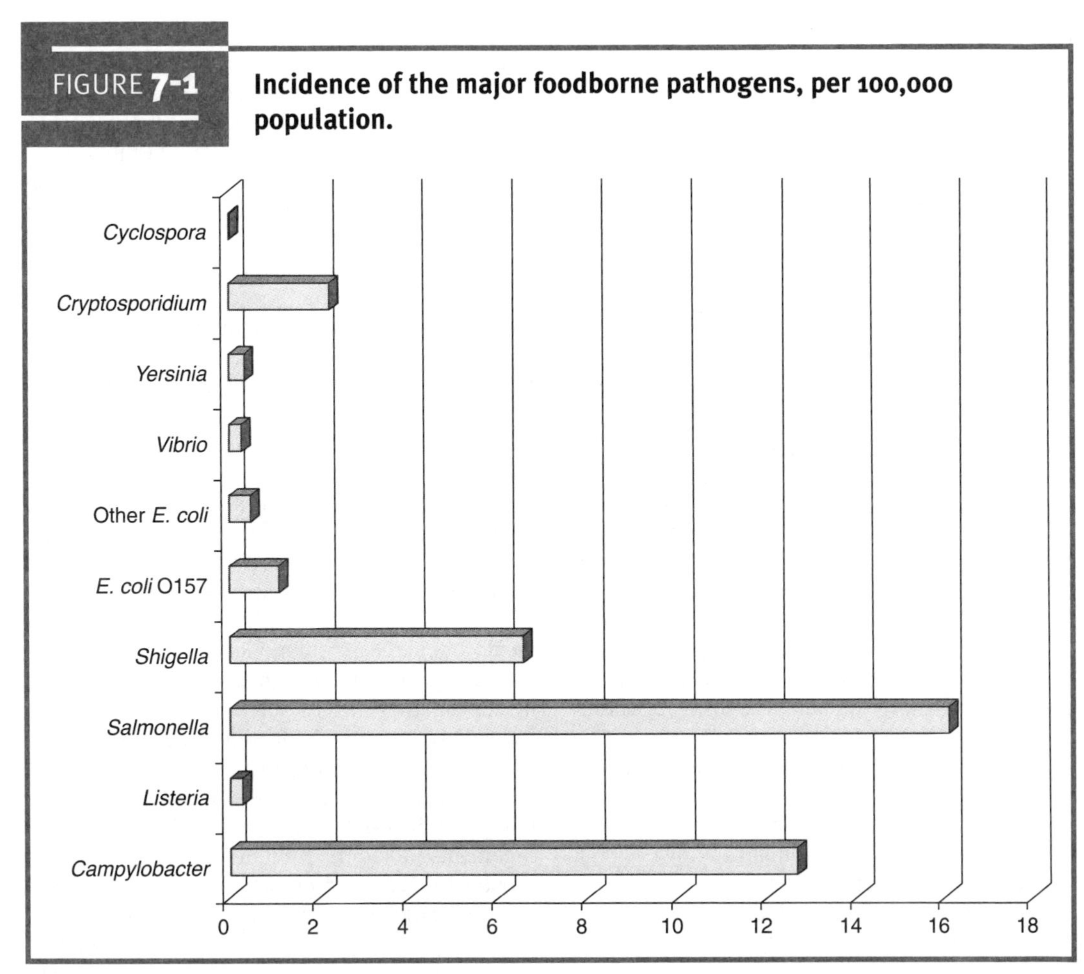

SOURCE: Data obtained from FoodNet for 2008.

After about 1850, these changes in society became evident. Clearly, an outside authority was needed to ensure that food products were unadulterated and safe. National food rules began to be created during this time.

In today's marketplace the range of foods is vast, and foods can originate from any part of the world. There is no practical way for the consumer to know where food came from or who the growers, handlers, shippers, or processors were. Literally millions of people are involved in the production, processing, transport, and sale of food. At any stage in the process a mistake can be made or an unscrupulous person may try to pass off substandard food. Today the role of government consists of overseeing the food industry, not of directing that foods are manufactured. This is a regulatory role. In fact, the body of law regarding food production is vast, and the government retains great authority over the processes that can be used, sometimes with a great deal of precise detail.

Inherent Toxicity

While this chapter is concerned mainly with the adulteration of food with microorganisms, some foods themselves are inherently toxic, or have the potential to be, for a number of reasons. For example, the Japanese pufferfish (also known as fugu, from its scientific name, *Fugu rubripes*) has a powerful toxin in its internal organs, typically the liver and ovaries. In fact, several species of related fish species are referred to as fugu, all of which are poisonous.

The fish contains tetrodotoxin, which has a LD50 (lethal dose in 50% of a population) of somewhere between 300 and 350 µg/kg of body weight. This makes it one of the most toxic substances known. It works by blocking nerve conduction, and the intoxicated individual typically dies from asphyxiation because the diaphragm no longer functions. Fugu's value as a delicacy apparently outweighs the potential health risk; it remains very popular. Even in Japan, where only licensed chefs are allowed to prepare fugu for consumption, two or three dozen poisonings occur every year, resulting in about half a dozen deaths.

Eating such a dangerous food is obviously unnecessary, yet some people actively pursue it as a delicacy. This desire cannot be considered rational outside of the known human propensity for curiosity and for risk taking.

Many people suffer from food allergies, and these allergies can be quite severe. Peanuts and other nuts, shellfish (e.g., clams, oysters, and mussels), wheat, and eggs are all typical foods that can cause allergic reactions. For the people who suffer from allergies, the malady can be quite debilitating. An allergy to peanuts, a common ingredient in many foods, can reduce the food choices of the allergy sufferer severely, requiring careful consideration of every food choice. Even peanut oil, used in some salad dressings and for cooking, can be dangerous.

Allergies are an aspect of the immune system's response to a perceived foreign material, usually a protein. Proteins in the food product are, for some reason, identified as invaders and are attacked by the immune cells. This is unusual because the human body takes in foreign materials every day in the form of food, but the food is processed in the gastrointestinal tract such that the potential allergen is destroyed before it encounters the immune system. Why this is not always the case is still a mystery. In severe cases, this phenomenon leads to an uncontrolled immune response called anaphylactic shock, which can be life-threatening. The incidence of allergies in the general population is increasing, and the reason for this trend is not known.

Ciguatera (briefly discussed in Chapter 4 in the section on harmful algal blooms) is a true microbial illness because it occurs as a result of the production of toxins (e.g., ciguatoxins, scaritoxins) by certain dinoflagellates during certain times of the year. These microorganisms are in turn eaten by reef fish, which are themselves eaten by predator fish. At each level, the toxins bioaccumulate, and if the predator fish are eaten by a human, a substantial dose of toxin is also ingested. In some rare cases, the effects can be lethal. The production of toxins is probably due to environmental conditions, but these are poorly defined, and much additional research is needed.

Another fish-related illness that can be traced to microorganisms is scombroid poisoning. The scombroid fishes include several large game fish, such as tuna, mackerel (also called scomber, hence the name of this class of fish), swordfish, and mahi-mahi (technically not in the Scrombroidea family). All of these fish have in common a high level of the amino acid histidine in their tissues. These substances help the fish fight off microbial infection. After the death of the fish, the microorganisms start to decompose. If the fish is kept on ice, the rate of decomposition is low, but if the fish tissue warms slightly, the microorganisms utilize the proteins in the fish tissue to propagate. Certain microbial species, such as *Proteus* and *Photobacterium*, have the enzyme histidine decarboxylase, which converts histidine to histamine (**Figure 7-2**). Histamine, of course, is the biochemical associated with the degranulation of the mast cells of the immune system, which are involved in allergic reactions. Consumption of large amounts of histamine will mimic one of these allergic episodes, causing headache, nausea, vomiting, a skin rash or hives, and tachycardia. In asthmatic people, pronounced wheezing may occur.

True Contamination of Food

Although specific types of contamination of food by microorganisms are of primary concern, a distinction must be made between food that has been contaminated and food that has become contaminated solely by water. Water is used extensively in food production both for rinsing food and for cooking. However, waterborne sources of food contam-

FIGURE 7-2 Conversion of the amino acid histadine to the immune activator histamine.

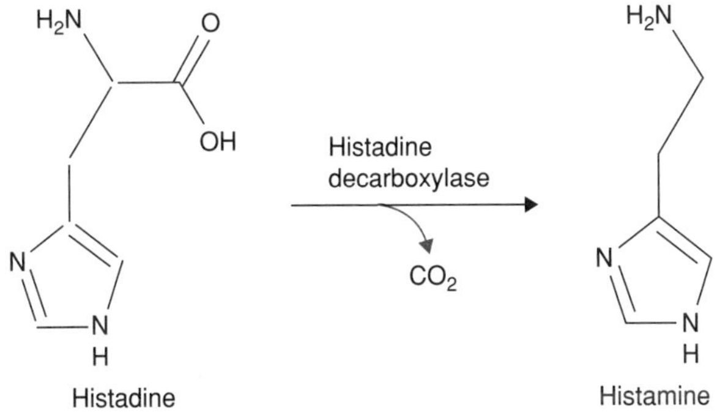

ination are principally a problem of water contamination and are only secondarily a food problem. Therefore, such problems are best treated as water problems (see Chapter 4).

Food has certain disadvantages in terms of pathogen contamination. Many of our common foods come from warm-blooded animals (i.e., mammals and birds), and humans are warm-blooded. Therefore, the pathogens on these animals may be identical to the pathogens that can infect humans. If allowed to grow, the pathogens can increase from low levels to infectious-dose levels in a relatively short period of time. Food then becomes a good growth medium for bacteria (although not for viruses). Other food sources bioaccumulate pathogens, such as the filter feeders (e.g., oysters and clams). Of course, many of our other foods are plants that do not normally have any human pathogens on them. If pathogens are present, however, it is likely that they have been deposited there in contaminated water from fertilizer application or rinsing during processing. Of course, all types of food are susceptible to rotting, and this must be combated as well.

Our food supply does have a few advantages, however. As long as the food animal is alive, the protective barriers of the animal (skin) and its immune system work in our favor, keeping any pathogens from invading and multiplying. The best option is thus to keep the animal alive as long as possible and to kill it and process the carcass as fast as possible.

In considering food contamination, the most likely contaminants are bacteria. Of these, some bacteria produce toxins and therefore cause intoxication. Toxins can be quite potent, as seen in the botulism toxin or the staphylococcal toxin. Of course, toxins do not multiply in the body as microorganisms do, so the dose is more comparable to a toxic dose of a chemical. Other bacterial species enter the body through food and then cause an infection, typically through the gastrointestinal tract. These are infectious agents, and in some cases the illness can spread throughout the body. Pathogens such as *Salmonella*, *Campylobacter*, and the pathogenic *E. coli* strains fall into this category.

Certain viruses are associated with food, but often the host range of viruses is so restricted that food is an unlikely source of distribution. Some viruses are known, such as Newcastle virus (which can pass from contaminated meat to the eye after splashes) and Norovirus (formerly Norwalk virus or Norwalk agent), which is usually associated with contaminated water but appears to be transmitted through food as well.

Fungi are usually not a problem as infectious agents in food, but they cause a serious problem in terms of food spoilage. Parasites (e.g., trichina worms and tapeworms) are a major issue in many parts of the world because they are spread in contaminated meat. As discussed in Chapter 8, they are not a major concern in the United States, and control is not overly difficult.

How Does Food Become Contaminated?

Four levels of food handling must be considered. **Production** is the first level: plants are grown and harvested, and animals are raised for eggs, milk, or meat. A number of potential problems can occur. Animals must be cared for properly, and sick animals must be effectively treated or culled from the herd to protect the rest of the animals. Crops are fertilized, and if a natural fertilizer is used, care must be taken to ensure that any pathogens do not remain on the produce as it is harvested.

After production, the next step is **wholesaling** or **food processing**. Companies receive food from many sources and prepare it for retail sale or process it into different types of food. For example, the production of a canned stew or soup requires many different foods, a cooking step, and a canning step. At any of these steps, contamination may be introduced, but in practice the system is tightly controlled, and few mistakes are made. If a mistake does occur, the potential for illness is widespread because the wholesalers distribute products to many retailers.

The **retail step** involves the grocery store or supermarket or a restaurant where food is prepared for consumption. The retailer is responsible for keeping the foods under appropriate conditions so that they do not spoil or, if they do go bad, for removing them from sale. Many foods are packaged, and there is little that can go wrong at the retail sites. However, many supermarkets have their own butcher shops, where the potential remains for contamination during handling of the meat.

Restaurants are of increasing concern in terms of food safety because eating meals out has become so prevalent. An estimated 43% of household food budgets go to restaurants (Collins, 1997). This does not mean that 43% of meals are eaten in restaurants; these meals are considerably more expensive than dining at home. This figure equals about four meals out per week, a believable estimate when the vast number of fast food meals consumed is considered.

Consumption is the final step. This includes preparation of raw foodstuffs, and at this level, many things can go wrong. In fact, the vast majority (perhaps 50%) of all food contamination problems occur at consumption. In many millions of households in the United States, food is prepared for meals. Different people have different levels of skill at food preparation, storage, and handling. The mitigating factor is that households typically serve only a handful of people, so the effects of mistakes are generally limited to the number of people in the household.

Overall, a downward trend in foodborne illness is occurring, and this can be attributed to federal oversight of food handlers and to an effective campaign to handle food appropriately in the home. This is a continually changing picture, as new concepts in food preparation and packaging are constantly changing. However, the trend toward dining out is a factor that must be considered. It should be remembered that "dining out" includes not only regular restaurants, but also "fast food" restaurants and even snack-style establishments like ice cream parlors and soda machines. There have been a number of reported disease outbreaks that have been traced to unhygienic workers at restaurants or to poorly maintained facilities (including the presence of vermin). Because the meals are prepared by strangers, it is difficult to guarantee that the conditions are always acceptable. For oversight, the effectiveness of local restaurant inspectors is vital. Other significant social trends include movement of populations, either through tourism (bringing exotic diseases to new locations) or through immigration (bringing people and their food habits from other areas). Sometimes these food preferences are inherently dangerous, such as a preference for cheese made from unpasteurized milk.

PRODUCE

Fruits, vegetables, and nuts are generally safe to eat. While plants do have their own pathogens, those pathogens do not harm humans. Spoilage can be a problem, although these cases may or may not involve true pathogens. Ordinarily, spoilage leads to gastrointestinal distress that is not life-threatening. A diet of plant material can be quite healthy as long as sufficient attention is paid to a balance of nutrients: vitamins, minerals, carbohydrates, proteins, and fats and oils. Nuts are an excellent source of protein, and vegetable oils can supply a person's needs. The vegetarian diet has distinct advantages.

Fruits and vegetables grown for personal use are highly unlikely to make consumers ill as long as they are prepared using common-sense principles: thorough washing before eating and standard caution in canning and preserving. After all, soil contains spores of *Bacillus* species that can cause food poisoning through toxin production and that also cause spoilage. Also, spores of the obligate anaerobes *Clostridium perfringens* and *C. botulinum* are often present in soil, and improper canning can result in botulism poisoning if the later microorganism is present. Every year a handful of botulism cases are attributed to home canning problems, usually due to insufficient washing of the food to remove soil particles. The fact that soil and vegetation contain *Listeria monocytogenes* has been well established. However, these pathogens are usually found in low concentrations and only become a problem if the produce is stored improperly and for an extended period of time.

Supermarket produce has different problems. Once spoilage is noticed in a supermarket, the produce is discarded, and "sell by" dates are usually followed. However, contaminations of produce on large, commercial farms have resulted in sickness in consumers. Contamination at the point of food production is especially dangerous because the food travels to any number of processors, which in turn send them on to wholesalers and retailers. The affected population can be quite large.

Contaminated produce is usually not caused by typical soil microorganisms. It is usually due to fecal contamination, which may originate from a natural fertilizer (manure) or from contaminated water (i.e., from a sewage source). Other potential points of contamination include the workers, who might not have good personal hygiene or access to reasonable hand-washing facilities on the job. In these cases, the contaminants are typical pathogens from the fecal–oral route of transmission: pathogenic *E. coli* strains (including O157), *Salmonella*, *Campylobacter jejuni*, parasites (e.g., *Giardia*, *Cryptosporidium*, and *Cyclospora*), and a number of viruses (e.g., Norovirus, Hepatitis A virus, Rotavirus) (Beauchat and Ryu, 1997). In other countries, the range of pathogens may reflect local health problems or the unreliability of the water supply.

When artificial fertilizers are used, there is less risk of fecal contamination. Natural fertilizers are sometimes used because the supply is readily available. Manure naturally contains high concentrations of microorganisms, and if the animal is sick, high concentrations of pathogens are shed in the feces. This does not need to be a problem, however. Manure is not applied as fresh fecal material. When it is composted before distribution, the odor is reduced, as are the number of pathogens (through cell death and predation by other species). Also, the manure should be applied to a farm field only at certain times during the growing season. If it is applied soon after planting, there is plenty of time before harvest for the bacteria to die (although cysts and spores may take far longer). If the manure is applied within a month of harvest, there is far greater likelihood of fecal contamination of the produce.

Farms in the United States can be monitored on a regular basis. Their water supplies are usually identified clearly and tested for contamination. If the farm field is close to a cattle grazing area, dust containing the pathogens might blow over the produce and contaminate it. Subsequent irrigation of the produce might allow the microorganisms to infiltrate the tissues of the plant material, where they will be very difficult to remove. Wild birds defecate wherever they land, and local wild animals (e.g., insects, mammals, and reptiles) may do the same when they wander onto a farm field. Any of these animals may deposit pathogens on the future food supply, and they may have acquired pathogens from feeding on contaminated land areas, such as waste dumps. Greenhouses probably offer the best protection for food growth, although all manner of rodents may still manage to invade. In short, it is not practical to isolate the food supply from outside influences.

Modern transportation and refrigeration means that worldwide production of food is possible, providing fresh produce year-round, which creates a better diet for everyone. It also allows different regions to specialize in local food production, which benefits the economy and results in lower food costs overall. However, food produced in other countries may not meet the high standards for hygiene that we expect. This is particularly true when considering the water source. Water easily contaminated with fecal waste might be used on farm fields or for washing harvested produce. Many documented disease outbreaks have been traced to this type of error.

Water purity is a major requirement for safe food production. When chlorinated water is used for washing produce, soil particles and many of the microorganisms on the surface are eliminated. But this process is not absolute; any microbes that become lodged in the interior of the produce are extremely difficult to remove.

Microorganisms remain alive on produce and foods of all kinds much longer than expected. This includes both the produce and the plants from which they are harvested; opportunities for contamination are significant. It is unwise to use a contaminated water supply on a farm field with the expectation that, by the time the produce is harvested, the microorganisms will have died. This may not be true at all.

Limiting factors relate to the physiology of the microbes themselves and to the environmental factors that determine the degree of physical insult to the cells. Spore-forming bacteria and cyst-forming parasites may remain viable for months. Even vegetative cells may last from weeks to months as long as they do not desiccate or are not directly exposed to the ultraviolet radiation of sunlight. Under these circumstances, the bacteria do not actively grow, but instead they go into a prolonged quiescent state in which there is little physiological activity. Monitoring of plant surfaces under these conditions may not reveal the presence of bacteria. These bacteria are said to be "viable but not culturable," or VBNC, because they are relatively fragile and might not readily grow on artificial media; they are not the same as dead cells. In fact, under the right conditions, VBNC bacteria can begin to grow again (e.g., when the produce is processed and different ingredients have been added). This problem can be considerable because high concentrations of VBNC might not be detected by standard monitoring methods.

Microorganisms always remain on the surface of produce, whether pathogens or spoilage microorganisms. In 2009, an outbreak of *Salmonella* was traced to alfalfa sprouts (Safranek et al., 2009). Surprisingly, the source of the contamination appeared to be the irrigation water of a seed company in Wisconsin. The contaminated seeds were sold to a grower, who germinated the seeds into sprouts, which were then sold for consumption in a multistate area. The bacteria survived throughout the process and caused a substantial outbreak of disease.

After harvest is complete, produce must be quickly brought to market before it overripens. Refrigerated trucks and rail cars are used to both retard bacterial growth and to limit ripening, which extends the shelf life of the produce. Alternatively, the food may be processed (cooked and canned) on site before shipping. In all scenarios, food-processing equipment must be cleaned on a regular basis because an accumulation of food material on the machines is unavoidable, creating an environment in which microorganisms can certainly grow.

Spoilage

Spoilage refers to the loss of food because its quality has been degraded. This may come about by overripening, by desiccation, by infestation of insects, or by consumption by animals. It can also be caused, of course, by the growth of microorganisms on the food that produces disagreeable odors, flavors, and toxins by those microorganisms. For example, fungal growth on strawberries by *Phytophthora cactorum* produces a disagreeable odor that has been compared to watercolor paint (Jelen et al., 2005). Many similar instances where microbial infection led to foul odors and tastes could be cited. Spoilage is not usually associated with pathogens, but is still very important because it decreases the general food supply.

PATHOGENS

Many pathogens can be found on foods, but those listed in the following sections are some of the most common. Some are also found among the waterborne pathogens, but here they are explained in terms of foods that are commonly contaminated. The pathogens described here can be found in a variety of foods (see Chapter 8 on meats and Chapter 9 on dairy products). The microbiological analyses of the many different foods and food processes comprise a vast area of research methods. For the latest methods, see Downes and Ito (2001). A typical worksheet for the identification of enteric pathogens is shown in Appendix II.

Clostridium botulinum

Clostridium botulinum is the causative agent of botulism, which is an intoxication with the botulinum toxin, one of the most powerful toxins known. *Clostridium* species are obligate anaerobes, but they are also spore formers. The spores persist for extended periods of time and may infiltrate food products. Canning used to be a major concern for botulism, especially if soil was not completely removed from the food before canning. However, current industrial canning procedures are more than sufficient to kill all spores. An average of 145 cases of botulism occur every year in the United States. Some of these are cases of wound botulism, in which *Clostridium* spores enter a wound and germinate under anaerobic conditions. But about 21 cases are foodborne botulism, mainly from mistakes in home canning operations. The rest, about 90–100 cases, occur in infants.

Importantly, botulism toxin is the danger, not the *C. botulinum* microbe. Cases of infant botulism have been reported in which the baby was fed food with *Clostridium* spores (e.g., honey has been mentioned often). The spores germinate in the baby's gastrointestinal tract and multiply there because the baby does not yet have fully formed intestinal flora. The *Clostridium* cells produce the toxin, which spreads systemically and can kill the affected child. In an adult, normal flora prevent the spores from establishing residency in the intestine, and the spores pass out through the feces.

The toxin interferes with normal nerve transmissions, the muscles no longer contract properly, and a flaccid paralysis sets in. If enough toxin is present, death results from failure of the muscles of the diaphragm, which leads to asphyxiation. In infants, the symptoms are related to the intestinal location of the infection. The infant may become constipated, which is very unusual in young children, who usually have relatively loose stools. However, the toxin causes peristalsis to cease, which then causes fecal matter to collect and harden in the bowel. Another sign in an infant is a weak sucking action on the mother's breast, which is caused by the toxin's effect on the muscles of the mouth.

Adults may also succumb to botulism poisoning, and it takes surprisingly little toxin to be lethal. Death may occur in a few hours or may take several days. Signs of intoxication in adults vary, but in all cases they can be described as interruptions of nerve impulses that are observed as lack of muscle response. A history of eating home-canned produce or suspicious meat (sometimes from hunting) has been associated with botulism.

A very rare case of botulism in commercially canned food occurred in 2007 (Ginsberg et al., 2007). Cans of hot dog chili sauce were not treated properly on the production line, and they contained *Clostridium* spores. The resulting botulinum toxin sickened a number of people in several states before the product recall was announced. Prior to this case, the most recent episode of commercial canning producing botulism occurred in 1974.

Clostridium perfringens

C. perfringens is the causative agent of gas gangrene. It is a spore former and thus has the capacity to persist for long periods of time in soil, dust, vegetation, and so on. Ordinary cooking might not be sufficient to kill all spores. Therefore, the potential exists for germination of spores in cooked food, and then multiplication of the bacteria if the food is warm for long periods of time. Fortunately, the ungerminated spores do not easily invade a healthy gastrointestinal tract, and proper canning procedures are more than sufficient to eliminate these spores. Mild infections are not uncommon; they result in diarrhea. The symptoms typically subside in a few days.

Shigella

Four species of *Shigella* are known pathogens: *S. dysenteriae*, *S. flexneri*, *S. boydii*, and *S. sonnei*. This microorganism is also passed primarily through contaminated water, but it may be found in food. *Shigella* produces diarrhea that is often mixed with blood, and this fact frequently encourages people to seek medical help. However, most cases resolve without complication, frequently with only supportive therapy.

Shigella can be found in the human gastrointestinal tract, so poor personal hygiene by food handlers must be suspected during an outbreak. This microbe is primarily considered to be passed through the fecal–oral route, and water transmission is the major source. However, in many countries with poor sanitation, it is transported via flies that land in untreated fecal waste and then land on food. The infectious dose has been estimated at 10–100 organisms, making a simple transfer by a fly sufficient to make a person ill. The damaging elements of *Shigella* are the powerful Shiga toxins. These toxins are also found in toxigenic *E. coli* strains. It causes severe dysentery that includes blood and mucous. Symptoms usually resolve but may take 2 weeks. Like *E. coli*, *Shigella* can also cause hemolytic uremic syndrome (HUS). Most of the fatalities that occur due to this microbe are in children.

Staphylococcus aureus

Contamination with this microbe occurs easily because it is present on the hides of many animals. *Staphylococcus aureus* (**Figure 7-3**) is a true

FIGURE **7-3** ***Staphylococcus aureus* on a blood agar plate.**

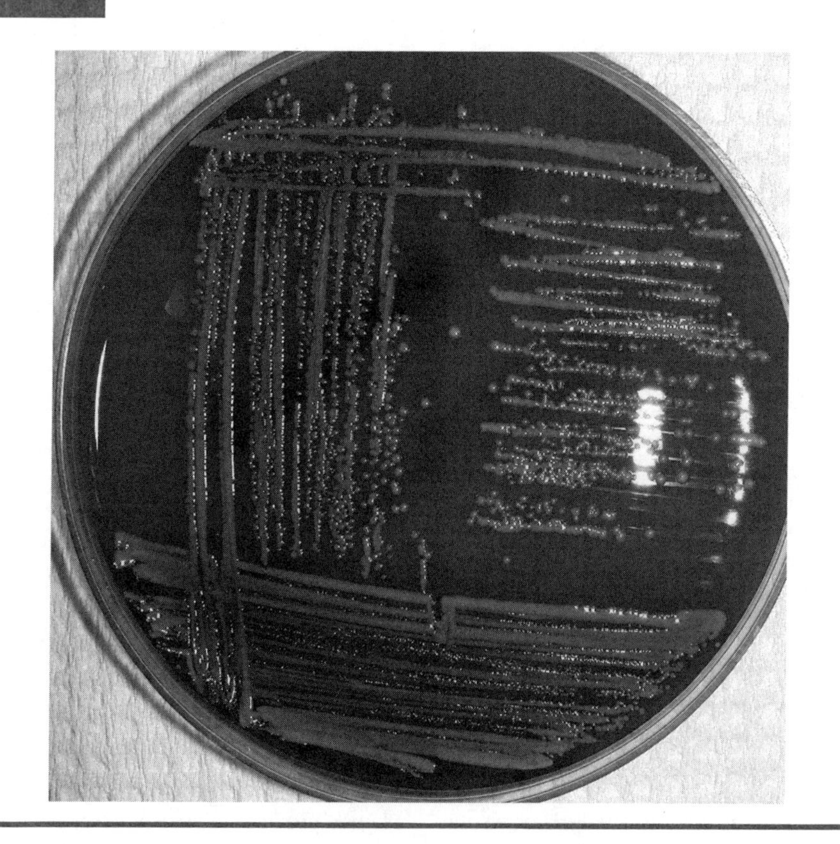

pathogen and causes a variety of superficial infections (e.g., pimples, boils, and sores) that require care but are not sufficient to cause a worker to call in sick or for an animal to be excluded from slaughter. As a foodborne pathogen, *S. aureus* is most well known for its role in staphylococcal food poisoning, an intoxication. Staphylococcal species inhabit human skin, and workers easily pass the bacteria from nasopharynx or hands to food. The skin usually harbors *S. epidermidis* (**Figure 7-4**), although the much more dangerous *S. aureus* can colonize the skin transiently.

Staphylococcus aureus makes an enterotoxin that is heat stable and thus withstands cooking temperatures. If ingested, it produces a violent reaction, usually within 1 to 6 hours. The symptoms of this infection typically include severe diarrhea and vomiting that may result in dehydration, especially in younger children. *Staphylococcus* has a growth advantage on salted foods (e.g., salt pork) because of its high tolerance for saline environments. A cow with mastitis sheds *Staphylococcus* in its milk. If the food has been cooked, then the bacteria are dead and colonies cannot be grown on agar media. A test for the toxin can be performed, but this is rarely done. The course of the illness does not usually exceed 24 hours (i.e., the violence of the vomiting and diarrhea subside as the toxin

FIGURE 7-4 ***Staphylococcus epidermidis* on a blood agar plate. Note the similarity to *S. aureus*. Specialized agar plates and the coagulase test are used to distinguish these species.**

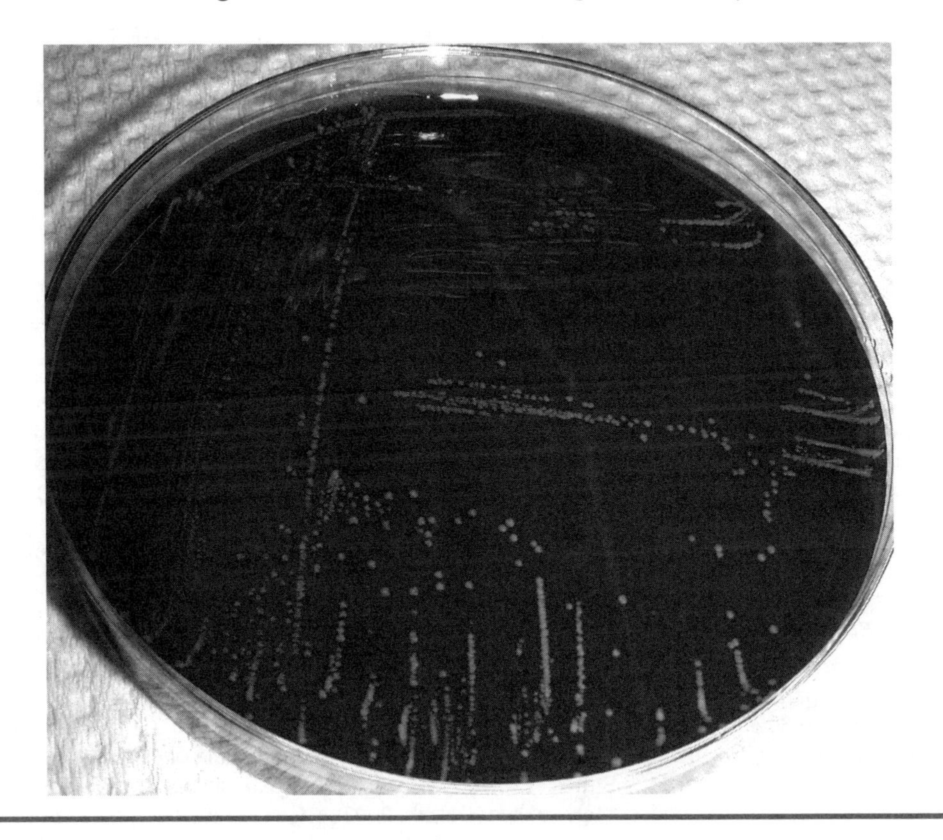

clears the gastrointestinal tract). The best treatment is supportive, which usually means fluids; intravenous fluids are given if severe vomiting prevents ingestion.

Bacillus cereus

Bacillus species are aerobic, Gram-positive rods that are ubiquitous in the environment. A stray piece of soil might easily contaminate a food product. These bacteria are spore formers and therefore can survive cooking.

The illness is usually either nausea and vomiting, or diarrhea. The microbe has two enterotoxins: one is heat stable and induces vomiting, while the other is heat labile and causes diarrhea. In either case, the illness rarely lasts more than 24 hours because the body expels the toxin quickly.

Hepatitis A Virus (HAV)

HAV is well known for its distribution in the water supply, and it can be transferred to food when contaminated water is used to wash the food. HAV is also passed through shellfish, which, as filter feeders, can accumulate the virus in their tissues. Cooking kills the virus, but consumption of raw shellfish puts the consumer at risk. One of the biggest outbreak risks is handling of food by restaurant workers who have the illness, especially an infected worker who has poor personal hygiene and contaminates the food just before it is consumed.

The bad news is that this virus causes an illness that can last from 2 weeks to 3 months. Symptoms are typical for hepatitis: fever, malaise, abdominal discomfort, and nausea. The good news is that the illness is usually self-limiting, and most people recover without complication. Fatality is rare, probably around 0.1%. Exposure to HAV by the general population is surprisingly high, with some reports that 33% of the population is seropositive for HAV. A vaccine for HAV is now in use and appears to be very effective.

Ascaris lumbricoides

Ascaris lumbricoides is the largest nematode (roundworm) capable of parasitizing the human intestine. Adults can grow to 30 cm in length. These infections are common in tropical and subtropical countries, but they are rare in the United States. They can be acquired by consuming inadequately cooked pork products or by bringing hands into contact with the mouth after butchering an infected hog. However, the more common pathway of infection is through soil that has been contaminated with feces. If a person brings contaminated hands to the mouth, the *Ascaris* ova may be swallowed. These ova are quite environmentally resistant. The best defense against these parasites is the use of modern sanitary practices for fecal material.

Trichuris trichiura

Trichuris trichiura is another intestinal parasite, also known as the whipworm. Much shorter than the roundworm, with a maximum length of 50 millimeters, it has much in common with the roundworm, such as environmentally resistant ova, passage through contaminated soil, and defense through sanitary handling of human waste. While the roundworm causes a gradual wasting disease, the whipworm causes colitis.

SELECTED CASE STUDIES

Cyclospora in the Wedding Cake

Two reports of a wedding made more memorable by the presence of the parasite *Cyclospora* in the food have been published (Ho et al., 2002; Murrow et al., 2002). Typically, wedding guests come from many places, congregate in one place for a time, and then return home. The incubation period for *Cyclospora* is about 7–10 days, so the symptoms begin long after the wedding. One case (Murrow et al., 2002) occurred in Georgia at a brunch for the bridesmaids. Among the foods that were served were fresh strawberries, blackberries, and raspberries. While investigating the outbreak, the berries came under scrutiny because *Cyclospora* has previously been associated with berries.

Of course, the food had long been discarded prior to the investigation, and so no direct analyses were possible. Epidemiological analyses of samples from the people at the brunch were generally unhelpful because most guests had eaten a wide variety of foods and most became ill. Tracing the origin of the berries only determined that the berries were from farms in either Guatemala or Chile.

A similar outbreak occurred in Philadelphia (Ho et al., 2002). Guests at a wedding became ill with long-lasting diarrhea, and the stool samples were positive for *Cyclospora*. Investigators again suspected berries as the vehicle for transmission; the wedding cake had been made with a butter cream filling and fresh raspberries.

In this case, the investigators were fortunate because the bride and groom had saved a portion of the cake in their freezer, with the intention of consuming it on their first anniversary. Of course, cake rarely lasts this long and is more safely consumed after returning from the honeymoon, but the important fact here is that the evidence was still available. The investigators acquired this cake and extracted DNA from the raspberry filling. PCR analysis demonstrated the presence of *Cyclospora*.

Tracing the origin of the berries in this case was not entirely successful until the investigators in Philadelphia conferred with those in Georgia. The only common source was a farm in Guatemala. However, an inspection of this farm by the FDA revealed no obvious source of contamination. *Cyclospora* can be difficult to isolate, and this type of contamination can be transient.

VIRUSES IN IMPORTED CLAMS

Shellfish are filter feeders that pass seawater over a specialized structure that captures particles of food. This tends to bioconcentrate certain undesirable substances, among them the viruses. In areas where untreated wastewater has come into contact with shellfish beds, the shellfish are no longer safe to eat. Thorough cooking will destroy the pathogens, but far too many people prefer to eat their shellfish raw or undercooked.

In 2000, an outbreak of both HAV and Norovirus occurred in New York after some clams were consumed raw. These clams came from China, where they were supposed to have been cooked and then canned, but apparently the cooking step had been omitted. China has an endemic HAV problem, and it would not be unusual for their clams to be contaminated.

As has been noted, when pathogen outbreaks occur, several things have gone awry at once. In this case, the clams should have been cooked but were not. The clams were served at a restaurant, and the chef there should have recognized that these clams were uncooked; the diners should also have noted that their food was raw, but apparently no one complained. These clams were also contaminated with Norovirus, and a coliform bacteria count was extraordinarily high: 93,000 per 100 grams of clam meat. Clearly, these clams came from a heavily polluted body of water. RT-PCR demonstrated the additional presence of HAV.

SICK AND IN JAIL

An outbreak of illness in a jail can be easily traced. After all, exactly where all the sick people have been and exactly what they have been eating are known. An outbreak of gastroenteritis in 2008 in a jail started with typical gastrointestinal complaints, starting just 8 hours after the previous evening's meal. The problem was quickly traced to a casserole that had been made with beef and turkey (Hsieh et al., 2009). The pathogen in this case was *Clostridium perfringens*; the enterotoxin from this microbe was found in stool samples. The casserole itself was highly contaminated, with 43,000 CFU per gram appearing on agar plates.

The more important question for this case study was how the microbe was transferred to the jail in the first place. Although the ultimate source of contamination remains unknown, the problem was exacerbated by handling the casserole items as leftovers. At improper storage temperatures, the bacteria may have multiplied significantly. Because *Clostridium* is a spore-forming microbe, initial cooking may not have been sufficient to eliminate the spores, and saving the leftovers allowed the spores to germinate.

O157 IN A DEER

Hunting is a popular pastime, and hunters often consume the game they kill. This meat has not been inspected by the USDA, of course, but this is not usually an issue. After all, humans have been hunting and eating their prey since prehistoric times. In one instance, however, the presence of a pathogen nearly proved tragic for the hunter's family.

In this case, a hunter in Connecticut shot a deer using a rifle (Rabatsky-Ehr et al., 2002). It is best to hit the deer in the vital organs (especially the heart) or in the head so that the deer dies quickly. In this case, the deer was shot in the gut, rupturing the intestine, and the deer ran off. The hunter tracked the wounded animal for 2 hours, finally finding it dead. He field-dressed the carcass (i.e., split open the belly and removed the organs) but did not wash the interior in the field (washing is preferred). Unfortunately, the intestinal flora had contaminated the meat, and this contamination included *E. coli* O157, a particularly deadly pathogen.

Once at home, the hunter split the meat into smaller portions and froze it. Venison that had been ground into hamburger patties was soon used in a family cookout. The hunter's young son had one of these hamburgers, grilled rare. The *E. coli* in the interior of the patty was not killed by this level of cooking, and the child soon developed bloody diarrhea. The O157 serotype of *E. coli* is known to cause HUS in young children, an illness that can easily be fatal. Fortunately, the child recovered completely. This case demonstrates that diseases associated with domesticated animals can also be found in wild animals and that each hunter takes on a great responsibility in killing, cleaning, and cooking a harvested animal appropriately.

QUESTIONS FOR DISCUSSION

1. Why do you think some pathogens are identified with a particular food, the way *Salmonella* is associated with chicken?
2. Why is *E. coli* O157 associated with ground beef instead of regular cuts of beef?
3. How might food become contaminated in the home? What are some basic safety measures that everyone should follow?
4. Why is contamination at the production stage so dangerous?
5. In the case study with the imported clams, HAV was found using RT-PCR. Why was this method used?

References

Beauchat, L.R., and J. Ryu. 1997. Produce handling and processing practices. Emerging Inf. Dis. 3: 459–465.

Collins, J.E. 1997. Impact of changing consumer lifestyles on the emergence/reemergence of foodborne pathogens. Emerging Infect. Dis. 3: 471–479.

Downes, F.P., and K. Ito (eds.). Compendium of methods for the microbiological examination of foods, 4th ed. Washington, DC: American Public Health Association.

Ginsberg, M.M., L. Granzow, R.F. Teclaw, L.K. Gaul, S. Bagdure et al. 2007. Botulism associated with commercially canned chili sauce—Texas and Indiana, July 2007. Morbid. Mortal. Weekly Report 56: 767–769.

Ho, A.Y., A.S. Lopez, M.G. Eberhart, R. Levenson, B.S. Finkel, A.J. da Silva, et al. 2002. Outbreak of cyclosporiasis associated with imported raspberries, Philadelphia, Pennsylvania, 2000. Emerg. Infect. Dis. 8. Available from: URL: http://www.cdc.gov/ncidod/EID/vol8no8/02-0012.htm

Hsieh, H., J. Archer, R. Heffernan, J.P. Davis, and C.F. Neilsen. 2009. *Clostridium perfringens* infection among inmates at a county jail—Wisconsin, August 2008. Morbid. Mortal. Weekly Rep. 58: 138–141.

Jelen, H.H., J. Krawczyk, T.O. Larsen, A. Jarosz, and B. Golebniak. 2005. Main compounds responsible for off-odour of strawberries infected by *Phytophthora cactorum*. Lett. Appl. Microbiol. 40: 255–259.

Kingsley, D.H., G.K. Meade, and G.P. Richards. 2002. Detection of both Hepatitis A Virus and Norwalk-like Virus in imported clams associated with food-borne illness. Appl. Environ. Microbial. 68: 3914–3918.

Mead, P.S., L. Slutsker, V. Dietz, L.F. McCaig, J.S. Bresee, C. Shapiro, P.M. Griffin, and R.V. Tauxe. 1999. Food-related illness and death in the United States. Emerging Inf. Dis. 5: 607–625.

Murrow, L.B., P. Blake, and L. Kreckman. 2002. Outbreak of cyclosporiasis in Fulton County, Georgia. Georgia Epidem. Rep. 18: 1–2.

Rabatsky-Ehr, T., D. Dingman, R. Marcus, R. Howard, A. Kinney, and P. Mshar. 2002. Deer meat as the source for a sporadic case of *Escherichia coli* O157:H7 infection, Connecticut. Emerg. Inf. Dis. 8: 525–527.

Safranek, T., D. Leschinsky, A. Keyser, A. O'Keefe, T. Timmons, et al. 2009. Outbreak of *Salmonella* serotype Saintpaul infections associated with eating alfalfa sprouts—United States. Morbid. Mortal. Weekly Report 58: 500–503.

CHAPTER 8

Meat Production

LEARNING OBJECTIVES

- Identify the pathogens that are associated with meats.
- List the principles for safe butchering of animals and storage of meat products.
- Identify the major factors in meat spoilage, including bacterial concentrations.
- Explain the process of sausage manufacture and what can go wrong.

Throughout history, people have hunted animals to provide meat in their diets, and animals have been domesticated for this purpose and for the production of other valuable products (e.g., milk, wool, and leather hides). Techniques have long been used to kill the animals and then butcher them for the valuable products. Hunters and some farmers still perform this task today, although it is more common for professional butchers to perform the service. Throughout most of human history, butchering was a valuable profession that required the customers' trust. The butcher had the duty to select only healthy animals for slaughter, to butcher the animal appropriately (sometimes in accordance with religious laws), and to sell the meat in a timely manner. A butcher who sold tainted meat or who passed off old meat as fresh could find himself outcast in his community.

For about 250 years of American history, the local butcher, or the individual providing for a family, was the dispenser of meat on a local basis. Customers went to the neighborhood butcher for the evening's meal because refrigeration had not yet been invented. Meat was thus very fresh. A farmer who butchered a hog might inform his neighbors, who would buy or trade for some of the meat. A single hog or cow would provide far more meat than a single family could consume, even assuming that some of it was processed into sausage for later consumption.

In the latter half of the 1800s, this system began to change. Most of the population of the United

States was concentrated in the northern states, east of the Mississippi River, roughly from Chicago to Boston. These were the industrialized areas of the country, and many cities grew because of the influx of people: Milwaukee, Cincinnati, Cleveland, Pittsburgh, and Detroit. Other cities had been very large for many years: New York, Boston, and Philadelphia. Local food could accommodate the needs of most of these people, but the demand for more and better food, coupled with improvements in transportation (railroads), ushered in the industrialization of food preparation.

A tremendous amount of food was available in the sparsely populated western states, notably in the form of beef cattle. The herds grazed on the grasslands where the buffalo had grazed for millennia, and the herds became vast. The only question was how to bring the relatively inexpensive beef from the west to the customers in the east. This began the era of the cowboy, roughly from 1865 to 1920. The job of the cowboy was to round up the cattle from the grasslands (where the range was considered "open," and cattle were identified by brands) and drive them to a railway station. In many cases, the railheads started as tiny towns—such as Abilene, Dodge City, and Wichita—that were located on the railroad line and that had ambitions to grow. These "cow towns" were small during much of the year, but when the cattle were driven to market, they increased in size dramatically. Corrals were constructed on the edge of town to accommodate the large herds. Saloons, hotels, and gambling houses catered to the cowboys, who were paid upon their arrival with the herd. And, of course, their patronage of brothels was not unknown (see Chapter 12 on sexually transmitted diseases), and occasionally a couple of cowboys would drink too much, argue about something, and then settle matters with a gunfight in the street. It was an iconic time in America.

Buying and selling cattle was a big business, and when the transactions were done, the cattle were shipped east in cattle cars. They were headed for the big meatpacking plants of Midwestern cities such as Chicago, Kansas City, and Omaha. Meat that originated in Canada was often sent to Wisconsin. One of the meatpacking companies of Green Bay, the Indian Packing Company, was instrumental in starting the local football club there, and in appreciation, the team adopted the name Green Bay Packers.

At the meatpacking plants, the cattle were slaughtered, their carcasses were processed, and their components were separated. Meat went into refrigerated rooms, while other parts of the animal were used for a variety of products. Initially, the process was very unsanitary; waste products were simply discarded into the local rivers. However, this brought tremendous criticism, and the meatpackers soon found that they could utilize waste parts of the animals profitably (e.g., for fertilizer, glue, gelatin, etc.). At the peak of the meatpacking industry, very little was wasted. Some of the meat was cooked on site and then packed into cans. Sides of beef were loaded onto refrigerated rail cars for the trip east (the first cars were cooled with blocks of ice). There was a steady demand for meat in the big eastern cities, and the meatpackers earned fortunes during these years.

Unfortunately, the industry developed faster than the technology that was needed to keep the products safe. Greed became a factor, and the meatpackers cut many corners to maximize profits. As has been described previously, the customers and the suppliers never met each other, and so the supplier was able to sell meat that he would not eat himself. The branch of the U.S. Department of Agriculture that regulates food practices, the Food and Drug Administration (FDA), existed during this time as the Bureau of Chemistry, but it had no regulatory powers. Conditions in the meatpacking houses became deplorable and more generally known, but it took the publication of a work of fiction to bring them to the attention of the public.

The Jungle, a novel by Upton Sinclair, was first published in serialized form in 1905 and then as a book the following year. This book had a rare and enormous influence on society—it was partially responsible for shaping the Pure Food and Drugs Act (1906) and the Meat Inspection Act (1906) (see Chapter 10 for more detail). Readers of Sinclair's book were treated to descriptions of the slaughterhouses like this one, in which the character Antanas describes how the chemical pickling solution was recovered and reused:

> *For they had set him to cleaning out the traps; and the family sat round and listened in wonder as he told them what that meant. It*

seemed that he was working in the room where the men prepared the beef for canning, and the beef had lain in vats full of chemicals, and men with great forks speared it out and dumped it into trucks, to be taken to the cooking room. When they had speared out all they could reach, they emptied the vat on the floor, and then with shovels scraped up the balance and dumped it into the truck. This floor was filthy, yet they set Antanas with his mop slopping the "pickle" into a hole that connected with a sink, where it was caught and used over again forever; and if that were not enough, there was a trap in the pipe, where all the scraps of meat and odds and ends of refuse were caught, and every few days it was the old man's task to clean these out, and shovel their contents into one of the trucks with the rest of the meat! (Sinclair, 1995)

It is hard to believe that conditions like these once existed, or to imagine that they did not cause massive outbreaks of disease and death from all manner of pathogens. Clearly, oversight of the meat industry was needed to eliminate the worst abuses and to ensure a healthier food supply for everyone.

PREPARING MEAT FOR SALE

Meat refers to the muscle tissue of any of several animals, both domesticated and wild. This includes cattle (beef), swine (pork), lamb, poultry (chicken, turkey, duck, etc.), deer (venison), fish of many types, and other aquatic animals (shellfish, lobster, squid, and octopus). Generally, the term *meat* refers to any animal that is typically harvested as food. Some operations are industrialized, such as large beef cattle ranches; others are not, such as open-ocean fishing. Technological improvements are always changing the industry, such as the development of fish farming. As with any new industry, some problems cannot be anticipated, but vigilance should prevent any major problems.

Foodborne illnesses are partially or completely responsible for about 5000 deaths every year in the United States. Illnesses associated with meat and dairy products are covered separately because, as products of warm-blooded animals, they pose specific and significant health concerns. Major difficulties are associated with animals as food. First, the flesh is a rich nutrient source, containing a large amount of protein as well as many vitamins and other complex biochemical compounds. After slaughter, however, it must be protected against bacterial attack. Second, and specifically in mammals, animal pathogens may be transferrable to humans. The examples abound: *Staphylococcus*, *Streptococcus*, the many gut microflora, *Campylobacter*, and *Listeria*.

As long as the animal is alive, the animal's own physiology keeps the future meat supply untainted. Skin is resistant to microbial invasion. Animals also have immune systems of some type, and mammalian immune systems are comparable to our own. Obviously, animals get sick, but these animals can be recognized and removed from the rest of the herd until they recover.

A few general principles should be remembered with regard to harvesting meat from animals:

- Speed is essential. The time from the death of the animal through processing of the carcass to the storage of the dressed meat in a freezer should be as short as possible.
- Slaughter should be performed as aseptically as possible. Although true aseptic conditions are impractical, unnecessary contamination is avoided. Dust should be kept to a minimum, and entrails of the animals should be removed cleanly and without spillage of contents.
- Stressed cattle should not be used. Cattle are accustomed to ambling around in pastures or feedlots. Some of them become agitated when they are transported by truck. The reasoning behind this principle is detailed in **Box 8-1**.
- Cattle are typically not fed for the 24 hours before slaughter. This reduces the fecal contents, which is a major source of contamination during the slaughter process.
- Obviously, sick cattle should not be slaughtered. Sickness is often defined as an inability to stand. This can be caused by a number of illnesses or injuries, including broken bones, nerve damage, severe mastitis, or neurological diseases like bovine spongioform encephalitis (BSE), also known as mad cow disease (discussed in more detail in the following section).

BOX 8-1 Meat Appearance

Stressed cattle are not immediately slaughtered because they will form "dark, firm, dry" (DFD) meat. Ordinarily, when death occurs, glycolysis stops and lactic acid formation increases. This results in a decrease in pH. Stressed cattle deplete their stores of muscle glycogen. If glycogen is low, then no lactic acid builds up in the tissues, and the pH remains near neutrality. Meat in this condition appears darker than usual. At neutral pH values, spoilage can occur more rapidly, and bacteria such as *Pseudomonas* and *Shewanella* can take advantage of the high protein content. In DFD meat, the pH remains high, usually over 6.0, and spoilage is noticeable at lower concentrations, perhaps $10^6/cm^2$. Stressed hogs form "pale, soft, exudative" (PSE) meat. Again, accelerated glycolysis leads to a lower pH accompanying changes in the tissue after death. The meat may be of a lower grade for sales purposes but does not seem to spoil any faster than normal.

Animals should be killed in the most humane manner possible. Methods for cattle slaughter include physical trauma (the use of a captive bolt pistol or stunner) or electric shock. The stunned animal is then killed by severing the carotid artery, resulting in massive exsanguination. Poultry are killed by controlled atmosphere killing, which is simply asphyxiation with a gas such as carbon dioxide or nitrogen.

In cattle carcass processing, the head is removed, along with the hooves and hides. Contamination is most likely to occur during this step. As can easily be imagined, the hooves and hides might be caked with mud and fecal material. Cattle spend their time in pastures and feedlots and are not particular about where they step or lie down. When these body parts are removed, especially the hide, much dust can be kicked up that lands on all nearby surfaces. Among the organisms of concern are both pathogens and spoilage microorganisms (e.g., *Micrococcus*, *Staphylococcus*, *Pseudomonas*, and many fungal species). In fact, this dust is the greatest source of contamination for the surface of the underlying meat. Interestingly, cleaning the carcass before slaughter appears to have absolutely no effect on mold and bacterial counts during the processing step (Biss and Hathaway, 1996).

Scalding is the technique used on hog carcasses instead of skinning. The problem of stripping off the hide and shaking dust particles into the air is avoided, but the scalding process itself does not sterilize the surface. A significant bacterial and fungal population remains. This is true for poultry as well; the bird carcasses are scalded to facilitate removal of the feathers. Poultry carcasses are washed in a bath on the processing line, which eventually becomes contaminated with microorganisms. To sanitize the water, weak acids may be added (e.g., acetic, citric, or lactic acid [2–5% solution]); hydrogen peroxide or chlorine compounds may be used as well.

Next, the carcasses are opened along the midline, and the viscera are removed. This includes the complicated digestive tract of cattle. If this technique is performed correctly, the opportunity for contamination is minimal; however, an accident at this point that ruptures the bowel will spill fecal contents, thus contaminating the entire area. The lymph nodes are also removed and are then inspected for signs of disease.

The sides of meat are carved down to the desired sizes, and the meat is then flash frozen. The multitude of utensils and grinders used in this process may not be cleaned as often as they should be, but there are practical limitations on how often this can be done. Ultraviolet (germicidal) lights inside the meat lockers are effective because they illuminate the meat surfaces where the bacterial contamination occurs. For further preparation, the meat can be treated with the preservative sodium nitrate (see Chapter 10, section on additives). At the packing house, the meat can be cooked and canned (e.g., for canned ham or deli meats) or shipped in a refrigerated truck to market. At the market, meat cuts are placed in nonsterile containers, which may be the biggest source of surface bacteria.

Flash freezing kills parasites effectively; few incidents of parasitic infections such as trichinosis or taeniasis are contracted through food in the United States. In developing countries, this is not always the case, and it is always wise to cook meat thoroughly.

To produce a delicacy, beef is sometimes aged. Freshly killed meat has a distinctive flavor, and supermarket meat (aged less than a week) has a different flavor. Aged meat has a "gamey" quality that is often highly valued. Aging means that the meat is hung in a refrigerated space for an extended length of time (as long as 3 weeks). Enzymes within the beef work to tenderize the meat during this time. The meat dries and loses a considerable amount of weight, and between the extensive processing and the weight loss, the cost of a pound of steak produced in this way can be quite high. Notably, this beef develops a crusty exterior that contains mold species that can grow at reduced water activity (a_w) in cold temperatures. The mold produces enzymes that further tenderize the beef. The process gives the meat a distinctive flavor, but before this aged beef is consumed, the crusty portion should be cut away and discarded.

After processing, it is not unusual to find meat surface concentrations of bacteria at 10^2–10^5 per square centimeter. Notably, this represents only the surface, and the underlying meat is usually microbe free. This becomes an important distinction when ground meat is considered (**Box 8-2**).

BOX 8-2 The Problems of Ground Meat

Recalls of ground beef (sometimes ground pork) are not uncommon. Recalls of intact meats, such as steaks or chops, are much less common. Ground (or "comminuted") beef is used in a variety of food dishes, such as spaghetti sauce, chili con carne, and tacos; however, it is the mainstay for the production of hamburgers. As the name states, this beef is ground in a machine. Typically, only pieces of larger cuts of beef are included, the so-called trims and ends. In commercial, industrial meatpacking settings, they are all tossed together from different beef sources, which is the first problem. If a single piece of beef is contaminated with bacteria, then tossing the small pieces into the grinder spreads it among a number of meat packages. Another significant source of contamination is lymph nodes, which may occasionally be ground up in comminuted meats. Furthermore, all parts of the beef carcasses are handled by humans, who might pass on their bacteria to the meat.

The next problem is also obvious. The grinding is physically damaging to the cells of the meat. As cell contents are released, far more water becomes available. This moisture creates a suitable growth medium for any bacteria that are present. In addition, the meat is thoroughly mixed. Whereas previously the contamination was probably confined to the meat surface, it then becomes mixed throughout the meat. This creates small, protected pockets inside the masses of meat where the bacteria can multiply and where facultative anaerobes, like *E. coli*, may have an advantage.

Bacterial contaminants, such as the destructive *E. coli* O157, can be killed by sufficient cooking. However, cooking burgers to a rare or medium rare state might not kill the bacteria in the interior of the burger. The U.S. Department of Agriculture recommends that hamburger should be cooked to a temperature of 160°F to kill pathogens.

In fact, it is not at all unusual to find small numbers of pathogens on meat in the supermarket. *Staphylococcus aureus* is the microbe responsible for most cases of meat-associated food poisoning, and *Micrococcus* is the major contaminant found on chicken (Geornaras et al., 1998). A survey of retail chicken, turkey, beef, and pork from several grocery

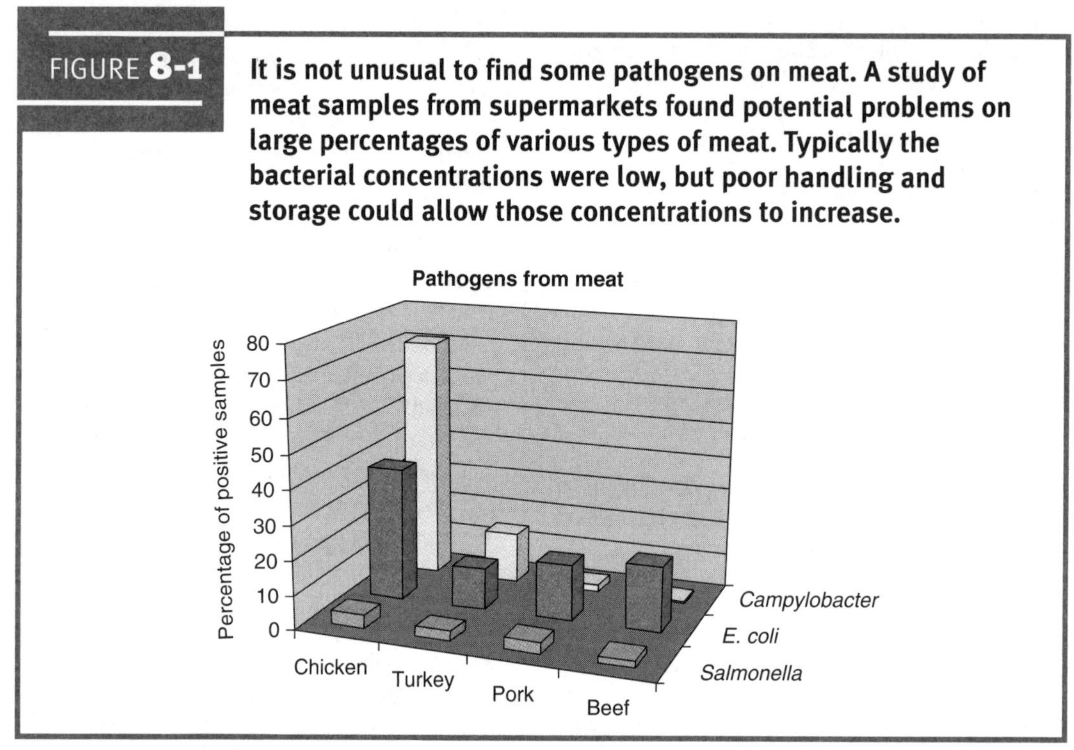

FIGURE 8-1 It is not unusual to find some pathogens on meat. A study of meat samples from supermarkets found potential problems on large percentages of various types of meat. Typically the bacterial concentrations were low, but poor handling and storage could allow those concentrations to increase.

SOURCE: Data from Zhao et al. (2001).

stores found *E. coli*, *Salmonella*, and *Campylobacter jejuni* (Zhao et al., 2001). The most prevalent was *Campylobacter* on chicken (70.7%), while the least prevalent was *Campylobacter* on beef (0.5%). The three main pathogens were found on at least some of the samples of each type of meat (**Figure 8-1**). Interestingly, only 4.2% of the chicken samples showed evidence of *Salmonella*, despite its general association with chicken. Also, the prevalence of *E. coli* on chicken (38.7%) was higher than on beef (19%). This may reflect the increased vigilance of the meat industry.

SPOILAGE

Meat spoilage often involves Gram-negative rods because they are facultative anaerobes that may originate from the intestines. However, a number of genera are implicated in spoilage: *Shewanella putrefaciens*, *Brochothrix thermosphacta*, *Pseudomonas*, *Lactobacillus*, *Moraxella*, and *Acinetobacter* can cause spoilage under refrigeration conditions, especially on meat (Gill, 1986). Meat has a high level of water activity (a_w = 0.99; see Chapter 10), and this fact combined with the high nutrient value of the meat means that many different microorganisms are able to grow quickly on it. The smaller molecules, glucose and amino acids, are the early targets of spoilage microorganisms. Free amino acids are next, followed by lactate. The pH will probably be around 5.6 at this point. Once these small molecules are consumed, typically when bacterial populations reach 10^6–10^8 bacteria per square centimeter of meat surface, the polymers of the tissue (i.e., proteins and lipids) are attacked. The rich source of protein can enable growth to the very high level of 10^8 bacteria per square centimeter of meat surface (Dainty et al., 1975). When lipids are decomposed, an unmistakable odor of rancidity emerges from the product. Bacterial growth results in the utilization of sugars from the tissues, which lowers the pH. The resulting environment gives some types of bacteria a distinct growth advantage. In many cases, a single species, such as *Pseudomonas*, will predominate. Meat spoilage is due to production of compounds like hydrogen sulfide, ammonia, indole (from tryptophan), and amines and is revealed by organoleptic changes, such as taste, odor, and appearance.

If bacterial concentrations reach 10^7 per square centimeter, then spoilage has begun. Typical spoilage bacteria include *Lactobacillus*, *Leuconostoc*, and *Shewanella putrefaciens*. *Shewanella* often imparts a greenish color to the meat (hence known simply as "greening"). This is caused by the production of hydrogen peroxide and hydrogen sulfide after growth on proteins (utilizing the sulfur from cysteine). Hydrogen sulfide complexes with myoglobin, the oxygen-carrying molecule in muscles, form the compound sulphmyoglobin, which has a green appearance. A sign of spoilage on fish tissue is a characteristic phosphorescence, which is often caused by overgrowth of *Photobacterium*. Growth of *Streptococcus faecium* or *Enterococcus casseliflavus* on cooked meat results in a yellow color that is a type of carotenoid pigment (Whiteley and D'Souza, 1989; **Figure 8-2**).

> Fish and visitors smell in three days.
>
> —Benjamin Franklin, *Poor Richard's Almanack*, 1736

FIGURE **8-2** ***Enterococcus faecalis*** **on a blood agar plate.**

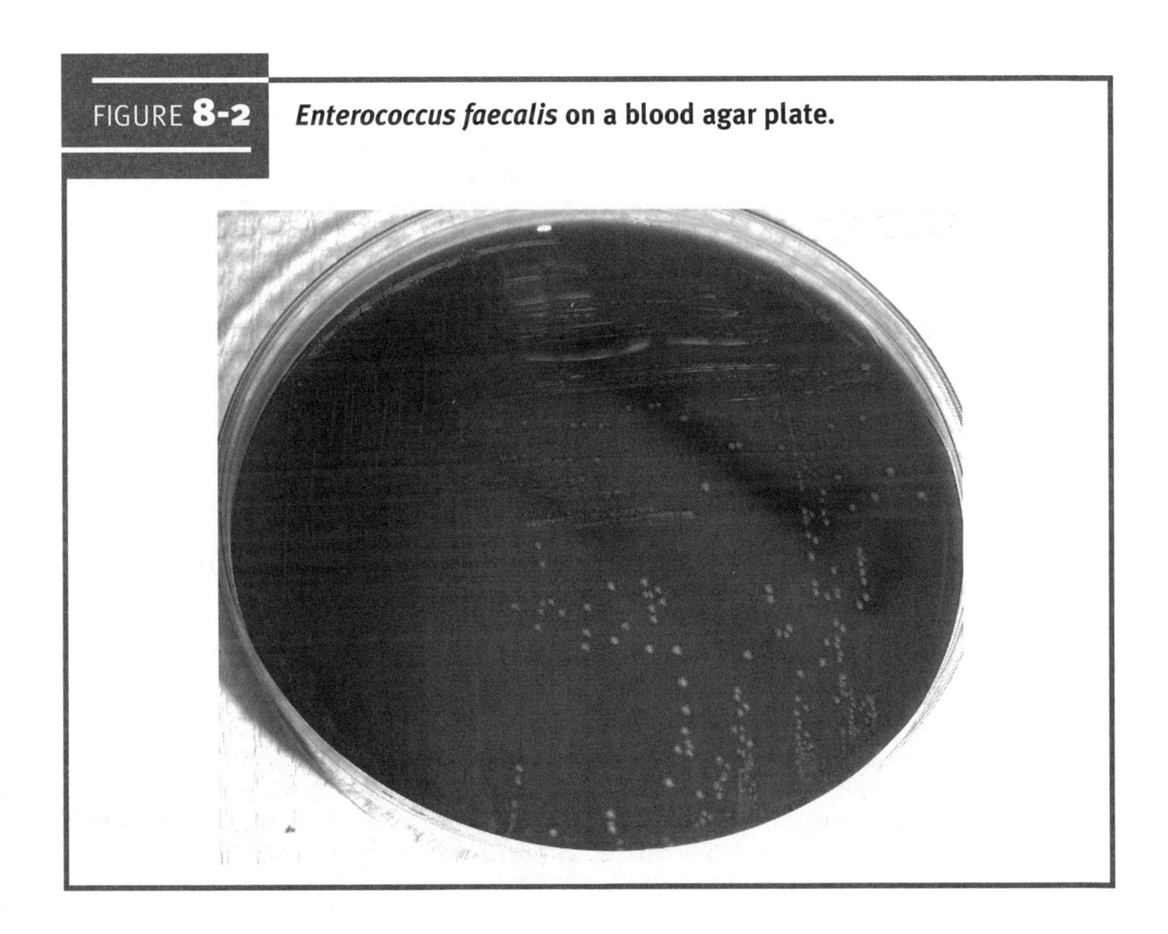

Molds are a constant problem because they grow under more extreme environments than bacteria. Fortunately, they also grow much slower. Yeast and molds are common in the environment and therefore are easily transferred to farm animals. During processing they are released into the air and may land on the raw meat. As long as the concentrations are low, this is not a problem; molds grow slowly. In a long-term test (Lowry and Gill, 1984), a sample of meat started with about 10 yeast per square centimeter, but after 20 weeks at a subzero (–5°C) temperature, the concentration was $10^6/cm^2$. While this concentration is certainly high enough to cause spoilage, it also required nearly 5 months of storage at nonfreezing (freezers are usually set at –20°C) temperatures to accumulate. As **Table 8-1** shows, many genera can be isolated at each step in the process. Notably, pathogenic fungal species are not typical at all; for example, many *Cryptococcus* species may be present, but not the pathogen *C. neoformans*. Mold growth is rare on ground beef; mold may be outcompeted for resources by bacteria during spoilage.

Interestingly, the major mold genera present in a farm environment are not the same as those present on meat surfaces. To a great extent this is due to the environmental conditions after meat processing. Meat is kept in a cold room, where psychrophilic molds will have a growth advantage. For this reason, *Candida* mold species are typically associated with spoilage. Xerotolerant mold species enjoy a growth advantage when the water activity (a_w) is low. *Eurotium* continues to grow until the a_w reaches the 0.62–0.70 range (Leistner and Rodel, 1976). The common molds *Penicillium* and *Aspergillus* grow until the a_w drops below 0.80. For this reason, these latter two mold genera can be found on sausages, which are ordinarily rather dry. This is sometimes observed as a yellow-green slimy film on the surface of the sausage. At these extremes of water activity, however, the molds do not form spores or mycotoxins. A slimy layer found on the outside of sausages may also indicate growth of *Lactobacillus* or *Enterococcus.* Souring found underneath the casings indicates *Bacillus thermosphacta,* which may result from excess acid production.

The surface of fish and shellfish are covered in microorganisms according to the waters in which they live. Therefore, the microflora of a fish from Cape Cod may be very different from that of a fish from the Caribbean. Bacterial concentrations are usually higher in warm waters, but other factors are also significant, such as the method of harvest. Line-caught fish have less surface contamination when compared to fish caught in nets. In all cases, the bacteria are ubiquitous on the fish but pose no real danger until the animal dies. These microorganisms, like those that live on our own bodies, are normal flora, but they start to cause spoilage soon after the death of the animal. High-pressure washes are effective at removing surface contamination on fish.

In general, fish and shellfish must be kept refrigerated from the moment they are harvested until they are prepared for consumption. Some types, such as lobsters and crabs, can be kept alive for extended periods, even until they are prepared for the table, but these are exceptions. Cooking will, of course, kill pathogens, but spoilage may occur quickly. Also, some fish and shellfish are consumed raw. Sashimi is a Japanese term for all types of fish that are consumed raw. Sushi actually means a food dish made of raw fish with other ingredients, such as rice and vegetables. Oysters are often consumed raw, and some consider them to be an aphrodisiac; it is difficult to see how this could be true. As filter feeders, oysters may bioaccumulate pathogens in the water supply such as the pathogenic *Vibrio* species (e.g., *V. parahemolyticus* and *V. vulnificus*). Shrimp can also be consumed raw, and when they are consumed whole, the organism's gut is consumed with them. Shrimp themselves are very small, but there are substantial concentrations of bacteria in the gut; some may be pathogenic.

WHICH DISEASES ARE FOUND IN ANIMALS?

Domesticated animals are prone to diseases just like other animals and humans are. Some of these diseases are caused by microorganisms that have a narrow host range, meaning that the microbe is unable to infect humans. However, examples abound in which animal pathogens are easily transferred to humans. These examples include bacteria and parasites. Passage of viruses is not common, probably due more to their host range, and fungi

TABLE 8-1 Mold Species That Are Associated with Stages of Meat Preparation

Genus	Environmental	Fleece and hides	Meat processing	Isolated from meat
Acremonium			X	
Alternaria			X	
Aspergillus			X	
Brettanomyces				X
Bullera	X			X
Candida	X	X		X
Citromyces				X
Cladosporium			X	
Cryptococcus	X	X		X
Curvularia			X	
Debaryomyces	X	X		X
Dipodascus				X
Epicococcum			X	
Filobasidium				X
Fusarium			X	
Galactomyces		X		
Geotrichium		X		X
Hanseniaspora	X			
Hansenula	X			
Helminthosporium			X	
Hyphopichia				X
Kloeckera				X
Leucosporidium	X	X		X
Metschnikowia				X
Mucor			X	
Paecilomyces			X	
Penicillium			X	
Pichia	X	X		X
Rhizoctonia			X	
Rhizopus			X	
Rhodotorula	X	X		X
Saccharomyces	X			X
Schizoblastosporion	X			
Scopuloriopsis			X	
Sporobolomyces	X			X
Sporothrix		X		X
Thamnidium			X	
Torulaspora		X		X
Trichosporon	X	X		X
Yarrowia		X		
Zygosaccharomyces				X

SOURCE: Adapted from Dillon, 1998.

(yeasts, molds) are more commonly associated with spoilage than infection. Prions, on the other hand, are a genuine concern and require special consideration.

Campylobacter

Camylobacter are probably the leading cause of foodborne illness. *Campylobacter jejuni* is the major pathogen, and *C. coli* is also a contributor. They cause gastrointestinal disease that varies from mild diarrhea and gastric distress to severe life-threatening bloody diarrhea that may be accompanied by fever and nausea. Especially dangerous in newborns, the infection may take several days to incubate before symptoms appear, and illnesses lasting more than a week are not unusual. The U.S. Centers for Disease Control (CDC) estimates that *Campylobacter* is responsible for 2.4 million infections and more than 120 deaths every year. *Campylobacter* can be carried by a number of common mammals, including farm animals and pets.

These microorganisms can be found in mammalian intestinal tracts and thus are spread by the fecal–oral route. They have been found in a variety of foods and are linked to chicken and raw milk. Food handlers with poor hygiene should be suspected in an outbreak. *Campylobacter* can be found in contaminated water, although it appears to be more commonly acquired through food (including milk). *Campylobacter* has caused several neonatal deaths; babies and young children are susceptible to infection. For many years, the *Campylobacter* microbe was not identified. It grows only on special media, at reduced oxygen concentrations, and at a temperature optimum of 42°C. In fact, it is odd that such a prevalent pathogen should have growth requirements that are so different from that of the host.

A good example of the dangers of unpasteurized milk was reported by Hunt et al. (2009). A small community in Kansas suffered an outbreak of *Campylobacter* resulting in 68 illnesses. The vector for this illness was apparently a soft cheese that was made and consumed in the same day. The cheese was made from unpasteurized milk, and although *Campylobacter* could not be isolated from the cheese when it was examined a few days later, an epidemiological analysis found that the cheese was by far the most likely source of contamination. The microbe was isolated from fecal samples of the patients.

Part of the problem was that a soft cheese was made. Soft cheeses, such as cottage cheese and cream cheese, spread easily and are very popular. They do not have a long shelf life, however, even when refrigerated (probably no longer than a week). Soft cheese is made by adding rennet to milk and letting the milk curdle, which may take little more than an hour. Rennet is a mixture of enzymes that break down the proteins in milk; rennet is obtained from the stomach lining of the cow. After the curds (solids) are produced, the whey (liquid portion) is poured off and the curd is pressed into shape. The soft cheese is then ready to eat.

Soft cheeses have been so popular historically that they have even made their way into nursery rhymes:

Little Miss Muffet
Sat on a tuffet,
Eating her curds and whey;
Along came a spider,
Who sat down beside her
And frightened Miss Muffet away.

Just for clarity, a tuffet is a type of footstool.

In the case described previously, the soft cheese was consumed after it had been left out for several hours. It is likely that the milk contained some *Campylobacter* that multiplied during this time. Fortunately, all of the sick people recovered. Notably, some highly regarded soft cheeses are made in other countries specifically with unpasteurized milk. The savvy patron should be aware of this fact.

Listeria monocytogenes

Of several *Listeria* species, only *L. monocytogenes* and *L. ivanovii* are considered pathogenic, and the former is especially feared. This common Gram-positive bacillus can cause illness after it is ingested in contaminated food, milk and cheese, or water. *Listeria* can be found in intestinal tracts of a number of animals, but it can also be found on vegetation. *Listeria* is probably present in small numbers on produce, but it is not a problem unless it multiplies to infectious-dose levels. A variety of food types have been implicated in *Listeria* infections. Many meat and dairy products have been known to become contaminated.

Listeria is unusual in that it grows fairly well at refrigeration temperatures. The growth rate may be slow, but foods kept refrigerated for extended time periods run the risk of *Listeria* growth. The microbe is quite common and is found in many types of vegetation, so it is not unusual to find small numbers of them on a food product. It is also not unusual to find *Listeria* in stool cultures, because 5–10% of the population carries *Listeria* transiently without complications. *Listeria* is covered here under meat products, but it can easily contaminate milk and dairy products or produce. *Listeria* can be carried by a variety of birds and mammals as well as by humans. If allowed to grow in a food product, *Listeria* will produce an infectious dose. Unfortunately, the infectious dosage is still not known with any degree of accuracy.

This bacterium is so common that it is likely that everyone has been exposed to it, but not in concentrations that approach the infectious dose. As a foodborne infection, it causes a typical gastroenteritis, invading through the intestinal cells and multiplying in the cells of the Peyer's patches (Schuchat et al., 1991). From here the infection spreads to other nearby cells and then systemically. Fever, nausea, and diarrhea are typical symptoms, although many mild cases never come to the attention of health authorities.

Infections with *Listeria* can be severe, including an often-fatal meningitis. The microbe appears to have a tropism for nervous tissue and invades the central nervous system, causing symptoms of confusion, stiff neck, and severe headache. By the time the patient experiences convulsions, the meningitis is far advanced and life-threatening. Many people withstand high levels of exposure to *Listeria* without illness; many (not all) of those infected have deficiencies in cell-mediated immunity. In one especially unusual case (Douen and Bourque, 1997) a patient with *Listeria* developed encephalitis in the area of the brain known as the rhombencephalon, which includes the cerebellum, the pons, and the medulla oblongata. The patient was eventually cured, but he experienced musical hallucinations, often featuring tunes from the Big Band era (apparently, Glenn Miller music was not uncommon). This case demonstrates the unpredictable nature of brain infections such as meningitis and encephalitis.

Listeria can cross the placental barrier, infecting the developing fetus. While the mother will probably survive, the fetus often dies and is expelled by spontaneous miscarriage or is stillborn. Even if the child is born alive, an infected newborn has only a 50% chance of survival. The CDC estimates about 2,500 infections every year, with some 500 fatalities. Fortunately, a number of antibiotics are effective in killing *Listeria*.

Listeria will undoubtedly cause more unanticipated infections in the future. An outbreak in Massachusetts in 2007 was linked to milk from a particular dairy (Cumming et al., 2008). Interestingly, the milk had been pasteurized, yet it was contaminated after the process was complete and was passed on to consumers. In this outbreak, three people died from directly consuming the milk, while a fourth death attributed to *Listeria* was a child stillborn at 37 weeks gestation.

Public health authorities inspected the dairy and found *Listeria* in many environmental samples, but they found no problems with the processing equipment and could not identity the point of contamination. The remediation problem was so difficult, with such a high risk for more illness, that the owners of the dairy shut the facility down rather than attempt to clean it. This demonstrates how pervasive a contamination problem can be; the cost of closing the dairy must have been significant.

Yersinia

Two species of *Yersinia* are foodborne pathogens: *Y. enterocolytica* and *Y. pseudotuberculosis* (**Figure 8-3**). Infection with either species produces a gastrointestinal disease that resembles appendicitis and lasts about 2 weeks. However, about 25% of cases also produce bloody diarrhea, which is not a symptom of appendicitis.

Pork is the primary foodborne vector, especially if it is undercooked or if the person who has been slaughtering the animals infects himself directly. Assumedly, pet-to-person transmission is possible; cases of both humans and animals becoming ill in the same household have been documented. This remains unproven, however; the two may have shared the same food source. *Yersinia* species are also notable because they can grow slowly at refrigeration temperatures.

FIGURE 8-3 *Yersinia enterocolytica* on selective agar.

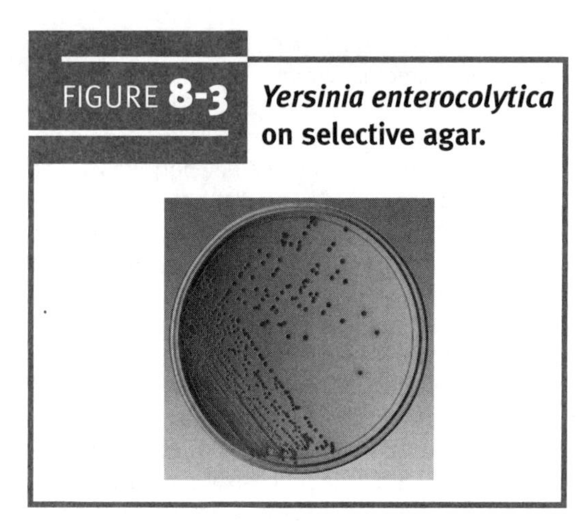

SOURCE: Courtesy of ThermoFisher.

Salmonella

Salmonella typhi causes typhoid fever, but it is rare in the United States. However, *S. enterica* is a fairly common pathogen associated with chicken and eggs, and outbreaks among poultry flocks are a constant threat to the industry. However, *Salmonella* can also be passed through pet feces. *Salmonella enterica* subtypes Enteriditis, Heidelberg, and Typhimurium are the most common pathogens. The CDC reports about 40,000 cases every year, although the actual number is undoubtedly much higher—other estimates are 30–100 times higher.

Salmonella is easily passed in the water supply, especially if wastewater contaminates fresh water. However, *Salmonella* is most well known as a contaminant of chicken and eggs. Chickens can acquire a chronic *Salmonella* infection in the oviduct, and this contaminates the eggs easily (**Figure 8-4**). Modern poultry farming is performed in confined spaces, so a single ill chicken has the potential to spread the infection quickly to the rest of the birds. Even mild infections can result in some *Salmonella* cells on the eggs. Although the eggs are washed to remove any fecal matter, a risk of contamination stems from this source as well (**Figure 8-5**).

Eggs are remarkable structures; they contain all that is needed for chicks to develop, and they protect the chicks from a hostile world. The hen, of course, provides warmth for the chick by sitting on the nest. The egg must allow gas exchange but must not allow any bacteria to enter, which could easily kill the chick. A distinct possibility exists, however, that the egg will develop very small cracks that are large enough for bacteria to pass through. Even in this case, the egg has defenses. Eggs contain large quantities of lysozyme, a powerful enzyme that destroys bacterial cell walls. They also contain avidin, an interesting compound that binds biotin so tightly that bacteria cannot gain access to it. In fact, the avidin–biotin binding capacity is so valuable that it has been used as a laboratory tool for separating specific compounds.

Invariably, some eggs are successfully invaded by bacteria such as *Salmonella,* and bacterial numbers increase within them. Eggs are a rich food. Because large numbers of eggs might be used in an industrialized food plant, eggs are pasteurized. Home use of eggs can also lead to infection, and food preparers should be mindful of contamination problems and the need for thorough cooking of egg dishes and refrigeration of products.

This microbe causes an acute gastroenteritis that is characterized by fever, diarrhea, nausea, and sharp stomach pain. Symptoms appear a day or so after ingestion, and they last for a day or two. Recovery is often accomplished with simple palliative therapy; in extreme cases, antibiotics are effective.

Escherichia coli

Some *E. coli* strains have *Shiga* toxins (known as STECs), which are essentially the same toxin genes that are found in *Shigella*, a close relative of *E. coli*. The most common of the STECs is serotype O157:H7, but many other types are common (**Table 8-2**). The "O" here refers to the antigenic type of the somatic antigen, which is part of the lipopolysaccharide of the cell wall in Gram-negative bacteria.

STEC strains are also dangerous because they have a relatively low infectious dose, perhaps as low as 100 microorganisms (Paton and Paton, 1998). About 8% of the O157 cases go on to develop hemolytic uremic syndrome (HUS), which is especially dangerous to children. Adults can get a similar syndrome called thrombotic thrombocytopenic purpura (TTP).

FIGURE 8-4 **Chickens can acquire persistent *Salmonella* infections when the microbe colonizes the oviduct.**

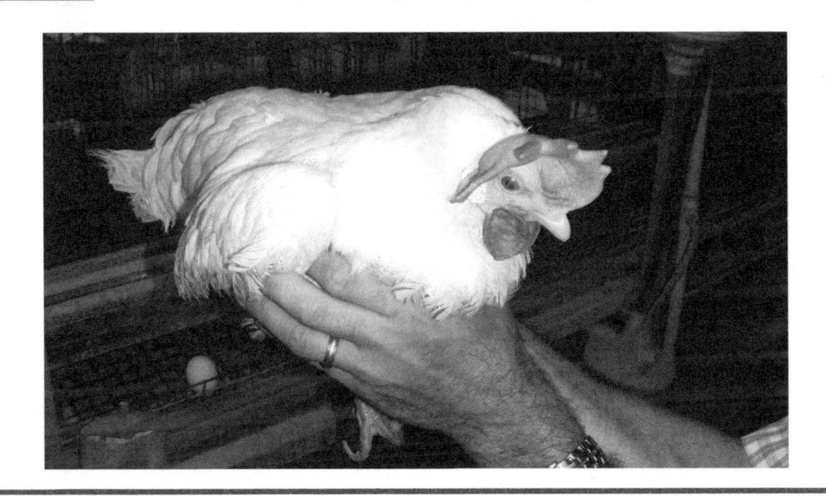

FIGURE 8-5 **Eggs do not arrive clean and pathogen free, as these examples demonstrate.**

TABLE 8-2 The Most Prevalent *Shiga* Toxin-Producing *E. coli* Strains (Non-O157), as Identified by Serotype of the O Antigen

O antigen	Number of cases	Incidence per 100,000 population
O26	68	0.15
O103	47	0.10
O111	35	0.08
O121	10	0.02
O45	8	0.02
O145	6	0.01
O124	5	0.01
O118	4	0.01
O69	3	0.01
O128	2	0.00

SOURCE: Information derived from FoodNet for 2009.

The identification of *E. coli* and its usefulness in tracking fecal contamination of food and water is documented in Chapter 3 (**Figure 8-6**). Because it is so common, it can be found in a variety of different foods. *E. coli* has been implicated in undercooked hamburger meat, although in recent years it has contaminated a variety of produce types (e.g., lettuce, cider, and milk) as well as water supplies. Because it is a common inhabitant of the human gastrointestinal tract, there is a substantial possibility of contamination by people with poor personal hygiene (i.e., if food handlers do not clean their hands thoroughly after defecating). *E. coli* is ubiquitous, causing approximately 20,000 infections every year.

FIGURE 8-6 *E. coli* on a blood agar plate.

Depending on the toxigenic strain, the *E. coli* microbe causes mild to severe gastrointestinal symptoms. Infections develop over a period of days, and the illness frequently lasts a week. The O157 strain is particularly dangerous because it can lead to HUS. This syndrome, which can be fatal in small children, results from damage to the kidneys during O157 infection.

A review of foodborne outbreaks (Kassenborg et al., 2004) found that risk factors included undercooked hamburgers, recent visits to cattle farms, and eating at a "table-service" restaurant. The fact that cattle farms are a risk factor is a reflection of the unsanitary state of the cattle themselves. Handwashing before meals, or ideally after leaving the cattle area, is sufficient to prevent exposure through this route. State fairs are one arena in which the general public is exposed to cattle. In fact, many such fairs now actually forbid contact between visitors and cattle. A "look but don't touch" policy is probably best.

Toxoplasma gondii

Toxoplasmosis is a greatly underappreciated disease caused by *Toxoplasma gondii*, a parasite that invades host cells as an intracellular pathogen. A large percentage of cases are asymptomatic, and seropositivity within the U.S. population is relatively high, indicating that the chances of encountering this organism are high. The CDC estimates that 22% of the U.S. population (about 60 million people) over age 12 has been infected with it (part of the NHANES III study). In some areas of the world the infection rate may be 95%. This is a great concern considering the possible effects.

The oocysts of *Toxoplasma* are resistant to environmental stresses and can be found in soil, but the greatest transmission threat appears to come from cat feces. Because other animals can carry these oocysts, it is possible that they will shed the oocysts in a garden area (where they may be grazing), and that the oocysts will cling to the produce grown there. Without adequate rinsing, the oocysts can become part of a meal. Transmission occurs after people handle contaminated materials and eat them without washing their hands, at which point the pathogen enters their gastrointestinal systems. The oocysts themselves are not infectious, but their sporozoites are. Eating contaminated and unwashed produce or contaminated and undercooked meat also poses a threat. For this reason, *Toxoplasma* is included in the section on foodborne illnesses in Chapter 7.

While a number of mammals and birds can carry the pathogen and excrete the oocysts in their feces, the domestic cat is the *Toxoplasma* vector for humans. Cats and other felines are the hosts in which *Toxoplasma* can complete its life cycle. The sexual cycle of reproduction of *Toxoplasma* occurs in the cat's intestine, which then produces the environmentally resistant oocysts. The cat will then excrete millions of oocysts over a period of weeks. Cats are notorious for killing field mice, other rodents, and birds and bringing the remains into the house, or at least as close as the homeowner will allow them. The infected animals pass the illness to the cat. The cat defecates in a litter box, which must be cleaned by a person. That person runs the risk of acquiring the infection if proper and timely handwashing is not used. Cats have clear preferences about where they wish to defecate, and might use a child's sandbox as the outdoor litter box. In these cases, the risk to the child is clear. Keeping the sandbox covered is an easy solution.

An active disease superficially resembles mononucleosis, with tiredness, muscle aches, and malaise that lasts for a period of weeks. Most people recover without complication, although some antiparasitic drugs are available for severe cases. After recovery, the cysts of *Toxoplasma* persist in the muscle and neural tissue, and the infection may reoccur.

In pregnant women, *Toxoplasma* poses an additional danger to the fetus. The parasite can cross the placental barrier and infect the fetus. If this occurs early in the pregnancy, it may cause a spontaneous miscarriage. If it occurs later in the pregnancy, it may cause neurological problems, often resulting in an abnormal head size (either hydrocephaly or microcephaly). Damage to vision or hearing occurs late in pregnancy and may not become apparent for many years. Drug treatment in pregnant women is problematic because drugs may also damage the fetus. For this reason, pregnant women are warned not to clean cat litter boxes. In fact, other potential sources of contamination, such as gardening, should probably also be avoided by pregnant women, and food should be thoroughly cooked to eliminate the risk of *Toxoplasma*. In immunocompromised persons, the toxoplasmosis can be acute, with neurological problems and a more pronounced systemic response.

Toxoplasmosis has an interesting link to schizophrenia (Torrey and Yolken, 2003). People with severe *Toxoplasma* infections can exhibit delusions and hallucinations that are similar to schizophrenic episodes. Also, studies of populations with mental disorders have revealed a high incidence of elevated antibody titers against *Toxoplasma*. Whether a true cause-and-effect is indicated here is unclear. If so, how does the disease severity develop? What is the age at disease acquisition? What are genetic predispositions? These questions only touch the surface of a host of other issues. However, this link is still very speculative, and it offers a fascinating glimpse into a possible complex relationship between an infectious disease and a mental disorder.

Vibrio

Vibrio vulnificus and *V. parahaemolyticus* occur in shellfish that bioconcentrate them from the brackish waters in which they live (**Figure 8-7**). Consuming raw shellfish, such as oysters, leads to infection. These infections have a high fatality rate, and death can be rapid; they are especially severe if the person has an underlying liver illness.

Trichinella spiralis

This parasite is the causative agent of trichinosis. The infectious larval form of the parasite lives in the striated muscle of the host, while the adult form lives in the mucosa of the intestine (**Figure 8-8**). *Trichinella spiralis* oocysts are passed to new hosts through consumption of the flesh of the animal if cooking has not been thorough enough to kill them. The environment of the stomach and intestine provide signals for the worm to excyst and for the larva to become infectious. Although the number of reported cases is very low, perhaps 12 cases per year, the number of people currently infected may be much greater but is expected to steadily decline.

This illness is typically associated with swine, and most cases probably originate from them. However, *T. spiralis* can be found in many mammals. Hunters are at risk if they consume the animals they kill, and many documented cases involve wild species such as bear, cougar, or wild

FIGURE **8-7** **Electron micrograph of *Vibrio vulnificus*.**

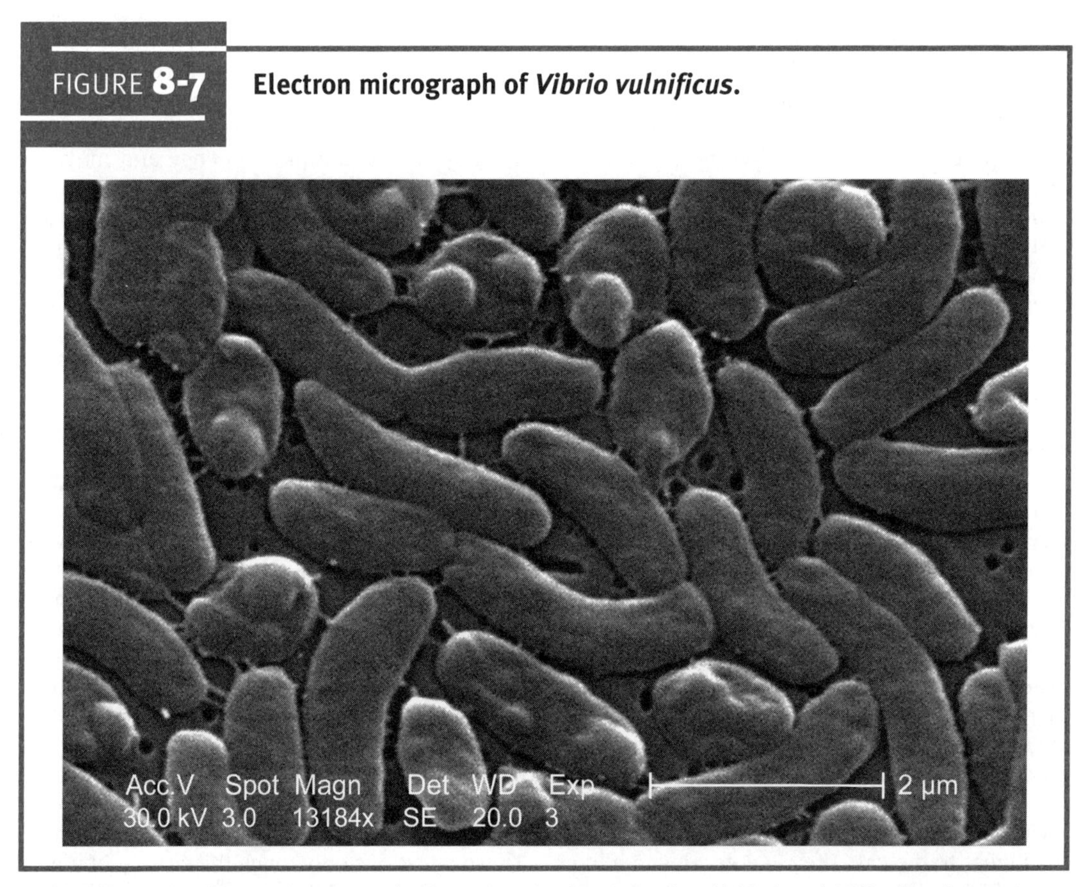

SOURCE: Courtesy of James Gathany, Centers for Disease Control and Prevention. (http://phil.cdc.gov/phil/details.asp). #7815.

FIGURE **8-8** ***Trichinella spiralis*** **cyst coiled in muscle tissue.**

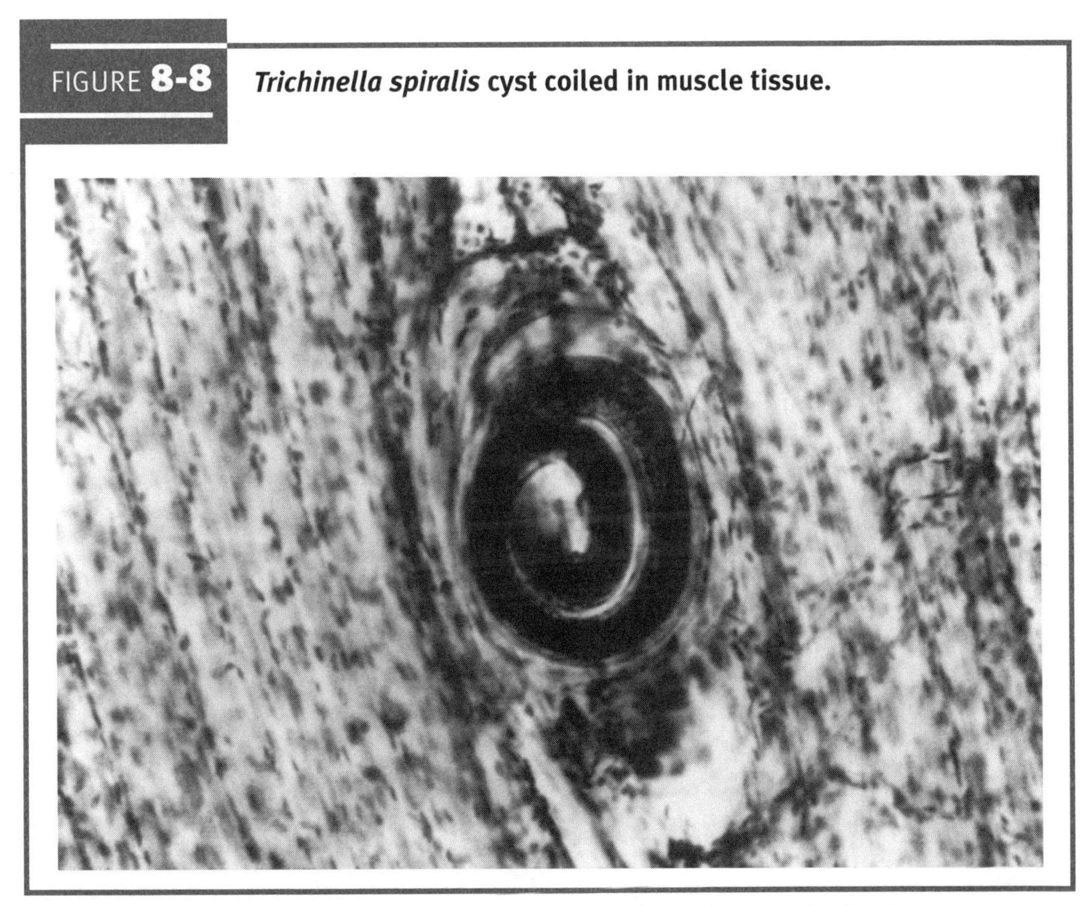

SOURCE: Courtesy of Dr. I. Kagan, Centers for Disease Control and Prevention. (http://phil.cdc.gov/phil/details.asp). #10180.

boar. In fact, *T. spiralis* is a significant problem in populations that consume walrus and polar bear flesh. The worms will travel through the body and lodge in muscle tissue. Severity of the symptoms of trichinosis depends on the extent of replication of the worms. Most cases are either asymptomatic or mild, with general discomfort or mild flu-like symptoms. Most extreme cases range from gastrointestinal distress to neurological symptoms (e.g., strokes, psychotic episodes, or encephalitis).

Trichinella is somewhat resistant to heat (cooking must reach the interior of the meat), but it is sensitive to cold. Quickly freezing pork after butchering, using a –40°C freezer, is sufficient to kill the parasites. Also, long-term storage (at least 20 days) at household freezer conditions (–17°C) is sufficient to eliminate *T. spiralis* oocysts.

Prions

Prions are not organisms at all but are instead rogue proteins. Accepting that a single protein could be responsible for a disease that looked and acted transmissible was difficult for scientists, but careful research over many years demonstrated that a single protein could cause the symptoms when introduced into a new host. Prions cause diseases known as transmissible spongiform encephalopathy (TSE) because they are transmissible to a new host under the right conditions, they cause a brain disorder, and the neural tissue takes on a "spongy" appearance on analysis (**Figure 8-9**).

FIGURE 8-9 **A photomicrograph of brain tissue from a patient with CJD. Note the empty spaces that give the tissue its spongy appearance.**

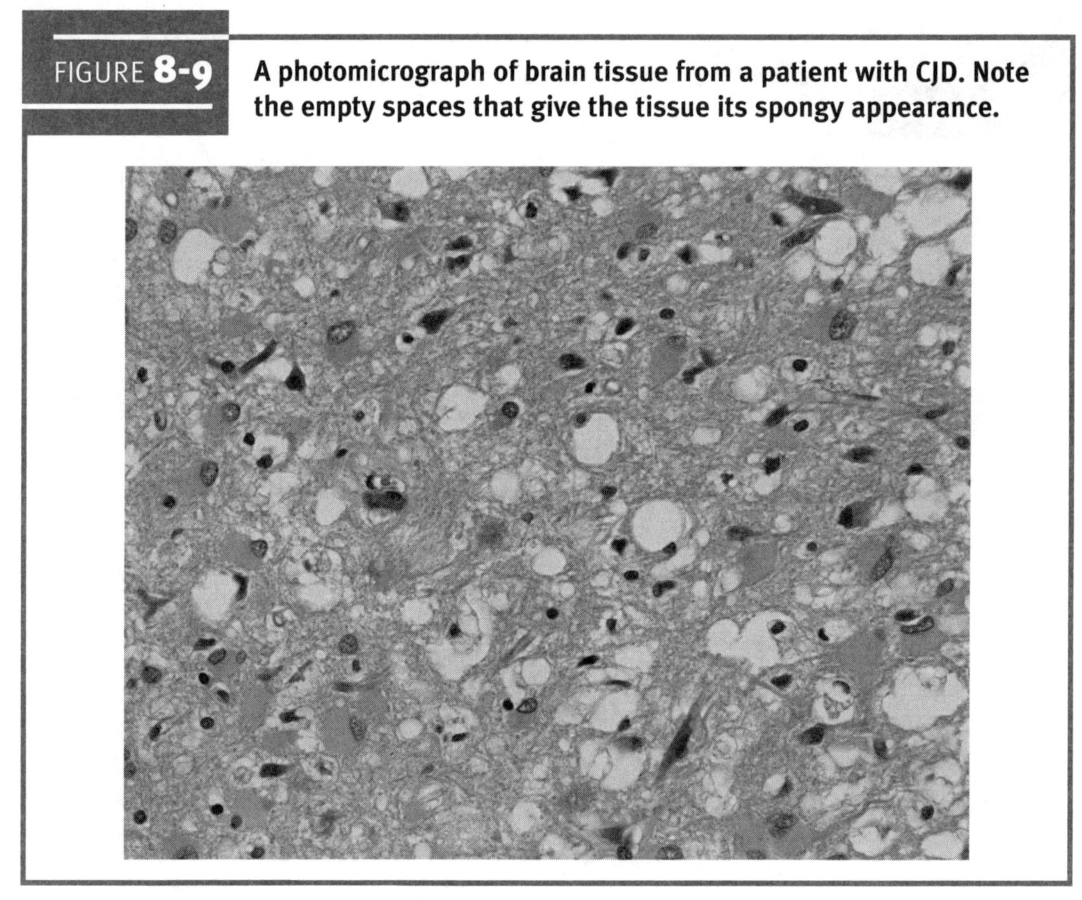

SOURCE: Courtesy of Theresa Hammett, Centers for Disease Control and Prevention. (http://phil.cdc.gov/phil/details.asp). #10131.

Three illnesses of humans are ascribed to prions; Creutzfeldt-Jakob Disease (CJD), a progressive neuropathy, is the most well known. In other species, prions are responsible for "scrapie" of sheep, so-called because the sheep will scrape themselves obsessively against objects until they are raw and bleeding, and in cattle the disease bovine spongiform encephalopathy (BSE) is better known as "mad cow disease." A number of other syndromes occur in wild animals, including some that are regularly hunted. For instance, chronic wasting disease has been described for deer.

Scrapie has been investigated extensively. The normal protein is called PrP^{C}, and when it has assumed its rogue form, it becomes PrP^{SC}. In this form, it is extremely resistant to all environmental insult, including strong disinfectants, proteases, and even autoclaving. For this reason, extreme measures are followed in the medical treatment of patients with a TSE. Instruments that touch infected tissues cannot be sterilized and must be discarded. This is the only safe alternative.

The transmissibility of TSE is a major public health concern. The disease can probably spread to humans through food, and it is unlikely that muscle tissue is involved. The practice in England for many years was to grind up dead and diseased animals and use them as part of cattle feed. This led to multiple BSE cases, and there was great concern that the number of TSE cases among the human popu-

FIGURE **8-10** **Although difficult for a layperson to discern, cows such as this one that display neurological symptoms can be suspected of having BSE and are separated from the herd.**

SOURCE: Courtesy of the U.S. Department of Agriculture and the Centers for Disease Control and Prevention. (http://phil.cdc.gov/phil/details.asp). #5437.

lation would rise dramatically. This has not occurred, however. More probably, the consumption of neural tissue led to the transmission of the prion. Until more is known about this agent, neural tissues are best avoided entirely, and of course, extreme care must be used when butchering a wild animal. Domestic animals that display any sign of neural impairment (so-called "downer" cows) are culled from the herd. This may include unusual nervousness or aggression or the inability to stand (**Figure 8-10**).

SAUSAGE

No discussion of food or meat safety can be complete without a discussion of sausage. On a superficial level this seems unnecessary: the world could get by without sausage. However, in the past sausages were a very important part of the diet in some countries; sausages last a long time and were used to provide meat through a long winter. Sausage was an excellent way to store the excess meat produced by the butchering of a large animal (e.g., cow or pig). Demand for sausage is still great. Small children want bologna (or baloney) sandwiches, pizza comes with either sausage or pepperoni toppings, breakfasts frequently include sausage and eggs, and so on. The additives also impart interesting flavors to the meat.

Yet sausage making is far from an aseptic procedure. In fact, the balance between normal microbial content and dangerous microbial levels (including pathogens and biotoxins) is critical to the safe production of this food.

Sausage is typically produced by first grinding up meat to a consistency suitable for packing into a casing. This is similar to the production of ground beef or ground pork in that the process damages the meat, releasing nutrients from the cells and potentially mixing microorganisms from the surface to the interior of the meat. The ground meat can be mixed with curing agents (e.g., salt, sugar, and nitrates) or a variety of spices or natural oils for flavor, and perhaps other colors and binders, extenders, and fillers. Sugars (e.g., glucose and sucrose) are used for flavor and to promote lactic acid fermentation. Nitrates and nitrites preserve the red color of meat and provide flavor. This process also inhibits spore germination of *Clostridium* species. Binders are proteins like sodium caseinate or soy protein, which bind water and therefore reduce water activity. Extenders are nonmeat additives, such as plant proteins, which are used to bulk up the product inexpensively. Carbohydrate fillers such as rice starch or wheat flour are also cheap and bind water. Whatever the mixture becomes, the final product is made by stuffing it into a casing. The casing can be natural or artificial, and a variety of packaging treatments can be done subsequently.

Salt is added to a final concentration of about 2%, and sugar makes up between 0.5% and 2.0%, depending greatly on the fermentation desired. Nitrates generally do not exceed 180 parts-per-million (ppm). Sodium ascorbate and sodium erythorbate ($C_6H_7NaO_6$) speed curing and reduce nitrosamine formation. They act as reductants and are added at a final concentration of 550 ppm. Other additives include phosphates, potassium sorbate, monosodium glutamate (MSG), hydrolyzed vegetable proteins, and lactates.

Sausage can be prepared in a number of final states. "Fresh sausage" is a mixture of raw meat and seasonings meant to be refrigerated until it is cooked by the consumer. Microbial spoilage of fresh sausage is a major concern, and these products should be cooked and consumed within a few days.

Fermented sausage also involves raw meat, which may or may not be cured (e.g., with salt, sugar, nitrates, etc.). The major difference between fresh and fermented sausage is the use of bacterial fermentation to extend the shelf life of the product. Fermentation results in lactic acid production, which lowers the pH to a level that inhibits the growth of many bacterial species. The pH typically does not get as low as 4.6, so these products are not regarded as high-acid foods (see Chapter 10). In the past, the source of fermentation culture was some of the previous batch that was used as a "starter." In industrial processes, fermentation is carefully carried out under specific conditions, but the overall result is the same. The bacteria ferment the meat product for about a week, depending on the carbon source used. The length of time of fermentation depends on factors like the sugar curing content and the water activity. The sausage is also dried to an extent to remove excess water, producing two types of fermented sausages: semi-dry and dry. Semi-dry sausages reach a pH of 4.8–5.2, but the water content is at least 35%. Dry sausages have a pH in the 5.0–5.5 range, but their moisture content is less than 35%, often below 30%. They have an extended drying time, which may extend from 10 to 100 days. The drying process is carefully controlled because the entire product must dry. If the outside dries too quickly, the moisture on the inside will be unable to escape. Both dry and semi-dry sausages have long shelf lives. So-called "summer sausages" are made in the winter and consumed in the summer. Sausages may also be smoked to improve shelf life. Smoking involves the addition of phenols, alcohols, organic acids, carbonyls, and hydrocarbons through the minute particles of wood smoke.

Some sausages are cured and smoked to retard bacterial growth, but are not fermented and depend on partial cooking to destroy bacteria and to reduce the moisture content. Other sausages are thoroughly cooked for sterilization and are then canned. These are ready-to-eat products. Finally, in some sausages, the meat has been fully homogenized through emulsification, making them easy to stuff into casings. They are typically pasteurized and kept refrigerated. The standard frankfurter ("hot dog") falls into this category.

The problem with sausage making is that many ingredients can introduce bacteria into the mix. In the past, this was favorable because lactic-acid-fermenting bacteria were used to start the whole process. However, the potential exists to introduce pathogens in these processes as well, including the deadly *Clostridium botulinum*. Bacteria can certainly be introduced on the meat itself, as outlined above. Extensive handling of the meat gives ample opportunity for the transfer of bacteria from workers' hands. Spices and condiments frequently have high bacterial content as well.

It is widely known that natural casings are made from some of the least attractive parts of the slaughtered animal. In fact, parts of the stomach (called the "maws") and intestine are routinely used. A pig's intestine may stretch over 60 feet long, while cattle intestines are twice as long, thus creating a large quantity of sausage casings. The organs are treated with a high-concentration salt bath and are then scraped to remove the tissue down to the submucosa layer, which is tough enough to withstand the stretching and filling procedure. Yet natural sausage casings have very high (10^4–$10^7/cm^2$) bacterial contents, mostly *Bacillus*, *Clostridium*, and *Pseudomonas*. This is not a problem with artificial casings, which are made of plant materials like cellulose. In preparing meat for sausage, typical contamination of the product must be considered. Processing of the product should incorporate methods for decreasing the bacterial content. The target in sausage production is a 5-log reduction in pathogens.

QUESTIONS FOR DISCUSSION

1. Have you ever eaten a meat product from an animal that was not mentioned here? Do any special conditions apply to that item?
2. After reading Chapter 10 on water activity, review the conditions under which meat is handled and determine why meat is sometimes treated with salt or sugar.
3. Why is ground beef such a problem for contamination?
4. Would you expect farm-raised fish to be more or less contaminated with bacteria compared to ocean-caught fish?
5. Of the pathogens mentioned here, which were also mentioned as waterborne diseases? What do the two environments have in common?
6. House cats are known to carry *Toxoplasma*. What additional precautions should be taken to avoid infection?
7. Will learning about *E. coli* O157 change your eating habits, such as how you prefer your hamburgers cooked?

References

Biss, M.E., and S.C. Hathaway. 1996. Effect of pre-slaughter washing of lambs on the microbiological and visible contamination of the carcasses. Vet. Rec. 138: 82–86.

Cumming, M., P. Kludt, B. Matyas, A. DeMaria, T. Stiles, et al. 2008. Outbreak of listeria monocytogenes infections associated with pasteurized milk from a local dairy—Massachusetts, 2007. Morbid. Mortal. Weekly Rep. 57: 1097–1100.

Dainty, R.H., B.G. Shaw, K.A. DeBoer, and E.S.J. Scheps. 1975. Protein changes caused by bacterial growth on beef. J. Appl. Bacteriol. 39: 73–81.

Dillon, V.M. 1998. Yeasts and moulds associated with meat and meat products. p. 85–177. *In* A. Davies and R. Board (eds.), Microbiology of meat and poultry. Thomson Science, London.

Douen, A.G., and P.R. Bourque. 1997. Musical auditory hallucinosis from *Listeria rhombencephalitis*. Can. J. Neurol. Sci. 24: 70–72.

Geornaras, I., A.E. de Jesus, and A. von Holy. 1998. Bacterial populations associated with the dirty area of a South African poultry abattoir. J. Food Protect. 61: 700–703.

Gill, C.O. 1986.The control of microbial spoilage in fresh meats. *In* A.M. Pearson and T.R. Dutson *Advances in meat Research*, Vol. 2. AVI Publishing, Westport, CT.

Hunt, D.C., M.C. Banez Ocfemia, D. Neises, and G. Hansen. 2009. *Campylobacter jejuni* infection associated with unpasteurized milk and cheese—Kansas, 2007. Morbid. Mortal. Weekly Report 57: 1377–1379.

Jay, J.M. 2000. Modern food microbiology, 6th ed. Aspen Publishers, Gaithersburg, MD.

Kassenborg, H.D., C.W. Hedberg, M. Hoekstra, M.C. Evans, A.E. Chin, R. Marcus, D.J. Vugia, K. Smith, S.D. Ahuja, L. Slutsker, and P.M. Griffin. 2004. Farm visits and undercooked hamburgers as major risk factors for sporadic Escherichia coli O157:H7 infection: data from a case-control study in 5 Foodnet sites. Clin. Inf. Dis. 38 (Suppl. 3): S271–S278.

Leistner, L., and W. Rodel. 1976. Inhibition of microorganisms in food by water activity. *In* F.A. Skinner and W.B. Hug (eds.), Inhibition and inactivation of vegetative microbes. Academic Press, London.

Lowry, P.D., and C.O. Gill. 1984. Development of a yeast microflora on frozen lamb stored at −5°C. J. Food Protect. 47: 309–311.

Paton, J.C., and A.W. Paton. 1998. Pathogenesis and diagnosis of Shiga toxin-producing *Escherichia coli* infections. Clin. Microbial. Rev. 11: 450–479.

Schuchat, A., B. Swaminathan, and C.V. Broome. 1991. Epidemiology of human listeriosis. Clin. Microbiol. Rev. 4: 169–183.

Sinclair, U. 1906/1995. The Jungle. Barnes and Noble Books, New York.

Torrey, E.F., and R.H. Yolken. November 2003. *Toxoplasma gondii* and schizophrenia. Emerg. Infect. Dis. 9. http://www.cdc.gov/ncidod/EID/vol9no11/03-0143.htm.

Whiteley, A.M., and M.D. D'Souza. 1989. A yellow discoloration of cooked cured meat products—isolation and characterization of the causative organism. J. Food Protect. 52: 392–395.

Zhao, C., B. Ge, J. DeVillena, R. Sulder, E. Yeh, S. Zhao, D.G. White, D. Wagner, and J. Meng. 2001. Prevalence of Campylobacter spp., Escherichia coli, and Salmonella serovars in retail chicken, turkey, pork, and beef from the greater Washington, DC area. Appl. Environ. Microbiol. 67: 5431–5436.

CHAPTER 9

DAIRY PRODUCTS

LEARNING OBJECTIVES

- Describe the need for pasteurization and its history.
- Identify the typical pathogens associated with dairy products and the diseases they cause.
- Describe the process of pasteurization, including the theory behind it.
- Explain how errors in the pasteurization process are commonly made and how the process is monitored.

KEY TERMS

- Pasteurization

Dairy foods refer to any product made with milk. The source of the milk could be, theoretically, any mammal (e.g., goat or yak), but cow's milk is by far the most common (**Figure 9-1**). The products that result from milk include the milk itself, cheese, ice cream, yogurt, cream cheese, butter, cream, cottage cheese, and even milk chocolate. All of these products have two things in common. First, they originate from a warm-blooded animal, which means they may harbor pathogens that can infect humans. Second, they are all high-nutrient products, which means they may provide a good medium for bacterial growth. One of the first tasks of public health departments was the monitoring of milk for pathogens (**Figure 9-2**).

Milk is far from sterile, despite efforts to disinfect the cow's udder before milking; it is easily contaminated with common hide and soil bacteria (**Figures 9-3**, **9-4**, and **9-5**). At the time of production, the contamination is usually low enough to make it potable, and it is not uncommon for dairy farmers to use some of their herd's milk as the household milk. However, bacteria grow quickly in such a nutrient-rich broth, and within a couple of days, the milk will become sour due to bacterial overgrowth (**Box 9-1**). The fact that milk from the grocery store will last a week in the refrigerator is due to the post-production processing of the milk, that is, pasteurization. In large measure, the study of dairy products is the study of pasteurization.

FIGURE 9-1 **Holstein cows on a dairy farm.**

FIGURE 9-2 **An early photograph of the public health department testing laboratory, showing bottles of milk ready for analysis.**

SOURCE: Courtesy of the Milwaukee Health Department.

BOX 9-1 **Bacteria Commonly Found in Milk**

Streptococcus
Lactobacillus
Achromobacter
Pseudomonas
Alcaligenes
Flavobacterium
Bacillus
Yeast

PASTEURIZATION

Pasteurization is the heat treatment of a product to kill or injure microorganisms; it does not typically sterilize the product (except ultra-high-temperature, or UHT, pasteurization). Instead, pasteurization achieves several goals: (1) it eliminates any pathogens in the product; (2) it reduces the total number of microorganisms in the product by killing or injuring them; (3) it increases the shelf life of the product; (4) it inactivates some of the mammalian enzymes that can give the milk an "off" taste; and (5) it achieves these goals without harming the taste or food quality of the product.

Pasteurization is used for more than just dairy products (**Box 9-2**). It can be used to protect a variety

FIGURE 9-3 **Milking time on the farm.**

FIGURE 9-4 The udder is washed off with a disinfectant prior to milking.

FIGURE 9-5 **An automatic milking machine does the work, and then detaches automatically when the flow of milk ceases.**

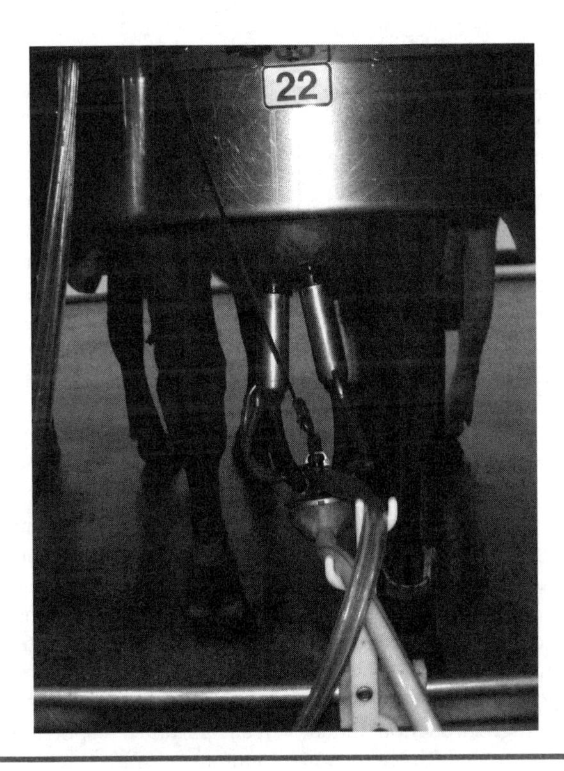

BOX 9-2 Pasteurized Goods

- Milk
- Wine
- Beer
- Fruit juices
- Cider
- Honey
- Eggs
- Sports drinks
- Canned foods

of liquids and even some solid products, such as almonds. In each case, a time and temperature are established for treatment that will produce the desired level of microbial eradication.

Fittingly, pasteurization is named after its inventor, Louis Pasteur. Pasteur was interested in making milk safer. Dairy cows were known to carry *Mycobacterium tuberculosis*, the causative agent of tuberculosis, and the organism was known to be passed through drinking the cow's milk. In contrast to the more well-known route of infection through aerosolized droplets, this route involved infection through the gastrointestinal system and infection of the major organs. Although milk could be made safe by boiling it, boiled milk changes significantly and does not appeal to many people.

BOX 9-3 Ultra-High-Temperature (UHT) Pasteurization

Ultra-high-temperature (UHT) pasteurization is done at 130°C for a 1-second exposure time. This can be accomplished using injected steam. This rapid method is appreciably faster than standard high-temperature, short-time pasteurization. In fact, the temperature is so high that less than a second is needed. This extreme condition sterilizes the milk. Products subjected to UHT are safe to store in sealed containers at room temperature, and they should last as long as any canned food.

Pasteur's solution was to heat the milk to a high temperature for a period of time, but nowhere near the boiling point. He settled on 62.8°C for 30 minutes, which is now considered the "classical" condition. Under this condition, *M. tuberculosis* was destroyed, although many other microorganisms survived. The result ensured a safe food product for untold numbers of consumers.

In the present era, an incubation time of 30 minutes is not usually practical; high throughput industries like dairies, which must process thousands of gallons of milk, could not operate with this constraint. Alternative conditions were investigated, and it was found that a temperature of 72°C could be used with only a 15-second contact time. This is called high-temperature, short-time (HTST) pasteurization and is the industry norm. Classical conditions are still sometimes used on high-value, small-quantity products, such as milk chocolate. (See **Box 9-3**.)

In each case, an elapsed time and a temperature are needed. This defines a thermal death time (TDT), which is produced for each constituent, which can refer to microbe death, enzyme inactivation, or loss of organoleptic quality. In a graph of time versus temperature, the slope is quite steep (**Figure 9-6**). Another way to state this is that, for a small increase in temperature, there is a corresponding large decrease in exposure time. Different components of the milk sample, such as flavors or enzymes, have different slopes. Different microbial targets have different slopes as well.

It is useful to define conditions on the basis of the *D* value and the *z* value. As has been mentioned elsewhere, the decimal reduction time *D* refers to a decrease in an order of magnitude, in this case an order of magnitude of the microorganism of interest. The *D* value is expressed as the conditions (time and temperature) needed to reduce microbial numbers by an order of magnitude. As should be clear now, a number of different pairs of time and temperature can be used to achieve the same result. Pasteurization aims for a 5 log (5D) reduction of viable microbes; that is, a decrease of five orders of magnitude. A single *D* value reduces the number of bacteria by 90%, leaving only 10% (**Figure 9-7**). The 5D result should be sufficient to kill all molds, yeasts, and many bacteria that contribute to spoilage. It should destroy all pathogens, such as *Mycobacterium tuberculosis* (TB) and *Coxiella burnetti* (Q fever). *Coxiella* may be the toughest of these, although other pathogens may be problematic. *Listeria*, an atypical microbe in milk that can cause mastitis, can resist normal pasteurization conditions (Doyle et al., 1987). Some organisms of interest are not completely eliminated, such as *Mycobacterium avium* subsp. paratuberculosis. This is the causative agent of Johne's disease in cattle and may be at least partially responsible for Crohn's disease in humans.

The *z* value is a little more definite. It is the temperature change needed to change the *D* value by an order of magnitude (**Figure 9-8**). To achieve this increase, the increase in temperature (°C) must decrease the time component from 20 minutes to 2 minutes.

It is useful to determine the *z* and *D* values for the important constituents of pasteurized products because many conditions can destroy the microorganisms without appreciably harming the vitamins or pigments of the food.

The modern dairy pasteurizes milk by sluicing it through a tube that has been heated to 72°C. The transit time of the milk is at least 15 seconds so that, upon exiting the tube, it is pasteurized. From there the milk goes directly into containers that are sealed and sent to a refrigerator until they are transported to the grocery store. This method works

FIGURE 9-6 Thermal death time (TDT) for *Mycobacterium tuberculosis* strain USDA 854 in milk. Note that the *y*-axis is in seconds and uses a log scale. According to these data, the same degree of killing can be obtained with 27 seconds at 64°C and with 3 seconds at 69°C.

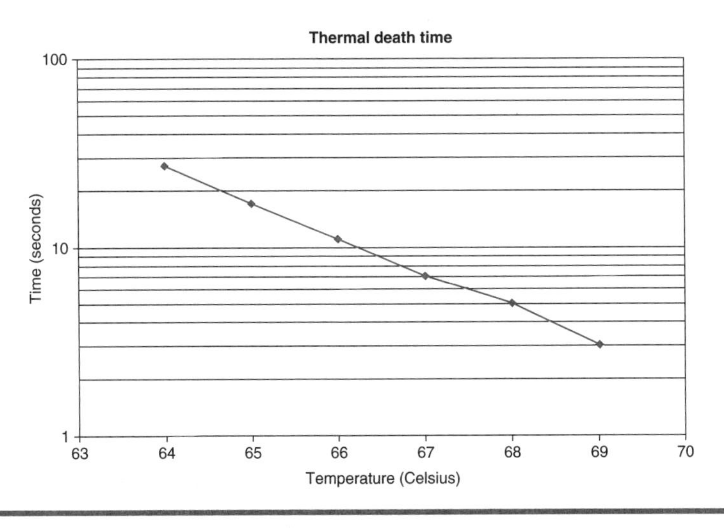

SOURCE: Based on data from H.R. Kells and S.A. Lear. 1960.

FIGURE 9-7 Graph showing the D values at a specific temperature. The dashed lines show the time required for a decrease of one order of magnitude of the target pathogen.

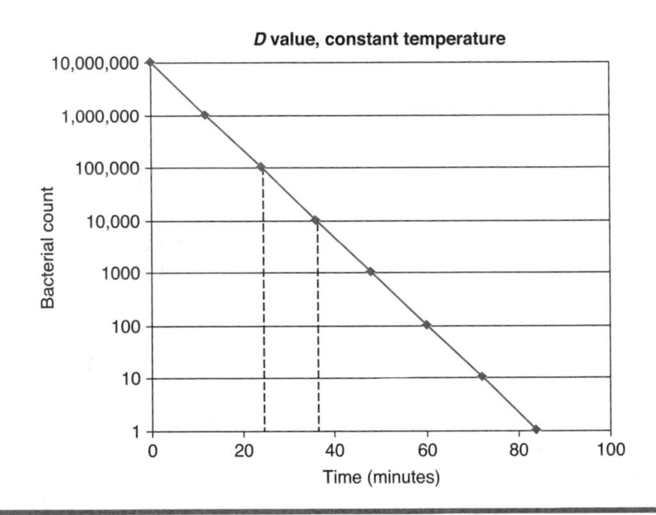

FIGURE 9-8 **Graph of the z value for a particular system. Note that the y-axis is a log scale, while the time change is a linear scale.**

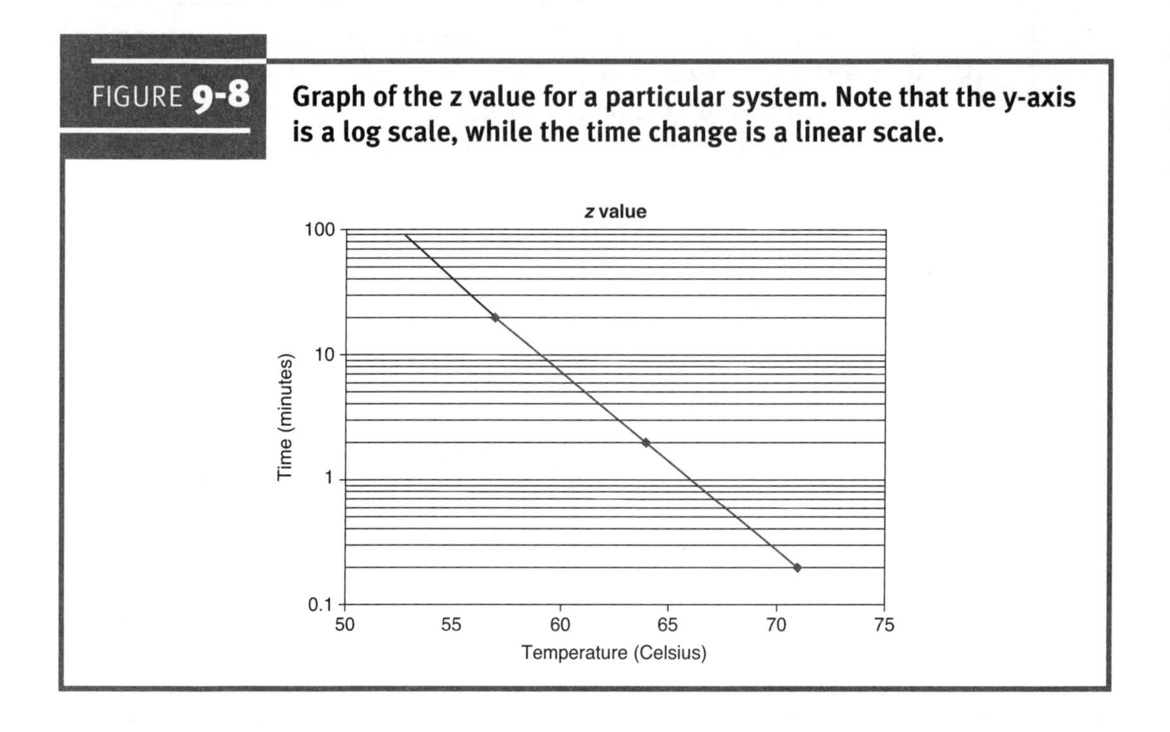

remarkably well; illnesses attributable to pathogens in pasteurized milk are very rare.

Batch pasteurization is used for high-value foods like chocolate and cream (**Figure 9-9**). Cold pasteurization refers to irradiation of milk at such conditions in which the same results are achieved. Pasteurized milk is then expected to last for 2 weeks at refrigeration temperatures.

Contamination of pasteurized products is always a concern because there is no obvious sign of contamination in the food product. Even in milk that has already been pasteurized there is a chance of mixing with unpasteurized milk. This happens because the pasteurization apparatus uses thin, stainless steel heat exchanger plates to warm the milk. The warm milk on one side of the plate is used to start the warming of the unpasteurized milk on the other side of the plate, thus conserving energy. Sometimes small holes in the heat exchanger plates develop and cross-contamination occurs. To determine whether this has occurred, a test for alkaline phosphatase is performed (Rocco, 2004). Unpasteurized milk has the enzyme, but pasteurization should have destroyed it completely. The test results in a colorimetric result (yellow color = positive) that is read on a spectrophotometer. If alkaline phosphatase is absent, the milk has been safely pasteurized.

One good example involves a case of contaminated ice cream that eventually sickened more than 200,000 people with *Salmonella enteriditis* (Hennessey et al., 1996). The ice cream was made and pasteurized as a pre-mix and was then shipped to customers in tanker trucks. However, the trucking company tried to maximize its trip efficiency by delivering other commodities on the return trips. In this instance, it had a contract to haul unpasteurized eggs on the return trip. The trucks were washed out between loads, but this was not sufficient to remove every trace of the eggs, at least some of which were contaminated with *Salmonella*. Because of this, the next load of ice cream pre-mix became contaminated, and a large population was exposed to the pathogen and became ill.

FIGURE 9-9 **A small-batch pasteurizer that is being used for pasteurization of cow's milk for feeding to calves.**

ANTIBIOTIC USE

Dairy cattle are prone to mastitis. A cow with mastitis will shed many pathogenic bacteria into its milk, rendering it unsafe. Pasteurization usually destroys these pathogens, but the best alternative is to take the cow out of production until the problem is alleviated. The cow must still be milked while it is treated with antibiotics, but the milk is simply discarded. In fact, a large percentage of all antibiotic use is in cattle (both beef and dairy). Antibiotics can be passed into the milk, where people consume them. The continued presence of antibiotics creates concern that antibiotic-resistant normal flora will evolve as a response to them, although this has not occurred thus far. However, there are regulatory limits on the concentrations of antibiotics present in milk and standard methods for determining concentrations (Bulthaus, 2004).

FERMENTING MILK

Streptococcus lactis and *S. cremoris* are found in milk. They utilize lactose and produce acid. As the pH drops, the proteins in the milk precipitate, forming curd, and the remaining liquid fraction is whey. In the production of cheese, this is useful because the water content (and water activity) is greatly decreased, meaning that the food will last longer. Some molds grow on cheese, but usually they can simply be scraped off. These conditions are too acidic for most bacteria that cause spoilage.

POTENTIAL PATHOGENS OF MILK

The pathogens that are specifically identified with milk or dairy products are described in this section. One classic contaminant of milk is *Corynebacterium diphtheria*, the causative agent of diphtheria. This formerly feared disease had a significant fatality rate, but vaccination against the pathogen has removed the threat. Other diseases that are normally passed by other means have been reported to be passed through milk or milk products. One such unusual case was the transmission of tick-borne encephalitis through goat cheese (Holzmann et al., 2009).

Methods for monitoring production of dairy products (Davidson et al., 2004) and for the detection of specific pathogens (Frank and Yousef, 2004; Henning et al., 2004) have been described. Predictably, the bioindicators for milk are also coliform bacteria. The tests used to monitor milk are similar to tests for waterborne pathogens, although the complex chemistry of dairy products can sometimes impede pathogen identification. Occasionally a cow harbors pathogenic *E. coli* strains, such as O157 (Shere et al., 2002). Without pasteurization, this contaminant is dangerous. A sample reporting form for milk analysis is shown in **Figure 9-10**.

A good example of the use of bioindicators is in the monitoring of ice cream, both soft-serve and hard ice cream. Soft-serve is the familiar swirled ice cream in a cone, while hard ice cream is scooped from a tub. Many people buy half-gallon containers of hard ice cream at the supermarket, but a trip to the ice cream shop is a fun activity. Hard ice cream is produced at a dairy facility and is pasteurized. It has a "sell by" date, but hard ice cream usually does not present health problems. Soft-serve ice cream comes to the retail outlet as a bag of ingredients that the ice cream machine whips into the familiar soft product. These machines become contaminated over time and need regular and thorough cleaning. Substantial handling of these products allows typical skin bacteria (*Staphylococcus*) and fecal bacteria from unwashed hands to invade the nutrient-rich dairy product.

Large corporations and franchise businesses are usually very good about establishing a schedule for cleaning the machines to minimize contamination problems. But schedules may be forgotten or precluded by activity. One of the functions of the local health department is to monitor these businesses periodically. A sample of ice cream is taken from the machines and brought to the laboratory; the sample is analyzed for the presence of typical contaminants, and if the sample proves positive for bacteria, the business owner is advised to clean the equipment routinely.

Brucella

Brucella is the causative agent of brucellosis, a severe systemic disease that may take several weeks or months to resolve. Brucellosis has also been called Malta fever and undulant fever; this last name refers to the undulating nature of the fever it produces. Because of the length of the illness, patients can be physically exhausted by the disease. Other symptoms include a general achy feeling, headaches, profuse sweating, and weight loss. In severe cases, the internal organs are attacked, which may lead to endocarditis (often the proximal cause of death). Depression is not uncommon in patients as they cope with this long illness. The fatality rate, if untreated, ranges between 2% and 5%.

Four *Brucellosis* species are known pathogens: *B. abortus*, *B. melitensis*, *B. suis*, and *B. canis*. Infection results from handling meat from an infected animal, probably through airborne droplets. The infectious dose appears to be very low. Infection is also associated with milk from infected animals and cheese made from that milk. The CDC reports about 100 to 200 cases per year, about half of which are attributable to food contamination. However, they also estimate that the true infection rate is about 14-fold greater and includes many unreported mild cases. If the infection is treated with antibiotics, the prognosis is good, although the course of treatment may require 6 weeks or more.

There are two major barriers to *Brucella* infection. The first is vaccination of cattle, for which the vaccine is highly protective. The second is the pasteurization of milk, which is sufficient to kill all *Brucella*. This addresses one current problem with continued infection: some people prefer unpasteurized milk as being "healthier." They can usually obtain unpasteurized milk (sometimes goat's milk) from small "hobby" farmers, whose animals might not be vaccinated. Outbreaks have been caused by cheese made from this milk.

Mycobacterium tuberculosis

Mycobacterium tuberculosis is a causative agent of tuberculosis (TB). The bovine strain is called *M. bovis*. Today it is far more common to acquire TB through airborne transmission, but in the past the

FIGURE 9-10 **An analysis sheet for the examination of dairy products, including ice cream.**

CITY OF MILWAUKEE HEATH DEPARTMENT — PRODUCT ANALYSIS

Dairy, Ice Cream Plant, or Mix Manufacturer ______________ Arrival Temp.

Restuarnt, Distributor, or Store licensee ______________ Agar _____ Water _____

Date Collected ______________ Collection Temperature ______________ Air _____ Petri dish _____

No.	Product	Size & Type	Code	Phos-pha-tase	Freezing Pt. (– °C)	Antibiotics	Coliform Count	Standard Plate Count	Dilutions		

SAMPLE COLLECTOR

Tested by:					
Read by:					

White –Bacteriology Laboratory
Canary –Consumer Protection Division
Pink –Dairy, Ice Cream Plant, or Mix Manufacturer
Golderod–District Sanitarian or Sample Collector
Blue –Restaurant, Distributor, or Store licensee
Green –Chemistry Laboratory

SOURCE: Courtesy of the Milwaukee Health Department.

microbe was passed through the milk of infected cows. For this reason, Pasteur experimented with disinfection of milk, finally producing the conditions that would kill this species and make milk safe to drink (see Chapter 13 on airborne diseases).

Mycobacterium is an unusual pathogen in that it cannot be effectively stained with Gram-staining reagents (technically it is Gram positive), but it can be stained well using acid-fast stain. If acquired through ingestion, the pathogen can invade any of the internal organs and create the primary infection. Later it will spread secondarily throughout the body. In the past, this resulted in a high rate of lethality, although it is rare today. In one notable outbreak (Winters et al., 2005), a young child in New York City died from *M. bovis* that caused a disseminated peritoneal tuberculosis. The source was suspected to be Mexican cheese made from unpasteurized milk. Several other cases in the city appeared to be attributable to Mexican cheese, either imported or sent as a gift. In the case that caused a fatality, the microbe colonized the peritoneum very quickly, and if there were other contributing factors to the death, they were not readily apparent. Dairy products should be made with pasteurized milk.

Mycobacterium avium complex (MAC)

Other *Mycobacterium* species are present in milk. In some cases, the HTST pasteurization conditions are not sufficient to kill all of them. Part of the problem is that the cells tend to clump together. A clump of cells is more heat resistant than individual cells in the milk, so HTST conditions may not be sufficient to get heat to the cells in the middle of the clump. Homogenization of the milk breaks up many of these clumps and improves pasteurization (Grant et al., 2005). At least a few cells are likely to survive current HTST conditions, and in some cases, significant numbers of cells may remain. More investigation is required to address this issue.

Until recently, these were considered to be nonpathogenic species and therefore harmless, but that view has been challenged. A pulmonary disease occurs sporadically due to MAC, but the most serious cases are in AIDS patients, where MAC causes a systemic disease.

The complex includes *M. paratuberculosis*, *M. avium,* and *M. intracellulare*. In cattle, these microorganisms cause Johne's disease, a progressive wasting disease. *Mycobacterium* can also infect sheep, goats, deer, and bison. It is surprisingly prevalent in cattle herds; one estimate indicates that 8% of all animals are infected. Because Johne's disease is contagious, other animals risk acquiring the infection. This disease superficially resembles Crohn's disease, an inflammatory disease of the large intestine in humans. It has been suggested that Crohn's is also caused by MAC, although definitive evidence is lacking. If *Mycobacterium* proves to be the cause, or at least a contributing factor, then new thinking on safe pasteurization will be required.

Coxiella burnetti

Coxiella burnetti is the causative agent of Q fever, which is usually included with the rickettsial diseases because it is an intracellular parasite and stains poorly with the Gram stain. *Coxiella burnetti* appears to have more in common with species like *Legionella*. In fact, in contrast to the pathogen of the rickettsial disease, *Coxiella* is quite environmentally resistant. Most infections today are due to airborne acquisition of this pathogen, but *Coxiella burnetti* can be transferred through the milk of an animal (e.g., cow, goat, or sheep).

The infectious dose is low, perhaps only 10 microorganisms. Symptoms of the disease are varied. Possibly, the vast majority of people who are infected are asymptomatic; definitive data are not available. Most of the symptomatic cases show nausea, vomiting, diarrhea, myalgia, severe headache, chest or abdominal pain, and fever with sweating. In particularly severe cases, a hepatitis syndrome may emerge. Among these cases, fatality can result (about 1–2% of overall cases).

Animal herds are typically vaccinated against *Coxiella*; it does not present a major problem in the United States. However, it has been suggested as a biowarfare agent because of the low infectious dose and the resistant nature of the cells. Eradication seems unlikely because this is a worldwide illness and because a number of wild animals also harbor the microbe.

QUESTIONS FOR DISCUSSION

1. A microbe has a *D* value of 9 minutes at 55°C. What does this mean?

2. How are D values different from z values?
3. What other products (besides those mentioned here) could be pasteurized?
4. How many dairy products can you list?
5. If you let milk sit for a while, the cream will rise to the top. Would you expect to find a difference in the microbial content between the milk and the cream?
6. Why is time management essential when examining milk products for microbial content?
7. Why is alkaline phosphatase tested? When should this take place?

References

Bulthaus, M. 2004. Detection of antibiotic/drug residues in milk and dairy products, p. 293–323. *In* H.M. Wehr and J.F. Frank (eds.), Standard methods for the examination of dairy products. American Public Health Association, Washington, DC.

Davidson, P.M., L.A. Roth, and S.A. Gambrel-Lenarz. 2004. Coliform and other indicator bacteria, p.187–226. *In* H.M. Wehr and J.F. Frank (eds.), Standard methods for the examination of dairy products. American Public Health Association, Washington, DC.

Doyle, M.P., K.A. Glass, J.T. Beery, G.A. Garcia, D.J. Pollard, and R.D. Schultz. 1987. Survival of *Listeria monocytogenes* in milk during high-temperature, short-time pasteurization. Appl. Environ. Microbiol. 53: 1433–1438.

Frank, J.F., and A.E. Yousef. 2004. Tests for groups of microorganisms, p.227–247. *In* H.M. Wehr and J.F. Frank (eds.), Standard methods for the examination of dairy products. American Public Health Association, Washington, DC.

Grant, I.R., A.G. Williams, M.T. Rowe, and D.D. Muir. 2005. Efficacy of various pasteurization time-temperature conditions in combination with homogenization on inactivation of Mycobacterium avium subsp. paratuberculosis in milk. Appl. Environ. Microbial. 71: 2853–2861.

Hennessey, T.W., C.W. Hedberg, L. Slutsker, K.E. White, J.M. Besser-Wiek, et al. 1996. A national outbreak of *Salmonella enteriditis* infections from ice cream. New England J. Med. 334: 1281–1286.

Henning, D.R., R. Flowers, R. Reiser, and E.T. Ryser. 2004. Pathogens in milk and milk products, p.103–151. *In* H.M. Wehr and J.F. Frank (eds.), Standard methods for the examination of dairy products. American Public Health Association, Washington, DC.

Holzmann, H., S.W. Aberle, K. Stiasny, P. Werner, A. Mischak, B. Zainer, M. Netzer, S. Koppi, E. Bechter, and F.X. Heinz. 2009. Tickborne encephalitis from eating goat cheese in a mountain region of Austria. Emerg. Infect. Dis. 15:1671–1673.

Kells, H.R., and S.A. Lear. 1960. Thermal death time curve of *Mycobacterium tuberculosis* var. bovis in artificially infected milk. Appl. Microbiol. 8: 234–236.

Rocco, R.M. 2004. Alkaline phosphatase methods, p. 341–362. *In* H.M. Wehr and J.F. Frank (eds.), Standard methods for the examination of dairy products. American Public Health Association, Washington, DC.

Shere, J.A., C.W. Kaspar, K.J. Bartlett, S.E. Linden, B. Norell, S. Francey, and D.M. Schaefer. 2002. Shedding of *Escherichia coli* O157:H7 in dairy cattle housed in a confined environment following waterborne inoculation. Appl. Environ. Microbiol. 68: 1947–1954.

Winters, A., C. Driver, M. Macaraig, C. Clark, S.S. Munsiff, et al., 2005. Human tuberculosis caused by *Mycobacterium bovis*—New York City, 2001–2004. Morbid. Mortal. Weekly Report 54: 605–608.

CHAPTER 10

FOOD PRESERVATION

LEARNING OBJECTIVES

- List the general methods used to preserve food, and when they are appropriate.
- Give examples where multiple barriers to microbial contamination can be used.
- Describe the specific conditions used in food preservation.
- Describe the equations used to measure food contamination.
- List the most important food laws of the last century and the reasons they were enacted.
- Explain the Ames test and why it was needed.

KEY TERMS

- 12D rule
- Aseptic handling
- Create a hostile environment
- Equilibrium Relative Humidity (ERH)
- High-acid foods
- Killing or injuring microorganisms
- Removal of microorganisms
- Thermal death time
- Water activity (a_w)

We are all descendants of people who learned how to *not* starve to death. This is not as easy as it appears, viewed from the comfort of a modern society with plenty of food in the refrigerator and at the supermarket, and with more food just waiting for delivery or take-out. In ancient times, starvation was a very real danger. Weather was unpredictable, and a drought could wipe out an entire harvest. People in ancient societies commonly worshipped gods of agriculture and the harvest; they appealed to them with offerings and sacrifices because the farmers of the time had so little control over the success of their efforts.

In tropical areas of the world, the growing season is year-round, so some types of food are always available. And, of course, domestic animals are kept as food-on-the-hoof. Although the food available might not be preferred, it is edible, and it will prevent starvation. In other areas of the world, the seasons play an important role, and crops are only available during part of the year. Fish and wild game are often available, and patterns of migration may be exploited by hunters and fishermen, but unknown factors might prevent successful harvest. The fact is that having food available takes a lot of planning, effort, and technology.

A few simple rules of food preservation, if applied correctly, keep food edible in a nutritious state for a period of time that is longer than normal. Additional goals are to limit losses due to other animals (e.g., vermin and insects), to slow the ripening process (i.e., caused by indigenous enzymes), and to reduce spoilage. Spoilage is the degradation of a food product, typically due to bacterial action, to a point that it is no longer healthy or desirable to eat. This last point is important because different cultural traditions may determine that a food is no longer desirable.

Interestingly, the general principles used today for food preservation are the same ones that have been used since antiquity. Individuals still employ them for household use, although the biggest change has been to apply these principles on an industrial scale. One of the most significant developments of the modern era has been the production of vast amounts of food in a form that is both safe to eat and that has a long shelf life.

The first general principle is **aseptic handling** of the food and the **removal of microorganisms** if they contact the food. In some cases, the food can be filtered to remove microorganisms. This is possible with nonviscous liquids, which are useful because they have the ability to remove the heat-resistant spores. An aseptic condition is usually a goal rather than an expectation because food comes from decidedly nonsterile environments where aseptic conditions are not possible. This principle refers to the development of methods that limit the introduction of additional microorganisms during the process of preservation. These include washing soil particles from fruits and vegetables, making sure food does not touch the floor, separating the digestive tracts of animals from the carcass during slaughter, and many other methods. Another general principle is **killing or injuring microorganisms**. Obviously, if microorganisms are killed, then they are no longer a problem, but even damaging them can significantly slow their growth rate and greatly increase the infectious dose. If it is not possible to kill the microorganisms, then **create a hostile environment** on the food so they cannot multiply. Pathogens may get on a food product, such as meat, but as long as the numbers remain below the infectious dose, the food is safe for consumption. Stopping growth, or alternatively stopping the expression of essential pathogenicity factors, can be achieved through a number of means.

This leads to the specific principles of food preservation.

KEEP IT DRY

All microorganisms need water to live. Removing the water from a food product means that even if microorganisms get on the food, they will be unable to grow. This stops bacteria from multiplying to an infectious dose level and curtails spoilage. A good example of this principle is pasta, a dried and shaped unleavened wheat flour (e.g., spaghetti noodles). Pasta is typically sold in a plain cardboard box that sits on the shelf in the pantry for long periods of time without spoiling. Pasta is, of course, extremely dry. Another example is raisins, which are partially dried grapes. The grapes would spoil within a week or so, while the raisins remain edible far longer. (See **Table 10-1**.)

This principle has been followed for uncounted centuries. It probably became known when a portion of meat dried in the sun before it had a chance to spoil, and hungry scavengers noted that it could be chewed to provide a satisfying meal. Dried foods were very important on long journeys, by land or sea, when regular supplies of food could not be obtained. A barrel of water (or other potable) was kept separate from the dried food, and the travelers chewed the dried food and washed it down with water. Mariners dried meat ("jerky") and biscuits (called "hardtack" or "shipbiscuit") and some fruits and vegetables as well. The early explorers would have starved to death at sea on their months-long journeys were it not for dried foods.

It is not sufficient to say that something is "dry." How dry it is determines how long it will last. The easiest way to determine dryness is to find the weight difference before and after the item is placed in a drying oven, which will drive off any moisture. However, this is not as useful as determining an item's water activity. **Water activity (a_w)**, in contrast to the moisture content of the item, is the fraction of water that is available for use, specifically by bacteria or fungi (i.e., the main agents of spoilage; **Table 10-2**). Much of the water in a food may not be available because it is bound to other chemical constituents of the food.

Calculation of a_w of a food is relatively simple. While the theory behind the principle is complex, in practice it is only necessary to measure the vapor pressure while the food is inside a small, enclosed chamber. Comparison is made to the vapor pres-

TABLE **10-1** **Selected Water Activities**

Food	Water activity
Raw meats	0.99
Fruit juices	0.97
Breads	0.95
Cheese	0.95
Syrup	0.87
Rice	0.85
Salami	0.82
Jams and jellies	0.80
Peanut butter	0.70
Dried fruits	0.60
Powdered milk	0.20

sure of the reference sample, which is pure water. The reference is assigned a value of 1.0, and the a_w is thus a simple ratio:

$$a_w = \frac{p}{p_0}$$

In this equation, p equals the vapor pressure of the food under study, and p_0 equals the vapor pressure of pure water.

Under these circumstances, the a_w is always a fractional value of 1.0. Measurements are very sensitive to temperature, so care must be taken to ensure that both the sample and the reference sample are at the same temperature. For instance, a fixed-temperature water bath can be used. Changes in temperature may lead to either increases of vapor pressure, or decreases, or no changes at all. This change is entirely dependent on the food sample, which is dependent on a variety of factors such as changes in water binding or the solubility of solutes in the food. Other important factors include the humidity of the air in the chamber and the moisture content of the food. If mixed foods (e.g, such as a bacon, lettuce, and tomato sandwich) are tested, the results become less predictable. Water may preferentially move from one type of food to another at different temperatures rather than become unbound and contribute to vapor pressure.

An alternative method of reporting a_w is to hold the food sample in a confined space and let the relative humidity come to equilibrium at a set temperature. This may take hours, but after equilibrium is reached, the equilibrium moisture content can be measured. The result is dependent on the relative humidity of the air inside the chamber. A series of equilibrium moisture-content values taken at different relative humidities can be used to construct a water sorption isotherm for the food under study. The **equilibrium relative humidity (ERH)** is measured as a percentage and then converted to water activity:

$$a_w = \frac{ERH}{100}$$

The determination of a_w has a practical benefit. It is a good predictive index of whether microorganisms will grow on a particular food. All cells have osmoregulatory mechanisms to protect themselves against a hostile environment. Bacteria, for example, typically will not grow when the a_w is less than 0.91. Molds are more tolerant of dry conditions, but they will only grow on foods with a_w values above 0.80. This explains why certain foods,

TABLE **10-2** **Water Activity (a_w) Levels for Microbes**

Microbe	Minimum a_w required
Clostridium botulinum	0.94–1.0
Penicillium patulin toxin production	0.95
Bacillus cereus	0.95
Listeria monocytogenes	0.92
Aspergillus toxin production	0.87
Staphylococcus aureus	0.86
Aspergillus	0.82
Penicillium	0.81
Halophilic molds	0.75

such as fruits, develop mold before succumbing to spoilage by bacteria. Below a_w of 0.80, few microbes will grow. Notably, certain metabolic reactions have preferred water activities. Most molds are dangerous because of the toxins they produce. In most toxigenic mold species, the toxin is only produced at a higher a_w than the a_w required for the growth of the mold. Therefore the mold may grow at low a_w, yet still not produce the toxins.

The a_w can be changed in other ways besides simply drying the item completely. Salting of meat is typically used; the salt creates a hostile environment for the bacteria by decreasing the a_w of the bacteria on the food. The higher osmotic pressure of the salted water will lead to osmosis of water from bacterial cells, resulting in their dehydration. However, the major reason that salt curing is used is that it changes the a_w at the surface of the food. Salting is often used for meats, such as salt pork, beef (e.g., pastrami), or fish. This occurs because the addition of salt changes the characteristics of the surface water in what are known as the colligative properties. That is, the properties of the solution depend on the *concentration* of solute particles in the solution, and not on properties of the solute itself. Examples include elevation of boiling point and depression of freezing point. One of the key changes caused by adding salt (usually to 1–2% concentration on the surface) is a reduction in the vapor pressure, which is the basis for determining water activity. Vapor pressure changes according to Raoult's law, which states that the vapor pressure of a volatile component of a system is proportional to the mole fraction of the component, or

$$P_A = X_A P_A^o$$

where P_A is the observed vapor pressure, X_A is the mole fraction of the volatile component (water), and P_A^o is the vapor pressure of a pure sample of the volatile component. Because, in any mixture, the mole fraction of any single constituent must be less than 1.0, there is always a fractional multiplier of the pure compound vapor pressure, and the result is a lesser value. Other components of the mixture interact with the water molecules, setting up hydrogen bonds, van der Waals forces, and ionic bonds, which influence the energy in the water.

This is also demonstrated by sugar curing, which is known through items such as sugar-cured hams. Ordinarily the idea of combining two highly caloric foods like sugar and meat would seem to create an ideal medium for growth of bacteria. In fact, the sugar is rubbed into the surface of the meat (usually to 3–4% concentration on the surface), where Raoult's Law ensures that a_w will decrease. This decrease protects the meat; not enough water is present to support bacterial growth. Even slight changes in a_w can disrupt the enzymes on which the bacteria depend.

Both salt and sugar are useful because they are edible. They improve the flavor of the food. Both substances are also used in creating sausages, which are stored for extended periods. Other substances could be used to take advantage of Raoult's law, but they might impart a bad flavor to the meat product.

Of course, if the food becomes wet (water activity increases) for any reason, the conditions are right for rapid growth of bacteria and fungi. The example of spaghetti noodles was used above as an example of a long-lasting dry food. Ordinarily this is true; however, in a very humid home (i.e., relative humidity above 60%) the moisture in the air forms a thin film on the surface of the noodles that has a higher a_w. Molds can sometimes grow under these conditions, leading to spoilage. For this reason, household humidity should be no more than 50%.

KEEP IT COLD

Refrigeration has been used to preserve food for long periods. During the winter months in northern latitudes, people learned that they could cut ice from the local lakes during the winter and store the ice underground packed with sawdust; the ice would linger even during the warm months. Food stored in these large refrigerated pits lasted longer. When industrial refrigeration facilitated the production of ice at any time, ice was sold to individual families to be stored in the icebox (the first family refrigerator), which allowed long-term storage of food in the household but lacked a certain convenience. As the ice melted, the water pan had to be emptied or it would leak all over the floor.

Refrigeration conditions are usually at 2–6°C. Refrigeration works not because it kills bacteria, but because it slows down the growth of the bacteria. This is an important distinction, because contaminated food that is put in the refrigerator still remains contaminated when it is taken out. Also, if the food is contaminated but not highly enough to cause an infectious dose, then storage in the refrigerator will not completely stop growth; the bacteria will be able to grow slowly and perhaps eventually produce enough to create an infectious dose.

The principle behind refrigeration is the concept of doubling time (t_d), or the amount of time the bacteria need to double in number. Bacteria reproduce by binary fission, and so one bacterium divides into two, two into four, four into eight, and so on. Growth is dependent on availability of the right nutrients, of course, but all things being optimal, the bacteria grow at a certain rate. This can be expressed by an equation:

$$N = N_0 e^{\mu t}$$

$$2.3Log\left(\frac{N}{N_0}\right) = \mu t$$

$$\text{when } N = 2N_0 \text{ doubling population}$$

$$\text{then } t = \frac{0.693}{\mu}$$

where N_0 is the initial bacterial population, N is the final bacterial population, t is the elapsed time, and μ is the growth rate.

When temperature decreases, enzymatic activity decreases, metabolic activity decreases, and doubling time increases. The rate of growth depends on the species of bacteria. The lowest temperature at which a bacterial species grows is also the one at which all other conditions for growth are optimal. Varying these other conditions further inhibits growth.

Mesophilic bacteria prefer temperatures that are higher than refrigeration temperatures, and their growth is severely inhibited by refrigeration. Psychrotrophic bacteria prefer lower temperatures and, while 4°C is not optimal for them, they will still grow slowly. Some important pathogens are psychrotrophic, such as *Listeria monocytogenes* and *Vibrio parahemolyticus*.

A freezer typically operates at –20°C. This is far below the freezing point of water, but ensures that food is sufficiently frozen at all times. Food begins to freeze at a few degrees below 0°C, but if the freezing process is slow, bacteria may multiply before the process is complete. The lower temperature facilitates rapid freezing. Freezing has two benefits. First, it is so cold that metabolism of bacteria stops. Second, any water in the food is frozen and as such is no longer available. That is, the a_w decreases. In reality, the food is both cold and dry. If water leaves the food, it is by sublimation rather than evaporation. In these cases, it causes the effect called "freezer burn."

Freezing also has the benefit of damaging any microorganisms that contaminate the food. Ice crystals that form inside the cell will damage cell membranes and enzymes. In fact, repeated freeze–thaw cycles are often used to rupture cells for recovery of cellular constituents in the laboratory. However, freezing–thawing also damages the food, rupturing cells and releasing the contents where bacteria might utilize them. Therefore, foods should not be repeatedly frozen and thawed.

Freeze-drying (lyophilization) uses low temperatures and a vacuum to quickly remove water from food by sublimation. This is often preferable because the process does not lead to movement of solutes inside the cells, and when the food is reconstituted, it assumes a more natural form. Grains are stored dry and cold to prevent any mold growth. In fact, the major safety issue is the creation of high dust concentrations in grain elevators, which have been known to spontaneously explode.

KEEP IT UNCOMFORTABLE

Bacteria have a number of requirements for growth, the specifics of which depend on the individual bacterial species. Some bacteria are known to be associated with certain foods, either as pathogens or as spoilage agents. If conditions are manipulated so that they are not optimal for certain species, then their growth will be either stopped or severely inhibited. Here again, storage is prolonged even if spoilage is ultimately inevitable.

The oxidation or reduction conditions (i.e., redox potential, or Eh) can be manipulated to inhibit certain bacteria. Aerobic microbes require

TABLE 10-3 Select List of Approximate pH Values of Foods

Food	pH (range)
Apple	3.3–4.0
Apricot	3.3–4.8
Artichoke	5.5–6.0
Asparagus	6.0–6.7
Banana	4.5–5.2
Beans	5.6–6.5
Beets	5.3–6.6
Blueberries	3.1–3.3
Bread (white)	5.0–6.2
Bread (rye)	5.2–5.9
Broccoli (cooked)	6.3–6.5
Cantaloupe	6.1–6.6
Carrots	5.9–6.4
Celery	5.7–6.0
Cheese (American)	5.0
Corn	5.9–7.3
Cottage cheese	4.8–5.0
Crab meat	6.5–7.0
Cranberry juice	2.3–2.5
Gelatin dessert	2.6
Grapes	2.8–3.0
Honey	3.7–4.2
Ketchup	3.9
Lettuce	5.8–6.2
Maple syrup	5.2
Milk (cow)	6.4–6.8
Orange juice	3.3–4.2
Peaches	3.3–4.1
Peanut butter	6.3
Pineapple	3.2–4.0
Potatoes	5.4–5.9
Raspberries	3.2–4.0
Rice (white)	6.0–6.7
Shrimp	6.5–7.0
Soda crackers	5.7–7.3
Spaghetti (cooked)	6.0–6.4
Strawberries	3.0–3.9
Sweet potatoes	5.3–5.6
Tomatoes	4.3–4.9
Watermelon	5.2–5.6

Adapted from U.S. Food and Drug Administration, http://www.foodscience.caes.uga.edu/extension/documents/FDAapproximatepHoffoodslacf-phs.pdf

a positive Eh for growth, while obligate anaerobes require a negative Eh. Inclusion of 10% carbon dioxide in a package inhibits aerobic microbes. Vacuum packaging removes all atmospheric oxygen and thus inhibits aerobic microbial growth. The facultative anaerobes are more problematic because they are able to adjust to Eh conditions. In addition, Eh can determine the solubility of some nutrients, such as essential trace metals. Some species also have specific nutrient requirements, and if those nutrients can be removed or sequestered, the bacteria will not grow.

One of the most important conditions that can be changed is the pH of the food. All organisms have high and low pH limits for normal functioning of their cells (**Table 10-3**). Generally speaking, bacteria are sensitive to pH, although molds and yeast are more tolerant to pH. The benchmark is pH < 4.6. Foods with pH lower than this benchmark, either naturally or through manipulation of conditions, are "**high-acid foods**" and are especially resistant to spoilage. In addition, this level of acidity is sufficient to stop germination of bacterial spores.

It might appear that high-acid foods are rare in nature. But in fact, they are not at all unusual. Citrus fruits are an obvious example, but other fruits also have low pH values. Such low pH appears to be integral to the ripening process. Fruits are a high-energy resource and are therefore a good target for spoilage microorganisms. However, the skin of the fruit is a considerable barrier to microbes, and the pH of the fruit is an additional barrier. These conditions remain until the fruit has ripened, at which time the skin splits open and the conditions are altered. This also explains why molds are more likely to form on fruits; molds have a greater acid tolerance than bacteria.

Foods that are not naturally acidic can be turned into high-acid foods by adding a suitable acid, such as citric acid. Blanching (literally, "whitening") is a term that has been used for the surface treatment of foods. Steam blanching is effective at reducing bacterial counts on the food surface. Acid blanching (e.g., using 0.1–0.2% citric acid) has the advantage of treating the food surface while leaving a slightly acidic solution on the surface. Many foods acquire low pH through the fermentation processes.

Sauerkraut is made using lactic-acid-producing bacteria and shredded cabbage (similar to the Korean kimchi). In fact, "sauerkraut" literally means sour (or acid) herbs.

Pickling is another method for increasing acidity. There are two methods of pickling. The first involves adding acid directly to the food; for instance, vinegar (5% acetic acid) can be added to cucumbers to make pickles. The second method is called brining, and it involves a longer process that depends on certain bacteria in the solution that produce lactic acid that lowers the pH. In either case, the acidic taste is usually modified with a variety of spices or herbs and a little sugar to reduce bitterness.

While pickling of pickles sounds obvious, it is simply the archetype of the pickling industry. Many foods are preserved in this manner. Fish are pickled using vinegar, such as pickled herring and pickled mackerel. Eggs can be pickled. Often they are found in jars in bars, where, for some reason, they have become a traditional fixture. Pickled pork hock (the fleshy part of the leg) and pickled pig's feet are examples of domesticated animal products that are preserved by acidification. Vegetables such as green beans and asparagus have been pickled, as have mushrooms.

A variety of other inhibitory substances can be added to foods to aid in preservation. Natural inhibitors include sodium benzoate (found in cranberries), lysozyme (found in eggs), acids (such as propionic acid, a product of Propionibacteria, which is used to make Swiss cheese), alcohols (produced by bacterial fermentation, and which are seen in wine and beer), peroxides and antibiotics (which are produced by microorganisms and are not suitable as food additives). Some spices have traditionally been used as preservatives because they contain a variety of toxic chemicals, typically in low concentrations. Lysozyme is interesting because it attacks the cell wall of bacteria, weakening the structure until the cell bursts due to internal pressure.

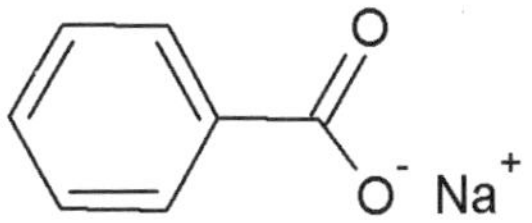

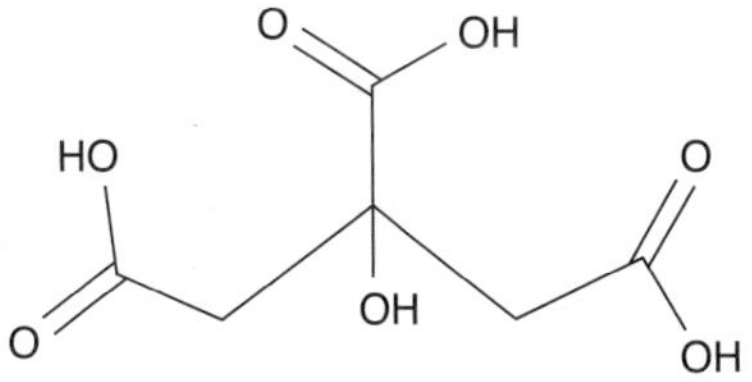

Citric acid

sodium benzoate

Artificial inhibitors have also been used extensively. Most people are familiar with smoked meats. Today the emphasis is on the type of wood used to "cure" the meat, which imparts a distinct flavor to the meat. However, smoking meat originated as a preservative. Wood was burned in a smoke house, and the particles of smoke would contact the surface of the meat, penetrating only a short distance into it. This was sufficient because typically meats are only contaminated on the surface, and the interior is sterile. The smoke particles carry a variety of toxic chemicals such as cresols, phenols, and formaldehyde. With such a toxic mixture on the surface, bacterial growth is greatly inhibited. Some people are concerned about the safety of eating meat that has been preserved in this manner because the chemicals identified are known carcinogens. However, they attach only to the meat surface, they can be removed, and the amounts are very small, assuming that normal portions of meat are consumed and that smoked meats are not a daily staple. The U.S. Food and Drug Administration (FDA) has identified other chemical additives that can be used as preservatives (**Table 10-4**). Again, the amounts used are small and do not appear to create a long-term risk for cancer. The FDA keeps a list of "Everything Added to Food in the United States" (EAFUS) at http://www.accessdata.fda.gov/scripts/fcn/fcnNavigation.cfm?rpt=eafusListing.

TABLE 10-4 Sample Classes of Food Additives

Class	Use	Example
Antioxidant	Inhibits oxidation of food	BHA, BHT
Emulsifier	A surfactant used to keep different liquids mixed	Lecithin
Spices, other natural seasonings, and flavorings	Distinct flavor	Ginger
Antifoaming (or defoaming) agent	Prevent foaming of foods during cooking	Aluminum stearate
Nutritive sweetener	Sweetens and is a substrate	Corn sugar
Chemical preservative	Retards growth of microorganisms	Erythorbic acid
Synthetic flavor	A non-natural flavoring	Ethyl vanillin
Color	Food coloration	Carrot oil

KILL MICROBES AND PREVENT THEM FROM RECURRING

Early humans discovered that cooking food, particularly meat, had many benefits. First, it imparted a better taste. Second, it made it easier to chew. Third, it destroyed any microorganisms on the food. Of course, they did not know what they were doing in this regard, but they must have noticed that they would not get sick as often eating cooked food. Finally, cooking preserved the food for a longer period of time. This is also due to the destruction of microorganisms on the food that would otherwise cause spoilage.

Over time, cooking (the application of moist heat) was employed in many formats to lengthen the time that food could be kept. The problem was always that, eventually, the food would reacquire microorganisms and then begin to spoil. If the food could be cooked and then placed in a sterile package, the spoilage was avoided. For most of human history, the concept of microorganisms was not understood, and therefore spoilage was not understood.

Cooking kills microbial cells by destroying their cellular constituents such as enzymes, structural proteins, and nucleic acids, and by rupturing cell membranes. Spores are much more heat resistant, and higher temperatures and/or longer times at higher temperatures are required to eliminate them. Spores may appear inert, but they contain enzymes that will be triggered under the right environmental conditions to begin the germination process. If these enzymes can be damaged, the spore is effectively inactivated.

A revolution in food storage occurred with the invention of canning, although the genesis of this durable technique came through an unexpected source. In the late 1700s, the emperor Napoleon Bonaparte needed to feed his army when it was on the march. In those days, armies typically "lived off the land," meaning that they bought or confiscated food from the people in the lands they passed through (or conquered). Armies thus had a certain maximal size because the land could only support a certain number of people and horses. It was also usual to go to war in the spring and summer months, when food supplies were more certain. Napoleon's armies were huge, and he did not wish to be constrained by the weather. In addition, his opponents might strategically destroy the food supplies as they retreated, making him dependent on long supply lines that were expensive to maintain and vulnerable to attack. Napoleon addressed this problem by offering a cash prize to the first person who could reliably preserve food.

Nicolas Appert won the prize by deducing that food could be cooked and then stored in a bottle in which air had been driven out, much as wine stayed drinkable after bottling. Because suitable containers were not available at the time, he settled on champagne bottles. Because these bottles had a narrow neck, they would not readily accept most food. But peas could be funneled into the bottles; the first canned food was thus cooked peas.

An Englishman named Peter Durand took the idea to the next stage by creating the first tin can. These containers withstood cooking far better than glass bottles, which occasionally would break if overheated. Today cans are usually made from steel or aluminum, but they are occasionally still referred to as "tins." Today, a "can" is hermetically sealed package of any type of material (i.e., metal, glass, or plastic).

Canning caught on quickly because of the obvious benefits. However, a new danger was introduced. If soil contaminated the food, then bacterial spores might be introduced. Spores survive cooking conditions and can then germinate. In some cases, the spore-borne bacteria would grow in the sealed can and spoil the food. This was usually discernable by the bad odor. In some unfortunate cases, however, the obligate anaerobe *Clostridium botulinum* was included in the can. These spores germinated, and the resulting colony produced the botulinum toxin, the causative agent of botulism. This toxin is one of the most powerful known in nature. It acts by interrupting nerve impulses that control muscular contractions, resulting in an inability of the muscles to properly contract. If enough toxin is ingested, the muscles that control breathing are affected and the person dies from asphyxiation. Contamination can occur before cooking and canning, and many bacterial toxins remain stable in high heat. In this case, the toxin was canned as well, sickening anyone who ate it later.

As an anaerobe, *C. botulinum* has no trouble growing in a sealed can. A single spore is sufficient to contaminate the contents and to produce enough toxin to kill a person. The LD50 is estimated at no more than 0.05 g/kg of body weight. In response to this problem, rules were established for proper canning. In particular, the **12D rule** is a requirement for the canning industry. This was first proposed by Esty and Meyer in 1922 and has become an industry standard. "D" is a shorthand term meaning an order of magnitude (see Chapter 9). The 12D rule specifies that the sterilization method used for canning must be sufficient to reduce the viable spores by 12 orders of magnitude. This depends on the concept of **thermal death time** (TDT), which is the length of time at a certain temperature that is required to kill a target population (in a water-based solution). Higher temperatures require shorter lengths of time to be effective. Canning temperature and time depend upon food acidity (i.e., high-acid foods require less time), density of the can contents, and heat-transfer properties.

TDT is specific for a particular microorganism; there is a big difference between TDT for *E. coli* and for spores. For *Clostridium* spores, the values are 2.78 minutes at 250°F (121°C). The concept of TDT is particularly relevant in pasteurization (see Chapter 9 for more detail). Many processed foods are pasteurized and stored semi-dry. This uses the two-obstacle approach, in which any of the techniques can be combined to ensure the safety of the food in storage.

The same principle holds true for energy sources other than heat, such as irradiation with gamma rays of Co^{60} or X-rays. These are powerful ionizing radiation sources, and the rays easily pass through metal cans. This means that the cans can be sterilized very quickly on a conveyor belt as the cans move past a high-energy gamma source. Radiation is measured in units of Grays, and *D* values can be determined for individual pathogens for specific food products and in specific packages. For example, one *D* value for *E. coli* is about 0.3 kiloGrays. To obtain a 5D (killing 99.999% of *E. coli*) result, 1.5 kiloGrays of radiation are used.

Because a high level of spore contamination is unlikely, the 12D standard has a generous margin of safety. In fact, problems with canned goods in an industrial society are quite rare. In the United States, approximately 150–200 cases of botulism occur per year; perhaps 20–30 are due to mistakes with home canning operations, and another 30–40 are due to wound infections. The remaining cases are infant botulism cases, which are discussed later in this chapter.

A number of other processes are being developed or are currently being used to kill or injure bacteria. The common element is the efficient transfer of energy to the food in some form to damage any living cells within it. Ultraviolet light has been used to damage cells in food production areas, at least where human eyesight cannot be affected. UV lights are used to knock down bacterial numbers in the air (e.g., dust or particle-related contamination) in filling and bottling operations and in food storage areas. The UV light damages bacterial cells by causing DNA damage.

FOOD LAWS

More law is probably written about food, food production and processing, and related areas than about any other area of law. The laws are comprehensive and specific, and they abound at the federal level and within each state (**Box 10-1**). A further layer can be found at local levels. Lawmakers depend heavily on food scientists and microbiologists to create effective rules that improve the safety of the food supply without overburdening the food industry.

Federal law trumps all other law in the United States. That is, state and local laws can make regulations that are more stringent, but state and local governments cannot make regulations that are less stringent than federal rules. The U.S. Code (USC) and the Code of Federal Regulations (CFR) describe the processes that must be used and the means by which quality is ensured. This is represented by a list of specific procedures that are approved by the government for use in food production, the Current Good Manufacturing Practices (CGMP, sometimes referred to only as GMP). These procedures are not merely suggestions; they are the only approved procedures that can be used because they have been determined to be safe and effective. New methods must undergo a series of steps before they are accepted, and these steps can be complex.

For defense against microbial contamination, the CGMP defer to the Bacteriological Analytical Manual (BAM), which is a collection of procedures and practices approved by the federal government

BOX 10-1 Code of Federal Regulations

The Code of Federal Regulations (CFR) lists the rules and regulations that are promulgated by the federal government through its various departments and agencies. The rules appear in the Federal Register and then in the CFR, which is updated frequently. The specifics of the laws can be found at the CFR website (http://www.gpoaccess.gov/CFR/).

Title 21 of the CFR involves laws about food and drugs. The rules cover much more than microbiological contamination of the food products. Other sorts of contamination are included, as are accepted practices, monitoring methods, and requirements for storage and transportation.

Other CFR titles are also relevant to this study. Title 7 is Agriculture; Title 9 is Animals and Animal Products; Title 50 is Wildlife and Fisheries. Also of note, Title 42 is Public Health.

The CFR is not the same as the U.S. Code (USC; http://www.gpoaccess.gov/uscode/), which is the compilation of all of the federal laws. Like the CFR, the USC is updated on a regular basis and is divided into various titles. Some of the USC titles match the CFR titles, such as Title 21, Food and Drugs. Others do not match. Title 9 CFR is Animals and Animal Products, while Title 9 USC is Arbitration.

that have proven effective. (See Appendix I for information about both CGMP and BAM.)

In the early days of regulation, the determination of safe and healthy food was relatively easy. A piece of meat was either normal or it was not. Over the years, the number of processed foods available has dramatically increased. Processed foods have been partially or completely prepared by a company to provide convenience to the consumer. For example, a can of beef stew has beef from one source, vegetables from several other sources, and spices from still other sources, any of which may introduce contamination. Cooking and canning steps must be performed without contaminating the product. The resulting processed food product becomes available in a form that can be reheated in minutes and then consumed.

Because of the intricacy of current food processing methods, it is no longer practical to have inspectors at every food processing company. Instead, the FDA depends on Hazard Analysis and Critical Control Points (HACCP). This method of oversight of the food manufacturing process identifies and monitors the likeliest points for contamination of the system, and over time, problems are corrected with new procedures and new technology (i.e., points in food production where loss of control would have serious health effects later). In this constantly changing industry, this type of flexibility is essential.

Federal Laws

The prominence of federal law was not always accepted. Only in the past 100 years or so has food regulation by a federal governmental entity been considered.

Federal law is considered to be dependent on the U.S. Constitution, Article I, Section 8, which says in part: "The Congress shall have power . . . to regulate commerce with foreign nations, and among the several States, and with the Indian Tribes. . . ." The idea here was that the colonial states would regulate their own internal economies, but the federal government could regulate interstate commerce. The federal government was considered to be a fair arbiter of interactions between the states. One area where this is evident is in the power to issue recalls of contaminated food (**Box 10-2**).

In colonial times, this was not a major issue, especially for food. Transportation was characteristically slow; horse-drawn wagons or riverboats were the most rapid means of transportation. Most food would spoil before it was transported to another state. One example of this problem is seen in the first major challenge to the federal government,

BOX 10-2 Recalls

Occasionally a food product is recalled because of microbial contamination (e.g., with *E. coli* O157). In fact, there are different classes of recalls. Quoting directly from 21 CFR 7.3(m) on the definition of types of recalls:

> *Recall classification* means the numerical designation, i.e., I, II, or III, assigned by the Food and Drug Administration to a particular product recall to indicate the relative degree of health hazard presented by the product being recalled.
>
> (1) Class I is a situation in which there is a reasonable probability that the use of, or exposure to, a violative product will cause serious adverse health consequences or death.
> (2) Class II is a situation in which use of, or exposure to, a violative product may cause temporary or medically reversible adverse health consequences or where the probability of serious adverse health consequences is remote.
> (3) Class III is a situation in which use of, or exposure to, a violative product is not likely to cause adverse health consequences.

the whiskey tax of 1791. The federal government implemented an excise tax on distilled spirits (chiefly whiskey). This angered many farmers of the frontier. They raised grains such as corn and used the excess to make whiskey. Whiskey had the advantages of being in great demand (colonial people were rather heavy drinkers, on average) and being much more portable than corn because whiskey did not spoil. Dissatisfaction with the tax led to the Whiskey Rebellion of 1794, which was quickly suppressed. If the farmers could have brought their corn to market faster, they would have avoided the tax.

Federal law was limited to importation of foreign food and drugs until the 1900s. As detailed in Chapter 8, the changes in transportation in the United States—the railroads and with rudimentary refrigeration—meant that food of all sorts could be transported long distances, including across state lines, giving the federal government authority over it. The first major national legislation was the 1906 Federal Food and Drug Act. It prohibited interstate transport of adulterated food or drugs, but notably, it did not regulate their manufacture. Over the next 100 years, federal law became more robust, until the extent of regulation became comprehensive. Today there is little opposition to this level of regulation; it is effective and efficient. Practically all foods are subject to federal food laws, even if the producer means it only for local or state consumption. The mobility of today's society, with individual cars and interstate trucking, means that anyone might travel to a state, buy a food product there, and carry it into another state in the same day. With the exception of roadside produce sales, practically every food item is subject to some regulation.

A number of federal departments and agencies are involved in food safety. The FDA, a branch of the Department of Health and Human Services, is responsible for much of the regulation and policing of the food industry. However, the U.S. Department of Agriculture is specifically responsible for meat and poultry inspection. The EPA oversees pesticide use. The Department of the Interior is responsible for fisheries. Many of the workers in these agencies have worked for the Public Health Service. Research is performed through the National Institute of Health.

The major legislative acts are outlined here.

The Biologics Control Act of 1902

This law was passed in the wake of a tragedy that claimed the lives of several children from a tainted antiserum. At the time, there was no diphtheria vaccine, and outbreaks were common. In severe cases, a diphtheria antiserum could be used. Unfortunately, a batch was contaminated with tetanus, and this proved fatal. The law required safety assessment of serums, vaccines, and other products that were used to treat disease.

The Food and Drug Act of 1906

This law was enacted in response to the publication of Upton Sinclair's *The Jungle* (see Chapter 8), which came to the attention of President Teddy Roosevelt and which detailed the unhygienic practices of the meatpacking industry. Among other things, it caused the creation of Title 21 of the Code of Federal Regulations. This law specifies that misbranded or adulterated food and drugs may not be used in interstate commerce. In addition, it defined dangerous food processing procedures and dangerous additives.

The Meat Inspection Act of 1906

This law was passed on the same day as the Food and Drug Act. It instituted cleanliness standards for meat processing plants and required federal inspectors to examine the animals being used and the meat after it was butchered.

The Federal Food, Drug, and Cosmetic Act of 1938

In 1937, a drug company created a sulfa drug called Elixir Sulfanilamide. It was dissolved in diethylene glycol, which has a lethal dose in the range of 0.5–5.0 grams per kilogram of body weight. Toxic doses of diethylene glycol were easily caused by ordinary doses of the elixir, and at least 100 people died. This event caused a new law to be passed to avoid future errors like this one.

As the name implies, oversight of cosmetics was awarded to the FDA. This law was more proactive than the 1906 laws; it outlawed some manufacturing practices. It established tolerances for poisonous components that could not be avoided. In addition, it allowed inspection of factories.

The Public Health Service Act of 1944

The Public Health Service (PHS) had a long history in the United States prior to 1944, but this act extended its range of disciplines to all of the known areas of health. The PHS now has a major impact on many areas of health in the United States, including training experts in food safety.

The Penicillin Amendment (1945)

Penicillin was one of the first wonder drugs, and this law established standards for the safety and effectiveness of penicillin. As new antibiotics were introduced, the law expanded to include them as well.

The Food Additives Amendment (1958)

During and after World War II, there was a revolution in the chemical manufacturing industry. A tremendous number of new chemical compounds were produced, some of which could be used in food manufacturing (e.g., pesticides and herbicides) and some of which could be used in food processing. Additives are not themselves a type of food, so they were not covered by previous laws, but they were consumed by people just as the food was. This amendment to the Federal Food, Drug, and Cosmetics Act corrected the additives oversight.

Notably, the Food Additives Amendment contained a provision that was added by its principal author, Congressman James Delaney of New York. The so-called Delaney Clause was inserted because of the concern that certain chemicals might, with repeated ingestion over a long period of time, cause cancer. Specifically, it stated the following:

> *No additive shall be deemed to be safe if it is found to induce cancer when ingested by man or laboratory animals or if it is found, after tests which are appropriate for the evaluation of the safety of food additives, to induce cancer in man or animals.*

The effect of this provision was that any food additive had to be tested for carcinogenicity before it could be used. Initially this included pesticides as well, because pesticide residues could be found in processed food. However, pesticides were removed from this law with the passage of the Food Quality Protection Act of 1996. The Environmental Protection Agency (EPA) took over regulation of pesticides and developed a standard of negligible risk, by which they meant less than one extra case of cancer per million people if exposed daily to the chemical at a certain concentration over a lifetime of 70 years.

The law was well intentioned, although Delaney focused on a single issue—cancer—as the rationale for the rule. He could have included other potential hazards, such as birth defects, nervous system disorders, immune system disorders, or endocrine disorders. But data were lacking on these other areas at the time, and the risk for cancer that was caused by chemicals was beginning to be appreciated.

The law was not especially well written, and parts of it are difficult to reconcile. In particular, Section 402 states that pesticide residues on processed food may not be higher than on the same raw food. This is sensible: no one wants the processing step to concentrate the pesticide. Section 408 specified that the amount of residue found on raw food would be subject to a threshold amount, based on a cost–benefit analysis. This is also sensible: pesticides are useful, but too much may be harmful, and a balance must be determined. However, Section 409 states that pesticide residues cannot be found in processed food if they are shown to cause cancer in humans of animals, regardless of the benefit. This clause was finally interpreted to mean a "zero risk" from additives to food.

Of course, zero risk is hard to prove. Even if the chemical compound causes one extra case of cancer in a billion people, that would disqualify its use under the Delaney Clause. And how could this extraordinarily low number be estimated? This was the question before chemical manufacturers after the passage of this law. Eventually it was addressed by two methods, the Ames Test (**Box 10-3**) and GRAS exemptions.

GRAS stands for "generally regarded as safe." Chemicals that have been used safely without incident for many years may get an exemption for the Delaney Clause. For example, salt has been used for hundreds of years as a preservative and as a flavor enhancer. Although too much salt is itself unhealthy, salt is not known as a carcinogen.

The Fair Packaging and Labeling Act of 1966

The descriptive labeling of ingredients on food packages had its origins in this act, which has been

BOX 10-3 The Ames Test

The Delaney Clause specified that any chemical that was known to cause cancer in humans or animals could not be used as a food additive. But how can this be proven? It is obviously unethical to test the chemical on humans. Testing on animals, such as mice or rats, is certainly possible, especially because they have relatively short life spans and thus can be studied as a surrogate for a human life span. However, the scale of such a study would be huge, and the cost would be prohibitive. Determining whether a standard can be met—such as no more than one extra case of cancer per million people—would cost millions of dollars.

A clever solution to this intractable problem was devised by Dr. Bruce Ames. The Ames test (Ames et al., 1973b) was predicated on the idea that carcinogenicity was related to gene mutagenicity. The more a chemical compound damaged the DNA of a cell, the more it would need repair. The more DNA repair in the cell, the greater likelihood of carcinoma. This can be seen in skin cancers (melanomas); people who damage their skin repeatedly through exposure to the sun have a higher risk of cancer.

Ames created several strains of *Salmonella*, each with a specific type of mutation in an easily monitored gene. Each strain was exposed to the chemical, and if DNA damage occurred, the gene mutation reverted to the wild-type phenotype, which was easily assayed on an agar plate. The numbers of colonies were easily determined, indicating the magnitude of the effect. In addition, a dose response was tested to determine whether there was a threshold effect for the chemical.

An early criticism of the test was that chemicals are often transformed in the body by the liver's enzymes. To account for this effect, the chemicals were exposed to macerated liver extracts and then used for mutagenicity experiments (Ames et al., 1973a).

These tests were used to easily and quickly screen new chemical compounds. Over the years, however, the test has been recognized to have significant limitations. If a high enough concentration of nearly any chemical is used, a certain amount of DNA damage occurs. Therefore only concentrations that are relevant to human exposure must be used. Of course, humans are exposed to small concentrations of chemicals over a period of many years. This cannot be duplicated with the bacterial test. Bacteria do not develop and age in the way that humans do. The other problem is that all food items contain some chemical compounds that are, in high concentrations, carcinogenic (or at least mutagenic when the Ames test is used). Of course, eating these foods and their small amounts of chemicals does not seem to present any certain difficulty for people in the real world. No one wants to outlaw ordinary foods (e.g., apples, potatoes, or beans) on the grounds that they contain miniscule amounts of dangerous compounds.

amended several times to provide clearer and more helpful information to the consumer about foods.

The Food Allergy Labeling and Consumer Protection Act of 2004

The incidence of food allergies has risen markedly in recent years. In some cases, the allergic reactions can be quite severe. This act requires food to be labeled clearly to indicate whether they contain any of the most common allergens: peanuts, soybeans, cow's milk, eggs, fish, shellfish, tree nuts, and wheat.

FIGURE **10-1** **A sample requisition form for food monitoring.**

Laboratory Requisition

City of Milwaukee Health Department
Public Health Laboratories

Public Health Laboratories
Frank P. Zeidler Municipal Building
841 North Broadway, Room 205
Milwaukee, WI 53202-3653

Phone: (414) 286-3526 Fax: (414) 286-5098

Submitting Division: ☐ CEH ☐ DCP ☐ HEH ☐ DNS

Collected By:__________ **Phone:**__________

Collection Address:__________
Patient/Client Name:__________
Address:__________
Date Collected:__________

Laboratory Division: ☐ Microbiology ☐ Chemistry ☐ Virology

Sample ID: **A** **Time Collected:**____ **Analyze For:**____ **Sample Information:**____	***Sample ID:*** **B** **Time Collected:**____ **Analyze For:**____ **Sample Information:**____
Sample ID: **C** **Time Collected:**____ **Analyze For:**____ **Sample Information:**____	***Sample ID:*** **D** **Time Collected:**____ **Analyze For:**____ **Sample Information:**____

Special Instructions/Comments:__________

SOURCE: Courtesy of the Milwaukee Health Department.

FIGURE 10-2 A sample form for a once-a-year festival.

MILWAUKEE HEALTH DEPARTMENT
Division of Disease Control and Environmental Health

Henry W. Maier Festival Grounds
Water Sampling Data Sheet

Sampler		Group Number		Date Analyzed	
Date Collected		Time Collected		Time Analyzed	

Station #	Location	pH	Temp °F	Total Chlorine	Total Coliform Count/100 ml	EHPC/ml
1	North Restroom					
2	Greek Village 7					
3	Leine's Restroom					
4	Saz's-Bldg 12					
5	Children's-Bldg 15					
6	Sport's Building-Bldg 19					
7	Harley Davidson-Bldg 28					
8	Venice Club-Bldg 29					
9	Miller Restroom					
10	Major Goolsby's-Bldg 44					
11	Pitch's Barbecue-Bldg 39					
12	Old South Restroom					
13	Forecourt					
14	North Concourse					
15	South Concourse					

Analyst ______________________ Date ________

HETEROTROPHIC PLATE COUNT TO FOLLOW

SOURCE: Courtesy of the Milwaukee Health Department.

State and Local Laws

State laws typically fine-tune federal laws to account for local conditions. That is, they may make the rules more stringent, but not less so. If local conditions suggest a local problem (e.g., because of the climate), they can take remedial actions. State laws also may establish statewide standards for industries. In Wisconsin, very specific instructions on cheese manufacture have been implemented because of the importance of that industry to the economy and the image of the state. Monitoring is performed on a routine basis (**Figure 10-1**) or for special occasions (**Figure 10-2**).

Local food laws respond to very specific concerns of a city or town. They may include additions to state and federal laws, but often there is no need. The most prominent role of local food laws is restaurant inspection, which is carried out by local Boards of Health. This can include local building codes, parts of which may be designed to limit contamination (e.g., resistance to vermin).

Local health departments, which in large cities often include a very impressive laboratory, are responsible for collecting a large volume of health information. This information includes not only data about infectious disease but many other illnesses as well.

QUESTIONS FOR DISCUSSION

1. What are the general methods used for food preservation?
2. How can these methods be combined to add extra layers of protection?
3. How has federal food law changed over the years, and what is the current extent of the law?
4. What are Current Good Manufacturing Practices?
5. If you want to try a new manufacturing practice for a food product, how would you get it approved?
6. What is water activity?
7. What is a high-acid food?
8. What effect did the Delaney Clause have on food processing?

References

Ames, B.N., W.E. Durston, E. Yamasaki, and F.D. Lee. 1973a. Carcinogens are mutagens: a simple test system combining liver homogenates for activation and bacteria for detection. Proc. Natl. Acad. Sci. USA. 70: 2281–2285.

Ames, B.N., F.D. Lee, and W.E. Durston. 1973b. An improved bacterial test system for the detection and classification of mutagens and carcinogens. Proc. Natl. Acad. Sci. USA. 70: 782–786.

CHAPTER 11

VACCINES

LEARNING OBJECTIVES

- Summarize how vaccines work.
- Describe the current vaccines and the diseases from which they protect us.
- Explain the types of problems that are encountered in vaccine use.
- Distinguish between killed and attenuated types of vaccines.
- Determine what the future needs are for vaccines.

KEY TERMS

- Herd immunity
- Killed
- Live attenuated

A LITTLE HISTORY

The creation of effective vaccines was one of the most significant medical developments in human history. Probably no other invention has saved more lives from deadly diseases than vaccination (**Tables 11-1** and **11-2**). Now a schedule of vaccines, most of which are given during early childhood, has been very effective in reducing morbidity and mortality.

The father of vaccination is Dr. Edward Jenner, an English physician who knew well the ravages of epidemic disease, particularly smallpox. Smallpox was one of the greatest killers in history, and during epidemics, the youngest people (birth to 3 or 4 years old) were often most afflicted and had the highest mortality rate. A procedure was available at the time to provide protection against smallpox. Variolation, named after the Variola virus, involved the use of pustular matter from the sores (pox) from a smallpox sufferer. If the material was taken when the person was in recovery, it included enough dead viral material to stimulate an immune response in another person in whose system the material was introduced. Typically this was done by scratching the material into the arm.

Variolation worked well if done properly, but there was always a significant risk that the procedure would actually infect the person with virulent smallpox virus. For this reason, it was not entirely trusted, and in some areas the procedure was outlawed. Jenner knew about this procedure and its

TABLE 11-1 Diseases That Can Be Prevented by Vaccines

Anthrax
Cervical cancer
Diphtheria
H1N1 flu (swine flu)
Haemophilus influenzae type b (Hib)
Hepatitis A
Hepatitis B
Human papillomavirus (HPV)
Influenza (seasonal flu)
Japanese encephalitis (JE)
Lyme disease
Measles
Meningococcal
Monkeypox
Mumps
Pertussis (whooping cough)
Pneumococcal
Poliomyelitis (polio)
Rabies
Rotavirus
Rubella (German measles)
Shingles (Herpes zoster)
Smallpox
Tetanus (lockjaw)
Tuberculosis
Typhoid fever
Varicella (chickenpox)
Yellow fever

limitations. He had also observed that people who had previously contracted the related illness cowpox appeared to be immune from smallpox infection. This discovery is commonly credited to his observations about milkmaids, who certainly had a close association with cows and who were often infected with cowpox. (Jenner made additional observations about swinepox.) However, his notes include descriptions of other people who had contracted cowpox in the past and who had emerged from smallpox epidemics without contracting the disease. He was a careful observer, and he used his data to hypothesize that a cowpox infection gave a long-lasting protection against smallpox.

Although cowpox has been assessed as an essentially benign infection, people who contract the disease can acquire blistering sores and skin ulcerations. This was not a pleasant experience, but if it could protect you from smallpox, it was worth the discomfort. Jenner tried his idea in 1796 by inoculating a young boy with cowpox from an infected milkmaid. Some time later, he tried several times to deliberately infect the same boy with smallpox. In every case, the disease did not appear, demonstrating that the boy had immunity from the disease. This type of experiment would not be allowed today, but it was certainly successful.

Jenner named his procedure "vaccination" because the source of the material was a cow (after the Latin word *vacca*, cow). He had no knowledge of viruses, which would not be described for another 100 years. But his success paved the way for the development of other vaccines, especially after microbiology became a science and specific pathogens were isolated and identified. Not all pathogens had convenient nonpathogenic analogs, but scientists such as Pasteur were able to produce vaccines for diseases such as anthrax and rabies.

The scientist Robert Koch also tried to develop a vaccine, in this case for tuberculosis, for which he had identified the pathogen. In this case, he produced a cell-free extract that he hoped would produce the desired effect. We recognize this extract today as a type of tuberculin, which is used now to identify cases of primary tuberculosis (i.e., an antibody response to the tuberculin protein). Unfortunately his tuberculin preparation was completely unsuitable as a vaccine, resulting in massive inflammation and, reportedly, some deaths. These results produced bad publicity, and his efforts to promote it damaged the reputation of vaccination for many years (Gradmann, 2001).

The pioneering work of Jonas Salk and Albert Sabin led to the development of successful vaccines against polio. Salk introduced his killed vaccine in 1955, and it had an immediate effect. Sabin's more effective, live, attenuated vaccine was introduced soon after, and in 1960, it was widely distributed in the United States. This vaccine was so

TABLE 11-2 Single- and Multiple-Target Vaccines

Part A: Single-Target Vaccines

Vaccine	Target	Type	Route of delivery	Manufacturer
Bacterial Vaccines				
BioThrax	*Bacillus anthracis* (anthrax)	Subunit (cell-free filtrate)	Intramuscular	Emergent BioSolutions
ActHIB	*Haemophilus influenzae* type B (meningitis)	Subunit (Hib polysaccharide conjugated to tetanus toxoid)	Intramuscular	Sanofi Pasteur SA
PedvaxHIB	*Haemophilus influenzae* type B (meningitis)	Subunit (Hib polysaccharide conjugated to Neisseria outer membrane protein)	Intramuscular	Merck
Hiberix (booster only)	*Haemophilus influenzae* type B (meningitis)	Subunit (Hib polysaccharide conjugated to tetanus toxoid)	Intramuscular	GlaxoSmithKline
Menactra	*Neisseria meningitides* (meningitis)	Subunit (Neisseria polysaccharide conjugated to diphtheria toxoid; types A, C, Y, W-135)	Intramuscular	Sanofi Pasteur SA
Menomune	*Neisseria meningitides* (meningitis)	Subunit (Neisseria polysaccharide types A, C, Y, W-135)	Subcutaneous	Sanofi Pasteur SA
Menveo	*Neisseria meningitides* (meningitis)	Subunit (Neisseria polysaccharide conjugated to diphtheria toxoid; types A, C, Y, W-135)	Intramuscular	Novartis
Prevnar	*Streptococcus pneumoniae* (meningitis)	Subunit (capsular saccharide serotypes 4, 6B, 9V, 14, 18C, 19F, and 23F conjugated to diphtheria protein)	Intramuscular	Wyeth

(continued)

TABLE 11-2 *continued*

Part A: Single-Target Vaccines

Vaccine	Target	Type	Route of delivery	Manufacturer
Prevnar13	*Streptococcus pneumoniae* (meningitis)	Subunit (capsular saccharide serotypes 1, 3, 4, 5, 6A, 6B, 7F, 9V, 14, 18C, 19A, 19F and 23F conjugated to diphtheria protein)	Intramuscular	Wyeth
Pneumovax 23	*Streptococcus pneumoniae* (meningitis)	Subunit (capsular saccharide serotypes 1, 2, 3, 4, 5, 6B, 7F, 8, 9N, 9V, 10A, 11A, 12F, 14, 15B, 17F, 18C, 19F, 19A, 20, 22F, 23F, 33F)	Intramuscular or subcutaneous	Merck
TICE BCG	*Mycobacterium tuberculosis*	Live attenuated *Mycobacterium bovis* (Bacillus of Calmette and Guerin)	Percutaneous	Organon
Mycobax	*Mycobacterium tuberculosis*	Live attenuated *Mycobacterium bovis* (Bacillus of Calmette and Guerin, BCG)	Percutaneous	Sanofi Pasteur SA
Vivotif	*Salmonella typhi* (typhoid fever)	Live attenuated	Oral	Berna Biotech
Typhim Vi	*Salmonella typhi* (typhoid fever)	Subunit (Vi polysaccharide)	Intramuscular	Sanofi Pasteur SA
Viral Vaccines				
Varivax	*Varicella zoster* (chickenpox)	Live attenuated	Subcutaneous	Merck
Havrix	Hepatitis A virus	Killed	Intramuscular	GlaxoSmithKline
Vaqta	Hepatitis A virus	Killed	Intramuscular	Merck
Engerix-B	Hepatitis B virus	Subunit	Intramuscular	GlaxoSmithKline
Recombivax HB	Hepatitis B virus	Subunit	Intramuscular	Merck

Part A: Single-Target Vaccines

Vaccine	Target	Type	Route of delivery	Manufacturer
Gardisil	Human papillomavirus (prevents cervical cancer)	Subunit (capsid proteins of types 6, 11, 16, 18)	Intramuscular	Merck
Cervarix	Human papillomavirus (prevents cervical cancer)	Subunit (capsid proteins of types 16, 18)	Intramuscular	GlaxoSmithKline
Afluria	Influenza virus A and B	Killed	Intramuscular	Merck
Agriflu	Influenza virus A and B	Killed	Intramuscular	Novartis
FluLaval	Influenza virus A and B	Killed	Intramuscular	GlaxoSmithKline
Fluarix	Influenza virus A and B	Killed	Intramuscular	GlaxoSmithKline
Fluvirin	Influenza virus A and B	Killed	Intramuscular	Novartis
Fluzone	Influenza virus A and B	Killed	Intramuscular	Sanofi Pasteur SA
FluMist	Influenza virus A and B	Live attenuated	Intranasal	MedImmune
Ixiaro	Japanese encephalitis virus	Killed	Intramuscular	Novartis
JE-Vax	Japanese encephalitis virus	Killed	Subcutaneous	Sanofi Pasteur SA
Ipol	Polio virus 1, 2, and 3	Killed	Intramuscular or subcutaneous	Sanofi Pasteur SA
Imovax Rabies	Rabies virus	Killed	Intramuscular	Sanofi Pasteur SA
RabAvert	Rabies virus	Killed	Intramuscular	Novartis
Rotarix	Rotavirus	Live attenuated	Oral	GlaxoSmithKline
RotaTeq	Rotavirus	Live attenuated	Oral	Merck
Zostavax	Herpes zoster (shingles)	Live attenuated	Subcutaneous	Merck
ACAM2000	Smallpox and monkeypox	Live Vaccinia virus	Percutaneous	Acambis
Dryvax	Smallpox and monkeypox	Live Vaccinia virus	Percutaneous	Wyeth
YF-Vax	Yellow fever virus	Live attenuated	Subcutaneous	Sanofi Pasteur SA

(continued)

continued

Part B: Multiple-Target Vaccines

Vaccine	Target	Type	Route of delivery	Manufacturer
ProQuad	Measles Mumps Rubella Varicella	Live attenuated	Subcutaneous	Merck
M-M-R II	Measles Mumps Rubella	Live attenuated	Subcutaneous	Merck
Daptacel	Diphtheria toxin Tetanus toxin Bordetella pertussis (whooping cough)	Toxoid Toxoid Subunit (cell-free extract)	Intramuscular	Sanofi Pasteur SA
Infanrix	Diphtheria toxin Tetanus toxin Bordetella pertussis (whooping cough)	Toxoid Toxoid Subunit (cell-free extract)	Intramuscular	GlaxoSmithKline
Tripedia	Diphtheria toxin Tetanus toxin Bordetella pertussis (whooping cough)	Toxoid Toxoid Subunit (cell-free extract, PT and FHA antigens)	Intramuscular	Sanofi Pasteur SA
Decavac	Diphtheria toxin Tetanus toxin	Toxoid Toxoid	Intramuscular	Sanofi Pasteur SA
Boostrix	Diphtheria toxin Tetanus toxin Bordetella pertussis (whooping cough)	Toxoid Toxoid Subunit (cell-free extract, PT and FHA antigens)	Intramuscular	GlaxoSmithKline
Adacel	Diphtheria toxin Tetanus toxin Bordetella pertussis (whooping cough)	Toxoid Toxoid Subunit (cell-free extract, PT, PRN, FIM and FHA antigens)	Intramuscular	Sanofi Pasteur SA

Part B: Multiple-Target Vaccines

Vaccine	Target	Type	Route of delivery	Manufacturer
Kinrix	Diphtheria toxin Tetanus toxin Bordetella pertussis (whooping cough) Polio virus	Toxoid Toxoid Subunit (cell-free extract) Killed	Intramuscular	GlaxoSmithKline
TriHIBit	Diphtheria toxin Tetanus toxin Bordetella pertussis (whooping cough) *Haemophilus influenzae* type b	Toxoid Toxoid Subunit (cell-free extract) Subunit (Hib polysaccharide conjugated to tetanus toxoid)	Intramuscular	Sanofi Pasteur SA
Pediarix	Diphtheria toxin Tetanus toxin Bordetella pertussis (whooping cough) Hepatitis B virus Polio virus	Toxoid Toxoid Subunit (cell-free extract, PT, PRN, FHA) Subunit Killed	Intramuscular	GlaxoSmithKline
Pentacel	Diphtheria toxin Tetanus toxin Bordetella pertussis (whooping cough) *Haemophilus influenzae* type b Polio virus	Toxoid Toxoid Subunit (cell-free extract) Subunit (polysaccharide conjugated to tetanus toxoid Killed	Intramuscular	Sanofi Pasteur SA
Twinrix	Hepatitis A virus Hepatitis B virus	Killed Subunit (surface antigen)	Intramuscular	GlaxoSmithKline
Comvax	Hepatitis B virus *Haemophilus influenzae* type b	Subunit (surface antigen) Subunit (Hib polysaccharide conjugated to Neisseria outer membrane protein)	Intramuscular	Merck

effective that polio has now been almost completely eradicated.

The list of effective vaccines is impressive, and more vaccines are available to protect animal herds that are important for the food supply. Notable exceptions, however, include a vaccine for the HIV virus, which remains elusive.

HOW VACCINES WORK

Vaccines present a particular structure, usually a protein, to the immune system. The structure is part of the pathogen. Under normal circumstances, the immune system responds to invasion by a pathogen by mounting the primary response. This takes some time, and during that period the person has the disease. Hopefully the immune response is sufficient to defeat the pathogen. Part of the response is the conversion of some of the immune cells into memory cells. If the person is ever reinfected with the same pathogen, the immune response is faster and stronger, and the pathogen is (usually) destroyed before causing the illness.

Vaccines work by introducing a molecule from a pathogen into the body, allowing the immune system to treat it like an actual infection. The memory cells are made, and if the person ever encounters the real pathogen, the immune response is swift and vigorous, defeating the pathogen before the disease can take hold. Ideally the vaccine stimulates both the cellular (i.e., T cell) and humoral (i.e., B cells producing antibodies) responses.

As tempting as it is to consider that vaccines might be developed against all pathogens (and even some nonmicrobial diseases, such as cancers and autoimmune diseases), it is not practical at this time. Generally speaking, it is best to restrict vaccination to diseases of some significance—that is, diseases that you really do not want to suffer from and yet have a real chance to acquire. Therefore, vaccines against the common cold are probably not practical, and vaccines against very rare diseases, such as Machupo virus, are probably unnecessary. Of course, if you are traveling in an area where Machupo virus is endemic, you might want the vaccine. But most people simply do not need this vaccine.

It is also helpful if the pathogen has only one or a few serotypes. Smallpox has only one serotype; polio has three. Combining all of the types in one shot protects the vaccinated person completely. For pathogens that have several serotypes, it may be sufficient to combine the most prevalent serotypes in one vaccine. The human papillomavirus vaccine incorporates the four most common or dangerous types.

Vaccines usually target a particular structure on the pathogen. Because the immune system detects invaders based on surface antigens, and because proteins make the best antigens, surface protein structures are the best subjects for use in vaccines.

Vaccines are produced by large pharmaceutical companies; they are expensive and time-consuming to develop and test for safety. The cost of an individual dose must be kept at a reasonable level, otherwise it might not be widely used. It is not essential that the entire population be vaccinated, even against common pathogens like measles and rubella. If most people in a community are vaccinated, the pathogen may infect a few individuals but will not spread widely. This is known as **herd immunity**, in which the majority of people comprise a protected population for the few individuals who are not vaccinated. However, when small communities—sometimes religious communities that do not accept vaccination—are infected by a visitor, people in the community may become infected because no "buffer" population is present.

TYPES OF VACCINES

The two general classes of vaccines are **live attenuated** and **killed.** The killed (or inactivated) vaccines use a form of the pathogen that is no longer viable. The antigenic structure—usually a surface protein—is still intact and can stimulate an immune response. This can be achieved by chemical inactivation with an agent such as formalin. This method is effective, although it usually does not result in a full immune response with both cellular and humoral immunity. A humoral (antibody) response is typical and might not include the very important secretory IgA antibody. For that reason, immunity is often short-lived and booster shots are needed. On the other hand, these vaccines are usually safe to use. After all, the pathogens are dead.

Killed vaccines may include whole-organism vaccines (e.g., the Salk polio vaccine) or specific components of the pathogen. For example, the vaccine for tetanus (caused by *Clostridium tetani*) uses the toxin molecule alone, which has been deactivated by converting it to a toxoid.

The live vaccines include both the attenuated strains of known pathogens and the genetically engineered vaccines in which the gene for the protein of interest has been cloned into another, safer vector microbe. The protein can then be expressed on the surface of the vector organism and can be used without fear of infection. Attenuation can be produced by growing the organism in an unusual culture medium. For example, the Sabin polio vaccine passes through chicken eggs several times before it becomes suitable for human use.

PROBLEMS WITH VACCINES

Vaccines are tested extensively before use, so large-scale deleterious effects are rare. While some bad reactions cannot be avoided, the rate has not been excessive. Even vaccines that are produced on an accelerated schedule, such as the Influenza A vaccine, are generally regarded as safe. A new vaccine is needed every year and is based on the previous year's strain. If a major change occurs in the antigenic properties of the new strain, the vaccine will have decreased effectiveness. While vaccine production has become more rapid in recent years, it is not possible now to create a new vaccine, to mass produce it, and to distribute it in time to blunt a new epidemic.

Other vaccines are developed and tested on a less hurried schedule, and the test results show that they are safe. However, human health comprises a wide spectrum, and in some rare cases, the vaccine might not have the intended effects. The attenuated polio vaccine, being a live vaccine, can occasionally produce a polio-like syndrome in vaccinated people. Named vaccine-associated paralytic poliomyelitis (VAPP), this illness appears to be caused by a revertant mutation in the virus that makes it act more like the wild-type virus. While the oral polio vaccine was being used, an average of eight people per year acquired VAPP (approximately 1 in 2.4 million vaccinations). The risk was much greater if the person receiving the vaccine was immunocompromised, especially with defects in immunoglobulin synthesis. Polio has been eradicated in the western hemisphere, so it is statistically more likely that a person will become infected by VAPP than by contact with the real wild-type virus. For this reason, vaccinations for polio, if they are needed at all, are likely to be performed with the very safe Salk vaccine.

The Bacille–Calmette–Guerin (BCG) vaccine, an attenuated strain of *Mycobacterium bovis*, has been used for many years as a vaccine against tuberculosis (TB). It is not used in the United States because the likelihood of acquiring a TB infection is very low. Instead, the population is monitored for an outbreak. But the BCG shot works reasonably well and is used in Europe, Russia, Japan, and other countries. BCG was tried without success in India several years ago. The problem appeared to be the overall poor health of the population. Apparently the vaccine is only effective if the health and nutrition of the person getting the shot are good, otherwise it is futile. This is unfortunate because many people in developing nations are therefore ineligible for this disease protection.

Other problems with vaccines are more mundane. For example, people get lazy about getting their shots. This happens constantly with tetanus boosters, which are needed approximately every 10 years. Fortunately, doctors have learned to inquire about the booster when they see a wound that might be contaminated with *C. tetani* and are quick to offer the shot. People also forget to have their children vaccinated, even though most school districts require the typical school-age series of vaccinations before a child can enroll at school (**Table 11-3**). Occasionally, a small outbreak of measles or whooping cough reminds people that the shots are important. Vaccines, especially live vaccines, can also be improperly stored and lose their potency.

One of the greatest challenges at this time is the protection of people who are immunocompromised. This group has been growing along with medical advances that extend the lives of many people suffering from previously fatal conditions. The group includes those with HIV, which attacks the cells of the immune system specifically and therefore destroys the very cells that the body needs to take advantage of a vaccine. This becomes an acute problem as HIV-positive individuals develop the full AIDS syndrome.

Other immunocompromised individuals include the elderly, who are not defined by age as much as

TABLE 11-3 **Immunization Program for Children**

Vaccine	Age at first inoculation	Subsequent inoculation ages	Notes
Hepatitis B	newborn	1–2 months, then 6–18 months	If mother is positive for hepatitis B, then immunoglobulin is also given to the infant.
Rotavirus	2 months	4 months, then 6 months	The 6-month shot might not be needed, depending on which vaccine is used.
Diphtheria, pertussis, tetatanus	2 months	Series: 4 months, 6 months, 12–15 months, 4–6 years	Tetanus boosters are needed periodically.
Haemophilus influenzae B	2 months	Series: 4 months, 6 months, 12–15 months	HIB vaccine, to avoid meningitis.
Pneumococcus	2 months	Series: 4 months, 6 months, 12–15 months	To avoid *Streptococcus pneumonia* meningitis. For children with severe health problems, the polysaccharide vaccine is used.
Inactivated poliovirus	2 months	Series: 4 months, 6–18 months, 4–6 years	The oral poliovirus vaccine is not used.
Influenza	2 years	Yearly during flu season	Inactivated vaccine can be given as early as 6 months.
Measles, mumps, rubella	12–15 months	4–6 years	MMR. Live virus; not given to immunocompromised people.
Varicella	12–15 months	4–6 years	Chickenpox virus. Often given with MMR.
Hepatitis A virus	12 months	23 months	The two shots should be at least 6 months apart.
Meningococcal	11–18 years	none	To avoid *Neisseria meningitides* meningitis.
Human papillomavirus	11–12 years	Second shot 1–2 months later, then third shot 6 months later	Females only.

Based on data from the Centers for Disease Control and Prevention (http://www.cdc.gov/vaccines/recs/schedules/default.htm).

by the health of the elderly person's immune system. At some point the immune system, like the other parts of the body, no longer functions as effectively as it once did. People receiving chemotherapy for diseases such as cancer are at a disadvantage because the toxic chemicals destroy the fastest growing cells of the body. In addition to the cancer cells, the cells that are derived from the bone marrow are greatly affected (e.g., red cells, platelets, and the immune cells). Finally, people who have received organ transplants must use immunosuppressive drugs to stop their body from rejecting the new organ. The balance of these drugs must be correct, or the drugs will so suppress the immune system that it cannot function properly. Even in an ideal case, the immune system is less than optimal.

CONTROVERSY

The value of vaccines is acknowledged by nearly everyone, although there is growing concern that vaccines bring a risk as well. In addition to the possibility of a bad reaction to a vaccine, an allergic response to a vaccine (especially those that are cultivated in chicken eggs), and the acknowledged risk of microbial contamination of a vaccine during production, other concerns about adverse reactions to vaccines are common. One prominent concern is the possibility that vaccination could lead to autism in some children. Because children receive multiple vaccinations, the idea that a shot could lead to autism probably discourages many people from getting their children vaccinated.

Autism is a developmental disorder that manifests as an inability to interact and communicate with other people. The degree to which this is true varies, leading to autism spectrum disorders (ASD). At one end of the spectrum are those with Asperger's syndrome, in which the child has reduced social interactions and a curious intensity with specific objects. At the other end of the spectrum are cases of severe mental retardation. ASD is generally considered to be an illness of the modern age; while isolated incidents may have been described in the past, they appear to have been quite rare. In contrast, Asperger's syndrome has only been considered as a diagnosis since 1981. The symptoms of autism usually begin to appear around the age of three years, which coincides with the age at which many vaccinations are given. The coincidence has been noted by many parents.

The incidence of autism has been increasing in the past few decades; the causes are inconclusive. A survey conducted by the Centers for Disease Control and Prevention puts the autism rate at 1 out of every 110 children—a startlingly high number (CDC, 2009). Autism occurs at a higher rate in boys than in girls. It is fair to speculate that this number has been climbing because physicians are now more familiar with the syndrome, and that in years past the children were diagnosed differently. Although this is certainly one factor, most researchers believe the increasing incidence is a phenomenon that is being driven by some other, contemporary factor of our society.

The first factor associated with autism risk was the presence of mercury in the vaccines. Thimerosal is an organic mercury compound used to inhibit bacterial growth when the vaccine was distributed in large bottles, from which many separate shots were obtained. The risk of contamination through the rubber cap of the vaccine vial was real, so the inhibitor was used. The concentration in the vaccine solution was low, but if multiple shots were given and the child was small, the overall mercury concentration delivered might have been significant. No one knows whether even a small dose of mercury at the wrong time in child development might damage a critical aspect of neurological development. In 2001, children's vaccines were reconfigured into single-dose delivery devices, and Thimerosal was discontinued. However, this compound can still be found in shots delivered mainly to adults, such as the seasonal flu vaccine. If mercury compounds were responsible for autism, then autism rates in children born since that time should have decreased. In fact, there has been no statistical difference in rates of ASD.

In recent years, the vaccine itself has been suggested to be to blame for autism, which strikes some children but not others. Evidence of a genetic component to autism has been presented, suggesting that some families may be prone to autism and may require only the trigger of the vaccination for autism to manifest. Determining whether vaccination itself leads to autism should be possible simply by comparing a population that received vaccination with a population that did not. Certainly some populations have poor vaccination coverage, but this usually also indicates that the level of medical care is poor as well, and a reliable diagnosis of autism might not be available. A definitive survey to test the hypothesis has not yet been performed and is certainly needed either to allay suspicions about vaccines or to provide data for effective policy changes.

SUCCESS STORIES

Some diseases are now under control thanks to vaccines (Dennehy, 2001). These include many childhood diseases and some of the most devastating diseases known to humankind. Smallpox (**Figure 11-1**) was one of the most devastating diseases known, yet the World Health Organization declared it eradicated in 1979. The last naturally occurring case of smallpox occurred in Somalia in 1977. Polio was also greatly feared for its ability to cause paralysis and death. After a worldwide vaccination drive and continued surveillance, it has nearly been eradicated. In 2008, only 1,652 cases occurred worldwide. Four countries are still considered endemic for polio: India, Afghanistan, Pakistan, and Nigeria. This pathogen will probably be wiped out completely because the vaccination campaign has extensive international support.

FIGURE **11-1** **A child with an active case of smallpox. The pustules will scar and may be disfiguring. In some cases the eyes scar as well, leading to blindness.**

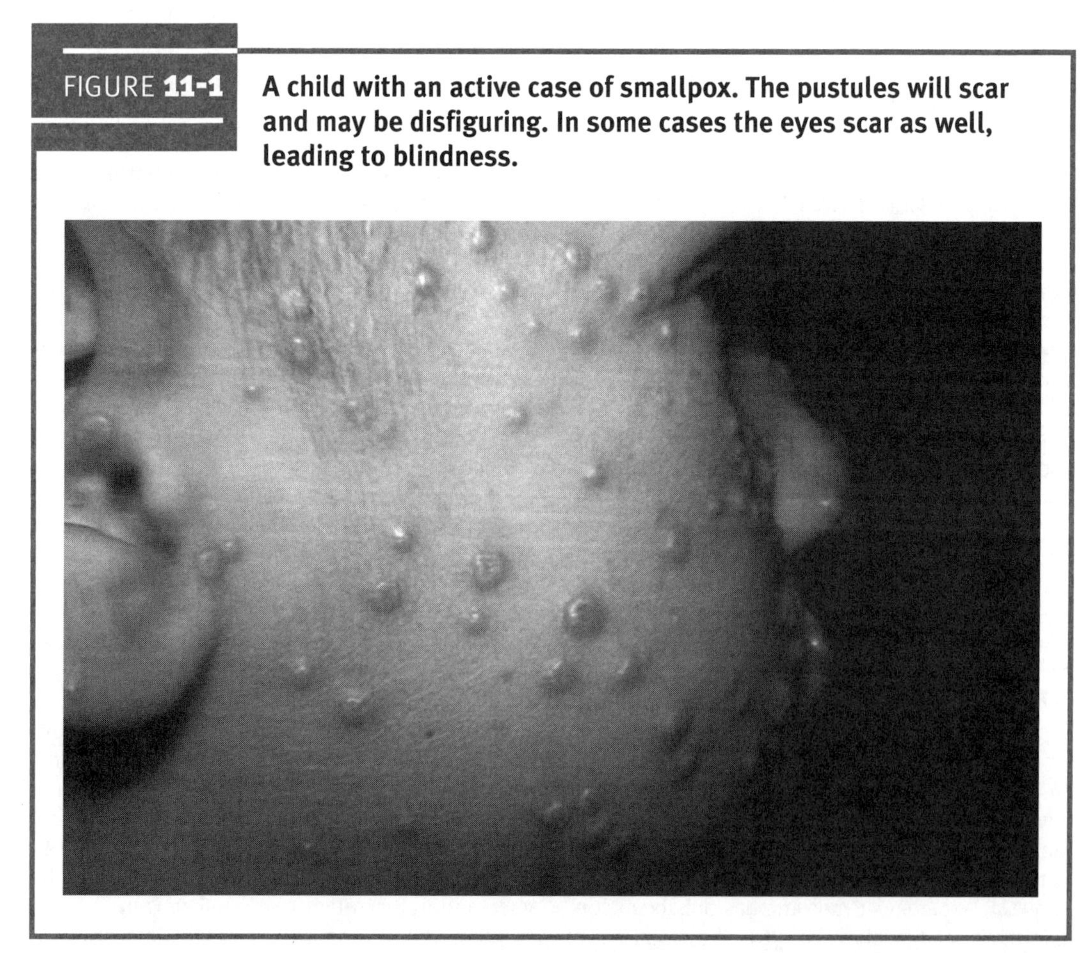

SOURCE: Courtesy of Dr. John Noble Jr., the Centers for Disease Control and Prevention. (http://phil.cdc.gov/phil/details.asp). #10481.

Measles–Mumps–Rubella

The measles–mumps–rubella (MMR) shot is one of the greatest vaccination success stories because these were all common childhood diseases only a few years ago. Measles was not a particularly debilitating disease, at least compared to other childhood diseases: it produced a distinct rash all over the body (**Figure 11-2**). Yet it was still responsible for many secondary infections (such as pneumonia) and had a fatality rate around 0.1%. The number of annual infections used to be 400,000–600,000 (Wood and Brunell, 1995), which means about 400–600 children died from measles-related causes every year. The peak year was 1941, with 894,134 cases. In recent years, the majority of cases in the United States were due to people visiting from other countries.

The mumps virus, however, was far worse. The hallmark of the disease was parotitis, an inflammation of the parotid salivary glands (**Figure 11-3**). Complications were many and frequent: orchitis (inflammation of the testicles), meningitis, encephalitis, and deafness. If encephalitis occurred, the fatality rate was about 1%. A sample mumps survey form is provided in Appendix III. Rubella ("German measles") produced a measles-like illness and was especially dangerous for pregnant women because the likelihood of birth defects in the fetus was much greater in these women. These defects included damage to major organs as well as mental retardation. (See **Figure 11-4**.)

The MMR shot drastically reduced the incidence of all three diseases (**Figure 11-5**). While there have been some spikes in their incidences over the years, the general trend has been consistently downward, along with the complications related to the diseases. For instance, in 2008 the total number of cases in the United States for measles was 140; for mumps it was 454; and for rubella it was only 16. This is a sliver of a fraction of the number that occurred just a few decades ago. Unfortunately, some outbreaks will still occur. An outbreak of mumps was seen in 2009 (High et al., 2009). A religious community that did not practice vaccination was affected after one of its members returned from an overseas trip where, presumably, the virus was encountered.

The same success can be reported for the vaccines against pertussis (whooping cough) and diphtheria, once dreaded diseases. The peak years for these pathogens were 1921 for diphtheria (206,000 cases) and 1934 for pertussis (265,000 cases). Both of these diseases have declined by 98–99% since then.

FIGURE **11-2** **The typical rash of measles.**

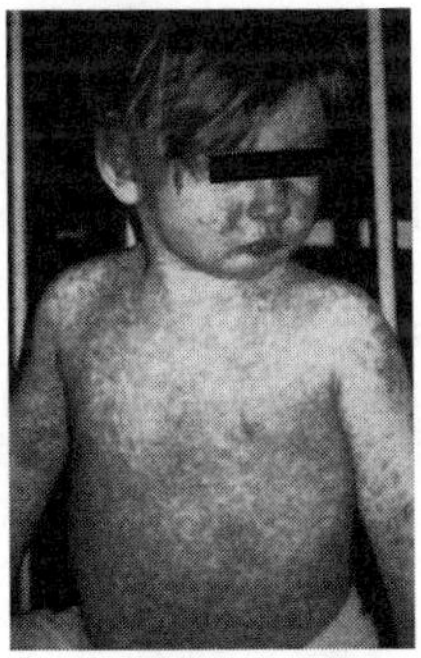

SOURCE: Courtesy of the Centers for Disease Control and Prevention.

Hepatitis A Virus

Much has already been written about HAV in this text, and a summary is not needed here. Because it can be passed through contaminated food or water, the victim probably has no idea of the danger. However, the introduction of the vaccine, along with improved education about hygiene and monitoring of food workers, have all contributed to a decrease in the total number of cases (**Figure 11-6**). This is a good example of how different facets of public health can all be brought to bear on a single problem and thereby can produce a very successful result. There is every reason to believe that HAV rates will continue to decrease in the future.

Varicella

Varicella is the vaccine for the chicken pox. Of the classical childhood diseases, chicken pox caused very little concern because the illness was almost always mild, with minimal discomfort (mainly itching) to the child (**Figure 11-7**). However, as

FIGURE **11-3** **A child with a swollen throat, characteristic of mumps infection.**

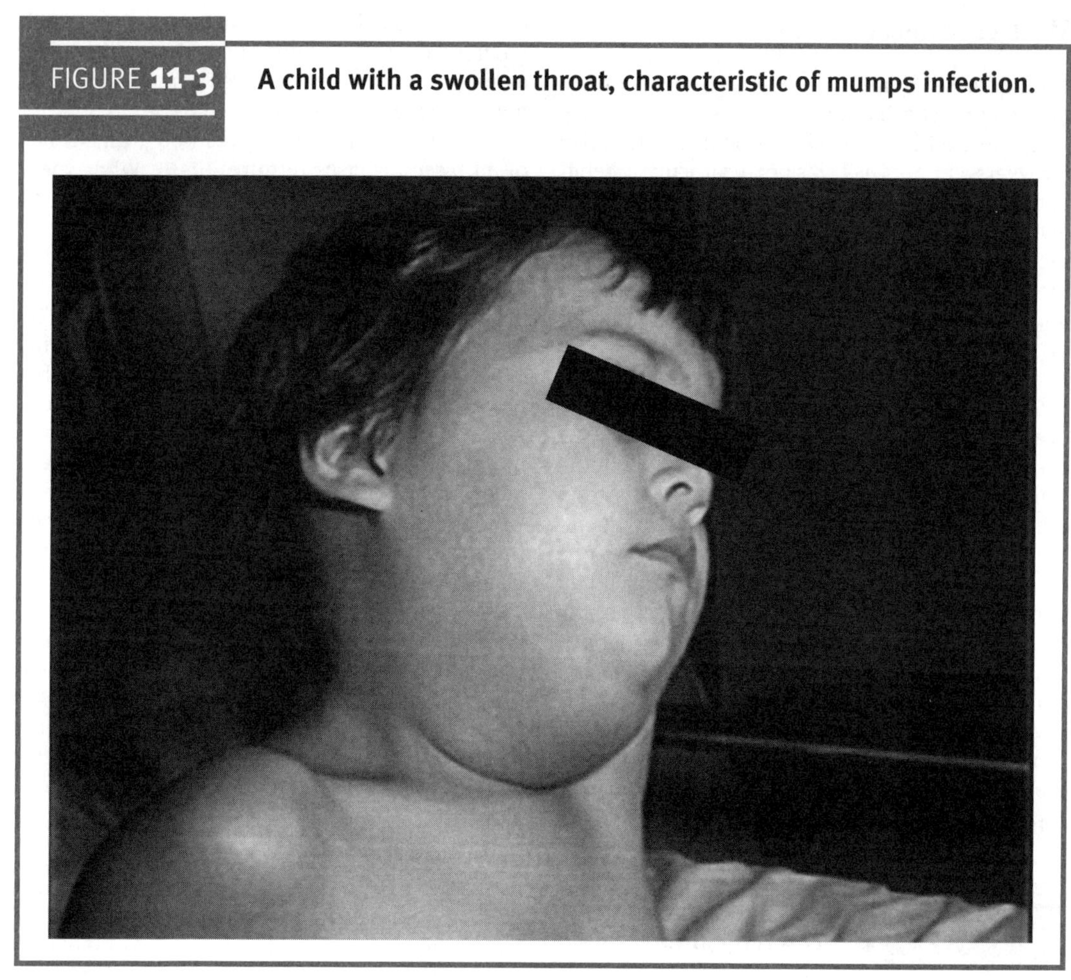

SOURCE: Courtesy of Barbara Rice, the Centers for Disease Control and Prevention. (http://phil.cdc.gov/phil/details.asp). #130.

a herpes virus, chicken pox can remain in the body for decades and can reemerge in adulthood as the disease shingles. Shingles is a painful disease worthy of avoiding by vaccination.

The varicella vaccine was introduced in 1995 and had a rather slow start. At the time, vaccines were distrusted because of fears that they were responsible for other syndromes—chief among them, autism. At this time, no evidence supports this belief; nevertheless, many parents skipped the vaccine and let the child get chicken pox the old-fashioned way. In fact, pox parties have been reported in which the parents take their children to the home of a child with the chicken pox so that their own children will get it and then get over it. Of course, this does not avoid the possibility of shingles in the future.

The introduction of the varicella vaccine caused chicken pox rates to plummet almost immediately (**Figure 11-8**). The safety of vaccines appears intact, and this example shows that even a mild disease can be a good target for an effective vaccine. With a little more than 30,000 cases of chicken pox reported in 2008, there is still room for improvement.

FIGURE 11-4 **The campaign for vaccination against rubella employed a number of different media types, including these posters.**

SOURCE: Courtesy of the Centers for Disease Control and Prevention.

FIGURE **11-5** **The number of cases of measles, mumps, and rubella has decreased dramatically. Note the spike during the 1989–1992 time period.**

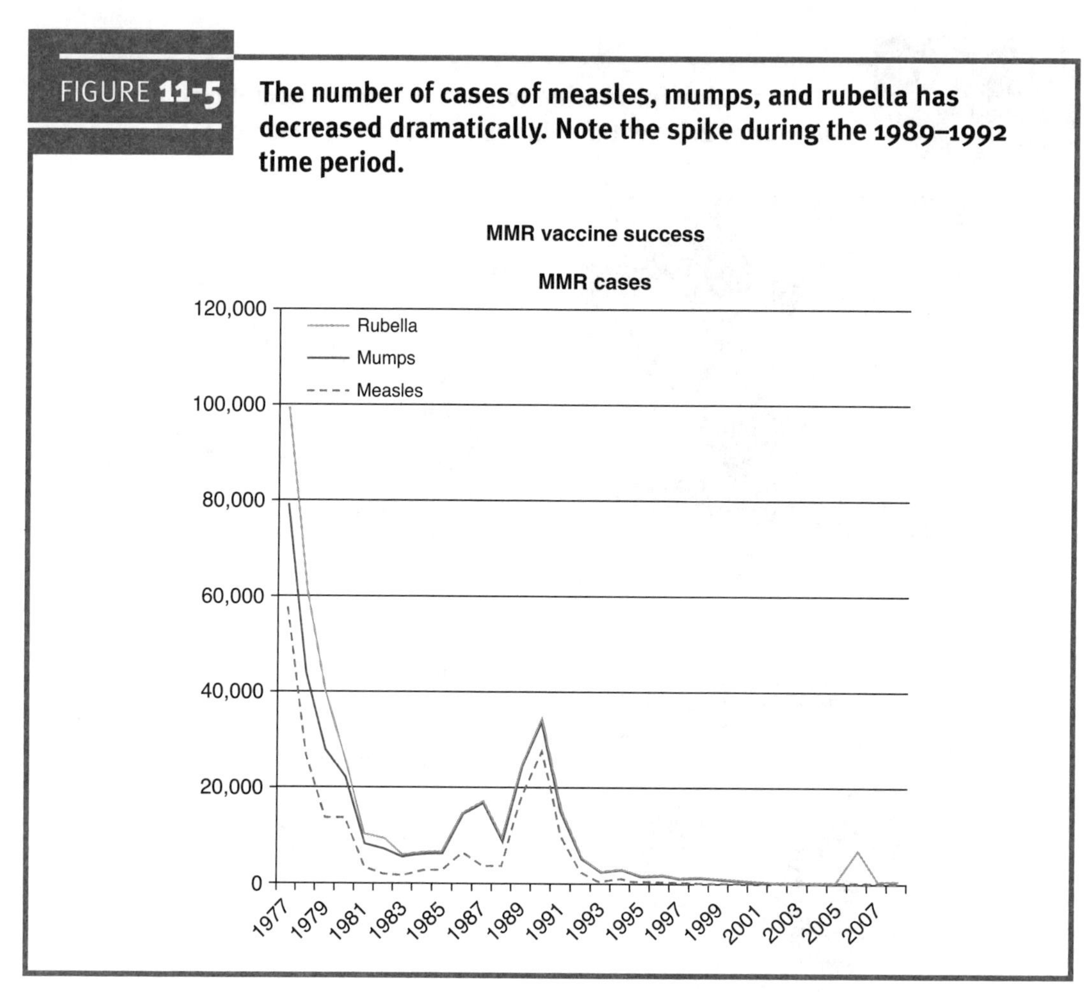

SOURCE: Based on data from the Centers for Disease Control and Prevention.

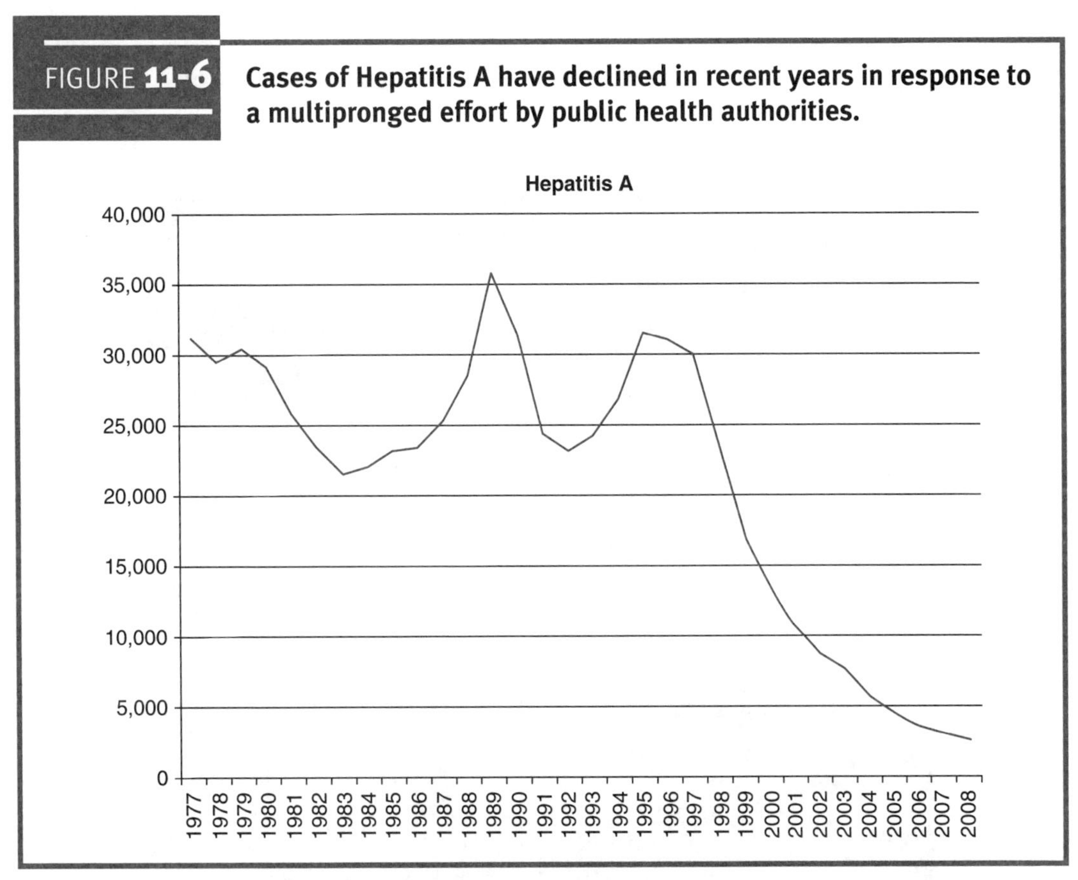

FIGURE 11-6 Cases of Hepatitis A have declined in recent years in response to a multipronged effort by public health authorities.

SOURCE: Based on data from the Centers for Disease Control and Prevention.

FIGURE **11-7** **A young boy with the characteristic rash of chicken pox (Varicella).**

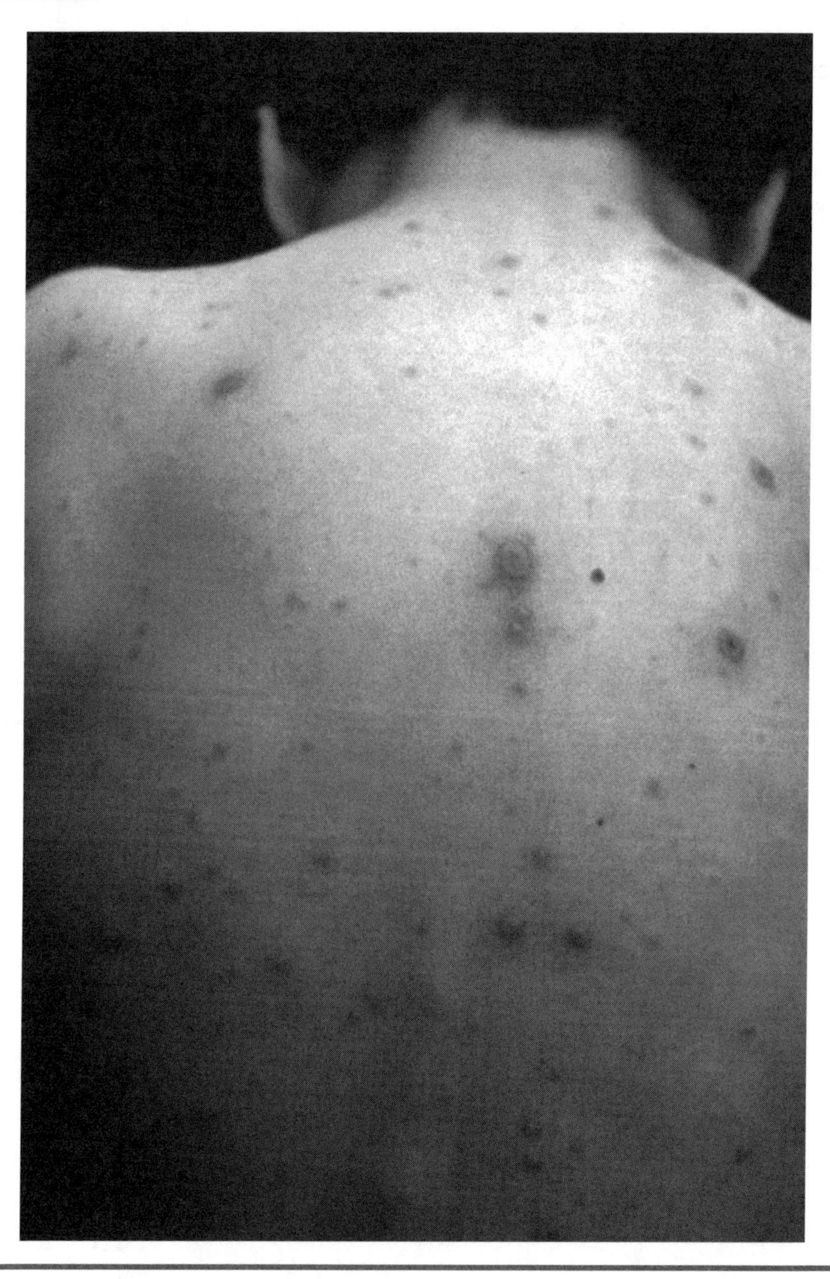

SOURCE: Courtesy of the Centers for Disease Control and Prevention.

FIGURE 11-8 **Since its introduction in 1995, the cases of chicken pox have decreased in the United States.**

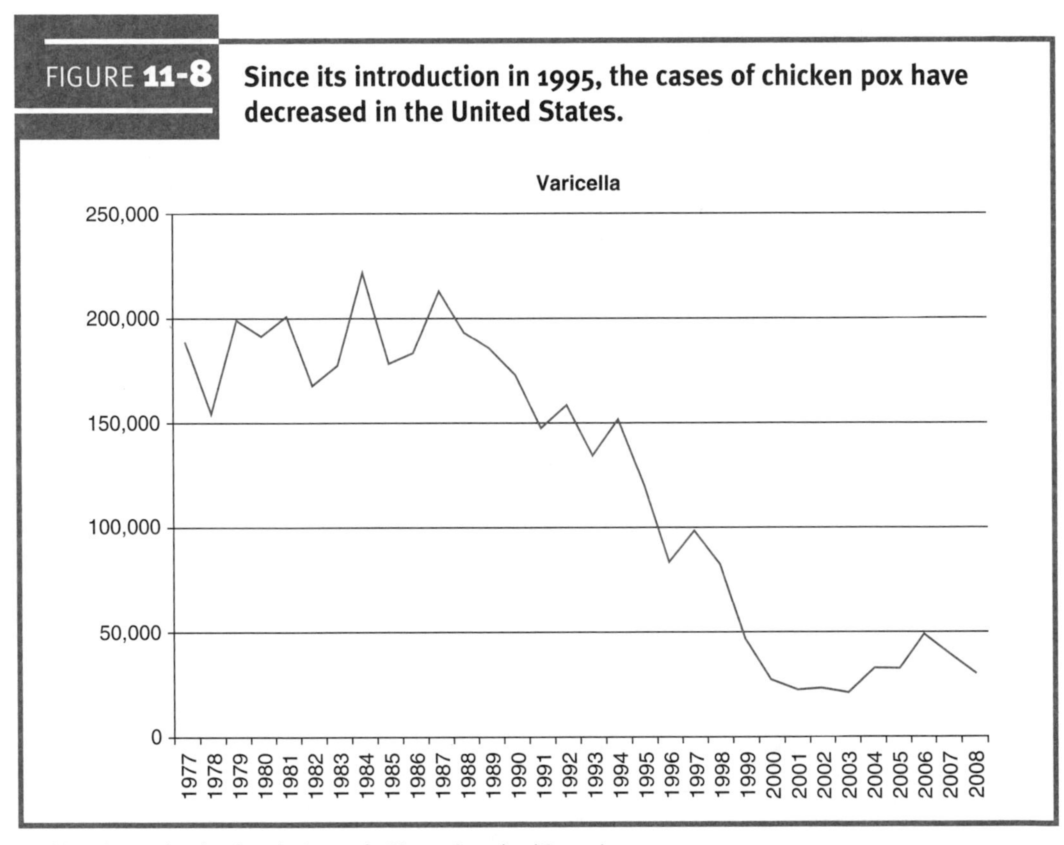

SOURCE: Based on data from the Centers for Disease Control and Prevention.

QUESTIONS FOR DISCUSSION

1. Which diseases still need an effective vaccine? What makes you think so?
2. Should people be compelled to vaccinate their children?
3. Should people be compelled to vaccinate themselves?
4. In your opinion, what is an acceptable level of risk (for illness or for death) of a vaccine?
5. What are the advantages and disadvantages of attenuated and killed vaccines?
6. Because vaccines are used on such a wide scale, cost becomes a major issue in their use. What is the best way to effectively protect a population while holding costs to a minimum? Would this change if you were considering a vaccine for a developing country?

References

Centers for Disease Control and Prevention, Autism and Developmental Disabilities Monitoring Network Surveillance. 2009. Prevalence of autism spectrum disorders—Autism and Developmental Disabilities Monitoring Network, United States, 2006. MMWR Surveill. Summ. 58(10): 1–20.

Dennehy, P.H. 2001. Active immunization in the United States: developments over the past decade. Clin. Microbiol. Rev. 14: 872–908.

Gradmann, C. 2001. Robert Koch and the pressures of scientific research: tuberculosis and tuberculin. Medical History 45: 1–32.

High, P., E.F. Handschur, O.S. Eze, B. Montana, C. Robertson et al. 2009. Mumps outbreak—New York, New Jersey, Quebec, 2009. Morbid. Mortal. Weekly Rep. 58: 1270–1274.

Wood, D.L., and P.A. Brunell. 1995. Measles control in the United States: problems of the past and challenges for the future. Clin. Microbiol. Rev. 8: 260–267.

CHAPTER 12

Sexually Transmitted Diseases

LEARNING OBJECTIVES

- List the sexually transmitted diseases and their etiological agents.
- Determine the incidence of each STD and any related epidemiological information.
- Describe the overall trends in disease incidence.
- Describe the reasons why eradication of STDs is difficult.

WHAT IS AN STD?

Sexually transmitted diseases (STDs) are diseases that are passed through human sexual contact, which includes the practices of vaginal sex, anal sex, and oral sex (**Box 12-1**). The term *sexually transmitted infection* (STI) is often used to highlight the fact that many people can be infected but remain asymptomatic carriers—this is one of the major difficulties presented by these pathogens. The range of effects runs from the almost ignored chancroid (*Haemophilus ducreyi*) to the greatly feared human immunodeficiency virus (HIV).

STDs create a unique problem for society. On one hand, how they are transmitted and how they can be eliminated is known. After all, these diseases do not survive in the environment, and there are no other hosts for the pathogens. On the other hand, the means to eliminate them is dependent on the sexual habits of the population—something that is not going to change easily. The range of human sexual preferences is vast, encompassing not only sexual practices that are likely to result in pregnancy but also practices that are solely intended for pleasure. Human interpretations of pleasure are based on individual choices, some of which do not appeal to the general population. For example, anilingus is the practice of oral–anal sexual stimulation (commonly referred to as "rimming"). As might be expected, the risk of infection through the fecal–oral route of transmission is rather high with this practice. The insertion

BOX 12-1 The STD List

Bacteria

- *Treponema pallidum*
- *Neisseria gonorrhoeae*
- *Chlamydia trachomatis*
- *Haemophilus ducreyi*
- *Calymmatobacterium granulomatis*

Viruses

- Hepatitis B virus
- Hepatitis C virus
- Herpes simplex
- Human papillomavirus
- Human immunodeficiency virus

Parasites

- *Trichomonas vaginalis*
- Scabies
- *Pediculosis pubis* (lice)

of foreign objects into the anus for sexual stimulation is well known in emergency rooms. The risk of perforation of the intestine is real, especially considering some of the objects that have been used: soda bottles, a salami, broomsticks, light bulbs, Christmas ornaments, a billy club, cucumbers, apples, knives, nails, and a trailer hitch (Agnew, 1986; Goldberg and Steele, 2010). As unappealing as some of these practices may seem, sexual habits can be a significant part of personal identity, and attempts to change that behavior can result in avoidance of or distrust of medical or public health authorities (**Figure 12-1**).

Those afflicted with STDs may not be the most personally responsible or educated individuals. Even after receiving antibiotic treatment, the patient might forget to take the medication or to stop taking it after feeling better, even though the pathogen has not been fully eliminated. This leads to antibiotic-resistant strains of bacteria, which is a growing problem. Methods to entice infected persons to return for additional treatment are greatly needed. Incentives for continued treatment—which may seem obvious—are actually needed in some cases. Better treatment of at-risk populations that are on the margins of society, such as prostitutes, has overall benefits for society.

At the present time, a group of measures is utilized to attack the STD problem from a social standpoint. Education about STDs, that is, about effective means of preventing them while engaging in sexual activities and about measures to take when infection is suspected, are all effective to a degree, especially if education efforts are concentrated on sexually "at-risk" populations. For people who are already infected, client-centered counseling is used, that is, a brief and individual counseling session on the importance of effective treatment and the need to institute safer sexual practices.

Contact tracing is the means of finding and informing people that they have been exposed to an STD. Trace-backs are difficult; most people react poorly to the news that they may have been infected. Notably, these contacts might not have the STD, and personal antagonism with the infected person may occur for no reason. (Another problem is that the affected person may not be able to ascertain the identities of all his or her contacts.) However, there is no way to know *a priori* whether the contact person has contracted an STD, so a procedure is followed for informing and testing him or her. (See **Figure 12-2**.)

WHAT ISN'T AN STD?

There are a number of limitations on what constitutes an STD. Sexual practices such as anal sex or oral–anal sex (which involve the mucous membranes associated with STD infection) can lead to infection with a number of common fecal bacteria. These have been referred to as sexually transmitted enteric infections. While they are certainly transmitted sexually, these bacteria are not specifically adapted for sexual transfer and growth on genital tissues. These infections can be thought of as opportunistic wound infections. Hepatitis A virus (HAV), which is usually acquired through contaminated food or water, is in the same general category. However, it is also passed by the fecal–

FIGURE 12-1 **Posters such as this one have been used by the U.S. Department of Defense to remind service men and women of the dangers of STDs. This one is from the World War II era.**

SOURCE: Courtesy of the Centers for Disease Control and Prevention.

oral route. Practices such as oral–anal sex can lead to acquisition of HAV from an infected partner. This is not a usual route of infection and therefore is not considered further in this chapter. Any disease that is typically spread by the fecal–oral route is regarded in the same way. That is, a microorganism might be spread by sexual means and might not be an STD. A good way to define STDs is that they

FIGURE 12-2 **Reported cases of bacterial STDs over time.**

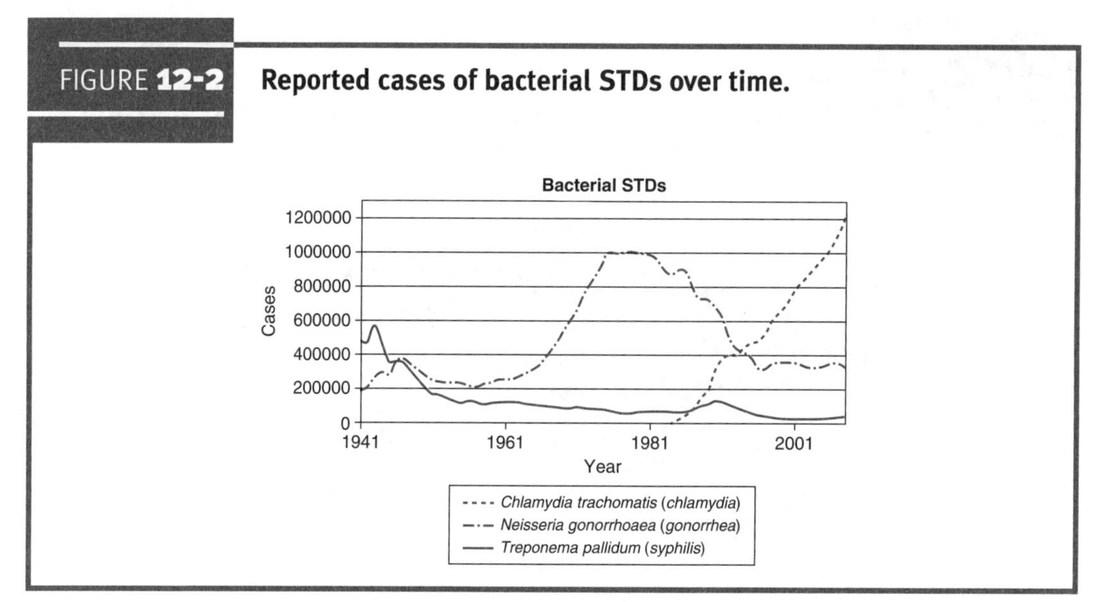

SOURCE: Based on data from the Centers for Disease Control and Prevention.

have little chance of spreading from one person to another *except by* sexual contact.

Some conditions of the genitourinary tract are certainly not due to an STD. Bacterial vaginosis (BV) is a poorly defined condition generally described as irritation of the vagina, which can occasionally be severe. Irritation may be caused by a true pathogen, but in many—perhaps most—cases, the cause is an imbalance of the normal flora of the vagina. Normally the vagina has a substantial growth of normal flora, dominated by *Lactobacillus*. These bacteria produce hydrogen peroxide, which inhibits the growth of other species. However, if normal conditions are disturbed, then other, minor species can begin to proliferate (**Figure 12-3**). In particular, overgrowth with anaerobes such as *Mobiluncus*, *Prevotella*, *Porphyromonas*, and *Peptostreptococcus* can occur. *Gardnerella vaginalis* has been implicated in producing BV with a discharge that has a fishy odor. However, all of these species can be found in normal flora in healthy women. Fungal overgrowth by *Candida* has also been demonstrated. In the vast majority of cases, this condition can be treated successfully.

Imbalances can be caused by having too many sexual partners, suggesting an underlying STD cause. However, the same symptoms can be found in celibate women, where it may be caused by excessive douching, hormonal changes, or even prolonged stress. For that reason, it is best referred to as a medical condition rather than a public health concern.

The aims of public health with regard to STDs are very similar to the aims of medical practice:

- The effective treatment of identified cases, as well as asymptomatic carriers.
- Identification of sexual partners and their notification of risk and treatment options. This varies with local laws.
- Appropriate handling of data; many STDs are reportable diseases.
- Identification of species, strains, and relevant characteristics, such as virulence and antibiotic resistance profiles.
- Education of the public about dangerous sexual practices and safe methods that people can use to limit their exposure to pathogens.

For many years, the treatment of STDs was hampered by taboos about the discussion of sex and sexual activities. In the United States, this changed

FIGURE 12-3 **Clue cells. These epithelial cells show adherent bacteria, suggesting a case of vaginosis.**

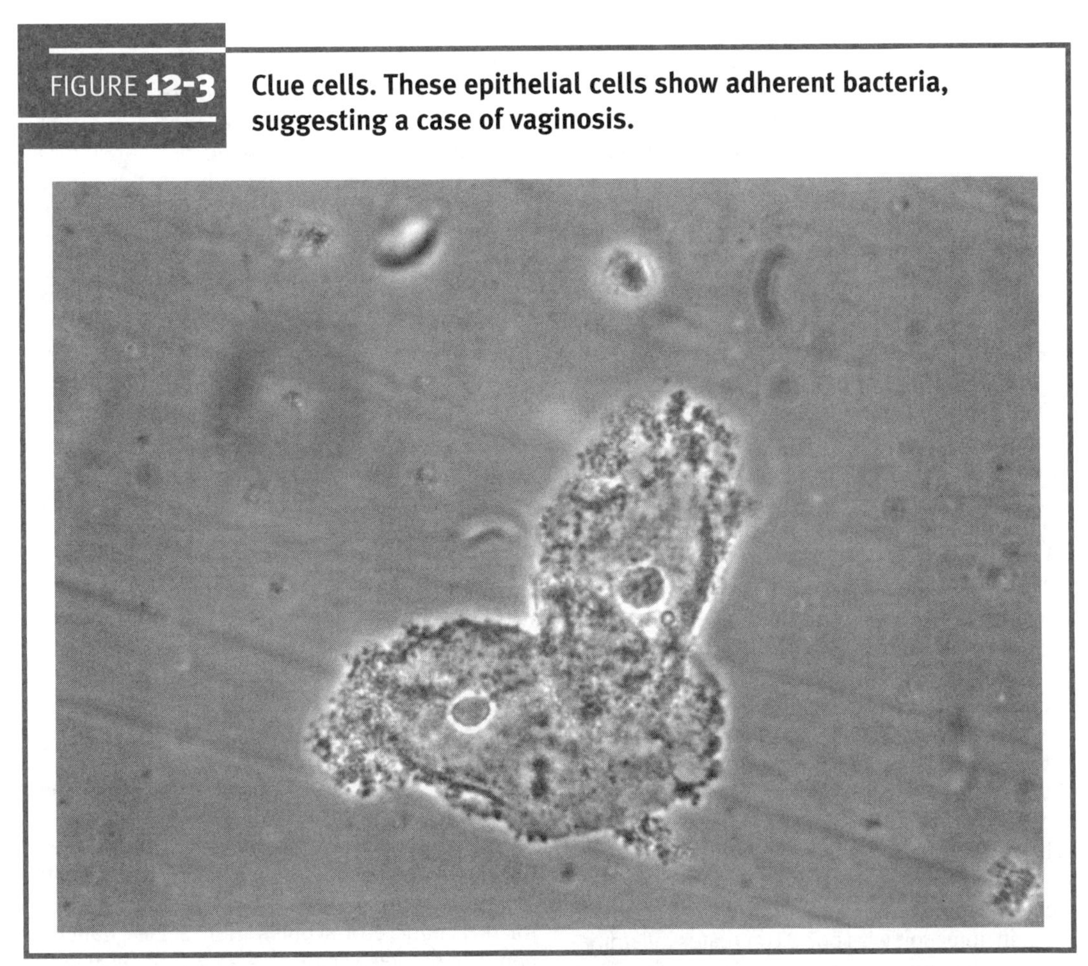

SOURCE: Courtesy of M. Rein, the Centers for Disease Control and Prevention. (http://phil.cdc.gov/phil/details.asp). #3720.

after World War II because many men, serving in uniform far from home, acquired STDs from prostitutes. The postwar years showed a dramatic increase in the incidence of STDs, leading physicians and public health officials to break the silence on the subject in the name of eradicating the diseases. Yet efforts to control sexual habits are likely to fail because people tend to have strong feelings about their sexual preferences. In particular, Americans have an individualistic streak that rebels at the idea of any authority telling them what to do in the bedroom. Thus a dynamic interaction between personal behavior and public safety has been created.

It is tempting to speculate that the reported numbers of these diseases are underreported because of the prevalent social stigma attached to them, not to mention the very real risk of the wrath of a person who finds out that he or she now has an STD because of a cheating partner. As with most information gathering, some degree of nonreporting exists, but overall the public health system does a good job of maintaining the maximum practical amount of anonymity. In most communities, the records are even protected from subpoena despite the fact that such records might have bearing on court cases. This is a measure of the importance that we attach to confidentiality with regard to this very sensitive subject.

One of the latest means of treating STDs is the use of expedited partner therapy (EPT) or patient-delivered partner therapy (PDPT). In these instances, the partner may have acquired an STD. (*Chlamydia* and gonorrhea are treated using these methods.) The patient is allowed to deliver the antibiotics to the partner without first seeing a physician. The value of seeing the physician is unquestioned, but the relative ease of obtaining treatment for a person who probably prefers a high degree of privacy is also important. These methods have merit. They are not usually recommended for men who have sex with men (MSM) because the presence of one STD may indicate others, such as HIV. Anyone in a group with a high risk from multiple STDs should be seen by a physician.

A number of known risk factors are associated with acquisition of an STD. These include sexual relations with several different individuals, specific sexual practices (especially anal sex), and the likelihood of multiple STDs being passed by one person. Infection with one STD has long been known to be a risk factor for others. Multiple STDs in a single person, who then has the ability to pass all of them to the next partner, are not uncommon. In addition, infection with one STD creates inflammation and lesions that increase the likelihood of another STD or other infection because lesions and ulcerations remove the protection usually afforded by the skin.

When information about the sexual history of a person with an STD is obtained, it is important to remember that major taboos still affect society. It is no longer advisable to ask men whether they are homosexual (or bisexual); many men who engage in homosexual sex do not consider themselves homosexual ("gay") or are unwilling to say it openly. In its place, the term MSM is used, for "men who have sex with men." This is an acceptable description that gets better compliance. Some people might not be eager to share such personal information or might be distrustful of authority. Some other cultures have even more stringent taboos on sexual practices. In these cases, the requirement to divulge sexual partners might get a poor reaction because sensitive information would become known and detrimental.

Partner notification is usually performed for all sexual partners the person has had within the previous 60 days. It is not acceptable to "wait until my wife shows symptoms of gonorrhea and then tell her." This might seem a convenient way to avoid a painful and embarrassing disclosure, but the fact is that many STDs are asymptomatic, and the unsuspecting person may still suffer from it. Many STDs lead to inflammation of the fallopian tubes, eventually resulting in scarring. This scarring can lead to infertility and occasionally to ectopic pregnancies, which are very painful and can be life threatening. Treating a partner covertly, such as slipping in some penicillin with her corn flakes, is also unacceptable. All persons have the right to know their medical conditions and which medications they are taking.

BACTERIAL STDS

Haemophilus ducreyi

Haemophilus ducreyi is the causative agent of chancroid, a notifiable disease (**Figure 12-4**). In 2007, only 23 cases were reported in the United States. The hallmark of chancroid is the presence of a painful ulcer(s) on the genitals. It is easily mistaken for the chancre of syphilis (hence the chancroid name), although a chancre is not especially painful. On the other hand, it is not at all unusual to see a chancroid–syphilis coinfection, and treatment for syphilis may well cure chancroid as well.

Chancroid is also seen in coinfections with HIV and genital herpes, and therefore thorough testing for these STDs should accompany treatment. As with other members of the *Haemophilus* genus, growth on artificial media is difficult and requires special preparation. The isolation of Gram-negative coccobacilli from a lesion is a good indicator. Antibiotic treatment is usually successful, and several drugs can be used.

Chlamydia trachomatis

Chlamydia, another notifiable disease, is an interesting and underappreciated pathogen although it

FIGURE 12-4 **A genital chancroid lesion due to *Haemophilus ducreyi*.**

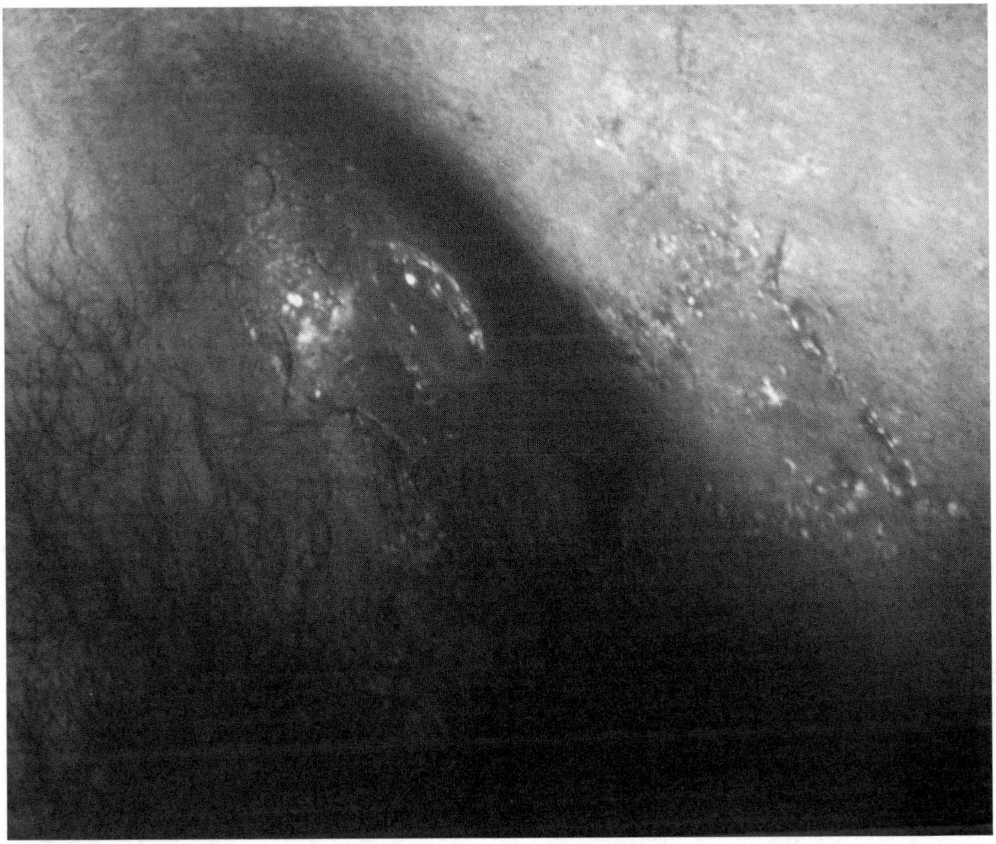

SOURCE: Courtesy of Susan Lindsley, the Centers for Disease Control and Prevention. (http://phil.cdc.gov/phil/details.asp). #5810.

is the most commonly reported of the notifiable diseases (**Figure 12-5**). In 2007, 1.1 million cases of *Chlamydia* were reported, far surpassing the second most reported disease, which happened to be another STD, gonorrhea. (See **Figures 12-6**, **12-7**, and **12-8**.)

A curious fact about this microbe is that it causes two distinct diseases. One is the STD, the other is trachoma, a serious eye infection. While it is certainly possible to acquire trachoma by receiving a bodily fluid in the eye, it is unusual. It is far more commonly acquired as a result of insects (flies) that deposit the bacteria directly on the eye as they swarm about the face. This is a relatively common infection in some parts of the world, especially the Middle East.

The STD was largely overlooked until the last two or three decades. *Chlamydia* is difficult to grow

FIGURE **12-5** **Inclusion bodies due to *Chlamydia trachomatis* are shown here on a monolayer of cells.**

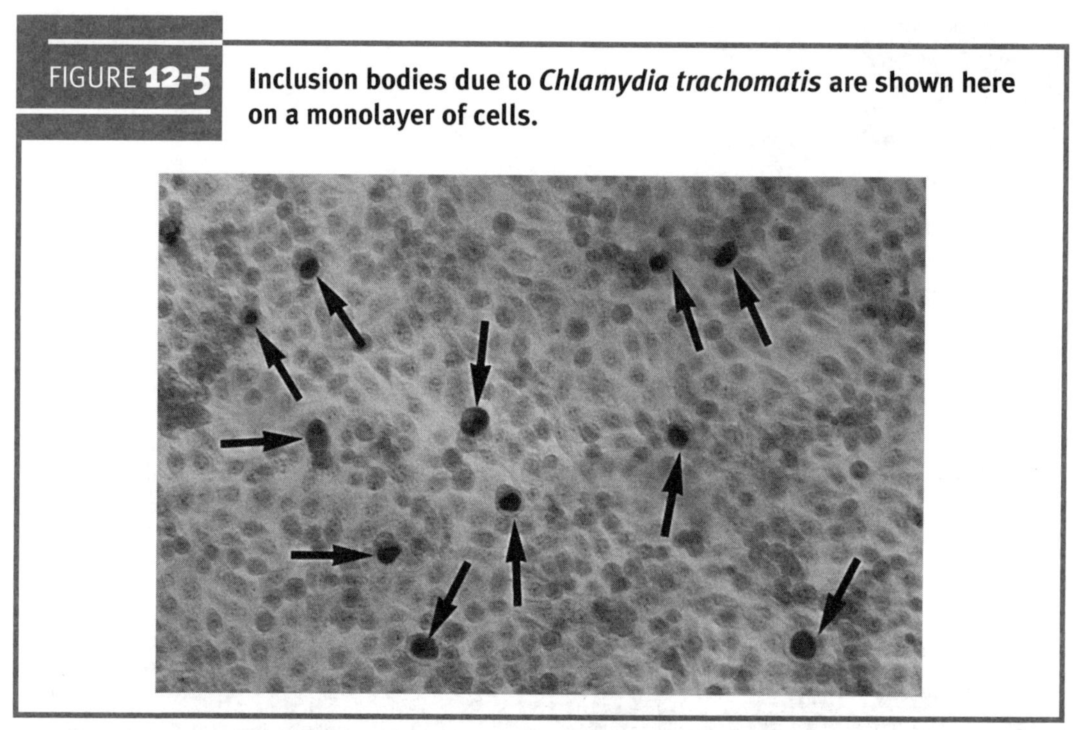

SOURCE: Courtesy of Dr. E. Arum, Dr. N. Jacobs/CDC.

FIGURE **12-6** **Cases of *Chlamydia* in 2008, broken down by age range and sex.**

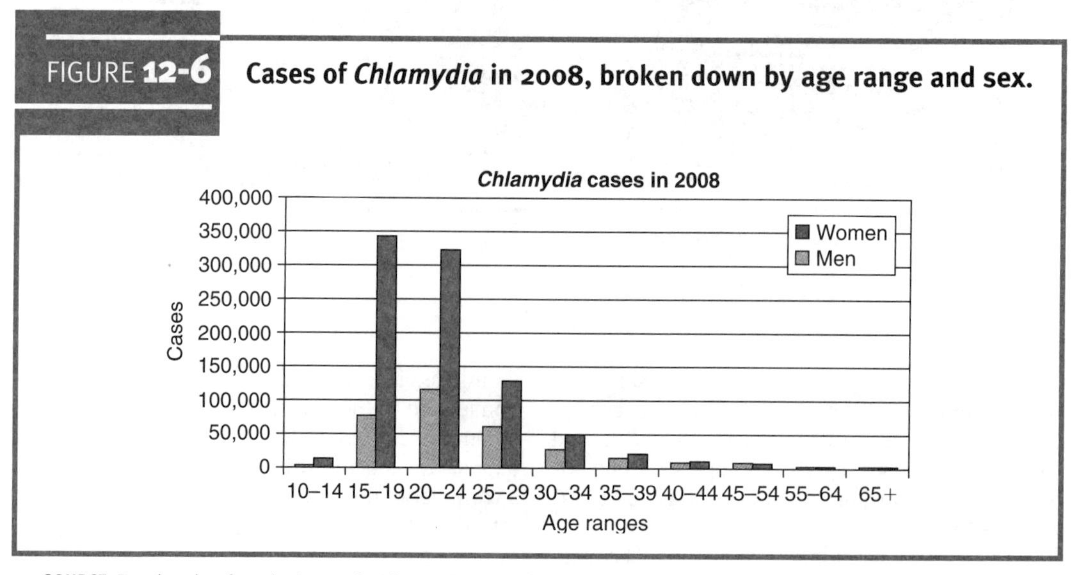

SOURCE: Based on data from the Centers for Disease Control and Prevention.

FIGURE 12-7 **Comparison graph of age ranges of men for infection with *Chlamydia*.**

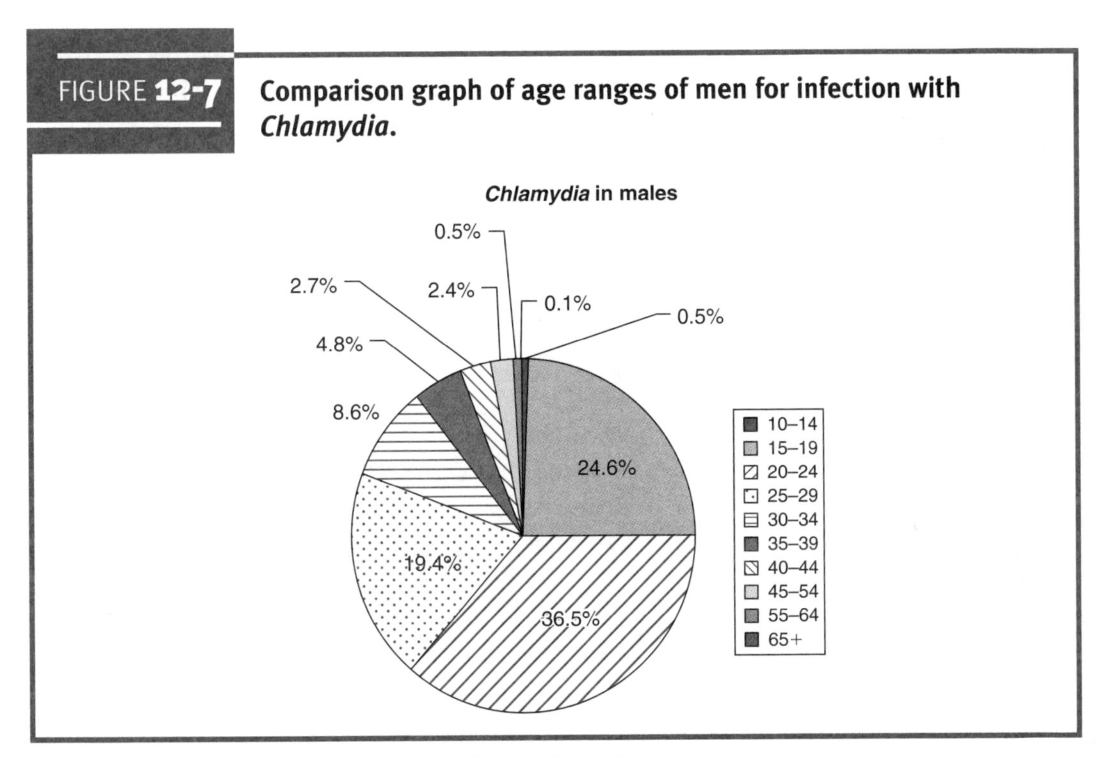

SOURCE: Based on data from the Centers for Disease Control and Prevention.

on artificial media, and obvious infections were often diagnosed as gonorrhea. This is understandable because the two diseases are often passed together. Unfortunately, treatment of gonorrhea might not effectively treat *Chlamydia*; however, numerous antibiotics are effective against *Chlamydia*.

In terms of therapy, *Chlamydia* has two distinct forms. One is the elementary body (EB), which is environmentally resistant and infectious but cannot replicate. The EBs infect host cells and then, once inside, convert them to the reticulate body (RB) form. The RBs are noninfectious and fragile, but they can replicate. The RBs go through many rounds of replication, aided by parasitizing the contents of the host cell. At a certain point, they convert back to EBs, and the host cell breaks open, freeing the *Chlamydia* cells to infect other nearby host cells.

In some cases, a chlamydial infection may become a case of lymphogranuloma venereum (LGV). These infections manifest as small ulcerations near the point of infection and can become systemic. If this happens, ulcerations may appear throughout the body, including fistulas in the colon and rectum. Painful buboes on nearby lymph nodes are not uncommon. The risk of secondary infections under these circumstances is very high.

Klebsiella granulomati

Klebsiella granulomati (formerly *Calymmatobacterium granulomatis*, and originally *Donovania*) is the causative agent of granuloma inguinale. These characteristic painless, dark red, raised lesions may become ulcerated, and they typically have a noticeably bad odor. These lesions occur at or near the

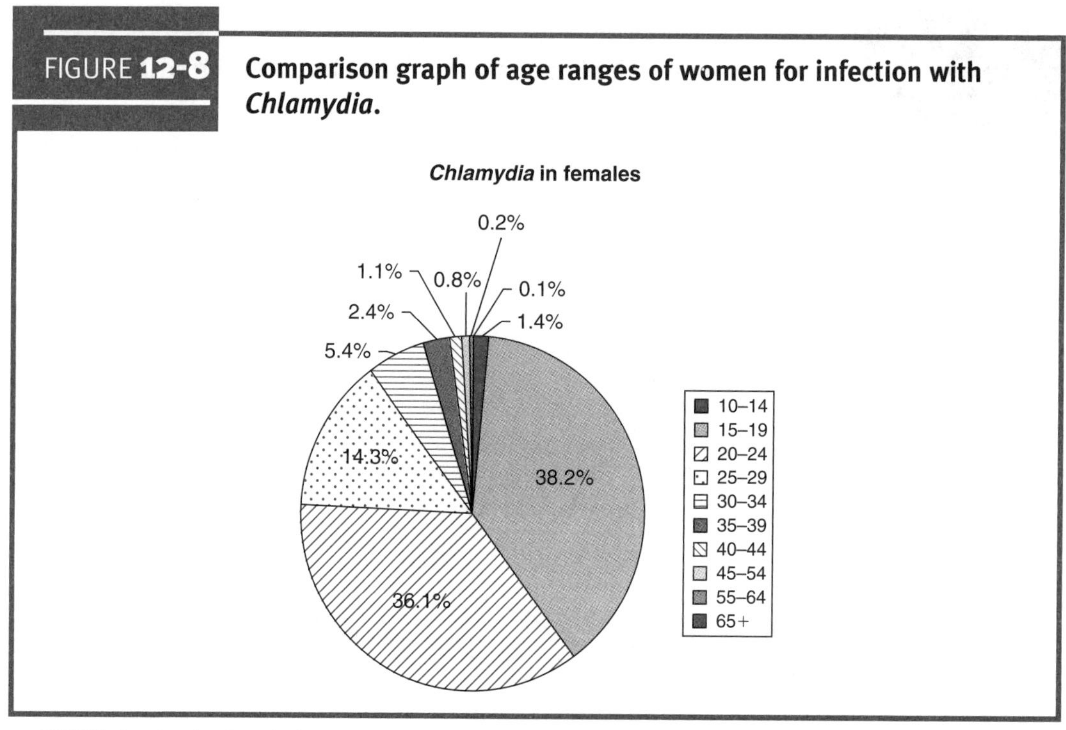

FIGURE 12-8 **Comparison graph of age ranges of women for infection with *Chlamydia*.**

SOURCE: Based on data from the Centers for Disease Control and Prevention.

site of sexual activity, which may include the penis, vagina, anus, and mouth (**Figure 12-9**). In particular, the warm, moist areas of the body are affected, such as folds of skin. The disease appears to be confined to tropical areas of the world. Granuloma inguinale is rare in the United States and is not even a notifiable disease.

Passage of the microorganism is probably facilitated by contact with ulcerated lesions. In addition, open sores are a potential route for invasion by other, more serious STDs. Only in rare cases does the microbe present a problem, but it can spread to the bloodstream, and an occasional death has been reported. Antibiotic therapy is effective at curing granuloma inguinale.

Klebsiella granulomati is much more difficult to grow than are other members of the *Klebsiella* genus. For this reason, diagnosis is made based on the presence of the red nodules and the microscopic analysis of macrophages, which are filled with characteristic Gram-negative rods. These are referred to as Donovan bodies and are found specifically in exudates of lesions.

Neisseria gonorrhoeae

Neisseria gonorrhoeae is the causative agent of gonorrhea, a well-known STD and a notifiable disease (**Figures 12-10** and **12-11**). In 2007, about 355,000 cases were reported in the United States. Rates of infection today are a fraction of rates that occurred just a few decades ago. Better education about STDs in general and a more open discussion about STDs have both contributed to this decline. However, gonorrhea has persisted and is still a major disease in developed countries. One reason for this is that women frequently have asymptomatic infections. An asymptomatic prostitute can infect many clients before being treated for the dis-

FIGURE **12-9** **Infection with *Klebsiella granulomatis* (*Calymmatobacterium granulomatis*), leading to an ulcerated penis.**

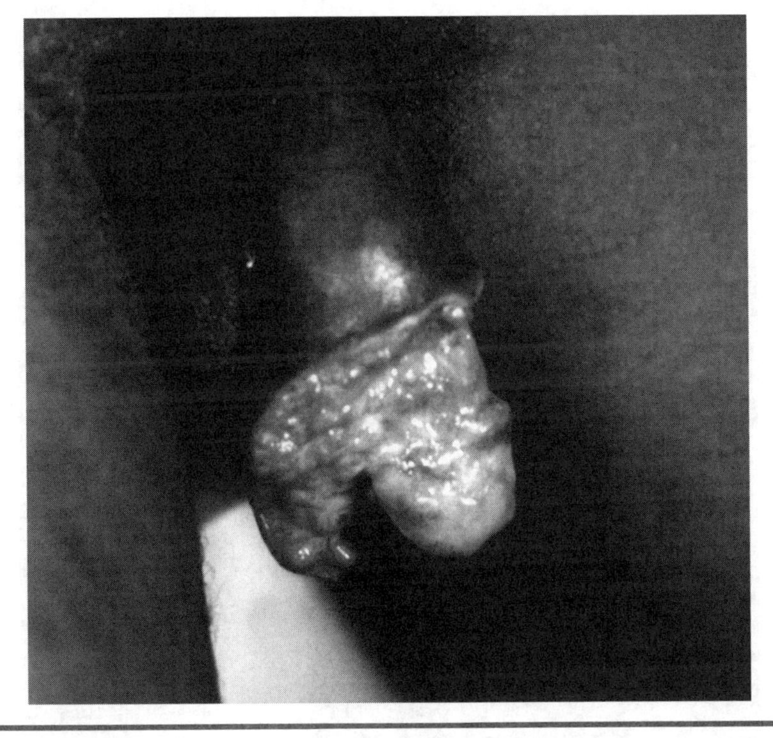

SOURCE: Courtesy of Renella Woodall, the Centers for Disease Control and Prevention. (http://phil.cdc.gov/phil/details.asp). #6523.

ease herself. The men may then go on to infect their wives or girlfriends, continuing the spread of the disease.

Antibiotic resistance has been growing among *Neisseria* strains. Even relatively new drugs like ciprofloxacin may be unsuitable for therapy. Coinfection with *Chlamydia* is also prevalent, which complicates the therapy choices. Presently, azithromycin and doxycycline are the most successful methods of treatment.

Treponema pallidum

Treponema pallidum is the causative agent of syphilis, a notifiable disease. Formerly a scourge of society, relatively few cases are now reported. However, if untreated, syphilis has the potential to cause terrible damage to the human body.

Syphilis infection occurs in several distinct stages. The primary infection results in an ulcer or chancre that gradually resolves. Usually this is a single lesion and is not painful to the touch. This might be easily dismissed by the sufferer, especially because the next stage of the disease might be significantly delayed. In the secondary infection, the symptoms are more general, with a skin rash, mucocutaneous lesions, and tender lymph nodes. Primary and secondary infections are often lumped together for reporting purposes. After the symptoms of secondary infection resolve, a period of latency occurs that may last several years. Finally,

FIGURE **12-10** **Gonorrhea can occur in other tissues besides the genitalia. Here an ophthalmic case is seen.**

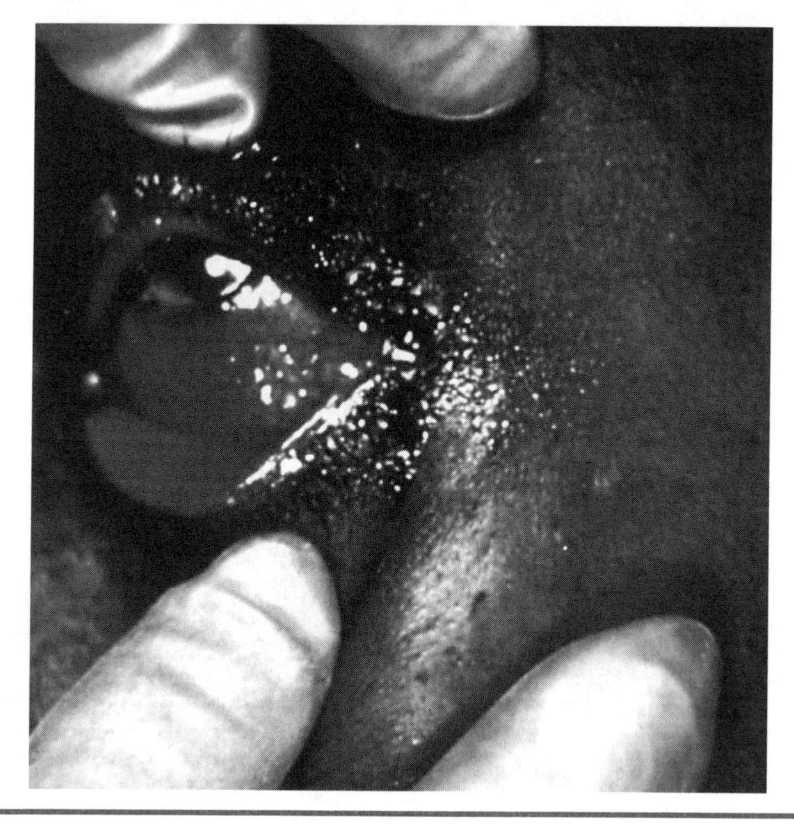

SOURCE: Courtesy of Renella Woodall, the Centers for Disease Control and Prevention. (http://phil.cdc.gov/phil/details.asp). #5175.

the tertiary infection arrives, bringing with it severe symptoms: extensive destructive lesions ("gummas"), lesions of the organs (e.g., heart, lungs, or eyes), and neurological symptoms (e.g., paralysis, numbness, or dementia).

Syphilis was the target of one of the earliest forms of antimicrobial treatment. A concoction known as Salvarsan was utilized early in the twentieth century. This arsenic-based compound was injected into patients. The idea was to kill the microbe before the patient died from the arsenic. This was a very real threat because the amount of arsenic used was quite high, and occasional fatalities occurred.

Serologic tests such as the VDRL and RPR tests have been used to diagnose syphilis infections for many years. Definitive testing is accomplished through a fluorescent antibody staining of the microbe. The treatment of choice for syphilis is penicillin G, which is effective and affordable.

Most states require a syphilis screening test for pregnant women. This is an important precaution because syphilis can cause devastating injury to a

FIGURE 12-11 ***Neisseria gonorrhoeae* as shown by Gram stain of a urethral exudates.**

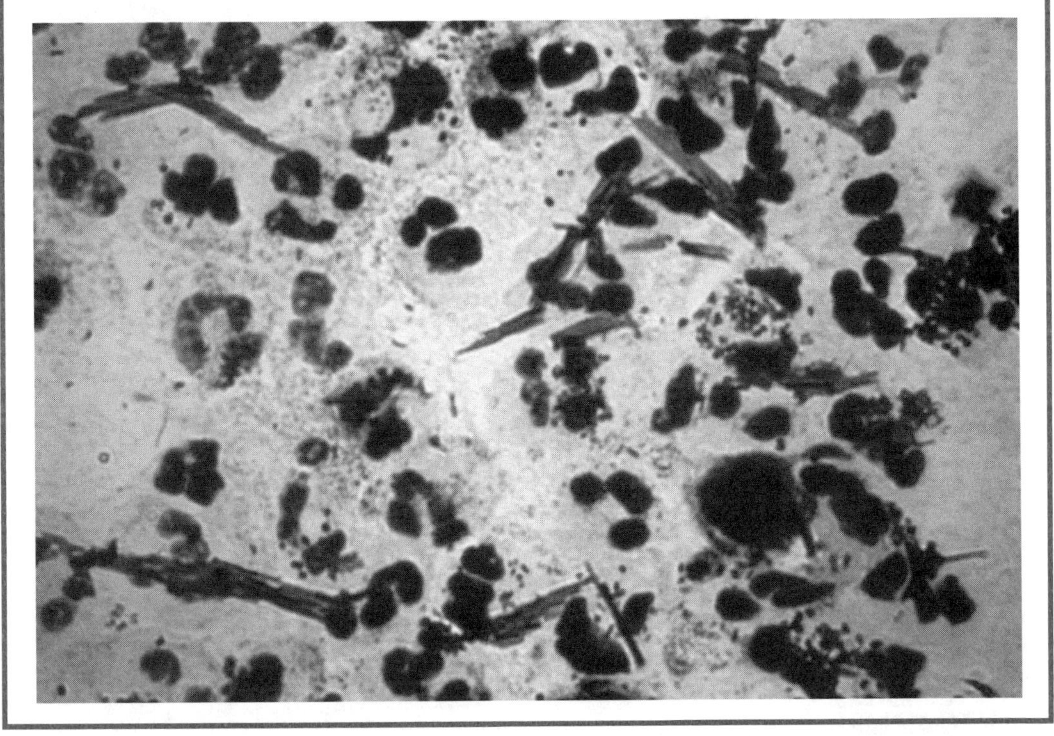

SOURCE: Courtesy of Joe Miller, the Centers for Disease Control and Prevention. (http://phil.cdc.gov/phil/details.asp). #5518.

developing fetus, including spontaneous abortion and neurological damage. Penicillin G is sufficient to cure both mother and infant.

FUNGAL STDS

Tinea cruris

The *Tinea* species are typical dermatophytic fungi that cause a variety of skin problems (**Figure 12-12**). This species stands out as a potential STD because it causes an irritation around the genitals that is commonly known as "jock itch." It is not exclusively an STD but is usually considered when a patient is being examined for other STDs.

Candida

Candida is a very common fungus that causes candidiasis, sometimes referred to as a yeast infection. Here again, *Candida* is not exclusively an STD, but it merits mention in the context of other STDs (**Figure 12-13**).

VIRAL STDS

Hepatitis B Virus (HBV) or Serum Hepatitis

HBV is a notifiable disease. The Centers for Disease Control and Prevention (CDC) estimate that approximately 43,000 cases occurred during 2007 if

FIGURE **12-12** **Photomicrograph of a dermatophyte, *Trichophyton soudanense*.**

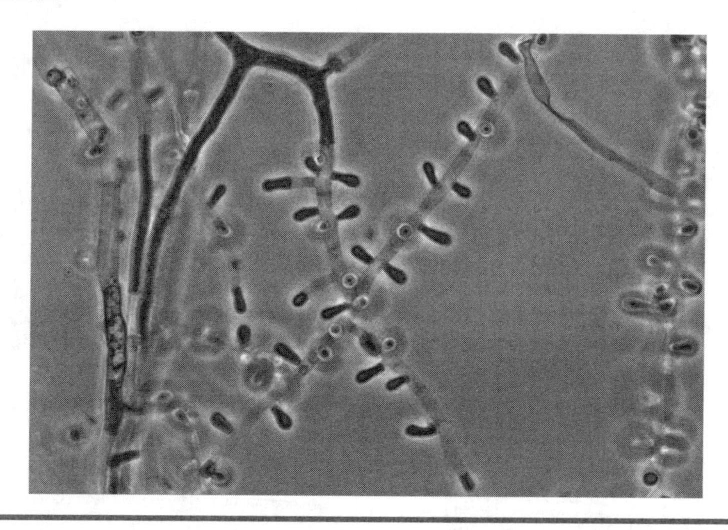

SOURCE: Courtesy of Dr. Libero Ajello, the Centers for Disease Control and Prevention. (http://phil.cdc.gov/phil/details.asp). #11008.

FIGURE **12-13** **Severe *Candida* infection of the pharynx in a patient with HIV.**

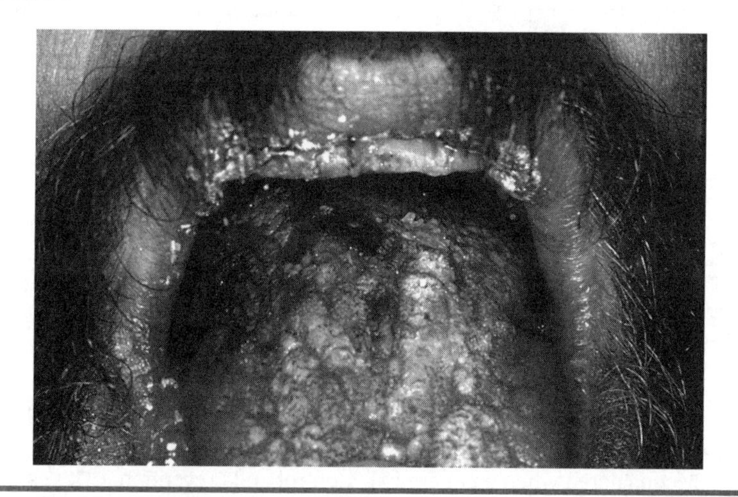

SOURCE: Courtesy of John Molinari, PhD, and Sol Silverman Jr., DDS, the Centers for Disease Control and Prevention. (http://phil.cdc.gov/phil/details.asp). #6067.

asymptomatic cases are included and undercounting is estimated. In contrast to hepatitis A virus, HBV can cause both an acute and a chronic hepatitis. Most people recover from acute cases, but chronic infections are both common and damaging. Perhaps 15–25% of chronic cases result in liver damage, including cancer (**Figure 12-14**). The risk of getting a chronic case is much higher when the person is younger. More than 90% of infants who are infected with HBV become chronically ill with the disease, while fewer than 50% of children aged 1–5 years and 6–10% of older children and adults develop chronic cases. In a typical year, there are 2,000–4,000 deaths attributed to HBV.

HBV is passed in blood products (i.e., red blood cells, blood plasma, or platelets), by accidental needle sticks, or by sexual contact. The HBV concentration can be very high in the blood but is also found in semen, vaginal secretions, and saliva. This is largely a problem for high-risk groups (e.g., MSM and intravenous drug users [IDUs]). In 2007, some 4,519 clinical cases were reported (**Figure 12-15**).

While about half of all infections are asymptomatic, the results for symptomatic cases are severe. About 1% of symptomatic cases result in a fatality. Symptoms include fever, fatigue, clay-colored stools, abdominal pain because of liver involvement, nausea, vomiting, and jaundice. A vaccine is available and recommended for people in high-risk groups, including healthcare workers who might get a needle stick accidentally (**Figure 12-16**). However, if an infection is acquired, there is no specific treatment for the disease.

Hepatitis B is especially virulent if coinfection with Hepatitis D occurs. Relatively little is known about Hepatitis D virus (HDV). It has little in common with Hepatitis B virus. HBV is a DNA virus, while HDV

FIGURE **12-14** **Massive hepatoma in a patient with chronic hepatitis due to Hepatitis B virus.**

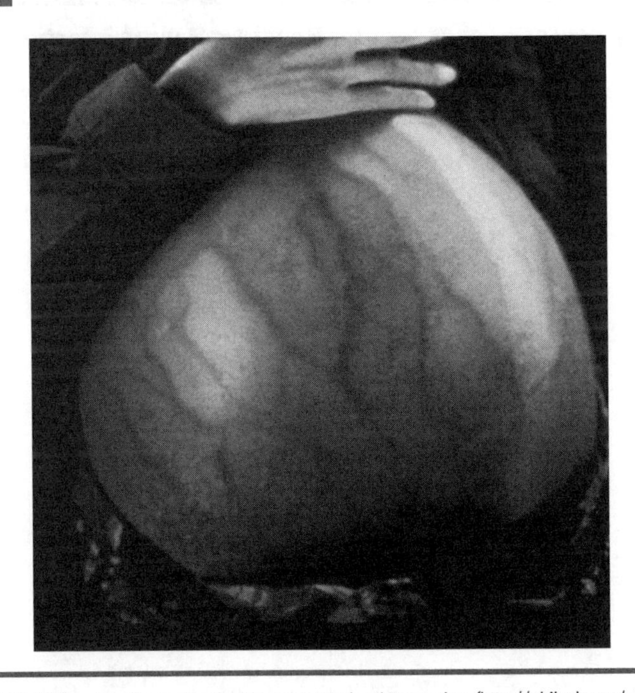

SOURCE: Courtesy of Patricia Walker, MD, the Centers for Disease Control and Prevention. (http://phil.cdc.gov/phil/details.asp). #5606.

FIGURE **12-15** Trends in reported cases of hepatitis over a 20-year period.

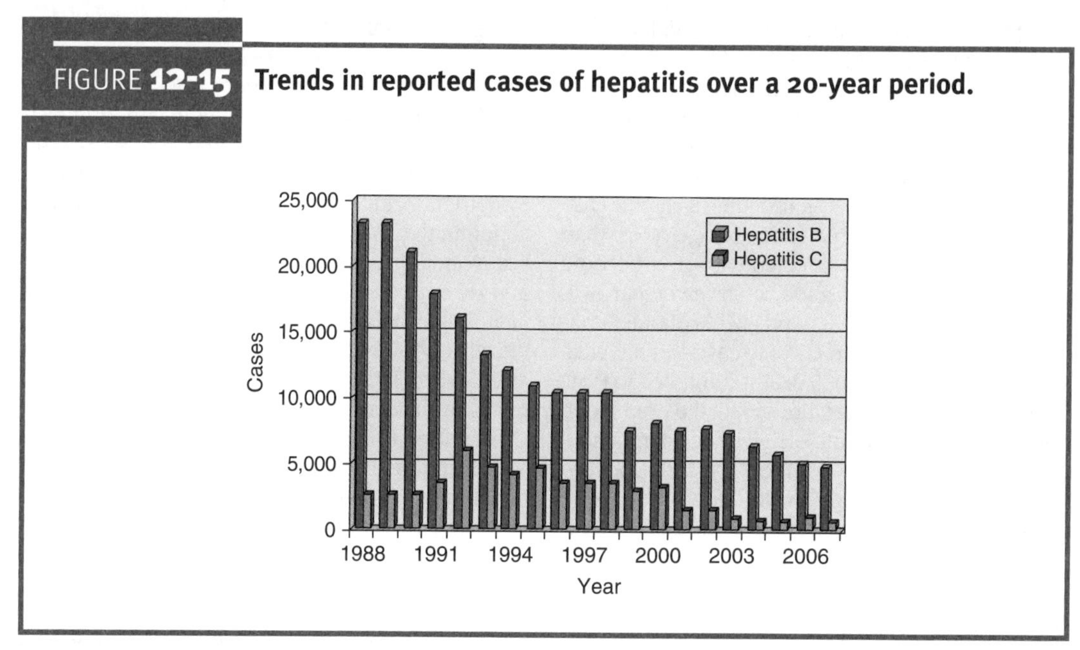

SOURCE: Based on data from the Centers for Disease Control and Prevention.

FIGURE **12-16** Routes of exposure for Hepatitis B virus infection.

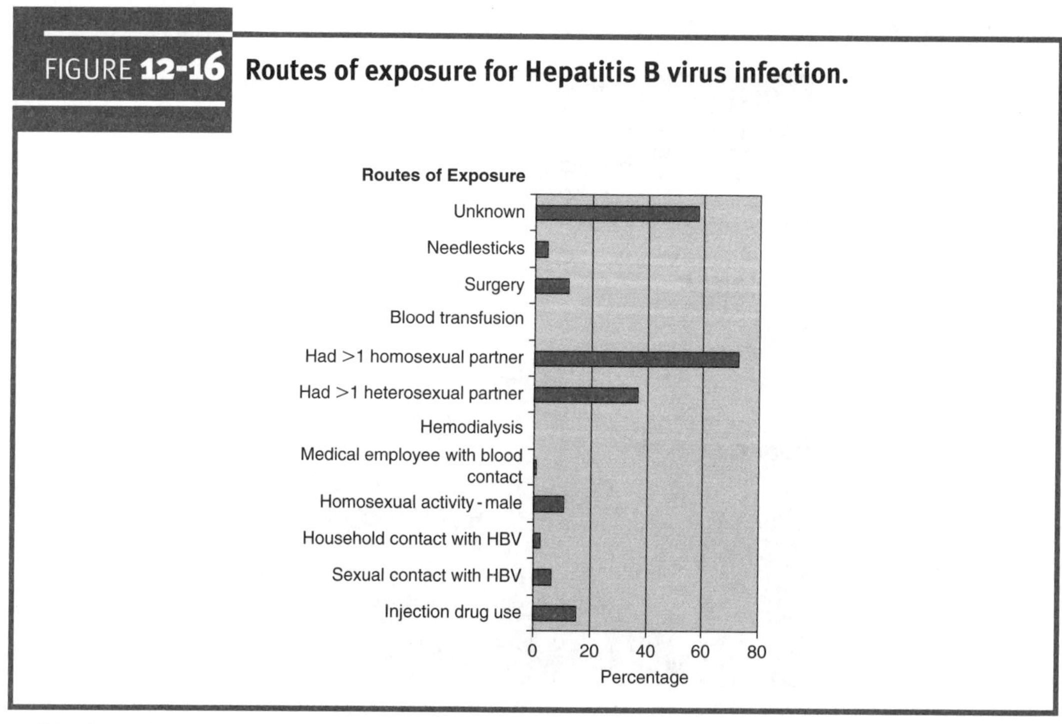

SOURCE: Based on data from the Centers for Disease Control and Prevention.

is an RNA virus. HBV causes disease by itself, while HDV does not. The transmission of HDV is not completely known; it may be more commonly passed through needle sharing than by sexual contact. However, passage by the fecal–oral route cannot be ruled out.

Hepatitis C Virus: HCV or Non-A, Non-B Hepatitis

Although HCV can be transmitted sexually, this does not occur with great efficiency. It is more likely to be transmitted through blood products and needle sticks, and therefore the IDU community is at greatest risk. One particularly unfortunate outbreak resulted from hemodialysis in a contaminated facility (Hallack et al., 2009). Even tiny amounts of blood or blood products may carry an infectious dose of HCV.

The CDC estimated that about 17,000 new cases occurred in 2007. The true figure is difficult to obtain because many cases are asymptomatic. Of the symptomatic cases, an acute illness is less common (no more than 30%). Symptoms are identical to those of HBV infection. Recovery from an acute case is unlikely to cause any liver damage. Acute cases are very uncommon. Chronic HCV develops in the majority (60–85%) of infections, and a high percentage of these cases result in liver damage. This, in turn, dramatically increases the risk for liver cancer. Only about 1 in 5 cases result in the clearance of the virus from the body. This means that the total number of HCV cases (new and continuing) is approximately 3.2 million. There is no vaccine yet for this virus. Antiviral drugs may be used in treatment; otherwise the treatment is supportive.

Herpes Simplex Virus (HSV)

HSV types 1 and 2 are responsible for a huge number (tens of millions) of cases of genital herpes in the United States (**Figure 12-17**). One of the problems with this disease is that it is chronic and, as yet, incurable. Because many cases are mild or asymptomatic, the infected person continues to spread the virus to other sexual partners. The virus manifests as small, superficial, painful ulcers. The virus can be transmitted even if visible blisters are not apparent. The lesions appear in clusters that are painful to touch. The base of the lesion is usually erythematous.

HSV-1 is usually but not exclusively associated with oral herpes infections. HSV-2 is responsible for the majority of recurrent genital herpes infections. The ulcerations that the virus causes on the genitals creates a greater opportunity for infection with other STDs, notably HIV. In very rare instances, HSV may produce encephalitis, which has a very high fatality rate. Until recently, there was little a physician could do about herpes infections. However, the acyclovir class of drugs is effective at limiting the severity of infection.

The herpes viruses can lie dormant for long periods of time. In both oral and genital herpes, a recurrent infection can be caused by a number of stresses on the body, such as another infection, exposure to toxic chemicals, or perhaps merely because of mental stress.

Human Immunodeficiency Virus (HIV)

Human immunodeficiency virus is the causative agent of acquired immunodeficiency syndrome (AIDS), a notifiable disease that has affected our society greatly in the past few decades and that is now having a worldwide impact (**Figures 12-18**, **12-19**, and **12-20**). It has been estimated that 95% of the people currently suffering from AIDS live in developing countries (Gayle and Hill, 2001).

The HIV infection may be asymptomatic at first, or it may present as flu-like symptoms approximately 2–4 weeks after infection. The virus infects the T cells of the immune system, those bearing the CD4 marker. This is the basis for the destructiveness of the virus, because it gradually eliminates these key regulatory cells of the immune system. Without the CD4 cells, the immune system cannot respond to any infection. Over time this means that the HIV+ patient will acquire more infections that will require longer recovery times. When the count of CD4 cells falls below 200 per microliter, the patient has AIDS and becomes very easily infected. One of these infections may prove fatal. In sub-Saharan Africa, the most common infection with HIV is tuberculosis, and the dual infection is especially dangerous.

In the early days of the AIDS outbreak, the virus killed its victims quickly, within weeks or months of diagnosis. Prognosis is much better today with the creation of highly active antiretroviral therapy (HAART), a cocktail of drugs that suppress the viral

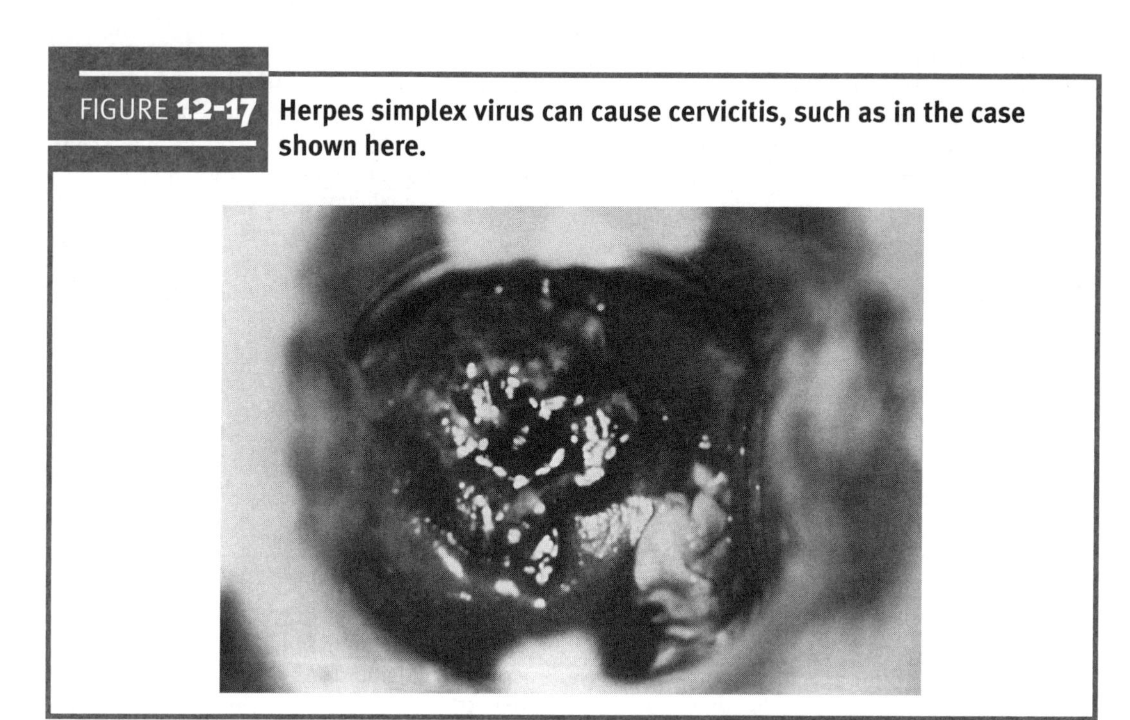

FIGURE 12-17 Herpes simplex virus can cause cervicitis, such as in the case shown here.

SOURCE: Courtesy of Dr. Paul Wiesner, the Centers for Disease Control and Prevention. (http://phil.cdc.gov/phil/details.asp). #6495.

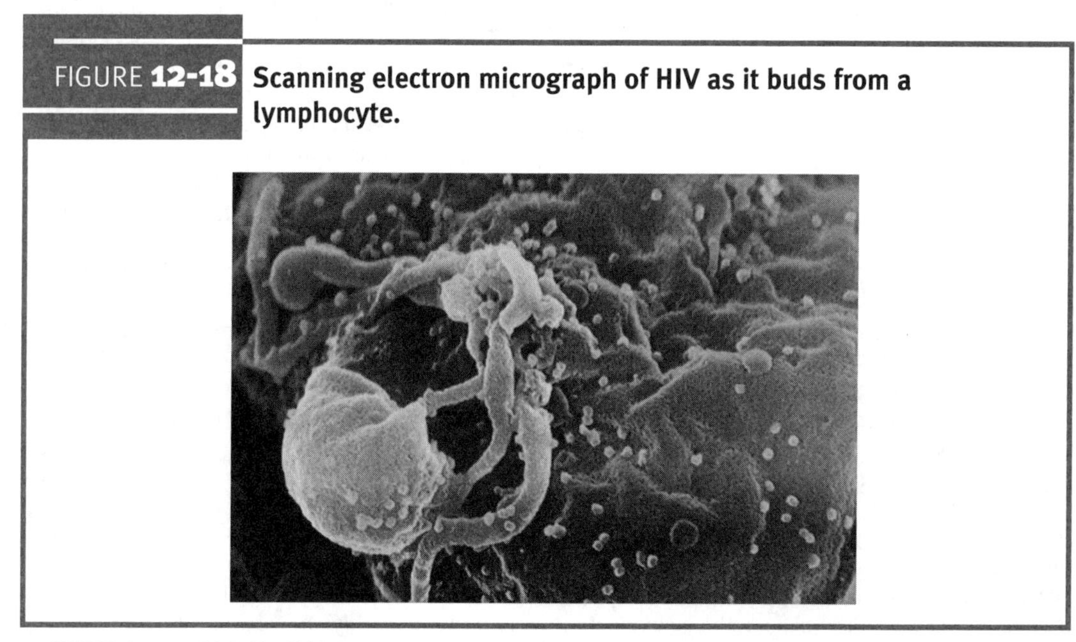

FIGURE 12-18 Scanning electron micrograph of HIV as it buds from a lymphocyte.

SOURCE: Courtesy of C. Goldsmith/CDC.

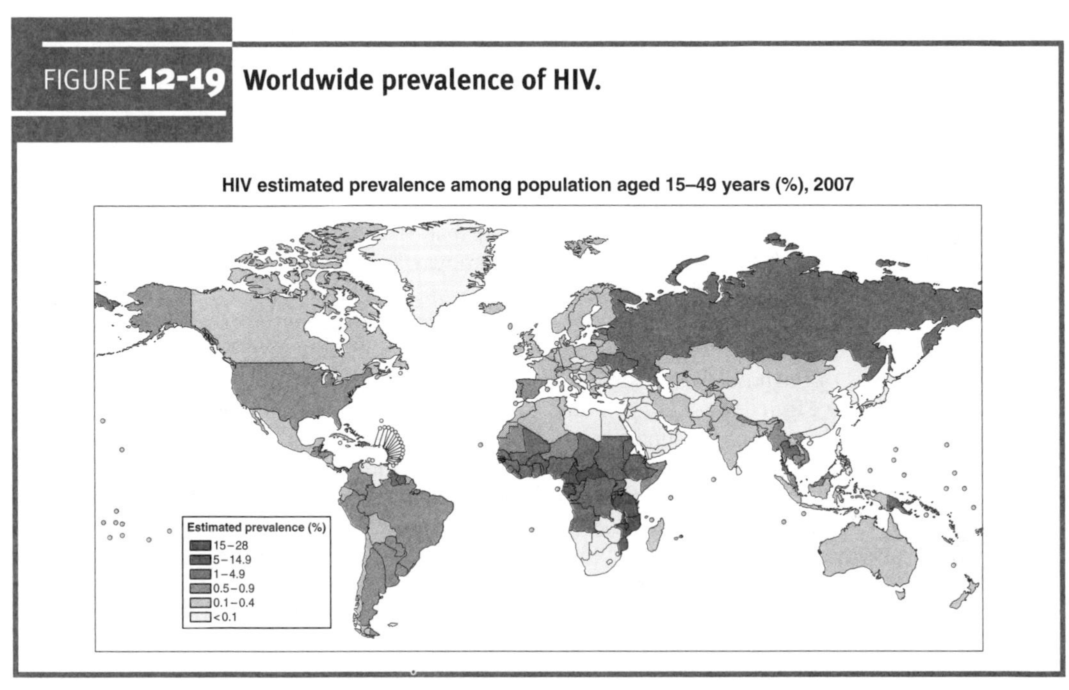

FIGURE 12-19 Worldwide prevalence of HIV.

SOURCE: WHO.

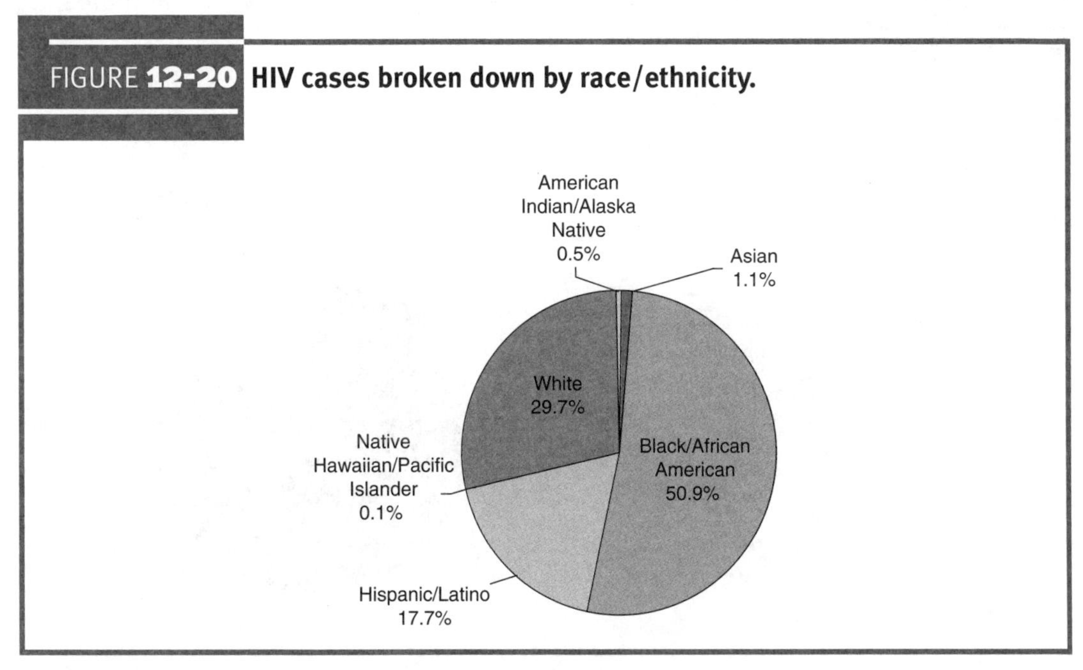

FIGURE 12-20 HIV cases broken down by race/ethnicity.

SOURCE: Based on data from the Centers for Disease Control and Prevention.

load. In the United States, the most common opportunistic infection in HIV positive people is *Pneumocystis carinii* (also called *P. jirovecii*). This fungus causes an atypical pneumonia found solely in HIV patients, often during the last stages of HIV attack. In contrast, the coinfection of HIV and tuberculosis is now one of the defining characteristics of morbidity in sub-Saharan Africa. HIV infection can activate latent TB infections, which are progressive and ultimately lethal. In fact, infection with HIV can raise the likelihood of activating latent TB by a factor of 10 (Gayle and Hill, 2001).

HIV is also passed through needle sticks (contaminated blood). (See **Figure 12-21**.) In many parts of the world, the donated blood supply is not adequately screened for pathogens. When a blood transfusion is needed, the choice of worrying about infection or receiving only a sterilized fluid can be difficult. HIV can also be passed through breast milk, and it has been estimated that in developing countries, as many as half of HIV infections in children are due to infection from the mother through breast feeding (Gayle and Hill, 2001). A particular risk factor appears to be the presence of mastitis, which brings additional immune cells to the breast and passes a higher viral dose to the child.

Human Papillomavirus (HPV)

More than 100 types of human papillomavirus (HPV) are known. The most common are Types 6 and 11. These types can produce genital warts, which are not especially troublesome. Usually these warts can be removed with any of the procedures that are acceptable for ordinary warts (i.e., cryotherapy or laser surgery). However, a few strains put the sufferer at greater risk for cervical cancer: Types 16, 18, 31, 33, and 35. A polymerase chain reaction (PCR) test is used to type the strain of virus that is causing the warts. The recent introduction of the Gardasil vaccine was welcome as a means to avoid infection by some of these high-risk strains, specifically Types 6, 11, 16, and 18.

FIGURE **12-21** **HIV infection in males and females results from different routes, as shown in these comparison charts.**

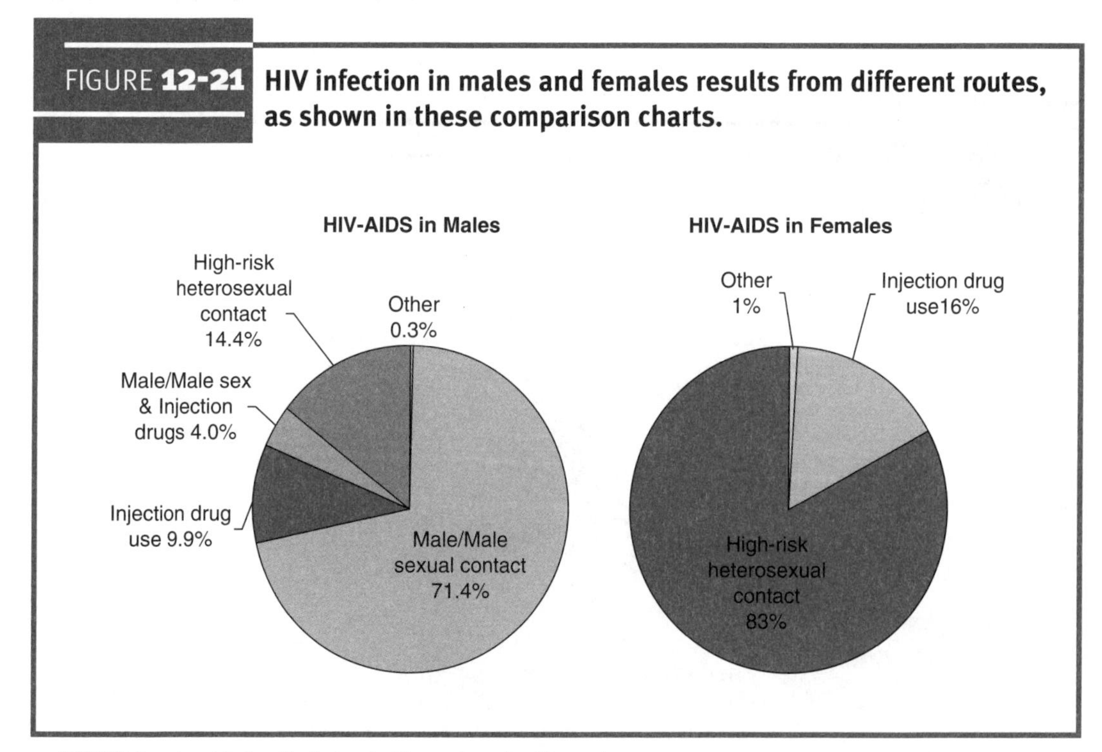

SOURCE: Based on data from the Centers for Disease Control and Prevention.

FIGURE 12-22 ***Molluscum contagiosum*** **is a poxvirus that occasionally causes a sexually transmitted disease.**

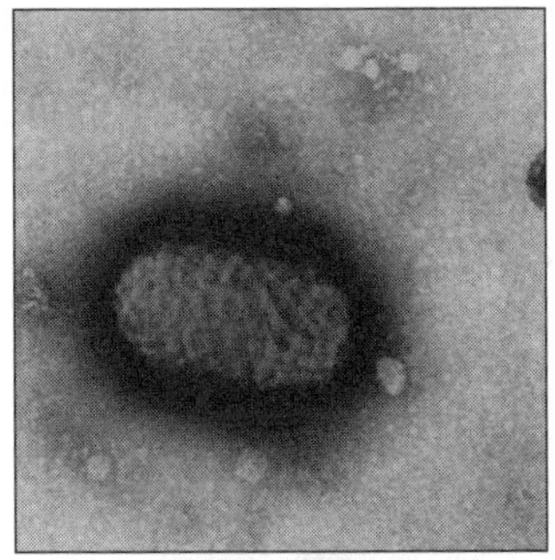

1. Orf virus "M" form

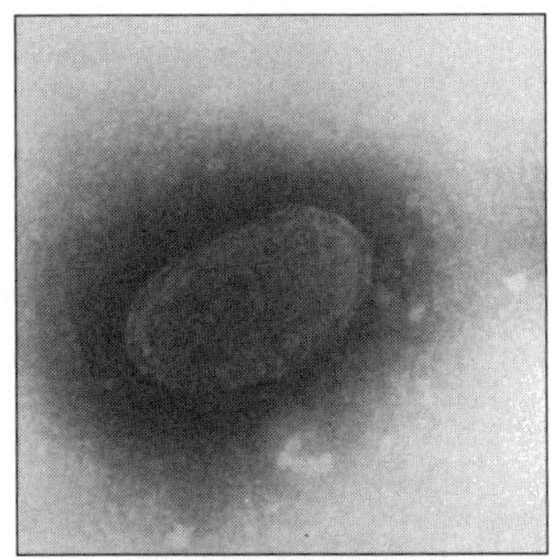

2. Orf virus "C" form

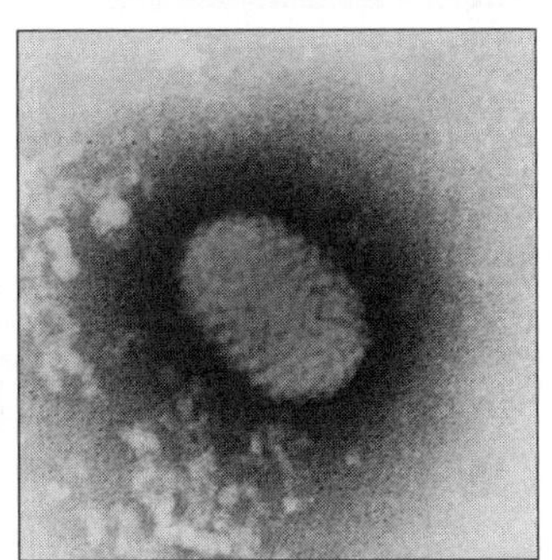

3. Milker's nodule virus "M" form

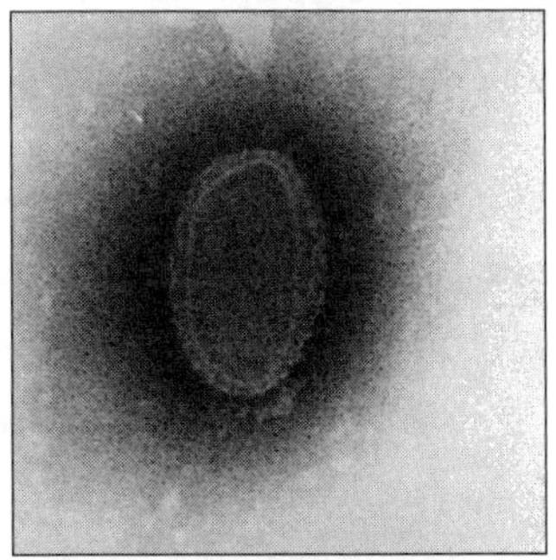

4. Milker's nodule virus "C" form

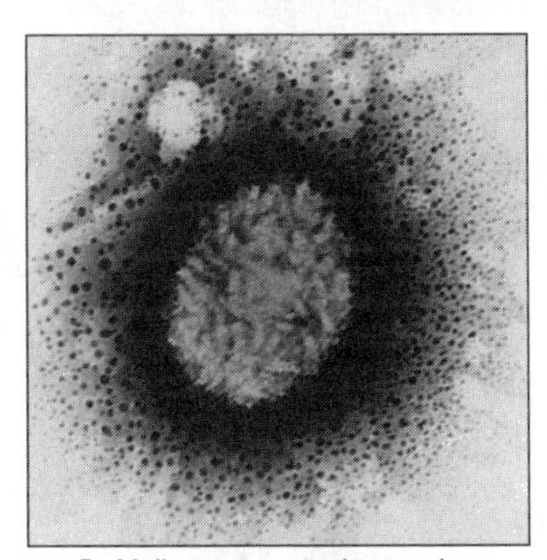

5. Molluscum contagiosum virus "M" form

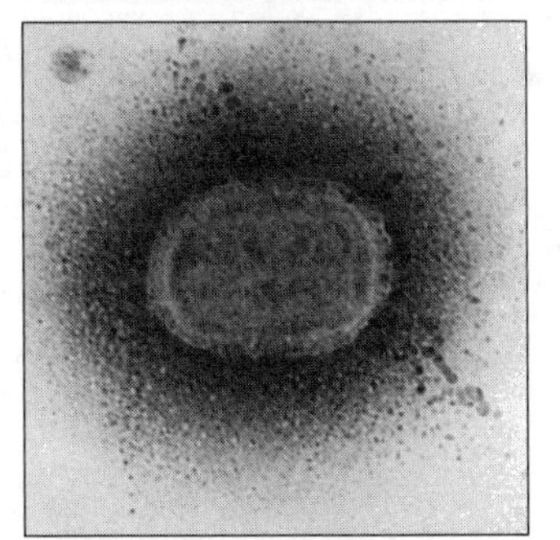

6. Molluscum contagiosum virus "C" form

Electron micrographs of orf, milker's nodule, and molluscum contagiosum viruses.

SOURCE: Courtesy of Dr. James Nakano, the Centers for Disease Control and Prevention. (http://phil.cdc.gov/phil/details.asp). #10198.

Molluscum Contagiosum Virus (MCV)

This is the causative agent of molluscum contagiosum, a disfiguring infection. The virus belongs to the poxvirus family and is spread by close contact, including sexual contact (**Figure 12-22**).

PARASITES

Crab Louse (*Phthirus pubis*)

Most people simply refer to these pests as "crabs." They are not the same as body lice, which are a different genus and which are able to transmit human disease (e.g., typhus). However, they cause intense itching in the pubic region and can spread to other people during sexual intercourse. Diagnosis is usually made on the discovery of pubic lice alone (**Figure 12-23**). A number of topical treatments are available to kill them.

Scabies (*Sarcoptes scabiei*)

Scabies are also known as the human itch mite. These tiny parasites burrow into the skin and cause intense itching (**Figure 12-24**). Scabies are passed through sexual contact, but they are also shared through personal products or even bedding. The lesions caused by these parasites may resemble burrows.

FIGURE **12-23** **The crab louse, *Phthirus pubis*.**

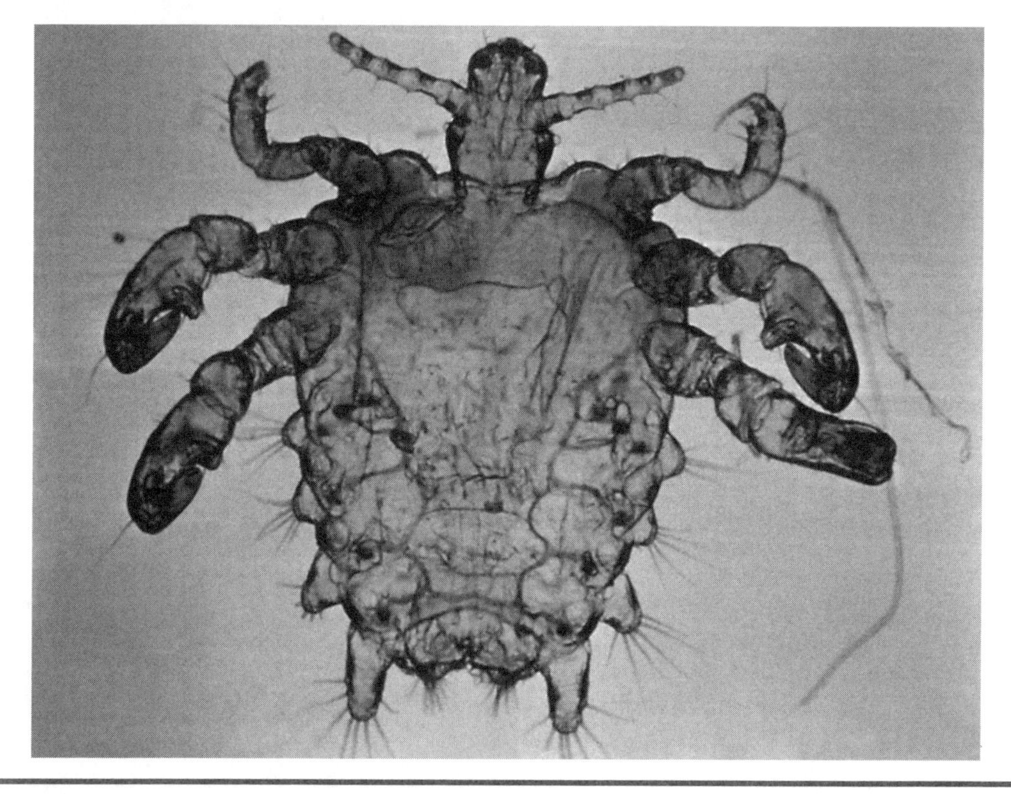

SOURCE: Courtesy of the World Health Organization and the Centers for Disease Control and Prevention. (http://phil.cdc.gov/phil/details.asp). #4077.

FIGURE 12-24 Scabies is shown here as a rash on the trunk of this person.

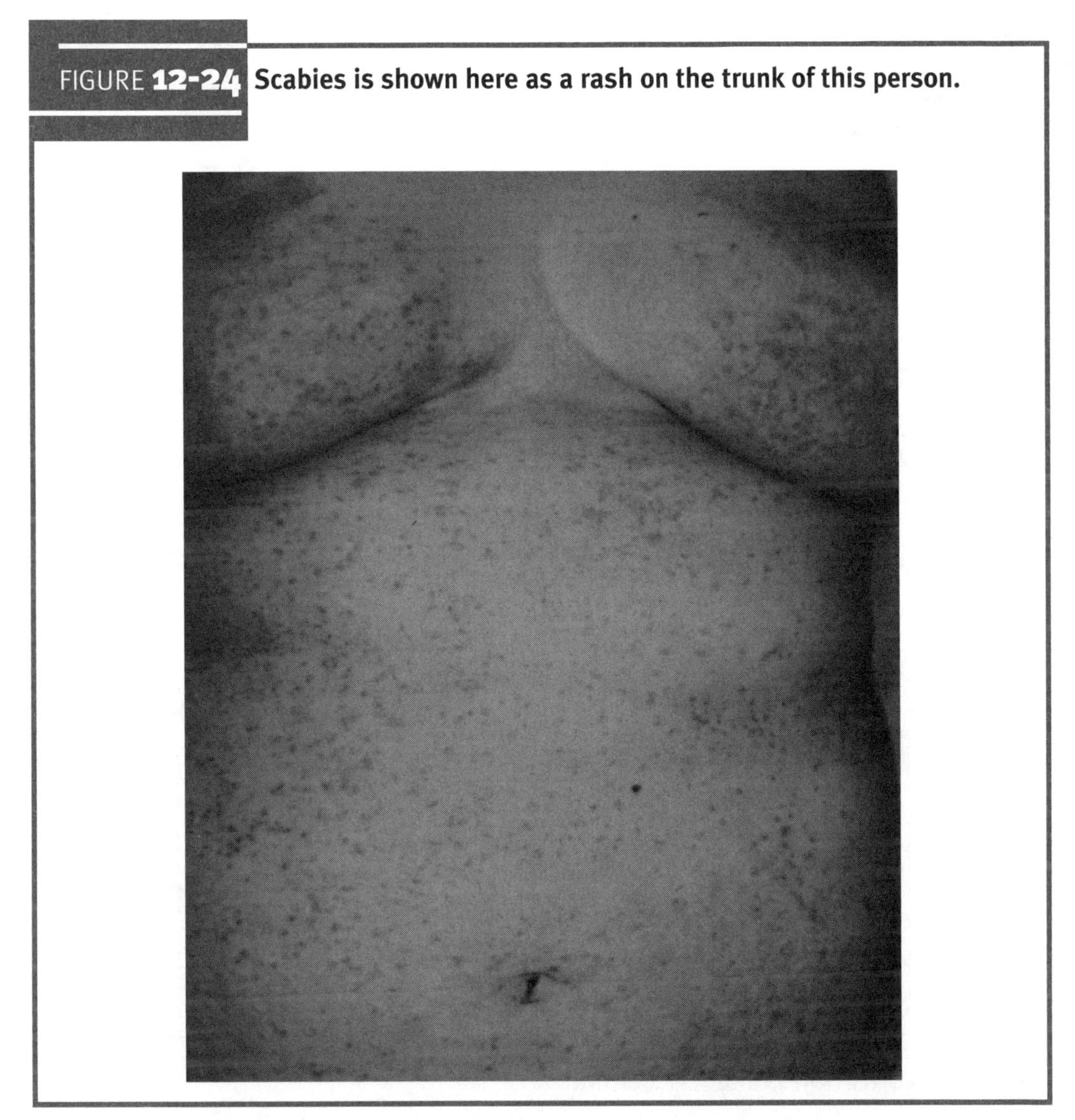

SOURCE: Courtesy of the Centers for Disease Control and Prevention. (http://phil.cdc.gov/phil/details.asp). #4799.

PROTOZOAL STDS

Trichomonas vaginalis

Trichomonas vaginalis is the causative agent of trichomoniasis (**Figure 12-25**). Large percentages of both men and women can have asymptomatic infections or relatively mild symptoms. In women, severe infections are accompanied by a foul-smelling, yellow-green, vaginal discharge. The infection can be successfully treated with metronidazole.

FIGURE 12-25 The life cycle of the parasite *Trichomonas vaginalis*.

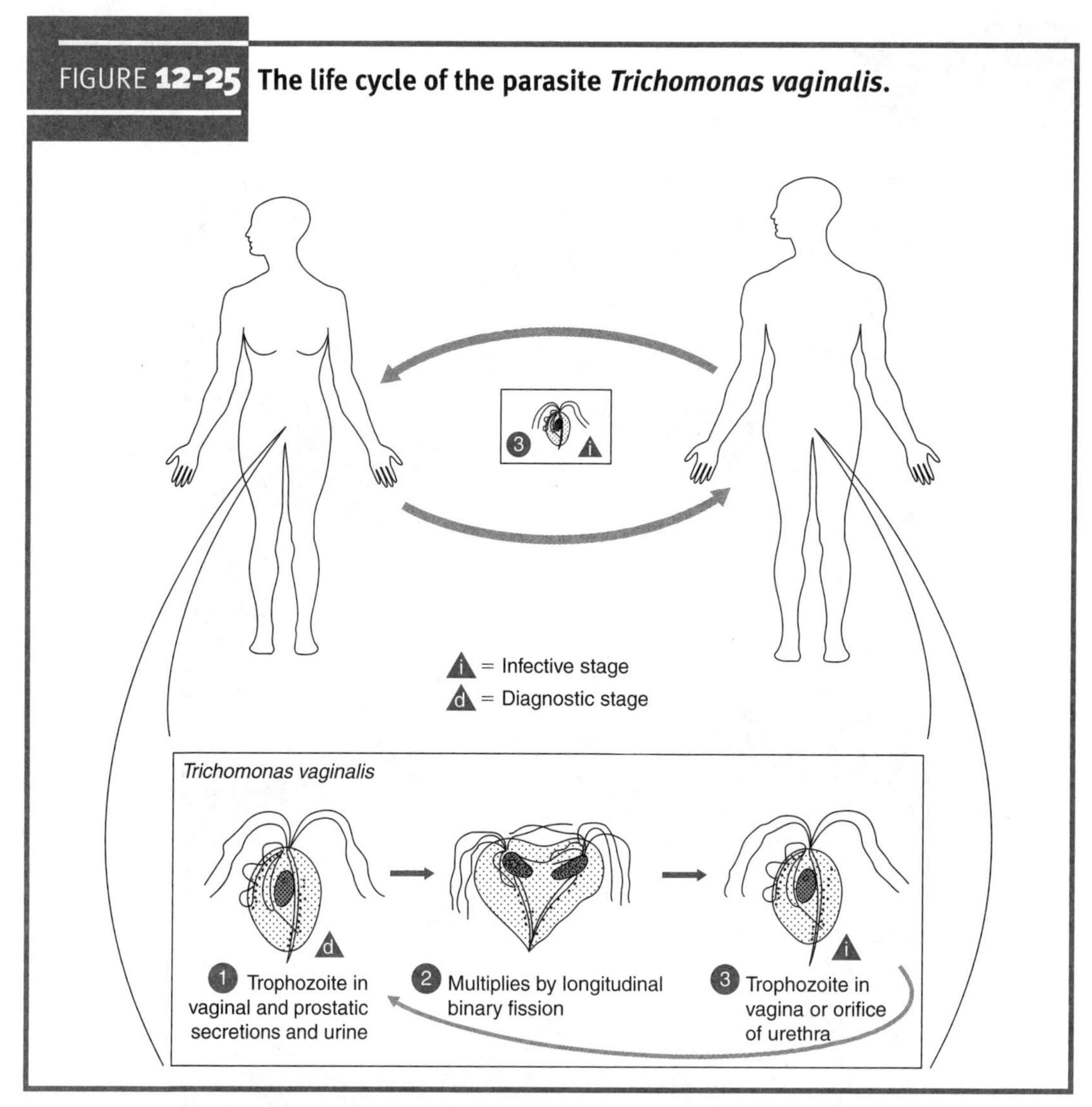

SOURCE: Courtesy of the Centers for Disease Control and Prevention.

QUESTIONS FOR DISCUSSION

1. What do you think are the main barriers to effective treatment of people with STDs?
2. How do infection rates in the United States compare with those of other countries? Are there any patterns?
3. Create a sketch for an advertising campaign to inform the public about STD risk.
4. Infection rates for STDs have fluctuated widely over the years. Why did this happen?
5. What role do the asymptomatic cases play in the continuation of STDs?
6. HIV is ultimately a deadly STD, and some countries have high rates of infection. What does this mean for those countries?

References

Agnew, J. 1986. Hazards associated with anal erotic activity. Arch. Sex Behav. 15: 307–314.

Gayle, H.D., and G.L. Hill. 2001. Global impact of Human Immunodeficiency Virus and AIDS. Clin. Microbiol. Rev. 14: 327–335.

Goldberg, J.E., and S.R. Steele. 2010. Rectal foreign bodies. Surg. Clin. N. Am. 90: 173–184.

Hallack, R., G. Johnson, E. Clement, M. Parker, J. Schaffzin, et al. 2009. Hepatitis C virus transmission at an outpatient hemodialysis unit—New York, 2001–2008. Morbid. Mortal. Weekly Rep. 58: 189–194.

CHAPTER 13

AIRBORNE PATHOGENS

LEARNING OBJECTIVES

- Determine which pathogens are primarily passed through airborne means and which are the most prevalent.
- Describe methods to limit infection by airborne infection.
- Explain the difference between quarantine and isolation.
- Explain how quarantine measures work and the legal basis for it.
- Describe the fungal species that can cause intoxications and the conditions that favor toxin production.
- List risk factors for tuberculosis and the origins of new cases.
- Explain what is meant by multidrug-resistant and extremely drug-resistant TB strains.

KEY TERMS

- Isolation
- Quarantine

INTRODUCTION

If you become ill sometime in the next year, it will probably be because of an airborne pathogen. This may be a simple cold virus that is best treated with bed rest and fluids, or it may be one of the influenza viruses. Every year, millions of illnesses are caused by these viruses. But there are many other airborne pathogens; many are viruses and are spread by droplet infection, especially the enveloped viruses. Some are now regarded as exotic diseases. The pneumonic variety of plague (caused by *Yersinia pestis*) is highly contagious through airborne droplets, and occasionally outbreaks still occur (Ogen-Odai et al., 2009). Most of these cases (90%) occur in Africa, most often in rural areas. The list of airborne diseases also includes airborne toxins that can be inhaled and enter the bloodstream. This is analogous to foodborne toxins that enter the body through the gastrointestinal tract.

As small particles, bacteria, viruses, and spores of bacteria and fungi are carried by the wind. All of these structures are denser than air, so they gradually fall back to earth, but strong wind currents can keep the particles airborne for long periods of time (depending also on the relative humidity). This may result in airborne pathogens traveling thousands of miles (Pedgley, 1982). In many cases, the spread of plant diseases can be mapped against wind patterns. This is especially true of fungal spores, which are more resistant to environmental insult. However, bacteria and virus particles are susceptible to ultraviolet light, which typically kills them. Coliform bacteria can be found as far as 100 m from wastewater treatment plants, but they are unlikely to go farther because of desiccation or damage by UV light.

Many airborne diseases are the formerly common diseases of childhood: measles, mumps, rubella, varicella zoster (chicken pox), and whooping cough. These are transmissible because of droplet transmission by coughing and sneezing (and often by direct contact as well), and they are highly contagious. To a great extent, they are under control through the use of vaccination, and they are described in Chapter 11.

THE AIRBORNE PATHOGENS

Legionella

In 1976, there was a meeting of the American Legion in Philadelphia. In attendance were many veterans of the U.S. military, many of whom were of advanced age and in poor health. They stayed at a number of hotels in the downtown area. After the convention they returned home, and many developed a persistent pneumonia that was very difficult to diagnose. Strangely for pneumonia, this illness also had gastrointestinal symptoms: nausea, vomiting, and diarrhea. Some 34 deaths resulted from this disease, and medical authorities conducted a thorough investigation to locate the causative agent.

The disease was caused by an organism that was later named *Legionella pneumophila*, and the disease was named Legionnaire's disease, in recognition of the outbreak that brought it to the attention of public health authorities (**Figure 13-1**). It is now more commonly called legionellosis and has become a common cause of atypical pneumonia in debilitated patients. *Legionella* stains poorly with Gram-stain reagents, although technically it is a Gram-negative bacillus. *Legionella pneumophila* grows on artificial media, but it grows poorly and has fastidious growth requirements. A black agar called BCYE (buffered charcoal yeast extract) is used (**Figure 13-2**).

The atypical pneumonia that *Legionella* causes is often misdiagnosed, leading to acute dangers for the patient. Surprisingly, *Legionella* is often the last possibility that some physicians consider when they see the symptoms. Most people have probably been exposed to *Legionella* at some point, but it is only dangerous to older people and the immunocompromised. Based on the information collected after the outbreak in Philadelphia, archived samples from a similar outbreak in Pontiac, Michigan, were examined and were associated with *Legionella* as well. Ironically, the affected people worked in the County Health Department building. This outbreak was dubbed Pontiac Fever, and this so-called "sick building syndrome" affected a number of office workers. However, these people were generally healthy, and no fatalities occurred. *Legionella* has been around humans for a long time, but it apparently did not cause any significant problems until the latter half of the twentieth century. *Legionella* can be found in freshwater lakes and ponds, where its preferred host is actually a protozoan called *Hartmannella vermiformis* (Abu Kwaik et al., 1998). *Legionella* invades this host and multiplies internally. Interestingly, it does the same thing in human phagocytes, which may be a coincidence or may indicate something more profound about species similarities. Either way, the fact is that *Legionella* has a distinct pattern of infection once it enters the human body.

The route of transmission in the vast majority of cases is airborne droplet (aerosol) transmission from a contaminated water source. If the aerosol particles are small enough, they are inhaled and get into the lungs, where the bacteria can infect alveolar macrophages. It is possible that people have become infected from natural water sources, such as ponds, but these cases would have been sporadic and would have been considered as atypical pneumonias of unknown origin. In today's world, aerosols can be created by fountains and misters, as well as sprinkler systems, showers, whirlpools, or splash fountains in recreational parks. Aerosol pathogens may also originate in the cooling towers of large heat exchangers, in water-cooled air conditioner units, and in hot water systems. In each case, the risk is greater if certain factors, such as the natural protozoan host, are found in the water source. *Legionella* also requires high concentrations of cysteine and iron. Because most cooling equipment is made of steel, this latter requirement occurs frequently. In water systems that are old and corroded, it is easy for dust and dirt

FIGURE 13-1 **A scanning electron micrograph of *Legionella pneumophila*.**

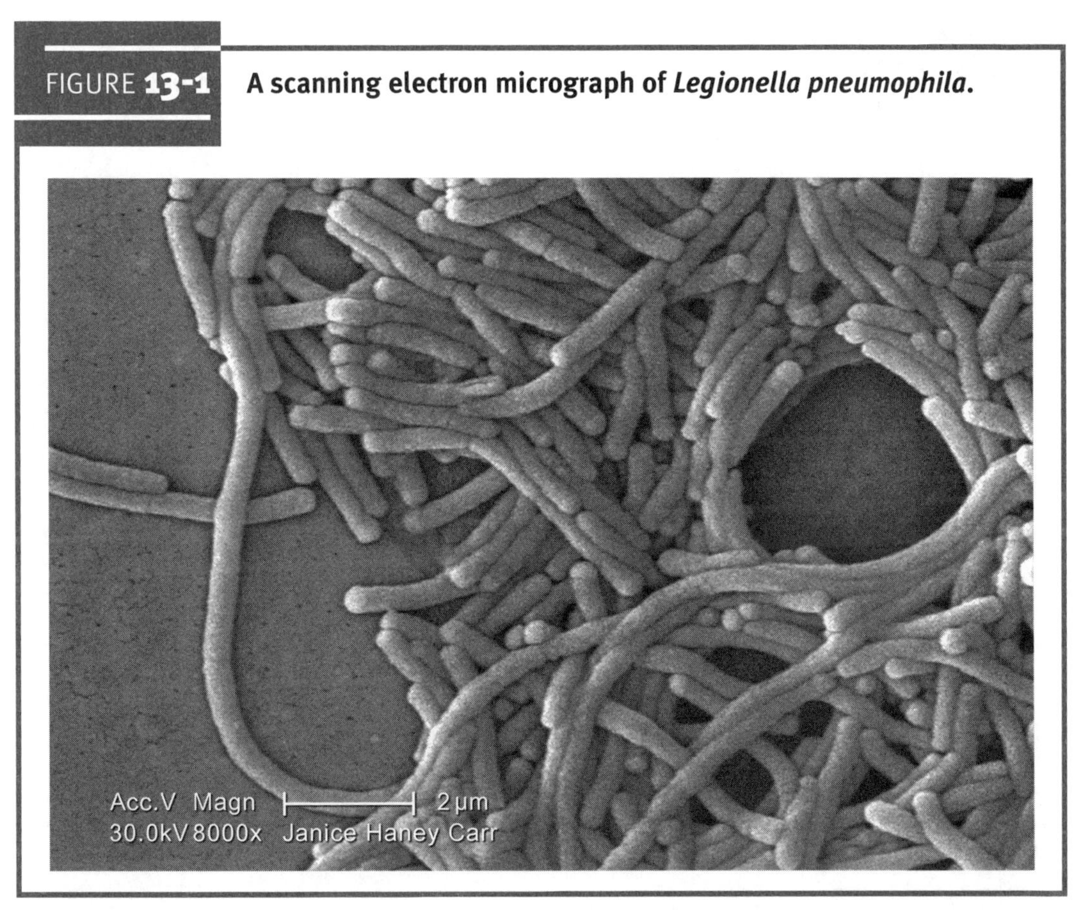

SOURCE: Courtesy of M. Williams, C. Lucas, andT. Travis, the Centers for Disease Control and Prevention. (http://phil.cdc.gov/phil/details.asp). #11152.

to enter the water supply (allowing growth of the bacteria), and aerosols easily exit minute cracks in the pipes. In a society in which air-conditioning units are common, a poorly maintained unit can occasionally cause a local outbreak to occur. Interestingly, the reservoir for *Legionella* usually seems to be the water distribution system, in which bacterial numbers originating at a different site, such as a cooling tower or evaporative condenser, are amplified. *Legionella* has been found in biofilms on the inner surfaces of many types of pipes, thus providing a substantial reservoir of microbes that may become significant if the right conditions appear for aerosol formation (Moritz et al., 2010). Control of biofilm formation by periodically flushing pipes to erode biofilms is a good preventive practice in some locations.

Bordetella pertussis

Bordetella pertussis is the causative agent of whooping cough. The disease takes its name from the symptoms, in which violent coughing expels the maximum amount of air from the lungs, necessitating a large gulp of air on inspiration that produces a whooping sound. In older children and adults, the disease ranges from a mild to severe respiratory infection, and whooping cough occasionally leads to a secondary pneumonia that can be problematic.

FIGURE 13-2 ***Legionella pneumophila* growing on a buffered charcoal yeast extract plate.**

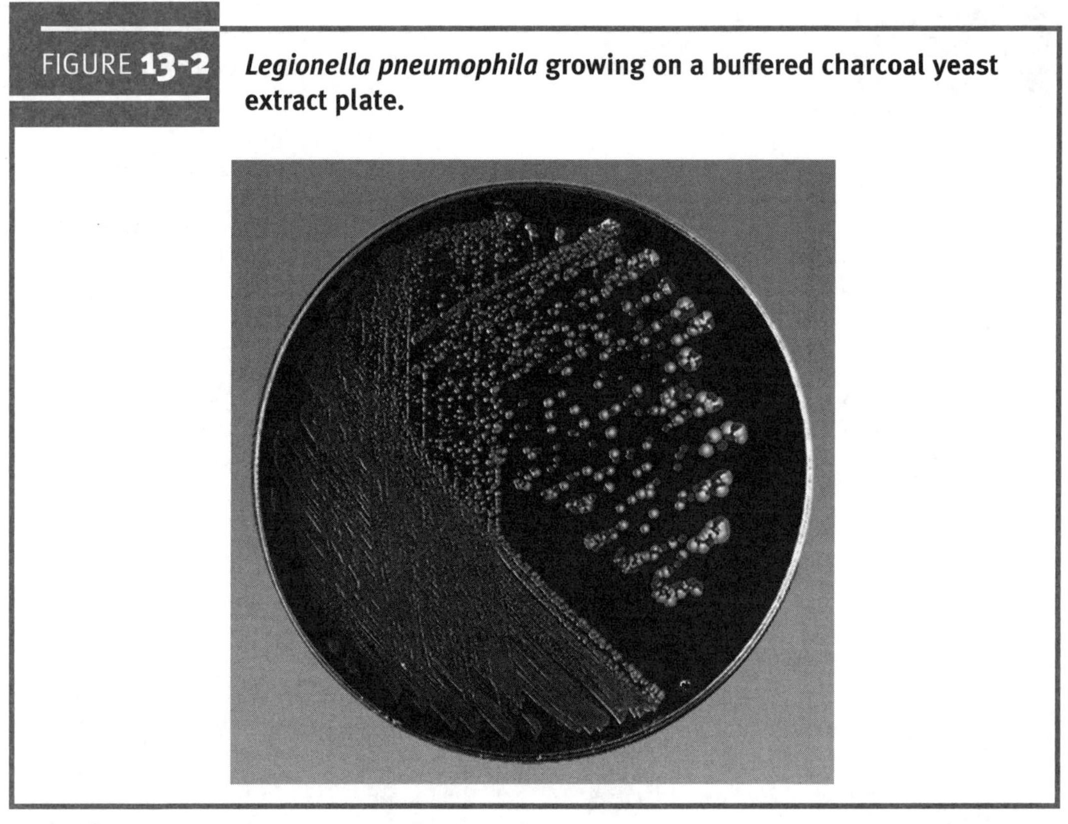

SOURCE: Courtesy of ThermoFisher.

However, the greatest danger is to small children, especially those less than 1 year of age, in which the vast majority of fatalities occur. In years past, whooping cough was greatly feared because of the significant mortality associated with the disease. In developing countries with populations that have other illnesses, poor nutrition, and no immunization, this disease still causes outbreaks and infant deaths.

An excellent vaccine for pertussis is available, and where it has been promoted extensively, the infection rate has fallen dramatically. Pertussis might be eradicated eventually because humans appear to be the only host for the illness. However, in the United States, the vaccination is commonly delivered as the DPT shot (diphtheria–pertussis–tetanus). A small but significant number of people are sensitive to the pertussis fraction of the shot and receive only a single dose. Other combinations of diphtheria and tetanus vaccines can be substituted. Some social communities do not vaccinate because of religious or personal beliefs, and these groups sometimes develop small outbreaks. Finally, parents sometimes neglect the DPT shot because no one really worries about whooping cough or diphtheria anymore, and because the tetanus shot is given after wounds occur. Small outbreaks serve as a reminder to vaccinate children. Typically, between 5,000 and 7,000 cases are recorded each year in the United States.

Mycobacterium tuberculosis

This microbe is the causative agent of tuberculosis (TB); it is an acid-fast microorganism that can be difficult to treat. *M. tuberculosis* causes a persistent infection that occurs in distinct stages. It is feared because, in the pre-antibiotic era, it caused a chronic and progressive disease that typically ended in death (**Box 13-1**). During the illness, the person's body would become noticeably emaciated, as though the body were being consumed by the microorganism. For that reason, the disease was also known as "consumption," as it is often referred to in literature of those times. The illness was known to strike anyone, regardless of social station, and many notable people suffered with or died from it. The Western icon John Henry "Doc" Holliday died from TB at age 36. Former first lady Eleanor Roosevelt developed tuberculosis in her bone marrow late in life. The writer Henry David Thoreau contracted TB at 18 and died from it at age 44.

BOX 13-1 Quarantine

Some illnesses should be avoided, if at all possible, such as communicable diseases that have serious health consequences and that can be identified in affected people. When these people are identified, it is possible to keep them away from other people and, hopefully, to cure them. Most people with such a disease would, one would think, wish to have medical attention and would want to be cured, especially at government expense. However, others do not see the importance of treatment in isolation. In these cases, the government reserves the right to use a form of coercion that is needed to protect society.

This coerced compliance takes two forms. One is **quarantine,** in which people who have been exposed to an illness or to an ill individual are separated from the general population until it is clear whether they have contracted the illness themselves. The other is **isolation,** in which an identified sick person is separated from healthy people. The difference has practical consequences. People in quarantine may chafe at the indignity and the inconvenience; after all, they might not be ill at all. People in isolation know that they have the illness—which is always a serious condition and can be a psychological burden—and they know that they must remain isolated until they are cured (with or without medical intervention) or until they die. A good historical example is that of Typhoid Mary, who lived during the early part of the twentieth century (i.e., the pre-antibiotic era). She worked as a maid and later as a cook. Unfortunately, at some point in her life she was infected with *Salmonella typhi*, the causative agent of typhoid fever. The pathogen colonized her gall bladder, and she became a carrier of the disease. As a cook, she spread the pathogen to several other people, some of whom died from the disease. Public health authorities confirmed that she was a carrier. Unable to cure her, they released her on the promise that she would not work as a cook. However, a cook's wages were higher than a maid's wages, and she returned to her former role. Once again the authorities stepped in, this time confining her to an island in New York harbor, where she stayed from 1915 until her death in 1938. Interestingly, she was never convicted of a crime: it is not illegal to be a carrier of disease. Nonetheless, she was confined against her will.

The U.S. government reserves the authority to take the necessary action to protect public health. The Public Health Service Act is codified in 42 USC 264, Subchapter II, part G, which states in part:

> The Surgeon General, with the approval of the Secretary, is authorized to make and enforce such regulations as in his

> judgment are necessary to prevent the introduction, transmission, or spread of communicable diseases from foreign countries into the States or possessions, or from one State or possession into any other State or possession. For purposes of carrying out and enforcing such regulations, the Surgeon General may provide for such inspection, fumigation, disinfection, sanitation, pest extermination, destruction of animals or articles found to be so infected or contaminated as to be sources of dangerous infection to human beings, and other measures, as in his judgment may be necessary.

This law gives the Surgeon General great authority over a wide range of products and the means to handle any communicable disease. It further states:

> Regulations prescribed under this section shall not provide for the apprehension, detention, or conditional release of individuals except for the purpose of preventing the introduction, transmission, or spread of such communicable diseases as may be specified from time to time in Executive orders of the President upon the recommendation of the National Advisory Health Council and the Surgeon General.

In other words, people cannot be detained without cause, but they can be apprehended and detained (and examined) if communicable disease is suspected. In fact, the Surgeon General has the authority to stop the entry into this country of communicable disease based on Section 265 of the same act:

> Whenever the Surgeon General determines that by reason of the existence of any communicable disease in a foreign country there is serious danger of the introduction of such disease into the United States, and that this danger is so increased by the introduction of persons or property from such country that a suspension of the right to introduce such persons and property is required in the interest of the public health, the Surgeon General, in accordance with regulations approved by the President, shall have the power to prohibit, in whole or in part, the introduction of persons and property from such countries or places as he shall designate in order to avert such danger, and for such period of time as he may deem necessary for such purpose.

Executive Order 13295, signed by President Bush in 2003, makes clear which diseases are covered by this act:

> (a) Cholera; Diphtheria; infectious Tuberculosis; Plague; Smallpox; Yellow Fever; and Viral Hemorrhagic Fevers (Lassa, Marburg, Ebola, Crimean-Congo, South American, and others not yet isolated or named).
>
> (b) Severe Acute Respiratory Syndrome (SARS), which is a disease associated with fever and signs and symptoms of pneumonia or other respiratory illness, is transmitted from person to person predominantly by the aerosolized or droplet route, and, if spread in the population, would have severe public health consequences.

In this area, the federal government gives great responsibility to the state governments. States are expected to isolate or quarantine people within state borders, while the federal government concentrates on individuals who cross state borders or who enter the United States from other countries. Civil rights are a concern in the application of this law. In theory, healthy people could be detained indefinitely as a possible carrier of disease. In practice, the government, both state and federal, has been very careful to test people with appropriate assays for specific pathogens and has treated people when they are ill.

The power of the Surgeon General in these cases is potentially great but is rarely exercised. This power is not all that different from many other federal departments. The head of the Food and Drug Administration has vast potential power to ban or recall food and drugs that are deemed a health hazard. The Secretary of the Department of Labor may declare himself (or herself) the mediator in labor disputes if a need is perceived. The Secretary of the Treasury has broad powers over the financial markets. The Secretary of the Transportation Safety Agency can, with a single order, ground every plane in the United States. This was seen on September 11, 2001, when several hijacked aircraft were used as weapons and no one knew how many more might be hijacked.

This great public trust must be used wisely and only for the public good. Abuse of such power would result in a lack of confidence in the federal government's collective wisdom and would probably lead to a decrease in their authority by an enraged electorate. That would make the government less able to react to future crises.

TB presents a small but persistent problem in the United States. In recent years, the number of annual cases has been approximately 15,000. However, this number is not as important as the reasons that the disease is resistant to eradication. The microbes invade the alveolar macrophages found in lung tissue, where they multiply. Free microbial cells are easily engulfed by activated macrophages, but the presence of the microbes inside host cells make them difficult to destroy. If the infection is large or if the host has trouble mounting an effective immune response, the microbes form a granuloma in an infected tissue, and the structure is surrounded with fibrin. In this form, the bacteria can persist for many years without noticeable effect. In later years or even decades, when the health of the host declines, the microbes can emerge from this latent state and cause another round of tuberculosis. In these cases, the disease may spread to any organ of the body.

Around the world, TB is a huge problem (**Figure 13-3**). On the African continent (particularly sub-Saharan Africa) it is often found in connection with people suffering from AIDS. The TB infection must be treated, but treatment invariably leads to antibiotic-resistant strains. When the virus weakens the immune system, there is no check on the bacterial infection, resulting in death.

In the past few decades, the incidence of TB in the United States has been declining steadily. In 2008, some 12,904 cases were recorded, making the case rate 4.2 per 100,000 people, an historic low (**Figure 13-4**). The question is this: What are the characteristics of the people who represent the new cases of TB? Part of the answer is shown in **Figure 13-5**, where case rates are broken down by race or ethnicity. The rates for Asians are by far the highest, with rates for Hawaiians/Pacific islanders being the next highest. Interestingly, the state with the highest case rate is Hawaii, with 9.6 per 100,000—more than twice the national average. This represents 124 cases in 2008, which is a relatively small number (New York had 1200 cases), but the population of Hawaii is also relatively small.

The reason for these high rates is contained in an analysis shown in **Figure 13-6**. Of the new cases of TB, more than half were in people born in another country (i.e., they emigrated to the United States). A decade ago, the fraction of cases attributed to U.S.-born persons was about half, but that has been decreasing steadily. Mexico tops the list of origins of emigrees with TB (with 23% of the foreign-born cases), followed by the Philippines (11%), Viet Nam (8%), India (8%), and China (5%). When case rates for the two populations are calculated, the U.S.-born TB case rate is about 2.0 per hundred thousand population, while the foreign-born case rate is 20.3.

FIGURE 13-3 **Tuberculosis is a worldwide problem, as evidenced by this estimation from the World Health Organization.**

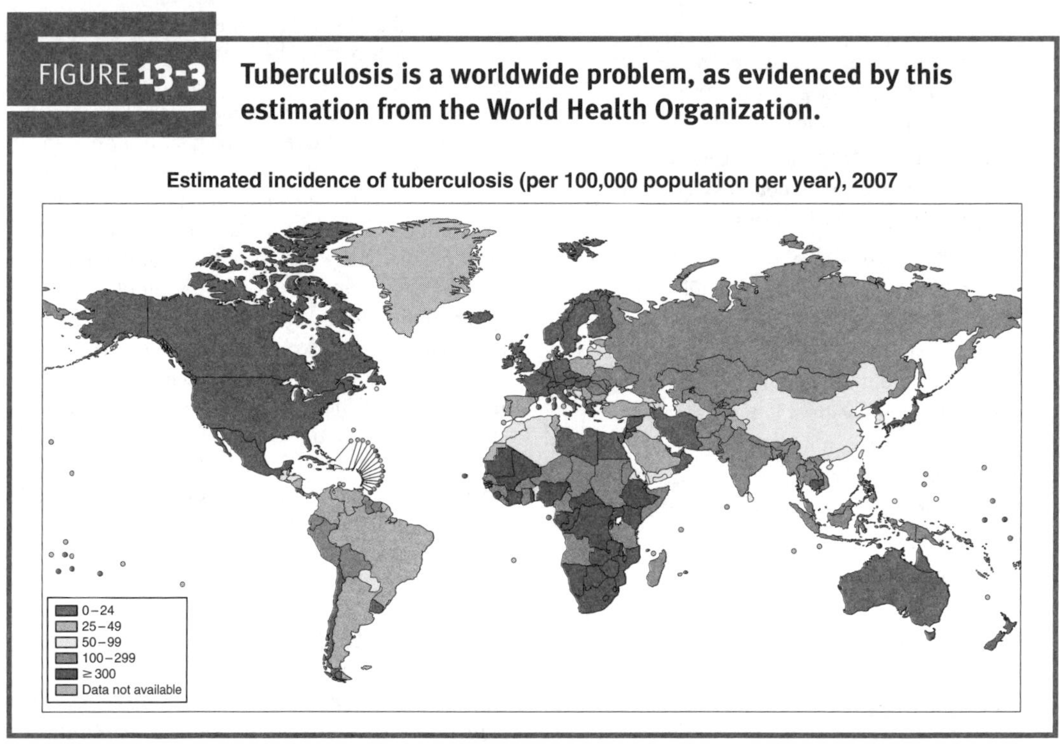

SOURCE: WHO.

FIGURE 13-4 **Rates of tuberculosis infections and deaths have fallen steadily over the last several decades.**

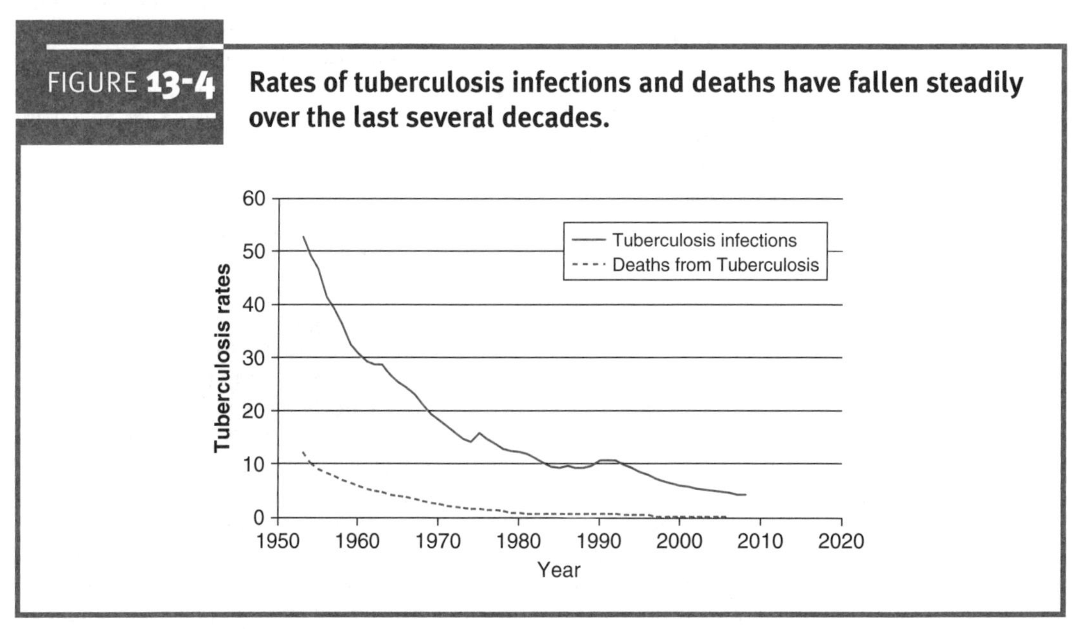

SOURCE: Based on information from the Centers for Disease Control and Prevention.

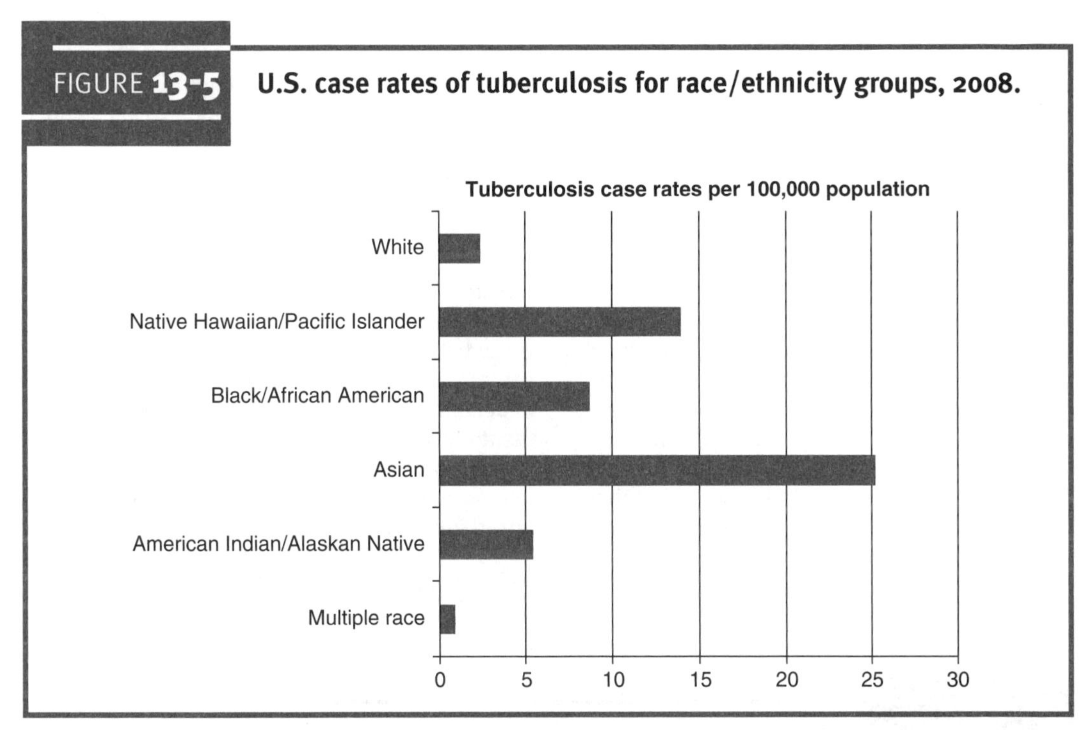

FIGURE 13-5 **U.S. case rates of tuberculosis for race/ethnicity groups, 2008.**

SOURCE: Based on information from the Centers for Disease Control and Prevention.

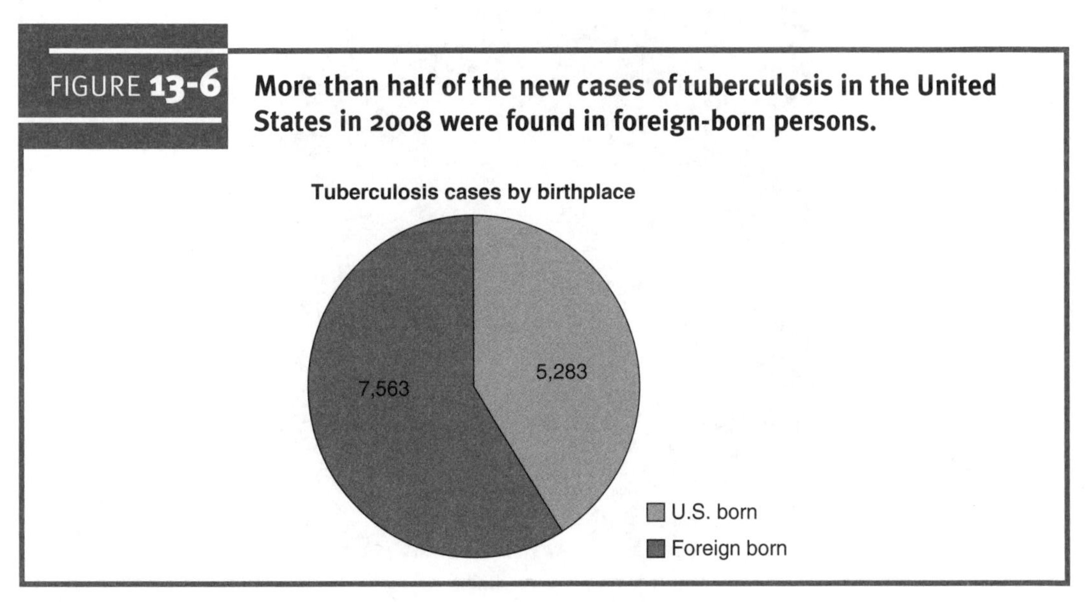

FIGURE 13-6 **More than half of the new cases of tuberculosis in the United States in 2008 were found in foreign-born persons.**

SOURCE: Based on information from the Centers for Disease Control and Prevention.

The large fraction of cases from Mexico is due to the large influx of people across a common border. The other countries are all Asian, where TB is endemic. People become infected in their home country and emigrate to the United States. If they had an active case of TB they might be spotted and treated, but if they have a latent infection they may not be identified. After living in the United States for decades (aging to the point of declining health), the latent infection may become active and then show up in the Centers for Disease Control and Prevention (CDC) statistics. In fact, the states with the highest case rates (i.e., California, Texas, New York, and Florida) are also the states with the highest numbers of immigrants (legal and illegal).

Notably, TB is endemic in some parts of the world; this fact affects the kind of screening that is ultimately performed. A comparison of the case rates for Pacific Ocean and non-Pacific territories makes this evident: Guam (51.1 per 100,000), Marshall Islands (197.9), Northern Marianas (39.3), and Republic of Palau (82.1). However, the rates for Puerto Rico (2.4) and the U.S. Virgin Islands (3.6) are much lower. The only exception here is American Samoa (4.6). Clearly, Pacific Islanders have a much greater chance of infection with TB.

In many countries, the population is vaccinated against TB using the BCG vaccine (see Chapter 11). However, the risk of individuals for acquiring TB in the United States is quite low, and therefore TB monitoring with the tuberculin test is used. This test determines whether TB has been contracted, and not whether the illness is active. The interferon gamma release assay is a relatively new test for TB infection. After an antigen-presenting cell contacts the antigen-specific T cell, the T cell then releases a significant amount of interferon gamma. The interferon can then be detected using an enzyme-linked immunosorbent assay (ELISA).

Tuberculosis (**Figure 13-7**) is treated with a cocktail of drugs, usually three different drugs at a time. Treatment is a long process, and switching to

FIGURE **13-7** ***Mycobacterium tuberculosis*** **growing on Lowenstein-Jensen agar slants.**

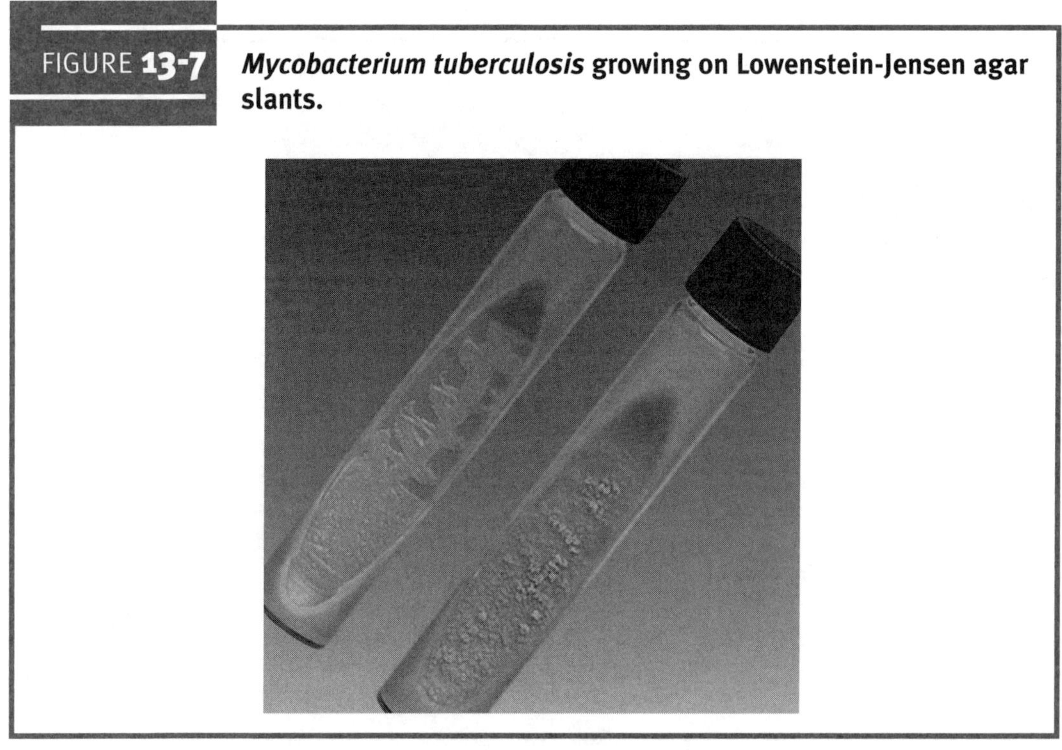

SOURCE: Courtesy of ThermoFisher.

a more efficacious drug is a likely outcome. Multidrug resistance (MDR) is therefore a major concern for TB treatment. Here MDR is defined as resistance to at least isoniazid and rifampin. Extensively drug-resistant TB (XDR) is defined as resistance to isoniazid and rifampin along with a fluoroquinolone drug and at least one of the second-line TB drugs (e.g., amikacin or kanamycin). Fortunately, these isolates are still rare (i.e., MDR accounts for about 1% of all cases), but this situation may change. The coinfection of some people with HIV and TB is especially troubling.

THE RESPIRATORY DISEASES

Many pathogens are passed to others through airborne particles. The "common cold" viruses are known to all and are suffered by all at one time or another. No vaccines are available for these pathogens, but they are usually not severe, and infected people almost always make a complete recovery. Collectively they are important, mostly because they are associated with a loss of productivity and the cost of over-the-counter medications. For the purposes of this text, they are largely ignored.

Influenza, on the other hand, is a very important disease. The pathogen is typically passed through coughs and sneezes, and probably also through fomites that have recently been handled by an infected person. Influenza is a member of the orthomyxoviruses; it is an enveloped, minus-strand, RNA virus (**Figure 13-8**). Notably, the genome has eight segments, and this provides a clue to the problems in dealing with influenza infections.

Many influenza viruses occur in the world, and they not only infect humans but also other species. Some infect birds (e.g., ducks or chickens), and others infect swine (e.g., "swine flu," which is transmissible to humans). Other known species susceptible to influenza viruses include whales, horses, and seals. The difficulty begins when two different viruses infect the same animal, creating a possibility that some of the eight segments are exchanged, effectively creating a completely different pathogen. Some (perhaps most) of these exchanges result in

FIGURE 13-8 **The general ultrastructure of the influenza virion is shown in this cartoon, with a cut-away section to show the interior.**

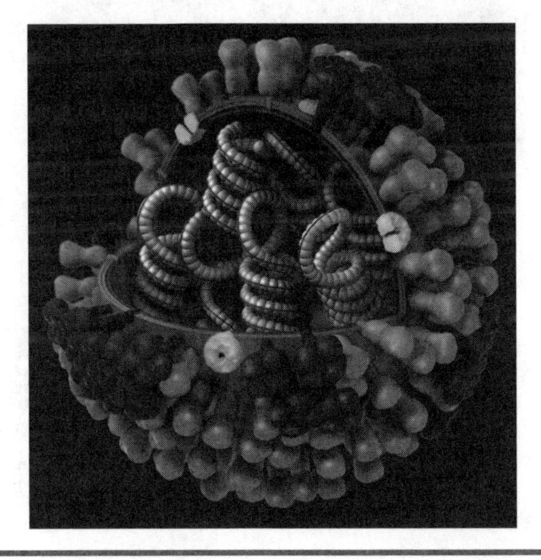

SOURCE: Courtesy of Doug Jordan, the Centers for Disease Control and Prevention. (http://phil.cdc.gov/phil/details.asp). #11878.

an ineffective virus, but others may create a more dangerous pathogen. In addition, small changes in the genome may result in different surface structures that are no longer recognized by the immune system, resulting in a worse disease. When humans live close to domesticated birds and swine, human acquisition of the infection becomes possible.

These types of molecular changes are impossible to address with a single vaccine for influenza. Usually a vaccine is made against the most recent isolate of virus, with the hope that the next virus will be relatively similar. Occasionally a substantial change in the virus renders the population unprotected. In this case, a pandemic can occur. The influenza pandemic of 1918 (the "Spanish flu") was one of the worst in history, with associated deaths ranging from 20 to 40 million worldwide. Many candidates have been identified as possible contributors to the next big influenza outbreak. An avian flu called H5N1 has circulated for several years (**Figure 13-9**). In 2009, several seasonal flu types circulated along with the H1N1 type. The severe acute respiratory syndrome (SARS) virus, a coronavirus, was first seen in 2003 and may recur if conditions become favorable (**Figure 13-10**).

Influenza is not a trivial disease (**Figure 13-11**). On the average, about 30,000 people die every year from the flu. In 2006, more than 56,000 deaths in the United States were attributed to it. Actually, only 849 deaths were directly attributed to influenza, but the balance were pneumonia-related deaths that followed an influenza infection, making influenza

FIGURE **13-9** **Electron micrograph of Avian influenza (H5N1).**

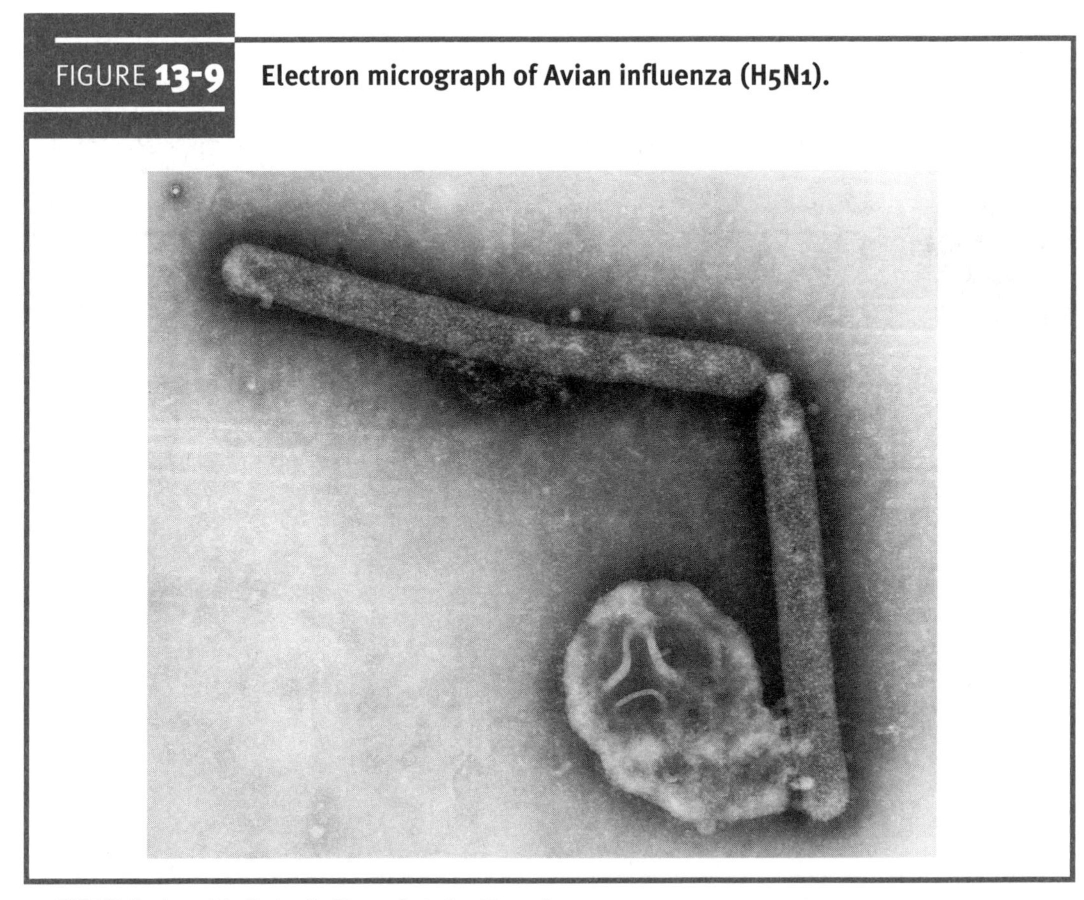

SOURCE: Courtesy of the Centers for Disease Control and Prevention.

FIGURE **13-10** **Scanning electron micrograph of the SARS virus, a potentially epidemic pathogen.**

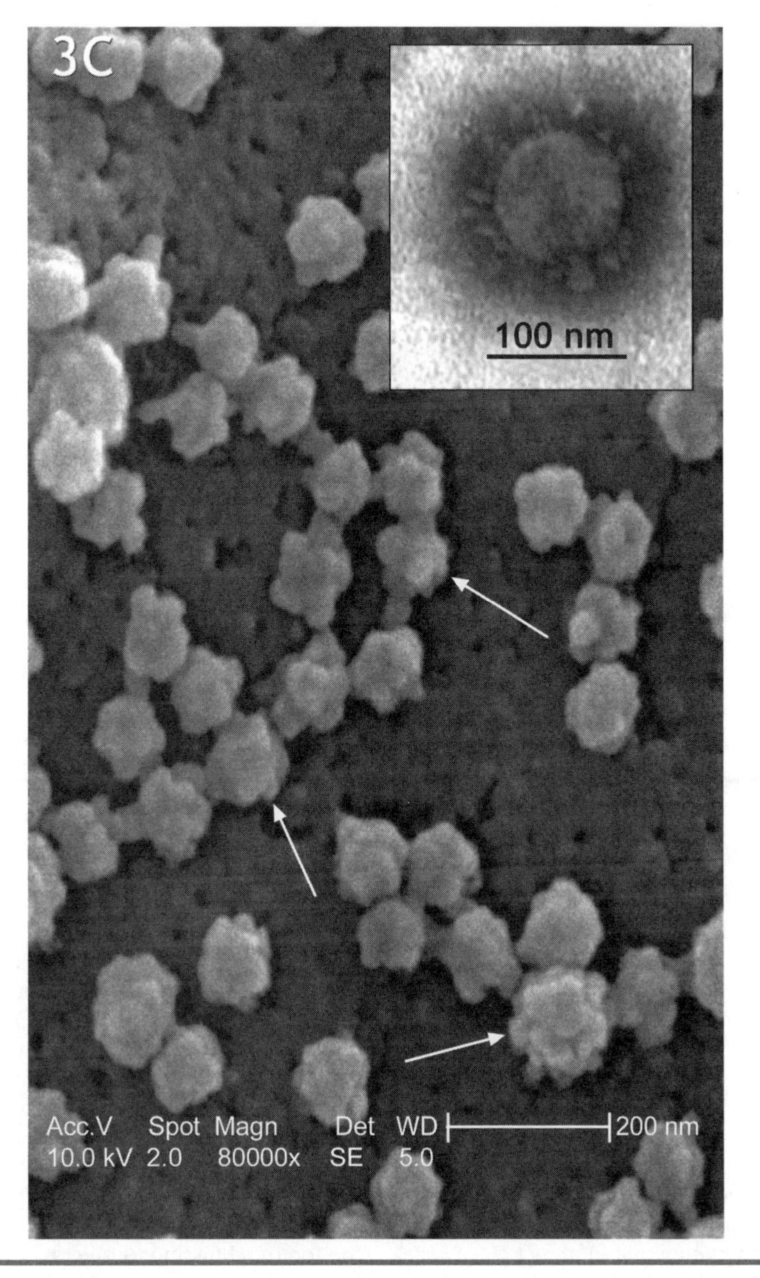

SOURCE: Courtesy of the Centers for Disease Control and Prevention.

FIGURE 13-11 The 1918 pandemic strain of influenza was recovered following a long search for specimens and is shown here in this electron micrograph.

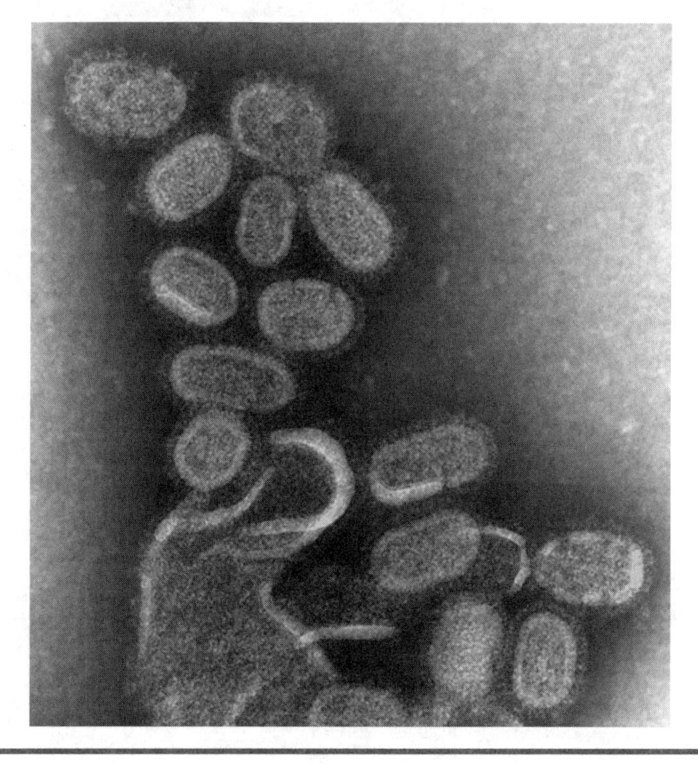

SOURCE: Courtesy of the Centers for Disease Control and Prevention.

the major factor. As expected, the vast majority of deaths occurred in people 75 years of age and older. This age cohort is very difficult to fully protect, making vaccination especially important for this group.

FUNGI

Histoplasma capsulatum

The causative agent of histoplasmosis is a fungal disease that starts when fungal fragments are inhaled, causing pneumonia in a fraction of cases (**Figures 13-12** and **13-13**). Most cases result in a mild flu-like illness, although for immunocompromised people, the results can be fatal. The fungus appears to be associated with droppings from birds and bats, and the risk of infection increases with proximity to large piles of these droppings. This can include chicken coops, so the timely removal of chicken guano before fungal growth can occur is advisable.

In one large outbreak (i.e., at least 18 people were affected), the cause was apparently the removal of a significant amount of bird guano from the top of a building (Stobierski et al., 1996). Bird guano has been implicated in other outbreaks of

FIGURE 13-12 ***Histoplasma capsulatum* has caused a lesion on the upper lip of this man.**

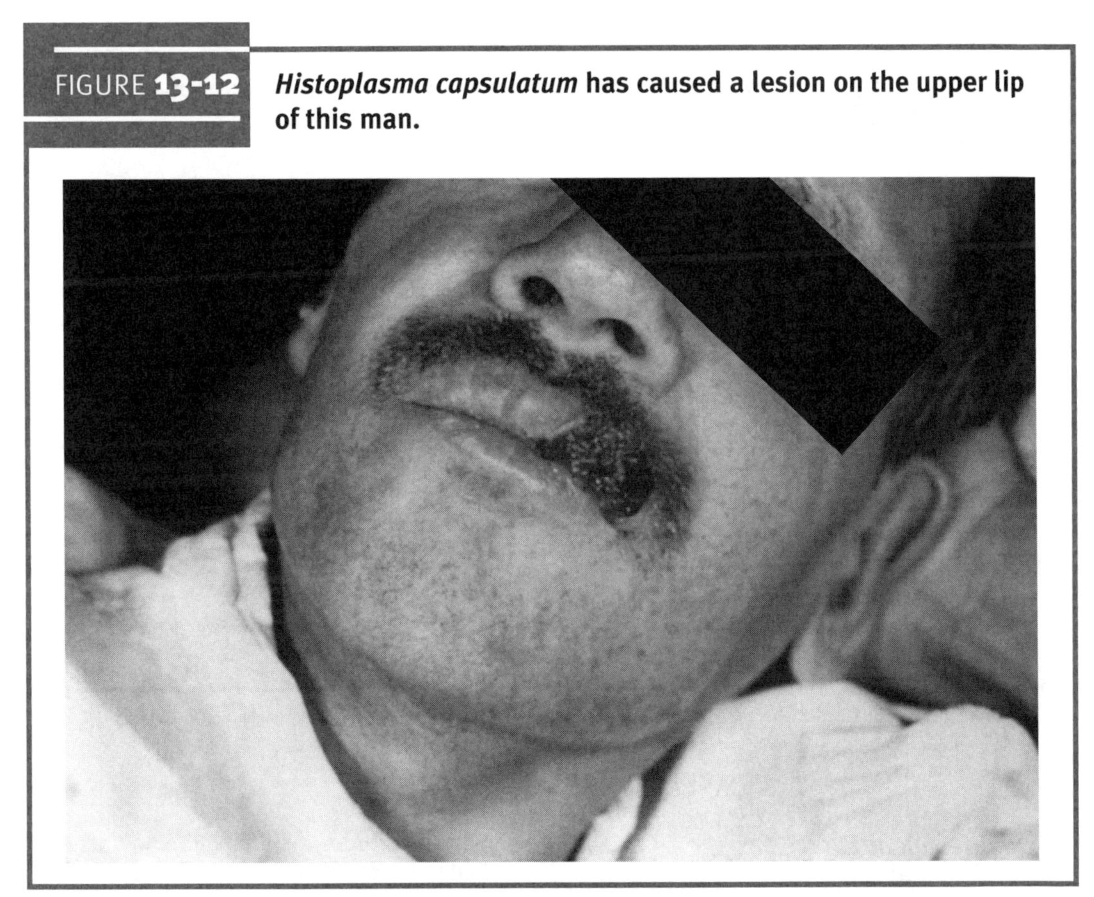

SOURCE: Courtesy of Susan Lindsley, the Centers for Disease Control and Prevention. (http://phil.cdc.gov/phil/details.asp). #6840.

this pathogen. The *Histoplasmosis* spores then drifted a short distance to an office building and caused illness among the workers. While it is possible that the wind picked up some dust and infected some unlucky people, the disturbance of massive amounts of guano increased the risk greatly. In addition to protecting the people who perform remediation activities, additional concern for others in the area is prudent.

Blastomyces

Blastomyces is the causative agent of blastomycosis, which can manifest as a severe systemic infection or as a cutaneous infection. *Blastomyces* grows as yeast in human tissue but as a mold under certain environmental conditions. The intermittent occurrence of these infections makes it difficult to determine the facts about its transmission. Even statewide analyses of cases are only partially conclusive (Baumgardner et al., 2005). This fungus is difficult to isolate from environmental samples, but it is assumed to play a role in the decay of plant material. This suggests that plenty of spores are released into the air, but infections are still relatively rare. It is probable that a variety of environmental factors must be present to cause the fungus to reach an infectious state.

FIGURE 13-13 ***Histoplasma capsulatum*** **growth at a magnification of 400X.**

SOURCE: Courtesy of Dr. Libero Ajello, the Centers for Disease Control and Prevention. (http://phil.cdc.gov/phil/details.asp). #10962.

Coccidiodes immitis

Another true fungal pathogen, *C. immitis*, is responsible for coccidioidomycosis, often called simply valley fever, after the San Joaquin Valley in California, where it is endemic. People who live there have probably been exposed to the fungus at some point and have immunity. The fungus is persistent because it survives below the soil surface and can become airborne when the wind blows dust into the air.

In some cases, infected people experience a slight fever, but most are asymptomatic. In other cases, however, a respiratory infection occurs. Most people recover easily, although in a small number of cases (usually involving an immunocompromised person), the disease can become systemic and can produce a life-threatening illness. Lesions and abscesses may appear on lung tissue, bone, skin, or central nervous system tissues. The fungus is transmissible from person to person only under highly unusual conditions. For example, three cases have been recorded of premature infants that acquired the infection from another person.

Candida

Several opportunistic species of *Candida* threaten public health; *C. albicans* is the most commonly encountered. *Candida albicans* can cause systemic infections, especially in the immunocompromised. If it infects the pharynx, a mat of whitish fungal hyphae appears. Therefore this infection is commonly called thrush, after the bird with the white neck (**Figure 13-14**).

Cryptococcus neoformans

This fungus deserves mention because it occasionally causes meningitis (**Figure 13-15**). Relatively little is known about susceptibility to this pathogen, although it is known to be associated with pigeon droppings. Infectious doses are probably encountered by those involved in cleaning large accumulations of pigeon droppings.

FIGURE 13-14 **Oral thrush caused by the fungus *Candida*.**

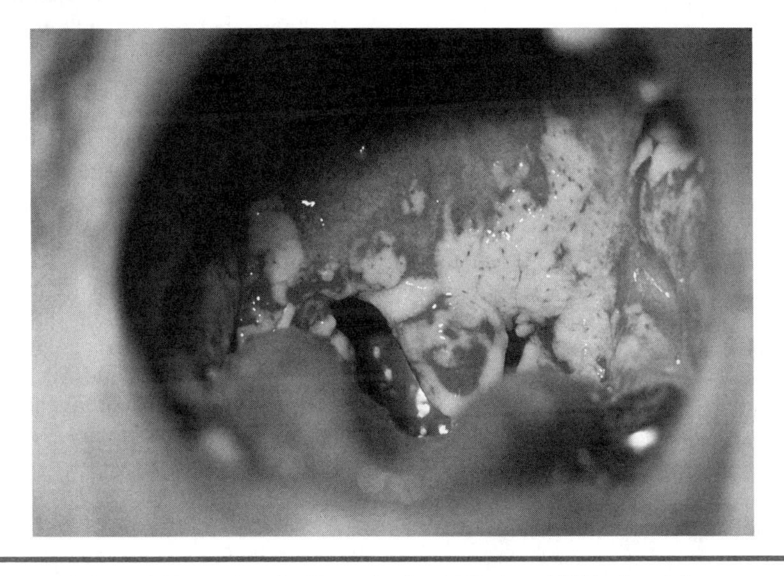

SOURCE: Courtesy of the Centers for Disease Control and Prevention. (http://phil.cdc.gov/phil/details.asp). #1217.

FIGURE 13-15 ***Cryptococcus neoformans*, which can cause meningitis, is shown on this photomicrograph. In the background are lung cells.**

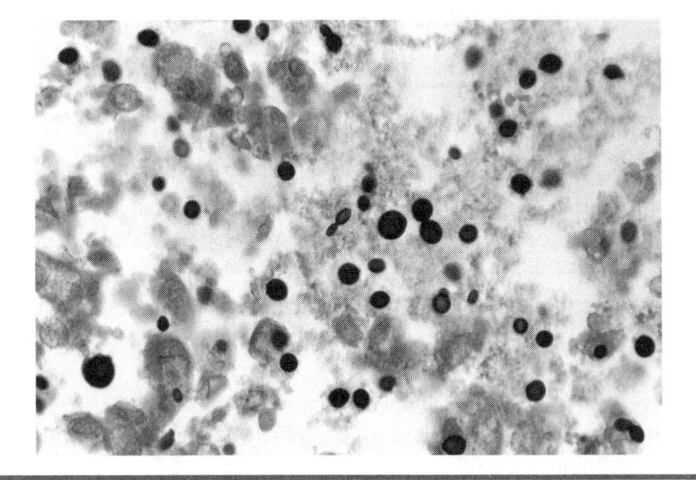

SOURCE: Courtesy of Dr. Edwin P. Ewing Jr., the Centers for Disease Control and Prevention. (http://phil.cdc.gov/phil/details.asp). #963.

Aspergillus

Aspergillus fungus is noteworthy because it can cause fungal pneumonia. In general, healthy people do not have a problem with pathogenic fungi, except for the dermatophytes (athlete's foot fungus). Problems arise in the immunocompromised population, people with chronic obstructed lung disease (COPD), or heavy users of antibiotics (in which the normal flora is destroyed and fungi enjoy a growth advantage).

Phialemonium obovatum

Phialemonium obovatum is one of the darkly pigmented fungi. It is considered to be an emerging pathogen, and little is known about it. One of the rare causes of human disease, it can be deadly, as demonstrated by a case of endocarditis in a 7-week-old baby (Gavin et al., 2002). So far, this fungus has only been a serious pathogen in immunocompromised patients (in this case, the baby was premature). However, the rising number of immunocompromised individuals in society makes this pathogen a concern for public health.

THE TOXIGENIC FUNGI

In addition to the pathogenic fungi, which cause an identifiable infection, toxigenic fungi also cause illness. Often referred to as mold, especially when found in an undesirable location, these fungi include organisms that are generally recognizable but have a variety of names. Fungi include the slime molds (known technically as Gymnomycota), the water molds (Mastigomycota), and the true fungi (Amastigomycota), which range from the microscopic forms (including yeasts) to the better known molds, mildews, rusts and smuts of important grain crops, mushrooms, puffballs, and other common fungi. Most are completely harmless and are more of a nuisance than anything else, such as the mildew that forms in shower stalls. Some fungi of interest are listed in **Table 13-1**.

Some diseases have been definitely described in terms of cause and effect. Alimentary toxic aleukia (ATA) is caused by any of several *Fusarium* species that may grow on grain and/or cattle feed. They produce the T-2 toxin, although it is uncertain whether this is the sole cause of all symptoms (Lutsky and Mor, 1981). Ingestion of small amounts results in gastroenteritis. All amounts of the mycotoxin are dangerous because they damage the bone marrow, leading to a decrease in the number of platelets, erythrocytes, and leukocytes produced. Because all of these cells are regularly replaced as a normal physiological function, the decrease means that fewer erythrocytes are available to

TABLE 13-1 Selected Toxigenic Fungi

Alternaria – these fungi produce more than a hundred secondary metabolites, many of which are known toxins.

Aspergillus – these fungi grow on a wide variety of food products, including grains (e.g., wheat, rice, sorghum, and corn), bread, cheese, and nuts. *Aspergillus* is very common, can be isolated easily from many environments, and can also cause systemic disease in the immunocompromised.

Penicillium – the original source for the antibiotic penicillin. This common fungus can be found in basements and other damp places. Although it produces a variety of other secondary metabolites, only one appears useful.

Fusarium – a very common fungus that can be found among decomposing vegetation. It is usually harmless, although it can synthesize a variety of potentially dangerous compounds, such as the fumonisins and the trichothecenes.

carry oxygen to the tissues, fewer leukocytes of all types are available to fight disease (leucopenia), and fewer platelets are available to repair damaged blood vessels. In severe cases, this results in a pancytopenia (loss of all new bone marrow cell types), and the resulting internal hemorrhaging poses a significant risk of death.

Molds are found in homes in many locations, typically where moisture levels are high (**Figure 13-16**). Molds produce compounds that have a deleterious effect on the human body. For many reasons, molds are also extremely difficult to study, especially in terms of health risk. Although mold growth in homes is regarded as unhealthy, no threshold values have been determined for mold growth or airborne spore concentrations. In all likelihood, people become ill at a wide range of concentrations, a range so great that establishing any one value as toxic is not possible. Deleterious health effects are seen when people are exposed to extremely high concentrations of fungal spores, such as farmers who must work around hay and straw that has become wet and is thus covered with mold. In these cases, toxicity occurs, but this is an occupational hazard and does not apply to a majority of people.

Classification of fungi is complex because various names for the same species are generally accepted. For example the important *Stachybotrys chartarum* is also known as *S. atra* and *S. alternans*. Often various names are all listed in the scientific literature so readers can recognize the species being discussed. In addition, fungi have complex life cycles in which sexual and asexual forms look entirely different. Because fungi are identified by their spore-forming structures, it is important to be able to identify such structures under the microscope.

As eukaryotes, fungi typically grow slowly. They form environmentally resistant spores that can remain dormant for long periods of time, especially under dark and cool conditions. As saprophytes, they grow on dead or decaying matter. For instance, during

FIGURE **13-16** **A typical mold conidium with spores.**

SOURCE: Courtesy of Dr. Lucile K. George, the Centers for Disease Control and Prevention. (http://phil.cdc.gov/phil/details.asp). #3961.

autumn they grow on fallen leaves and degrade them completely, eventually returning their nutrients to the soil. If a tree dies and falls to the ground, fungi are responsible for gradually consuming it; dying trees often support the growth of some "shelf fungus," which is difficult to degrade because of the tremendous amount of cellulose present (**Figure 13-17**). Were it not for the catabolic abilities of the fungi, dead matter would decay over a much longer period of time. However, this tendency also means that they will grow on many unusual types of carbon sources. The limiting factor is usually available water, and mold infestation is a problem whenever flooding due to broken pipes, leaks, or rainwater creates moist conditions. Fungi grow on old newspapers and magazines, photographs, cardboard, the paper backing on drywall, leather furniture and suitcases, or practically any other organic material (**Figure 13-18**). If the relative humidity is high enough, they will grow on exposed food and even the cardboard boxes in which the food is packaged (see Chapter 7 on spoilage). Active fungal species can be found in rugs and carpets, even in clean houses and in synthetic-fiber carpets because soil is tracked in from outdoors and the particles become lodged in the carpet fibers. A little moisture from rain-soaked shoes is all the fungi need to grow.

Fungi have an amazing metabolic capacity. A wide variety of different catabolic enzymes helps them degrade all sorts of complex compounds.

FIGURE **13-17** **A typical shelf fungus growing on the side of a tree. Fungi are found everywhere in the environment.**

FIGURE 13-18 **The walls and ceiling of this home are covered in mold, most likely due to a water leak somewhere nearby.**

SOURCE: Courtesy of Alonzo E. Scott, Jr./FEMA.

They also produce a wide variety of unique compounds (often referred to as secondary metabolites because they are not essential to the growth of the microorganism), but they give it some advantage. Many secondary metabolites are toxic compounds, or mycotoxins. By some estimates, more than 400 described compounds have been identified. The best explanation for the production of mycotoxins is that they aid the fungal spore to invade a niche; that is, the mycotoxin inhibits other microbes around the spore and allows the slower-growing fungus to compete effectively for resources. This is analogous to some antibiotic compounds produced by the bacteria, especially *Bacillus* and *Streptomyces* species. Fungi typically do not produce only a single toxic compound, but they may have the capacity to produce 20 or 30 different compounds. A clear example is the production of penicillin by the *Penicillium* fungus. This example should be kept in mind as the problems with mycotoxins are discussed. Some of the classes of fungal toxins are listed in **Table 13-2**.

Unfortunately, some mycotoxins also have a deleterious effect on humans. Apparently, most do not, and in the cases in which a mycotoxin does

TABLE 13-2 Selected Mycotoxins

Trichothecenes: Powerful mycotoxins produced by *Stachybotrys* and *Fusarium* species. The effects of alimentary toxic aleukia (ATA) are due to the T-2 toxin. Outbreaks of ATA occur when marginal grains are consumed, such as during times of famine. Vomitoxin is another type which, as the name implies, causes vomiting.

Satratoxin: Produced by *Stachybotrys*, and causes damage to blood vessels. They can also cause illness when consumed.

Aflatoxins: Compounds produced by *Aspergillus* species, which are quite common and grow on a variety of common grains and nuts. These compounds are also potent carcinogens, especially in causing hepatocarcinoma. However, hepatocarcinoma is usually caused by ingestion of significant concentrations (typically of aflatoxin B1) over long periods of time, and especially if the person is also suffering from Hepatitis B virus. Care must be taken to protect food from contamination, which usually means keeping grains dry enough to inhibit growth of fungi. The toxicity and/or carcinogenicity of the inhaled toxin are not known.

Aflatoxins are responsible for Turkey X disease, so-called because a mysterious ailment killed at least 100,000 turkeys in 1960. The cause was finally traced to Brazilian peanut meal contaminated with *Aspergillus* that had been fed to the turkeys. As it happens, turkeys and many other poultry species are extremely susceptible to aflatoxin.

Ochratoxin: Found in *Penicillium* and in *Aspergillus*. It is fluorescent, which makes it easier to detect when it is being produced. It is produced during fungal growth on a variety of foods and is toxic to the kidneys. It is also a probable carcinogen.

Ergot alkaloids: Produced by *Claviceps purpura*, which grows well on grains such as rye. Detection can be difficult, and in the past, they have been a source of much illness. During the Middle Ages in Europe, rye was used to make rye bread, a staple of the diet at that time. The ergot alkaloid compounds survived baking temperatures and were consumed by the local population. At that time, a small village might have only a single baker, and everyone would buy their bread daily from that one shop. It was possible, if contaminated grain was used, to afflict the entire town's population with ergotism.

If the concentration of mycotoxin is low, victims might feel a prickly sensation in the limbs, called St. Anthony's Fire. This symptom was thought to be cured by making a pilgrimage to the shrine of St. Anthony in France (Etzel, 2002). The local environment in that area did not usually support from *Claviceps* contamination. Because the supplicant was obliged to eat local food along the way (food preservation was rather crude), the afflicted person gradually flushed the mycotoxin out of his body, and the symptoms went away. In a sense, the shrine worked.

If higher concentrations were consumed, the symptoms could include convulsions and hallucinations. It has been suggested that many reported supernatural events of the Middle Ages, in which an entire village reported seeing religious figures or, alternatively, witches were due to the town becoming poisoned with ergotism. Apparently, people suffering from ergotism are very suggestible, and a person who sees an initial "vision" can easily convince others to see the same thing.

Fumonisins: Produced by any of the *Fusarium* species. They often contaminate corn and corn products, such as tortillas. These carcinogens may be responsible for increased rates of esophageal cancer. They are also teratogens; they create birth defects because of their role in inhibiting folic acid uptake by the developing fetus. This causes neural defects like spina bifida.

Fumonisin

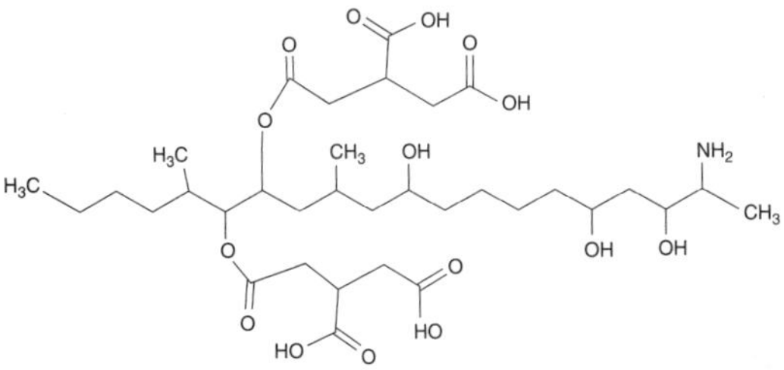

Satratoxin H

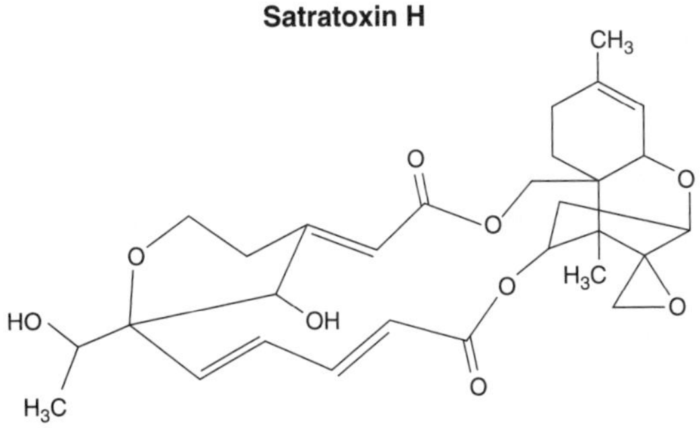

Aflatoxin B1

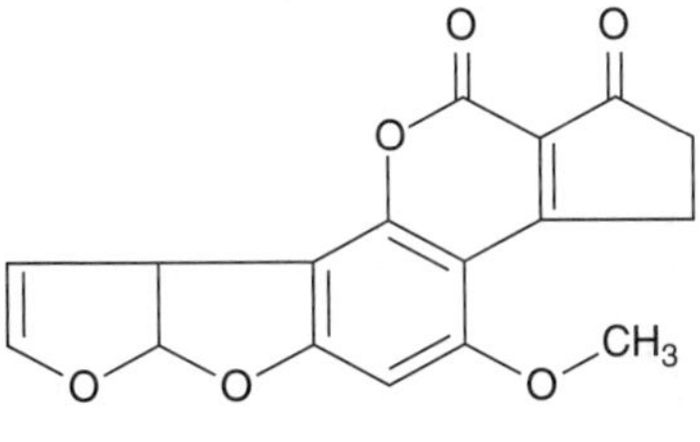

Vomitoxin

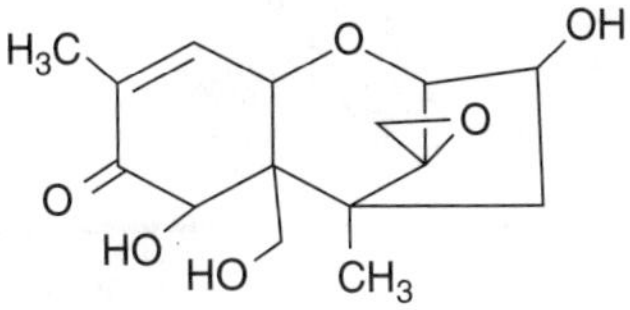

affect people, the effects are not great. However, for a subset of the population, the presence of certain mycotoxins, even in very low amounts, can cause health problems. Exposure can occur by contact (e.g., causing dermatitis), ingestion (i.e., a food-related illness), and by inhalation (**Figure 13-19**). The range of symptoms is quite broad. In mild cases, irritation, typically in the nasopharynx, is seen. Allergic reactions are common, with the most likely being allergic rhinitis (i.e., a runny nose, sinus headache, and scratchy throat). Mycotoxins are a likely source of asthma as well, or at least they exacerbate asthma. It is also possible to have severe hypersensitivity reactions to mycotoxins, including the life-threatening anaphylactic shock. This is seen in people who are hypersensitive to penicillin. Such severe reactions are not at all typical and are certainly caused by using the purified compound as an injectable or ingestible antibiotic. In past years, penicillin was greatly overused because it was considered to be a wonder drug for the treatment of many, many illnesses. In some people who overused the antibiotic, their immune systems became acutely responsive to penicillin, eventually producing an anaphylactic reaction when even tiny amounts were encountered. How these people respond to small amounts of penicillin in *Penicillium* mold in their homes remains to be determined (**Figure 13-20**).

An additional level of response to the toxigenic fungi is intoxication. In these cases, the illness is referred to as mycotoxicosis. The route of intoxication is airborne (hence their inclusion in this chapter), and the spores enter the moist environment of the lung, where the mycotoxin desorbs from the spore and enters the bloodstream. Unfortunately, this broad category of illness is difficult to define. To explain this, it is necessary to describe some of the parameters of the problem.

Fungi are known to produce many types of mycotoxins, but the conditions under which they are made, and the amounts they will make under those conditions, are difficult to determine with any

FIGURE 13-19 *Fusarium* species in a cross-section from a skin infection. Several fungi are known for these superficial infections.

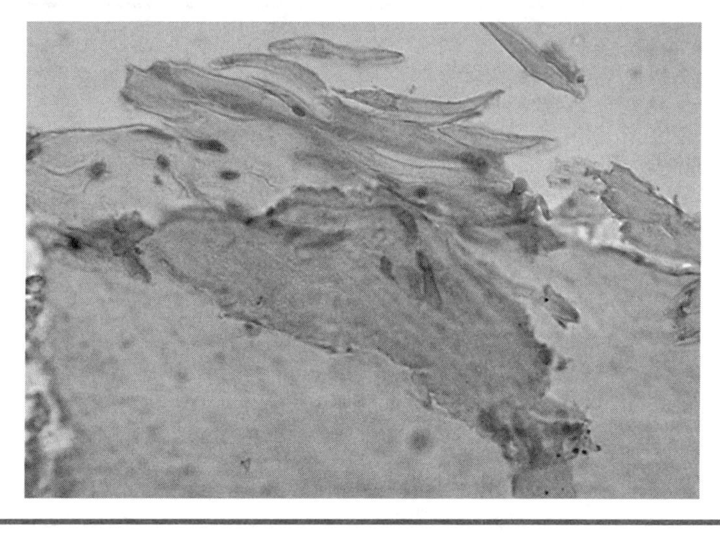

SOURCE: Courtesy of the Centers for Disease Control and Prevention. (http://phil.cdc.gov/phil/details.asp). #4323.

FIGURE 13-20 Conidia and spores from a *Penicillium* species.

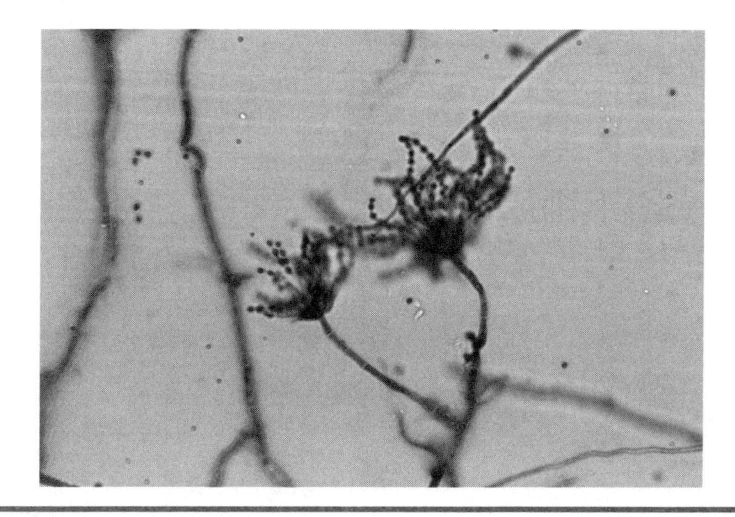

SOURCE: Courtesy of Dr. Lucille K. George and Dr. Arthur DiSalvo, the Centers for Disease Control and Prevention. (http://phil.cdc.gov/phil/details.asp). #4192.

degree of certainty. Mycotoxins are not made all the time, but only in response to certain environmental stimuli (e.g., water activity, temperature, presence of specific nutrients). There is no guarantee that the fungal cells will produce only the toxin that you are interested in, and perhaps the symptoms it causes are actually due to (or at least greatly enhanced by) the presence of other mycotoxins; that is, maybe a synergistic effect involving two, three, or more mycotoxins is at work. On the other hand, one compound might have an inhibitory effect on other compounds (Robbins et al., 2000). A patient molecular biologist could knock out the genes for all of the other mycotoxins, but this would require a tremendous labor (for which there is little grant support) and might produce a fungal species that no longer has normal regulatory control of the genes of interest. A reductive approach of this magnitude might yield a very different strain from the original.

Estimation of toxic dose based on the number of viable mold spores in a sample is not sufficient. Usually this is done by sampling a volume of air, in which the particles are concentrated onto the surface of an agar plate. After growth, the fungal colonies can be counted and identified. However, even dead spores can contain significant quantities of mycotoxin (because the compounds are very stable), and mold spores on surfaces always outnumber airborne particles. Perhaps the person suffering from mycotoxicosis cleaned out a dusty closet a few days before, for example, and wafted a tremendous number of spores into the air. By the time anyone searches for the spores, they will have settled again. In addition, it is not sufficient to simply count spores (living and dead) because the mycotoxins can be found in any part of the fungi. No standard method for the quantitative examination of spores or fungal growth on surfaces has been established, which makes comparisons and testing difficult.

A substantial time lag may elapse between the time a person encounters the mycotoxins and when the symptoms appear. This period may be several days, and the link between (for instance) encountering dusty conditions and feeling the effects of the mycotoxins may be lost in the interval. The sufferer may attribute the symptoms to some other effect, such as a cold or spoiled food or hay fever. These intervals differ greatly, but in cases of high-dose mycotoxicosis it is not unusual to see a delay of 3–9 days until flu-like symptoms appear. A few days later a leucopenia may be noted, especially if some other illness occurs and a blood count is performed. In especially severe cases, pancytopenia results, but only after 3–4 weeks. Recovery usually takes place after these episodes, although there is a window of vulnerability in which other illnesses may occur due to the poor health of the patient.

Even if a group of people is exposed to equal amounts of mycotoxins from the same source, their individual reactions may differ. This may reflect the fraction of spores that actually enter the lung and release their toxin to the bloodstream, or it may be simply a matter of different immune responses to the same mycotoxin, or there may be a genetic component that causes some people to be more susceptible. Even when symptoms are noted, they may differ from one person to the next: one may experience redness and itchiness, while another gets hives or shortness of breath. Even under the best of conditions, a cause-and-effect relationship is difficult to determine.

Most people are familiar with the term "black mold." This usually refers to the mold *Stachybotrys* or the related mold *Memnoniella*. These fungal species have a variety of significant toxins, such as satratoxins, trichothecenes, and stachybocins. In small amounts they can cause dermatitis or a bloody rhinitis due to damage to the blood vessels. Sometimes workers in grain elevators experience these symptoms. Higher amounts can cause flu-like symptoms. A large dose can cause an illness that actually resembles radiation poisoning: headache, diarrhea, fatigue, and hair loss. At these levels neurological symptoms can occur, such as irritability, dizziness, lack of concentration, mental fatigue, and difficulty sleeping (Fung et al., 1998).

Stachybotrys was implicated in the deaths of several infants during an unusual outbreak in Cleveland in 1994. The infants were all newborns who had only recently been brought home from the hospital. They suffered from a syndrome called infant pulmonary hemorrhage (IPH), in which blood vessels in the lungs (which are fragile in newborns) rupture and bleed. In these homes, evidence of *Stachybotrys* spores was found. The population sample was small, and the statistical analysis indicated that *Stachybotrys* presence in the home was a risk factor, but it was not the greatest risk factor in the study. The presence of smokers in the home was a greater risk. Obviously, the presence of smokers or smoking in the home is not the cause of IPH; in far too many homes, smoking is not associated with IPH. A reasonable inference would be a synergistic effect between cigarette smoking and *Stachybotrys* and the incidence of IPH. A subsequent case from Kansas City (Flappen et al., 1999) described a 1-month-old baby with IPH in which the airborne mold-spore concentration was estimated. However, this figure was only 420 spores per cubic meter of air, which does not appear particularly high. The home also had smokers, bolstering the argument that a synergistic effect is needed. At about this same time, a report described the first isolation of *Stachybotrys* from the lung of a patient (Elidemir et al., 1999). Initially this was an unusual case because the patient was 7 years old and came from a household with no smokers. Clearly, more remains to be discovered about this illness.

For reasons outlined above, establishing regulatory limits or actionable concentrations of mold spores is difficult, and currently no recognized threshold concentrations have been established. This is also true for the mycotoxins themselves. This creates a problem for public health authorities: they may note a very high level of airborne mold in a home, but no rule has been broken. Insurance companies have encountered this problem as well. Homeowners have sued contractors for faulty installation of building materials that resulted in mold infestation and subsequent illness in the household. The installation of windows is often a trouble spot, especially when they are not watertight and allow water to infiltrate and cause mold growth within the walls of the home. Because fungi are saprophytic, they have no need for sunlight and thrive in the absence of light. While the contractor can be sued for poor installation, the insurance company has been held accountable to remediate the mold problem (because the contractor rightly contends that there is no actionable level of mold, and therefore no one can prove harm). Individual juries may or may not agree about where the liability ultimately rests.

The effects that low levels of mycotoxins have on human health are debatable; they may be a part of what is termed "sick building syndrome" in which people occupying the building complain of a wide variety of symptoms: headaches, myalgia, dizziness, and inability to concentrate. Mycotoxins may be a factor in chronic fatigue syndrome, or at least in a subset of people who suffer from that poorly understood condition. Fungal toxins in general, and *Stachybotrys* in particular, may be responsible for some cases of the mysterious sudden infant death syndrome (SIDS). In SIDS, the cause of death is often completely unknown, and notably, mycotoxins are not generally suspected during such an investigation. Even if they were, the toxic dose for an infant has not been determined.

Household mold is not unusual, especially in damp areas. Even excessive humidity within the home can provide sufficient moisture for mold growth. The relative humidity should be maintained at 30–50%. Drier conditions may also cause problems, causing people's skin and mucous membranes to dry out (unrelated to disease). But a relative humidity in excess of 60% feels damp to most people and also allows mold to grow. In areas prone to flooding, the infiltration of water into basements leads to extensive mold growth, which may be difficult to eradicate. Along the Mississippi River Valley, many areas flood periodically, and mold-related health problems are more common there. After Hurricane Katrina caused extensive flooding in the New Orleans area, warm temperatures and wet conditions led to massive mold growth in many homes.

A substantial social element should be mentioned here. Those who suffer from mycotoxicosis are not solely from low socioeconomic status (SES) groups, but they may comprise a disproportionate number. The reason is quite simple: low-SES populations frequently do not own their own homes but rent instead. If their home gets a leaking pipe or a leaking roof, providing the water that mold needs to grow, then they must appeal to their landlord to fix it. Low-rent housing often means little money in the repair budget, and repairs may lag considerably. Even when low-SES individuals own their homes, repairs may be postponed because of budget concerns. In the meantime, the mold grows, producing mycotoxins and intoxicating the family members. The extent to which this occurs remains unclear, but it is easy to hypothesize that children who suffer from mycotoxicosis might not function well in school.

QUESTIONS FOR DISCUSSION

1. Which airborne pathogens would you think are seasonal in prevalence? Why?
2. Which groups are considered "at risk" for tuberculosis?
3. Why isn't the BCG vaccine used in the United States?
4. In your own home, where would you expect to find fungal growth?
5. What is the difference between isolation and quarantine?
6. What other diseases do you think could result in quarantine?
7. Why is it so difficult to study mycotoxins?

References

Abu Kwaik, Y., L. Gao, B.J. Stone, C. Venkataraman, and O.S. Harb. 1998. Invasion of protozoa by *Legionella pneumophila* and its role in bacterial ecology and pathogenesis. Appl. Environ. Microbiol. 64: 3127–3133.

Baumgardner, D.J., D. Steber, R. Glazier, D. P. Paretsky, G. Egan, A.M. Baumgardner, and D. Prigge. 2005. Geogrpahic information system analysis of blastomycosis in northern Wisconsin, USA: waterways and soil. Medical Mycology 43: 117–125.

Elidemir, O., G.N. Colasurdo, S.N. Rossmann, and L.L. Fan. 1999. Isolation of *Stachybotrys* from the lung of a child with pulmonary hemosiderosis. Pediatrics 104: 964–966.

Etzel, R. 2002. Mycotoxins. J. Amer. Med. Assoc. 287: 425–427.

Flappen, S.M., J. Portnoy, P. Jones, and C. Barnes. 1999. Infant pulmonary hemorrhage in a suburban home with water damage and mold (*Stachybotrys atra*). Environ. Health Perspect. 107: 927–930.

Fung, F., R. Clark, and S. Williams. 1998. *Stachybotrys*, a mycotoxin-producing fungus of increasing toxicologic importance. Clin. Toxicol. 36: 79–86.

Gavin, P.J., D.A. Sutton, and B.Z. Katz. 2002. Fatal endocarditis in a neonate caused by the dematiaceous fungus *Phialemonium obovatum*: care report and review of the literature. J. Clin. Microbiol. 40: 2207–2212.

Lutsky, I.I., and N. Mor. 1981. Alimentary toxic aleukia (septic angina, endemic panmyelotoxicosis, alimentary hemorrhagic aleukia). Amer. J. Pathol. 104: 189–191.

Moritz, M.M., H.C. Flemming, and J. Wingender. 2010. Integration of *pseudomonas aeruginosa* and *Legionella pneumophila* in drinking water biofilms grown on domestic plumbing materials. Int. J. Hyg. Environ. Health 213: 190–197.

Ogen-Odai, A., E.K. Mbidde, J. Lutwana, J. Wamala, A. Mucunguzi et al. 2009. Bubonic and pneumonic plague—Uganda, 2006. Morbid. Mortal. Weekly Rep. 58: 778–781.

Pedgley, D.E. *Windborne Pests and Diseases*. John Wiley & Sons, New York, 1982.

Robbins, C.A., L.J. Swenson, M.L. Nealley, R.E. Gots, and B.J. Kelman. 2000. Health effects of mycotoxins in indoor air: a critical review. Appl. Occup. Environ. Hyg. 15: 773–784.

Stobierski, M.G., C.J. Hospedales, W.N. Hall, B. Robinson-Dunn, D. Hoch, and D.A. Sheill. 1996. Outbreak of histoplasmosis among employees in a paper factory—Michigan, 1993. J. Clin. Microbiol. 34: 1220–1223.

CHAPTER 14

VECTORBORNE AND ZOONOTIC PATHOGENS

LEARNING OBJECTIVES

- Describe what a vector is and how they can spread disease.
- List several vectorborne pathogens and the diseases they cause.
- Recognize the various life cycles of the vector/pathogen pairs and how they are disrupted.
- Identify the diseases of highest incidence and the environmental factors that are relevant to dissemination.
- Identify the professions that are at greatest risk of vectorborne infections.

KEY TERMS

- Sylvatic
- Vectorborne diseases
- Zoonotic diseases

INTRODUCTION

Disease can be passed through contaminated food and water, or through direct contact with a sick person (either airborne or by sexual contact), but it can also be passed through an animal (i.e., another animal that suffers from the disease or perhaps is a mere carrier of the disease). **Zoonotic diseases** are passed to humans from animals, either by direct contact or by contact with excrement from the animals. Warm-blooded animals may host many of the same pathogens that can infect humans, so it is not surprising that transfer can occur if the animal is a reservoir of disease. One example is the disease rabies, which is transmitted through the bite of an infected mammal. Because it was so greatly feared, one of the first vaccines developed was for rabies. Some zoonotic diseases have already been mentioned in Chapter 8 on meat and in Chapter 9 on dairy (because diseases can easily be passed through an animal's milk). In addition, pathogens that are acquired primarily through the fecal–oral route were covered in Chapter 4 on waterborne disease because water is the most likely route of transmission. Zoonotic diseases that were not covered elsewhere are presented in this chapter.

Vectorborne diseases include an intermediate between hosts, often passing the disease from an animal to a human. Typically the pathogen enters a human through an insect bite that breaks the skin and injects the pathogen into the human circulatory system. Some of these diseases can be devastating, and understanding how they are passed and how they can be controlled is essential to preventing them. Death from a mosquito-borne illness is a sobering thought.

The biting insects are the major problem here. This includes the mosquito, the louse, the flea, and the tick. Among this group there are further important subdivisions: the vector for Lyme disease is the deer tick, not the wood tick. The common housefly is known to be a vector of many diseases, but only as a passive carrier. That is, a fly may land on fecal material, some of which will stick to its legs, and then land on a food item, transferring the fecal material to the food. If the food is left out under the right conditions, the bacteria multiply to infectious dose levels. In this regard, flies are vectors of disease, but they are functionally nonspecific. This chapter deals with specific vectors. Obviously, common-sense sanitary practices are also vital to public health. In the same sense, biting flies of all sorts may result in infections local to the bite, and bites from other insects may do this as well, but these are medical problems rather than public health issues. Most localized infections are relatively minor.

Many vectorborne diseases are caused by viruses, but some are caused by bacteria, including the very odd members of the *Rickettsiae,* which account for a large number of the diseases listed here. Parasites also account for some of the most important diseases, such as malaria. Fungi are atypical in this category.

THE ZOONOTIC DISEASES

Erysipelothrix rhusiopathiae

Erysipelothrix rhusiopathiae is a thin, Gram-positive bacillus found in animals (including fish) that causes the disease erysipeloid in humans. Erysipeloid appears as a cutaneous lesion and is not uncommon as an occupational hazard among people who handle animals (Reboli and Farrar, 1989). Workers who experience small cuts and abrasions on their hands and fingers during work are at risk for infection. The disease is usually self-limiting, although on rare occasions endocarditis can occur. In these cases, a heart-valve replacement may be needed. As an occupational hazard, *E. rhusiopathiae* is probably not completely avoidable, although the use of gloves during handling of animals helps greatly.

Chlamydophila psittaci (Formerly *Chlamydia psittaci*)

Chlamydophila psittaci is the causative agent of psittacosis, more commonly known as parrot fever. It got its name after an outbreak in the United States during the 1950s, when it was popular to have exotic birds as pets (e.g., parrots, parakeets, cockatiels, and macaws). Some of these birds carried the pathogen, and close contact with the bird was sufficient to transfer the pathogen. Some people teach their birds tricks, such as plucking a treat from their fingers. This close contact is sufficient to transfer the bacteria. Dust kicked up by the bird or from dried bird droppings may spread the pathogen. Infected people display fever, headache, a rash, and possible pneumonia. The symptoms mimic chlamydial pneumonia (*C. pneumoniae*) and can easily be misdiagnosed. Parrot fever is rarely fatal and can be cured with antibiotic treatment.

Imported birds are now inspected for disease, but it is entirely possible for an infected bird to get through the screening process because diagnostic tests are unreliable. This is a serious concern because the pathogen can infect chickens and turkeys as well, and this would have disastrous results. Tracing all persons who came into contact with a known infected bird is advisable, although difficult to do in practice. Bird smuggling is also a concern; the smugglers typically are not overly concerned with public health.

Leptospira interrogans

Leptospira interrogans is the causative agent of leptospirosis (**Figure 14-1**). It is a spirochete that can enter the body through the gastrointestinal tract or the mucous membranes (e.g., eyes or nose). *Leptospira* is difficult to categorize as a pathogen. It is spread through the urine of an infected animal, and many mammals can be infected. It might be considered a waterborne pathogen, but usually the urine does not reach the water supply. If it does, however, *Leptospira* can remain viable for months under the right conditions. Because it can infect rats and mice, food or dishware on which the vermin have urinated may pass the disease to an unsuspecting person, and therefore *Leptospira* may be

FIGURE 14-1 ***Leptospira interrogans*, the causative agent of leptospirosis, is shown here on a filter.**

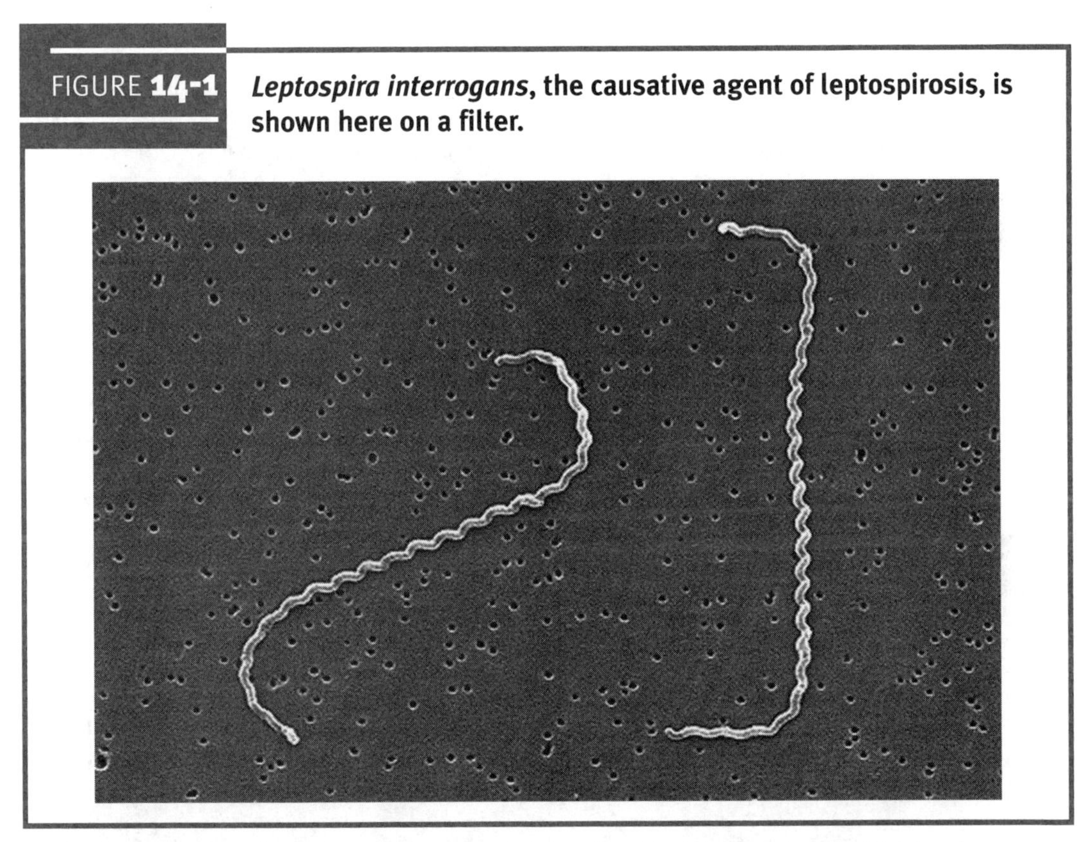

SOURCE: Courtesy of Rob Weyant, the Centers for Disease Control and Prevention. (http://phil.cdc.gov/phil/details.asp). #1220.

considered a food pathogen. Because most of the very few people who are infected are associated with animals through their occupations, it is considered here as a vectorborne pathogen.

Leptospira infection results in headache, fever, muscle aches (especially in the calves and thighs), and chills. Interestingly, it may also lead to vomiting and diarrhea, even though it is not a typical gastrointestinal disease. Numerous potential complications include jaundice, meningitis, myocarditis, and renal failure. Ordinarily the fatality rate is low, but when the more serious symptoms appear, the fatality rate may approach 20%.

Vaccination of animals helps to control infection. People who work with animals should avoid contact with urine as much as possible, particularly if the animal appears ill. Personal protective equipment also helps.

Burkholderia mallei

Burkholderia mallei is the causative agent of glanders, a disease associated with horses and mules. It can be transferred to humans who work closely with infected animals. Reportedly, *B. mallei* was used as a biowarfare agent during World War I (see Chapter 15), but against the horses that were depended upon by the army at that time, and not against humans. Glanders is extremely rare in the United States.

Spirillum minus

Spirillum minus is a Gram-negative spirochete that is the causative agent of rat-bite fever, also called sodoku. As the name indicates, the microbe is transmitted through the bites of rodents. The pathogen is also found in the urine and feces, so the presence of vermin around food or cooking utensils creates an opportunity for infection through ingestion. Maintaining good home hygiene is essential.

Another species, *Streptobacillus moniliformis*, causes an illness that is called by the same name and is essentially identical. The infection results in fever, headache, and muscle aches, as well as a rash that may become pustular. It responds well to antibiotic treatment. Both diseases are rare in the United States.

Hantavirus

The observation of hantavirus infection dates to the Korean War, when an unknown fever was described, then called Korean hemorrhagic fever. The site where it occurred was near the Hantaan River in Korea, and thus it was named. The causative agent was identified in 1993 after an outbreak in the southwestern United States (**Figure 14-2**). Here the virus was named *sin nombre* (Spanish, "the nameless one") apparently because no local group wanted a potentially deadly disease named after them or their home (**Figure 14-3**). Hantavirus is found worldwide but is called by many local names.

The distinguishing features of hantavirus are that it is acquired from rats and mice, that it causes a pulmonary infection, and that it is a member of the *Bunyaviridae*, making it an enveloped, minus-stranded, single-stranded RNA virus. The infected rodent excretes the virus in its urine and feces. Humans

FIGURE 14-2 **Incidence of hantavirus infection, by state.**

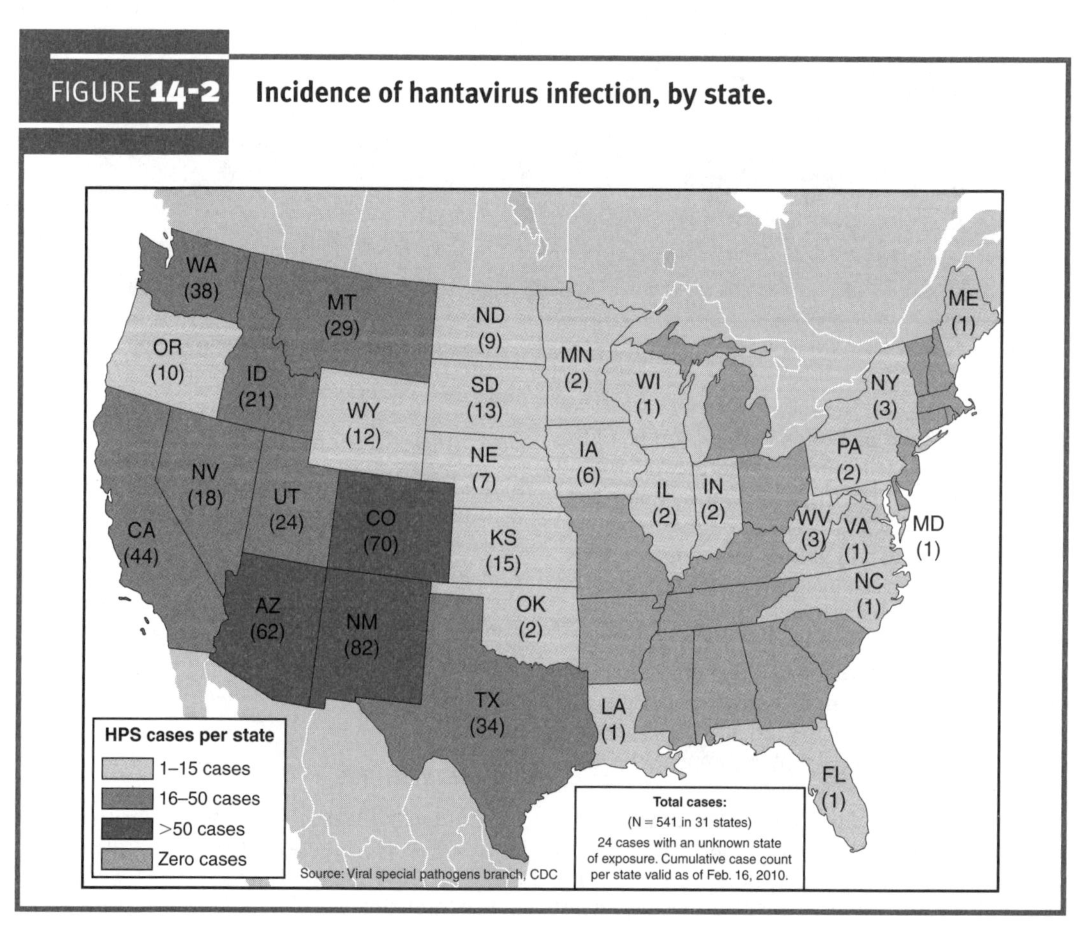

SOURCE: Courtesy of the Centers for Disease Control and Prevention.

FIGURE 14-3 **Electron micrograph of the virions of the *sin nombre* virus.**

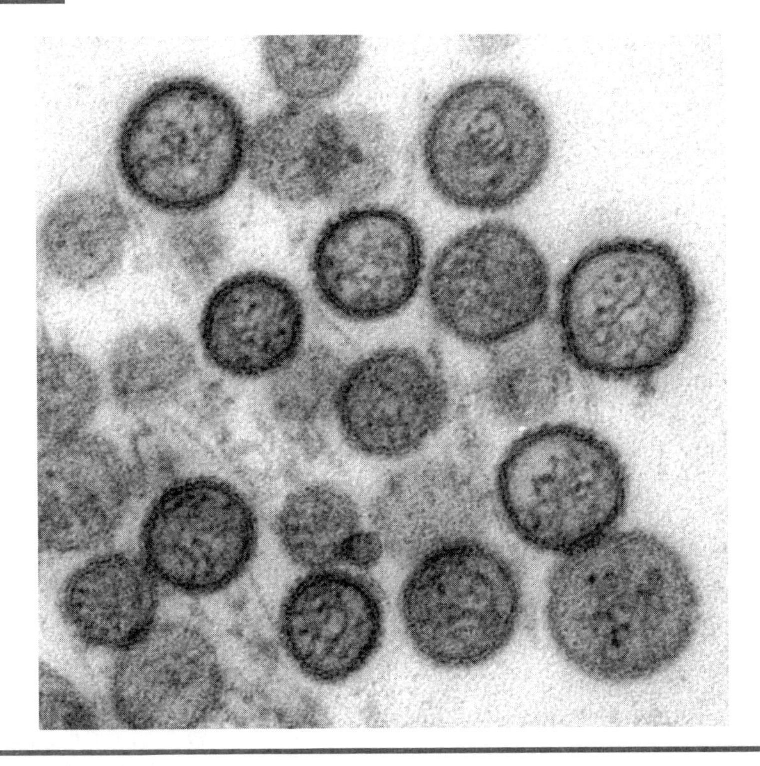

SOURCE: Courtesy of Cynthia Goldsmith and Luanne Elliott, the Centers for Disease Control and Prevention. (http://phil.cdc.gov/phil/details.asp). #1137.

acquire the virus through aerosolized particles. Obviously, maintaining sanitary conditions is the best preventative measure. It is easy to imagine that conditions during the Korean War allowed vermin to infiltrate military camps and infect the soldiers. The virus is also found in the saliva of the mouse, and it might be passed to humans who eat food that has been chewed by vermin (**Figure 14-4**).

The disease follows either a renal path or a cardiac path. The renal path starts as a flu-like illness that is compounded by gastrointestinal symptoms. This is followed by renal failure. At this point, it may be necessary to use dialysis to relieve the problem, or the case may prove fatal. After 3–7 days, the kidneys begin to work again, and the body then excretes high volumes of urine. This is the convalescent period of the illness. The cardiac path is very dangerous: it may lead to cardiac failure. Even with supportive therapy, the fatality rate may be as high as 35%.

Lymphocytic Choriomeningitis Virus (LCMV)

Lymphocytic choriomeningitis virus causes a disease that has some similarities with hantavirus, although this virus belongs to the arenavirus family. Mice host LCMV, and it appears to be passed to humans via urine and feces. Exposure to this virus

FIGURE 14-4 The deer mouse, a reservoir for hantavirus.

SOURCE: Courtesy of James Gathany, the Centers for Disease Control and Prevention. (http://phil.cdc.gov/phil/details.asp). #8358.

is not uncommon; in one study of seroconversion, 5% of those screened were positive. However, it only rarely causes a severe disease.

In most people infected with LCMV, no symptoms occur, or perhaps they have flu-like symptoms that resolve on their own. In a few cases, a meningitis-like or encephalitis-like illness occasionally also results in fluid accumulation in the brain. Partial paralysis has also been observed, and permanent damage to hearing is possible following nerve damage. However, the fatality rate associated with LCMV is very low.

Rabies

Everyone has heard of rabies, and everyone fears it. Symptoms are hideous and fatal if untreated. Rabies was one of the early targets for vaccination, and Pasteur succeeded in producing an effective vaccine.

Rabies is classically associated with dogs, and it is certainly possible for them to acquire the virus. The virus concentrates in the salivary glands of the animal so that virtually any bite will be infectious. Because dogs live close to humans, a rabid animal poses a huge risk. However, the campaign to vaccinate both dogs and cats has been remarkably effective in defeating rabies among household pets. A notable exception came from a dog that was brought into the country from Iraq as part of an animal rescue program (Mangieri et al., 2008). In dogs, the virus has an incubation period of about 10 days, meaning that the dog was infected shortly before transit. Usually a clear record of vaccination is required before an animal can be brought into the country. Notably, in many countries, vaccination of animals is not performed, or coverage is spotty.

Today most exposure to rabies is through wild animals, notably bats, skunks, and raccoons (**Figure 14-5**). The vast majority of infections is associated with these three animals. Other wild animals have also infected humans, such as woodchucks, coyotes, and foxes. Bats are a problem because they often nest in man-made structures like house attics and barns. Spelunkers (cave explorers) have become infected when exploring a cave full of bats. They easily kick up bat guano that may contain the rabies virus and then inhale the tiny particles, infecting the person. Skunks and raccoons often

FIGURE 14-5 **The raccoon, a possible vector for rabies transmission.**

SOURCE: Courtesy of the Centers for Disease Control and Prevention. (http://phil.cdc.gov/phil/details.asp). #2180.

scavenge for food around human dwellings. This puts them close to humans, who may surprise and panic them. Foxes and coyotes are shy creatures and normally avoid humans as much as possible. If traveling in a developing country, it is good to remember that many countries do not vaccinate their dogs, and dogs are the primary culprit in rabies transmission in such places. Approaching a strange dog is not advisable.

The rabies virus, a rhabdovirus, is an enveloped, minus-stranded, single-stranded, RNA virus (**Figure 14-6**) with an affinity for neural tissue. When a person is bitten and the rabies virus enters the body, the virus infects the local nerves and follows the nerves toward the central nervous system. As the disease progresses, hydrophobia (i.e, fear of water) often appears due to gradual paralysis of the muscles that control swallowing. The patient is actually quite hungry and thirsty but panics or starts choking when unable to swallow anything. If the virus infects the brain, there is essentially no hope, and the patient dies. However, the progression of the infection is slow, affording time for treatment.

The preferred treatment for known rabies infection is vaccination. The vaccine has time to induce the primary immune response before the virus invades the central nervous system and thus eradicates the live virus. Of course, this assumes that the bite was on the extremities, as most animal bites are. If the bite is closer to the central nervous system, the window of opportunity for vaccination is shorter. Combining vaccination with rabies immune globulin is especially recommended. Many frightening stories narrate the number and location of shots that are needed, but in fact, only four shots are needed (i.e., on the day of exposure and on days 3, 7, and 14), and the shots are intramuscular, typically in the arm.

At least one instance is known of a person surviving rabies without the vaccine (Willoughby et al.,

FIGURE 14-6 **Electron micrograph of the rabies virus, showing the characteristic bullet shape of the virion.**

SOURCE: Courtesy of the Centers for Disease Control and Prevention.

2004). It occurred in Wisconsin in 2004. A 15-year-old girl was bitten by a bat and did not seek medical attention until a month later. Because rabies has been known to take several months to infect the central nervous system, this was not a calamity, but by this time, the girl experienced neurological problems: double vision/blurred vision, tremors, and slurred speech. She was also making antibodies against the virus; therefore, the vaccine was no longer effective. To augment treatment, the doctors administered an antiviral drug, ribavarin, and induced a coma in the girl with drugs. After 7 days, she was revived from the coma on a gradual basis, and she made steady progress. Initially she suffered neurological damage, but she eventually resumed a normal life. However, this treatment has been attempted on other occasions without success (Pue et al., 2009).

If possible, a mammal that bites should be isolated and checked for rabies infection. Unfortunately, with most bites from wild animals, there is no opportunity to do so. In these cases, vaccination proceeds as a safeguard against the possibility of infection. Continued vigilance in vaccinating pets is essential to control of rabies.

Pox Viruses

The pox viruses are best known for smallpox, which does not properly fit into this chapter because it only infects humans. Because it has been eradicated from around the world, it troubles us much less, although it is covered as a potential biowarfare agent in Chapter 15.

Several other related viruses that infect other animals, however, may be transmitted to humans. Cowpox, which Edward Jenner utilized for his vaccine, still causes infections occasionally, but cowpox is not especially dangerous. Other pox viruses include camelpox and monkeypox; they are rare.

Hemorrhagic Fever Viruses

The severity of the diseases that hemorrhagic fever viruses cause makes them noteworthy. They all cause a syndrome in which extensive bleeding occurs; blood sometimes exits through the mouth, eyes, and ears, and sometimes blood is seen beneath the skin. All hemorrhagic fever viruses are enveloped RNA viruses. The group includes (with virus family) Ebola virus (filovirus), Crimean-Congo hemorrhagic fever (bunyavirus and a vectorborne illness), Lassa fever (arenavirus), and Marburg fever (filovirus). There are other, more obscure viruses like Rift Valley fever (bunyavirus), Bolivian hemorrhagic fever (arenavirus), and Venezuelan hemorrhagic fever (arenavirus).

Fortunately, all of these viruses are very rare, but the death rate during outbreaks can be shockingly high. The 1976 Ebola outbreak in Zaire had an 88% death rate. In all cases, no vaccine or effective treatment is available. Only supportive treatment is used, and isolation procedures for the patient must be followed precisely to limit exposure of healthy people to any fluids from the infected person.

Lassa fever has a reservoir in rodents, which shed the virus in urine and feces. It passes to humans in contaminated food or water, or possibly as an aerosol. Marburg virus apparently has a reservoir in African fruit bats (as does Hendra virus; see the section on equine encephalitis viruses). The virus infects primates, and people handling sick monkeys are at great risk of acquiring the virus, probably from secretions. The primary means of transfer from the fruit bats may be through aerosols; rabies can be spread from bats in this manner. Ebola (**Figure 14-7**) can also be acquired from monkeys, and the true reservoir of the illness is still unknown.

FIGURE **14-7** **Electron micrograph of Ebola virus.**

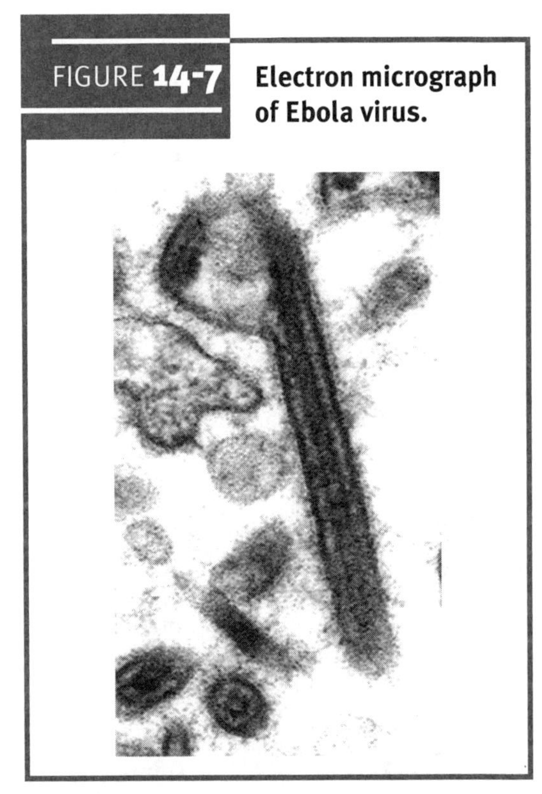

SOURCE: Courtesy of Cynthia Goldsmith, the Centers for Disease Control and Prevention. (http://phil.cdc.gov/phil/details.asp). #1835.

These viruses have not yet shown the ability to leave the small geographic area where they now exist. The Marburg virus was named after Marburg, Germany, which is not the endemic region, but a laboratory there was the site of a recognized outbreak. The virus is not found in Germany otherwise. However, within endemic areas, the incidence may be significant; it is estimated that 5,000 people die of Lassa fever every year. Any unexplained outbreak outside of the endemic area would be cause for great concern.

Schistosomiasis

Schistosoma are parasitic worms (carried by snails) that can infect people through direct contact with a freshwater source. Schistosomiasis (also known as Bilharzia) is extremely rare in the United States, but these parasites are a big problem in other areas of the world. Interestingly, the worms can penetrate the skin, which is an unusual trait.

Toxocara

Toxocara parasites are commonly known as roundworms. They are passed to humans through infected dogs and cats. They spread to humans via the fecal-oral route, usually as a result of close contact with a family pet. It can cause two serious syndromes, one called visceral larva migrans and one called ocular larva migrans.

Visceral larva migrans occurs when the roundworms infect the major organs. Ocular larva migrans is an infection of the eyes. Both are relatively rare, despite the high incidence of roundworms in pets. *Echinococcus multilocularis* parasites are also passed in this manner. This type of tapeworm migrates to the liver, where it can cause substantial damage.

Ancylostoma

Ancylostoma is commonly known as the hookworm. They are produced in various animals and excreted in their feces. They then enter the human body through the skin, typically when bare feet come into contact with contaminated soil. These worms cause disease in humans, including a syndrome known as cutaneous larva migrans. This disease causes skin eruptions. The worms can also migrate to the intestines and colonize them, causing intestinal symptoms. These infections can be effectively treated with anti-parasitic drugs.

VECTORBORNE DISEASES

Bacterial Diseases

BARTONELLA

Bartonella is a Gram-negative bacillus that was formerly known as *Rochalimaea*. The diseases *Bartonella* causes result in the appearance of a rash of small papules, which result from proliferation of blood vessels in a localized area. This is referred to as bacillary angiomatosis and is notable because the papules bleed easily when struck.

Bartonella hensalae is the causative agent of cat scratch fever (Regnery and Tappero, 1995). The illness was long considered a mystery until the organism was identified in the lymph nodes of a patient (English et al., 1988). *B. hensalae* is harbored by common house cats, which do not show signs of illness but can transmit the illness in either bites or scratches. Cat feces also harbor *Bartonella*. Fleas and ticks (**Figure 14-8**) from the cat are also able to transmit *Bartonella*, and it has been suggested that the bites and scratches actually allow flea feces into the wound to start the infection. It is difficult to predict which cats harbor the organisms, and estimates predict that 40% of house cats are infected. Therefore, it is important to clean wounds from cats immediately with soap and water.

The typical infection consists of swollen, tender lymph nodes, headache and backache, chills, and a general feeling of malaise. The condition can persist for 3 or 4 weeks. Most infections resolve without treatment. In some cases, particularly with immunocompromised patients, the illness becomes much more severe and can affect the liver, eyes, or heart (causing endocarditis). Occasionally the illness is fatal. If treatment is needed, a variety of antibiotics can be used successfully.

Bartonella quintana is the causative agent of trench fever, which earned its name during the trench warfare of World War I. Sanitary conditions in the trenches were appalling, and the soldiers often suffered from body lice, which carry this pathogen (**Figure 14-9**). Encountering a case of trench fever is now a rare event, but it might recur under the crowded conditions of a refugee camp.

Symptoms of trench fever include a headache lasting for about a week, pain that seems to be located behind the eyes, body aches, and the bacillary angiomatosis rash that covers the arms and trunk of the body. Most cases resolve without complications. If an outbreak occurs, it is important to segregate patients with body lice and to delouse them thoroughly before letting them rejoin the general population.

Bartonella bacilliformis is the causative agent of Carrion's disease, which is native to certain parts of South America, particularly in the Andes. The vector is the sandfly. The disease is aggressive, resulting in hemolysis (leading, in some cases, to immunosuppression) and high fever. Neurological symptoms are serious signs of the advanced stage of the illness. Untreated, the disease can easily be fatal, but Carrion's disease responds well to antibiotic treatment. Verruga peruana is a chronic form of this disease.

Several other species of *Bartonella* are known and are suspected to cause disease. However, evi-

FIGURE 14-8 **An electron micrograph of the common flea, which is capable of carrying several diseases.**

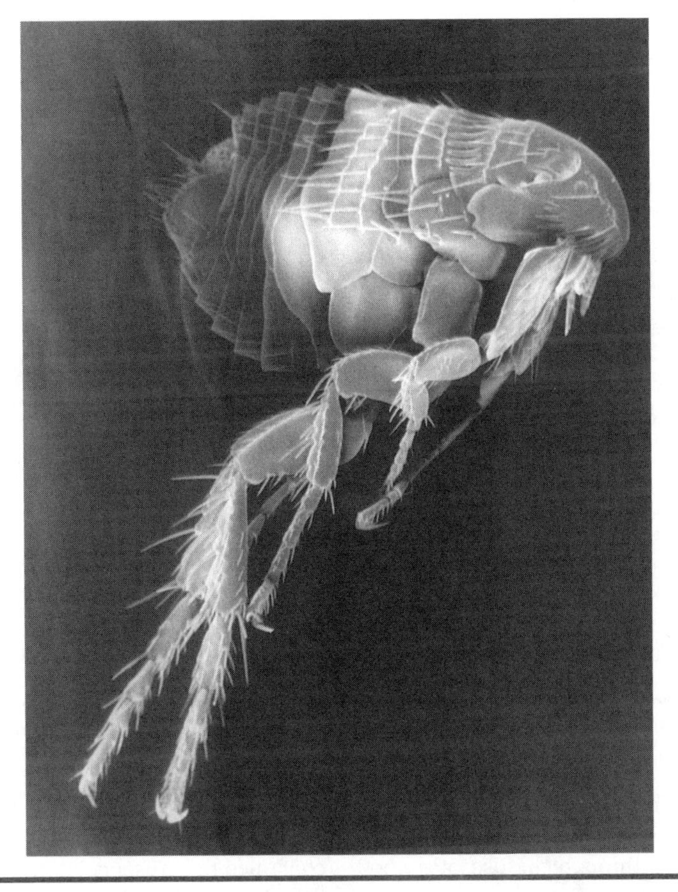

SOURCE: Courtesy of Doug Jordan, the Centers for Disease Control and Prevention. (http://phil.cdc.gov/phil/details.asp). #11878.

dence for pathogenesis and for a possible vector is lacking.

Borrelia

Borrelia are spirochaetes. This genus includes *B. burgdorferi*, the causative agent of Lyme disease, and *B. recurrentis*, the causative agent of relapsing fever. *Borrelia burgdorferi* is spread by ticks, while *B. recurrentis* can be spread by ticks and, rarely, by body lice.

Relapsing fever is rarely reported, with no more than 25 cases per year on record. Those at risk include campers and hunters, especially if they camp in cabins where the ticks live. The fever is recurrent because the spirochete changes the antigen structures on its surface and thus evades the immune system to a great extent. As the microbe is recognized by the immune system, the fever appears and antibodies are created. Then the antigens shift, and the bacteria begin to multiply again.

FIGURE **14-9** **The common head louse.**

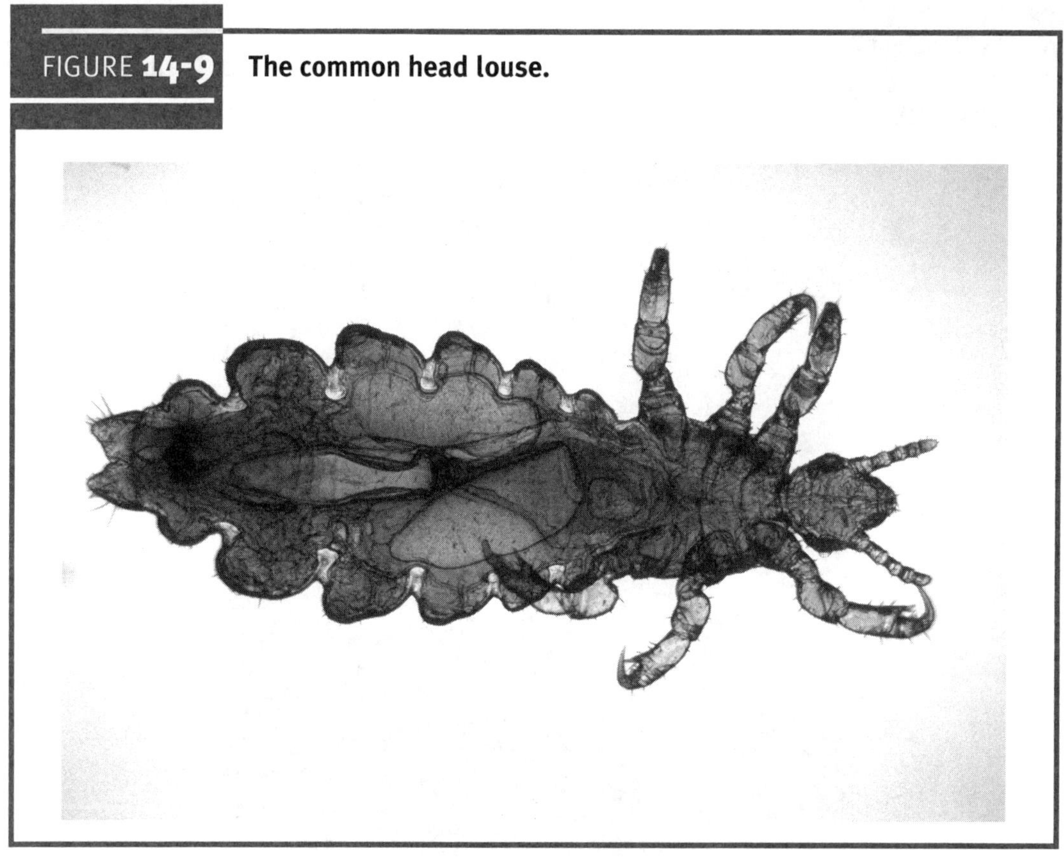

SOURCE: Courtesy of Dr. Dennis D. Juranek, the Centers for Disease Control and Prevention. (http://phil.cdc.gov/phil/details.asp). #377.

Infections can be dangerous if untreated. Prevention is accomplished through the use of a good insect repellant.

Lyme disease has an interesting history. It is named after Lyme, Connecticut, where it was first identified. Several children in the town became ill with an odd cluster of symptoms. The diagnosis at the time was juvenile rheumatoid arthritis, which is a real but rare condition. The idea that so many cases could be found in a small area suggested that something else was the cause. The mothers of the victims petitioned the Centers for Disease Control and Prevention (CDC) for an analysis of the outbreak, and they responded with a team of scientists that gradually put the puzzle together, finding the pathogen and the mode of transfer. Their investigation confirmed the presence of the microbe in the deer tick (Burgdorfer et al., 1982; **Figure 14-10**).

The key finding was that the deer tick (*Ixodes*) was the vector, not the common wood tick (**Figure 14-11**). Deer (specifically white-tailed deer) serve as a reservoir for the microbe and then pass it to deer ticks. Humans are an accidental host, but in areas where deer and humans are in close proximity, deer tick bites are common.

The hallmark of Lyme disease is the appearance of a red ring around the site of the tick bite, which is called erythema migrans (**Figure 14-12**). The ring is apparent in many, but not all, infections. Other symptoms include fever, headache, and joint pain

FIGURE **14-10** **The deer tick, which is capable of carrying *Borrelia burgdorferi*, the causative agent of Lyme Disease.**

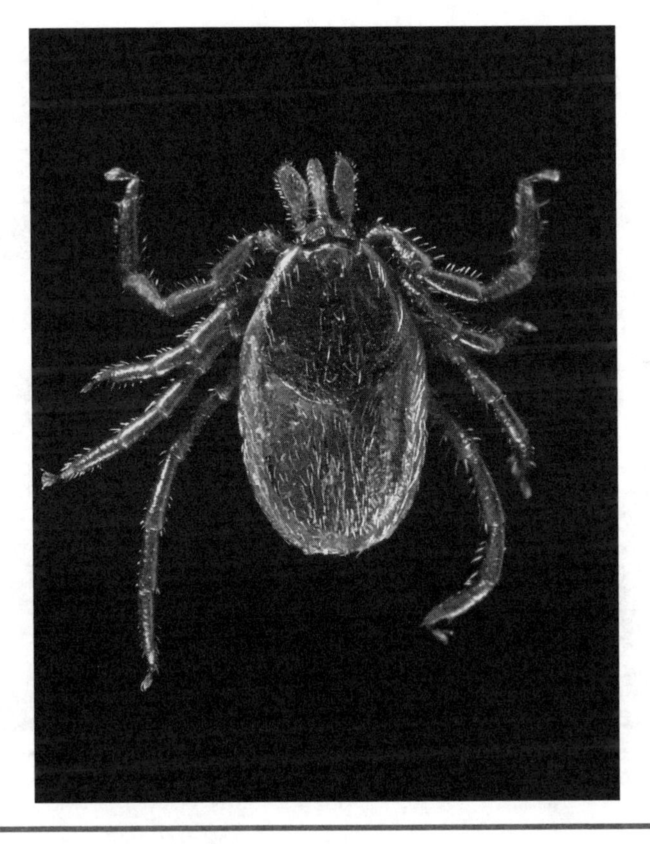

SOURCE: Courtesy of the Centers for Disease Control and Prevention.

(thus the mistaken diagnosis in Lyme). If not recognized and treated at this stage, the microbe disseminates throughout the body and can cause more severe symptoms. These include splenomegaly and cardiac abnormalities. Especially significant are the neurological symptoms, which may include forgetfulness, confusion, and a general feeling of dullness.

Most tick bites occur during the summer months, when people are outdoors and wearing lightweight clothing. The best prevention is to wear long pants and sleeves when in the woods, if possible, and to use an insect repellant. If any ticks are found, they must be removed as soon as possible. The longer the tick is attached, the greater the chance of infection.

Yersinia pestis

This is the causative agent of plague, also called the black plague. The vector is the common flea, which is often found on rats. The rat may also succumb to the pathogen, and the fleas from the dead rat may

FIGURE 14-11 **Comparison of wood ticks and deer ticks.**

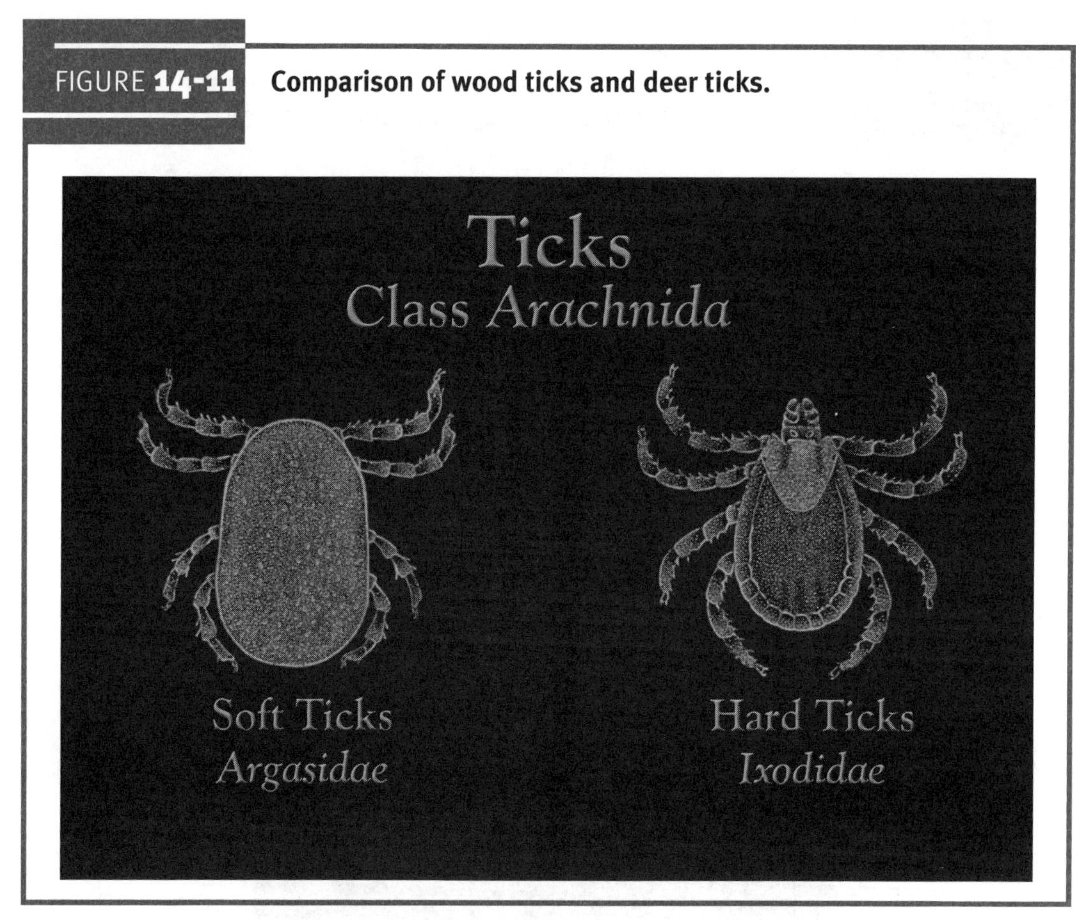

SOURCE: Courtesy of the Centers for Disease Control and Prevention. (http://phil.cdc.gov/phil/details.asp). #5993.

jump onto a person who is removing the dead rat. Plague has been used as a biological weapon and is covered in more detail in Chapter 15.

FRANCISELLA TULARENSIS

Francisella tularensis is a short, Gram-negative bacillus that is the causative agent of tularemia, also called rabbit fever (**Figure 14-13**). There are a number of ways to become infected with this pathogen: it can be transmitted by fleas, ticks, biting flies, and mosquitoes. However, a general outbreak of tularemia has never occurred as a result of insect bites, despite the fact that there are plenty of all of these insects in the world. It does appear to spread easily through the direct handling of tissue from an infected rabbit, and this probably indicates a low infectious dose. Ingestion of insufficiently cooked contaminated meat results in a gastrointestinal illness.

People who work directly with rabbit carcasses are at the greatest risk. People who handle animals, such as veterinarians, may acquire superficial ulcerations on the hands or arms (**Figure 14-14**). Ingesting the pathogen may result in ulcerations of the pharynx along with diarrhea and vomiting. If acquired through inhalation, *F. tularensis* causes pneumonia. If not treated with antibiotics, the lethality for the pneumonic variety of infection can

FIGURE 14-12 **The "bull's-eye" rash (erythema migrans) of Lyme Disease.**

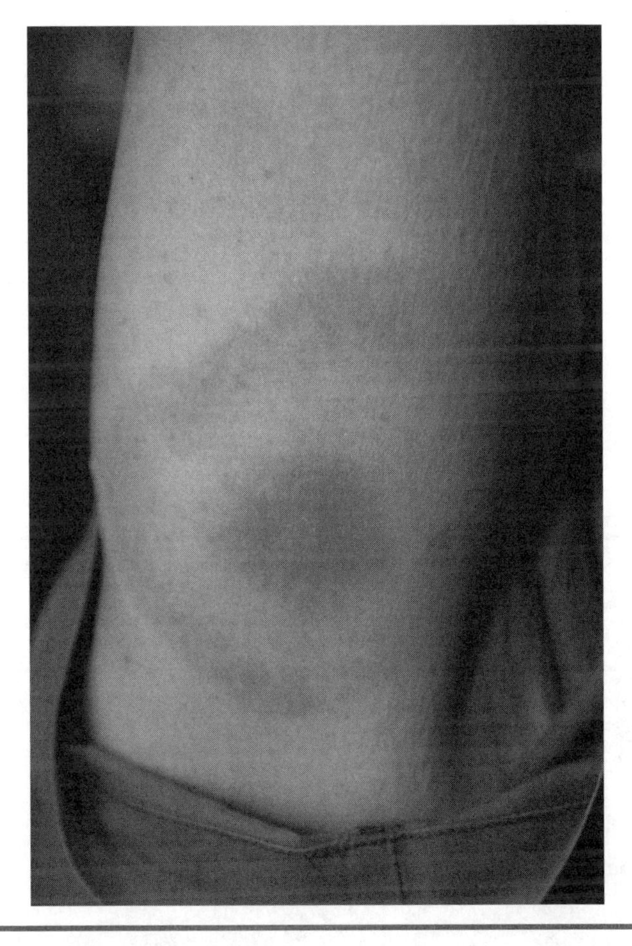

SOURCE: Courtesy of the Centers for Disease Control and Prevention.

reach 60%. The disease is not widespread and affects mostly animal handlers. Personal protective equipment is essential to avoid exposure.

Rickettsia rickettsii

This is a rickettsial microbe, an unusual bacterial species that is an obligate intracellular pathogen passed from animal reservoirs to humans via the tick. It causes Rocky Mountain Spotted Fever (RMSF). Ticks of the genus *Dermacentor* are the vectors in the United States, while other ticks indigenous to other countries can transmit the disease as well.

RMSF is an accurate description; the disease results in a high fever and a spotty rash that covers the body (**Figure 14-15**). The disease is not limited to the Rocky Mountain area, and in fact, it

FIGURE **14-13** ***Francisella tularensis* on a chocolate agar plate.**

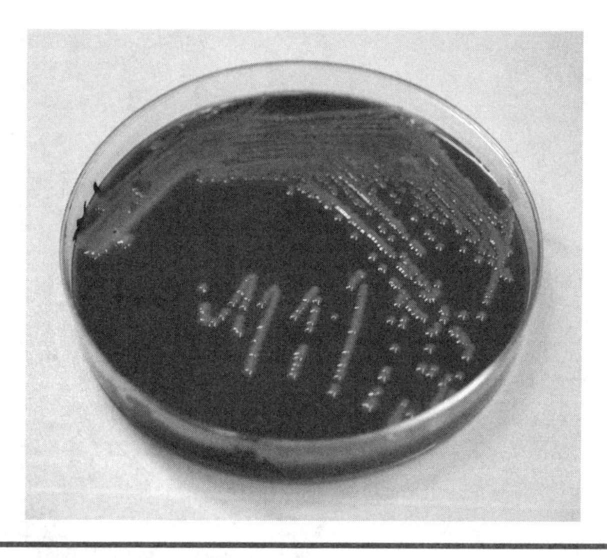

SOURCE: Courtesy of Megan Mathias and J. Todd Parker, the Centers for Disease Control and Prevention. (http://phil.cdc.gov/phil/details.asp). #11765.

FIGURE **14-14** **Superficial wound on the hand caused by *Francisella tularensis*.**

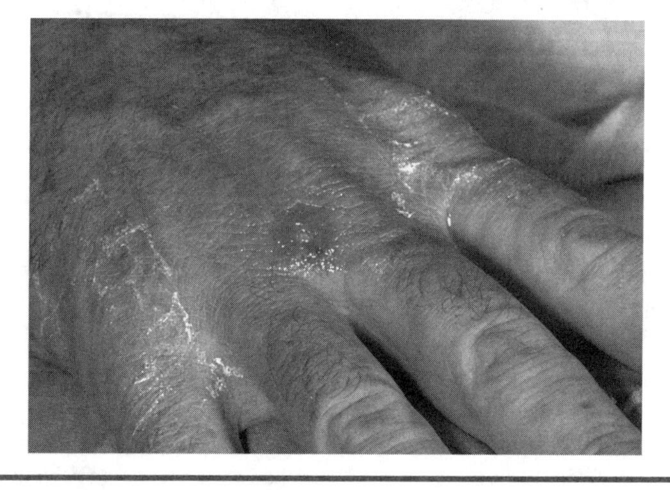

SOURCE: Courtesy of Dr. Brachman, the Centers for Disease Control and Prevention. (http://phil.cdc.gov/phil/details.asp). #2032.

FIGURE 14-15 **The spotted rash characteristic of Rocky Mountain Spotted Fever.**

SOURCE: Courtesy of the Centers for Disease Control and Prevention. (http://phil.cdc.gov/phil/details.asp). #4476.

is far more prevalent in the eastern United States, particularly the Carolinas. The incidence of RMSF is high during the summer months, when people are engaged in outdoor activities that bring them into contact with ticks (**Figures 14-16** and **14-17**). The disease can be very dangerous; without treatment, the fatality rate is about 20%. Even with antibiotic treatment, the fatality rate is still about 5%.

RMSF is difficult to avoid entirely, although the case rate is very low. Strict attention to tick avoidance is the best strategy. If a tick is found, it should be carefully removed, and the bite should be washed thoroughly with soap and water.

Similar diseases are caused by different Rickettsia species. *Riskettsia akari* causes rickettsial pox, a milder form of RMSF. The common mouse is the reservoir for the pathogen; it is passed to humans via the mouse mite. This disease threatens urban areas rather than the forested and mountainous areas where RMSF is found. The rash can certainly be frightening, but most cases resolve without complication. Fortunately, *R. akari* has not proved to be a major concern because eradication of vermin from households is an effective means of control.

R. conorii causes Boutonneuse fever, also known as Mediterranean fever. This is a worldwide disease, although not an epidemic one. The reservoir is the dog, and the vector is the tick. The disease resembles RMSF except that some of the spots can be black as a result of local tissue necrosis. *Rickettsia felis* causes cat-flea rickettsiosis. As the name implies, the reservoir is the cat (as well as rodents), and the vector is the common flea found on both dogs and cats. In addition to the typical symptoms of rash, fever, headache, myalgia, and so on, the patient also suffers from photophobia. Worldwide, a number of other spotted fever diseases are caused by Rickettsia species. Generally speaking, they are spread by ticks.

FIGURE 14-16 The incidence of Rocky Mountain Spotted Fever is much higher during the summer months. Data are from 2002.

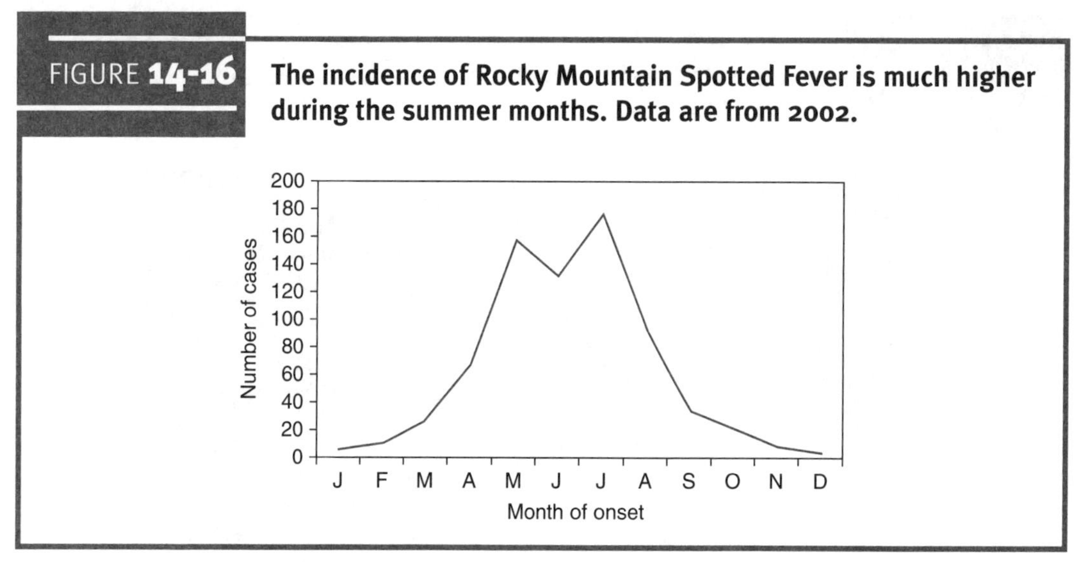

SOURCE: Courtesy of the Centers for Disease Control and Prevention.

FIGURE 14-17 Annual incidence per million population for Rocky Mountain Spotted Fever by state for 2002.

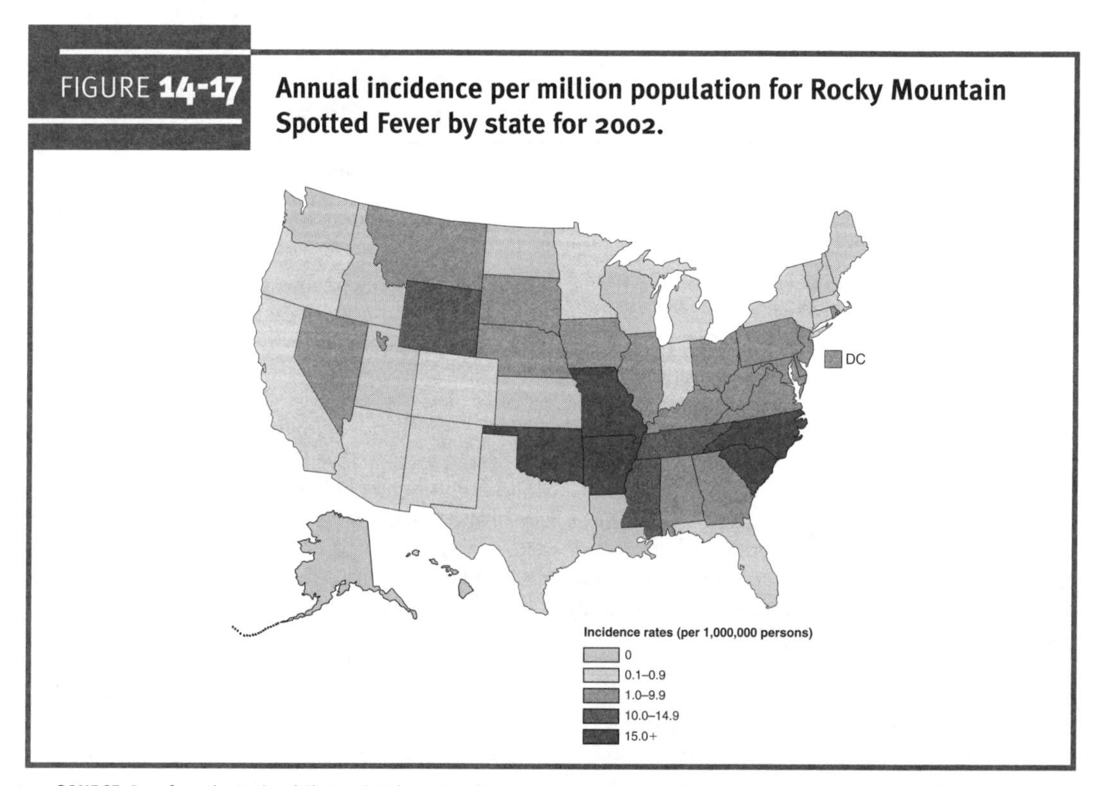

SOURCE: Data from the National Electronic Telecommunications System for Surveillance. Courtesy of the Centers for Disease Control and Prevention.

Typhus

Typhus is a complicated disease because several types are caused by different organisms and are transmitted by different vectors. In all cases, the symptoms are largely the same: fever, myalgia, headache, and a rash. All are treated in the same way as well.

Murine typhus is caused by *Rickettsia typhi* (not to be mistaken with *Salmonella typhi*, the causative agent of typhoid fever). Here the reservoir is rodents (i.e., rats and mice), and the vector is the flea.

Epidemic typhus is caused by *R. prowazekii*. The reservoir has a **sylvatic** (woodland) source in flying squirrels; otherwise humans are the reservoir. It is passed through the body louse vector, and this explains the epidemic nature of the disease. It is passed from person to person quite easily under crowded and unsanitary conditions, like those found during a war or in a refugee camp or after a natural disaster.

Scrub typhus is caused by an organism in a different genus, *Orientia tsutsugamushi*. Here the vector is the chigger, and the reservoir is rodents.

Ehrlichia chaffensis

This rickettsial microbe is responsible for human monocytic ehrlichiosis. The reservoir is the large mammal population: deer, elk, and many domesticated hooved animals. It is spread through the tick vector.

Anaplasma phagocytophilum

Until recently, this was seen as another species of *Ehrlichia*, but it is different enough to deserve its own genus. *A. phagocytophilum* causes human granulocytic anaplasmosis (formerly human granulocytic ehrlichosis). It shares similar disease symptomology, reservoir, and tick vector with *Ehrlichia*.

Viruses

The vectorborne viruses are divided into arthropod-borne viruses (arboviruses) and rodent-borne viruses (roboviruses).

Equine Encephalitis Viruses

This category includes Eastern equine encephalitis virus (EEE; **Figure 14-18**), Western equine encephalitis virus (WEE), and Venezuelan equine encephalitis

FIGURE **14-18** **Eastern equine encephalitis in the United States. Data show cumulative cases from 1964 to 2008.**

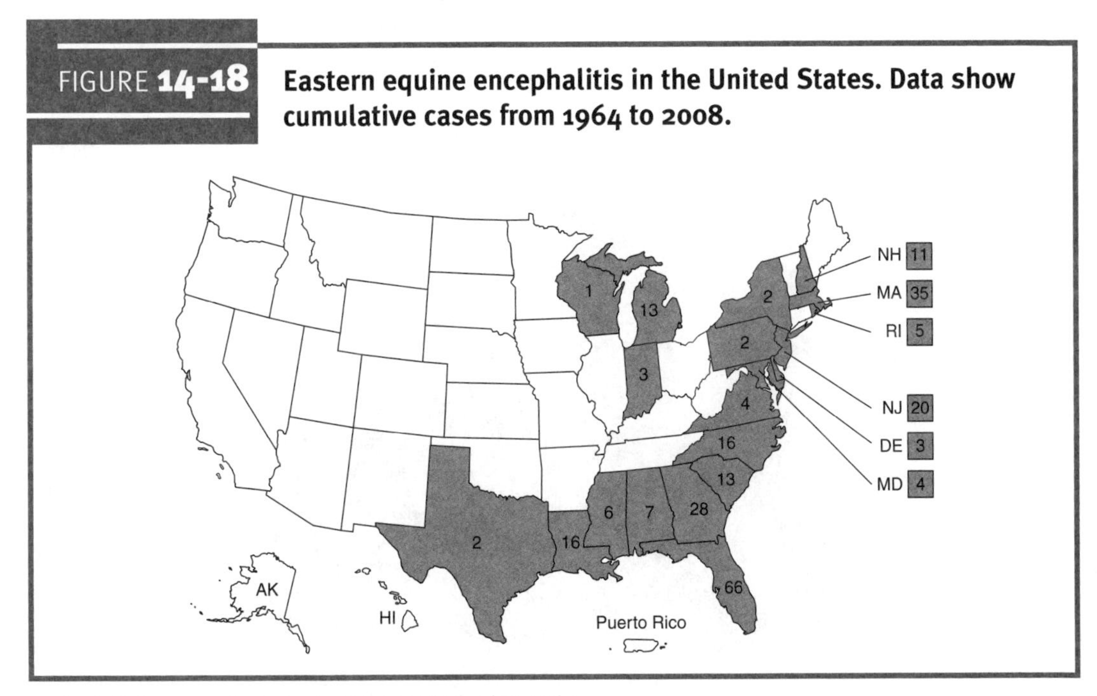

SOURCE: Courtesy of the Centers for Disease Control and Prevention.

virus (VEE; **Figure 14-19**). These viruses are found in horses and are transmitted by mosquitoes. People who work directly with horses are at much higher risk; a suspected case of any of these viruses quickly prompts a medical history question about exposure to horses. If the horses have been sick as well, equine encephalitis is highly likely. Other reservoirs for these viruses have been postulated, but the sporadic nature of the illness makes an epidemiological study difficult.

Equine encephalitis viruses are all togaviruses, which are enveloped, plus-stranded, single-stranded, RNA viruses. Infection with one does not appear to confer any antibody protection against the others. Despite bearing the name of a South American country, VEE outbreaks have occurred in the United States. As the name indicates, in some infections VEE produces an encephalitis that is life threatening. Horses can develop aggressive infections that result in high titers of virus in their blood. Mosquitoes biting them pick up the virus and transmit it to humans when they bite. Humans do not develop these same high titers, but the illness can still be severe.

FIGURE **14-19** **Electron micrograph of the virions of Venezuelan equine encephalitis virus.**

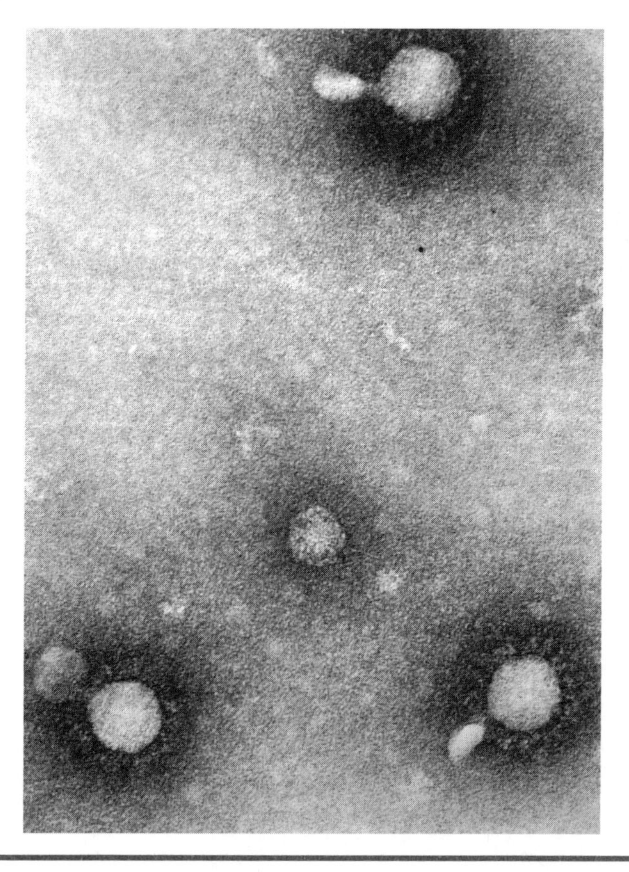

SOURCE: Courtesy of the Centers for Disease Control and Prevention.

In general, the illness for all three viruses looks the same. General, nonspecific symptoms are followed by a worsening condition of vomiting and diarrhea. This then progresses to neurological symptoms, such as spastic paralysis, tremors, seizures, and coma; deaths are not unusual (Vilcarromero et al., 2010).

Another virus of interest is Hendra virus, a paramyxovirus. It is also found in horses, and infections in humans can cause encephalitis. The reservoir is probably fruit bats and flying foxes that are native to Australia and Southeast Asia. How the horses are infected is not clear. It passes to humans who come into close contact with the infected horses and their secretions. Therefore this illness might be more accurately regarded as a zoonotic disease. In humans, the illness appears as a flu-like disease that can become encephalitis. Because this virus is in the paramyxovirus family, it can be treated with certain antiviral drugs such as ribavarin. However, the illness has a high fatality rate.

A related virus, called Nipah, infects swine but otherwise acts the same as Hendra virus. It is found in Singapore and Malaysia. Again, the transmission of the virus from fruit bats to swine is not understood.

West Nile Virus

A few years ago, this virus would have been almost ignored by this text because West Nile virus (WNV) was considered to be confined to distant parts of the world. All that changed in 1999 when an outbreak occurred in New York State. There it caused encephalitis in a number of animals and was found in birds, which it was also known to infect. From there it has spread to locations across the United States, including some of the coldest states (**Figure 14-20**).

Until 1999, WNV was thought to be found in warmer regions such as Uganda (where it was first isolated), Egypt, and India. A few cases had been identified in Mediterranean countries, but these

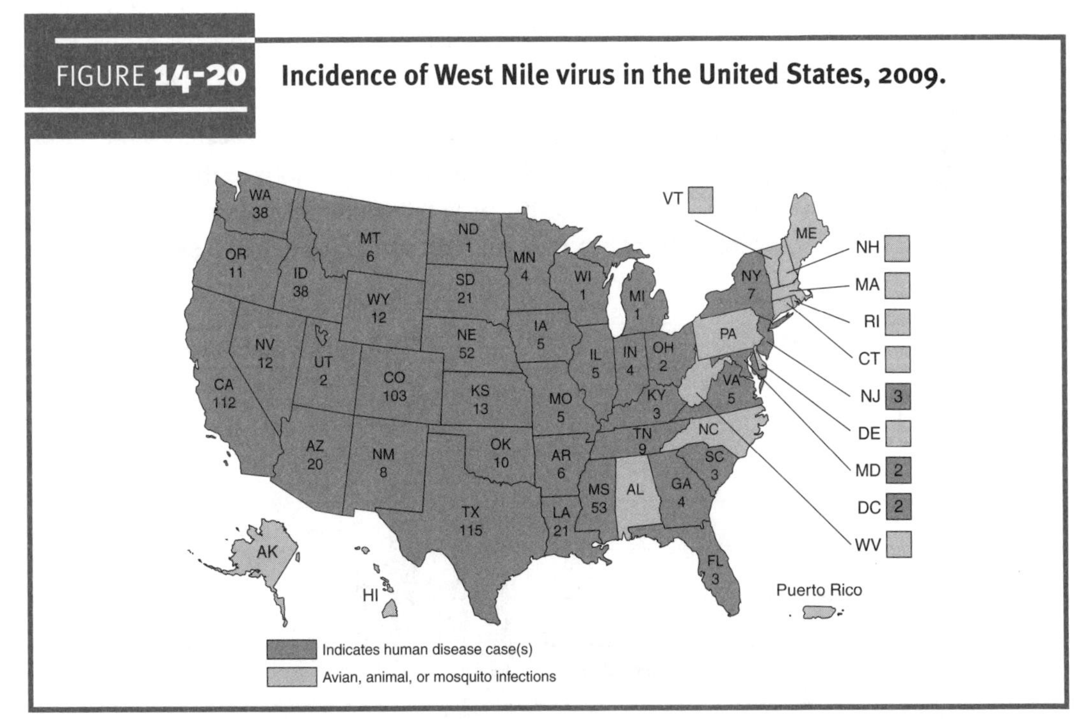

FIGURE 14-20 **Incidence of West Nile virus in the United States, 2009.**

SOURCE: Courtesy of the Centers for Disease Control and Prevention.

were relatively infrequent sightings. In the United States, encephalitis viruses certainly occur, such as St. Louis encephalitis virus and La Crosse virus. But this was clearly the WNV and did not originate in North America.

The virus is suspected to have arrived either by a mosquito or an infected bird. Mosquitoes live for a period of weeks, which is too short a time to cross an ocean on a sailing ship but plenty of time to hitch a ride on an airplane. Once in the country, it may have been passed along from one infected animal to another by the local mosquito population. Such hitchhikers may have arrived over the years, and continue to do so, but only in this unlucky instance did the disease gain a foothold. This is a cautionary tale about the spread of exotic diseases and the speed of air travel.

Mosquitoes spread WNV through their bites. Most people who are exposed to the virus are asymptomatic. Perhaps 20% develop flu-like symptoms along with a rash and swollen lymph nodes. In less than 1% of the infections, the patient experiences severe, debilitating neurological problems. These problems can include paralysis or numbness, coma, convulsions, and vision problems. Death is rare but not unknown. (See **Figure 14-21**.)

Yellow Fever Virus

This is a flavivirus and therefore is an enveloped, plus-stranded, single-stranded, RNA virus. It is primarily a jungle virus, residing in primates and transmitted by mosquitoes (**Figure 14-22**). Humans are accidental hosts. In the past, the range of yellow fever was much greater, with an outbreak recorded in New Orleans as late as 1905. However, effective control of the mosquito population has eliminated the threat in the United States.

Its impact on history is unquestioned; yellow fever has been a major killer for many years. Due to yellow fever and malaria, the building of the Panama Canal cost thousands of lives. The mosquito vector was first suggested by Dr. Carlos Finlay, who identified the vector species as *Aedes*. The control of the mosquito population became a paramount objective based on his work. Dr. Walter Reed later proved Finlay's theory.

Mosquito breeding must be limited as much as possible. For travelers to these areas, an effective vaccine often provides good protection. Because treatment for the virus is still only supportive, it is far preferable to avoid the virus if possible. While most cases are relatively mild, a number of cases progress to hepatitis, creating the yellow

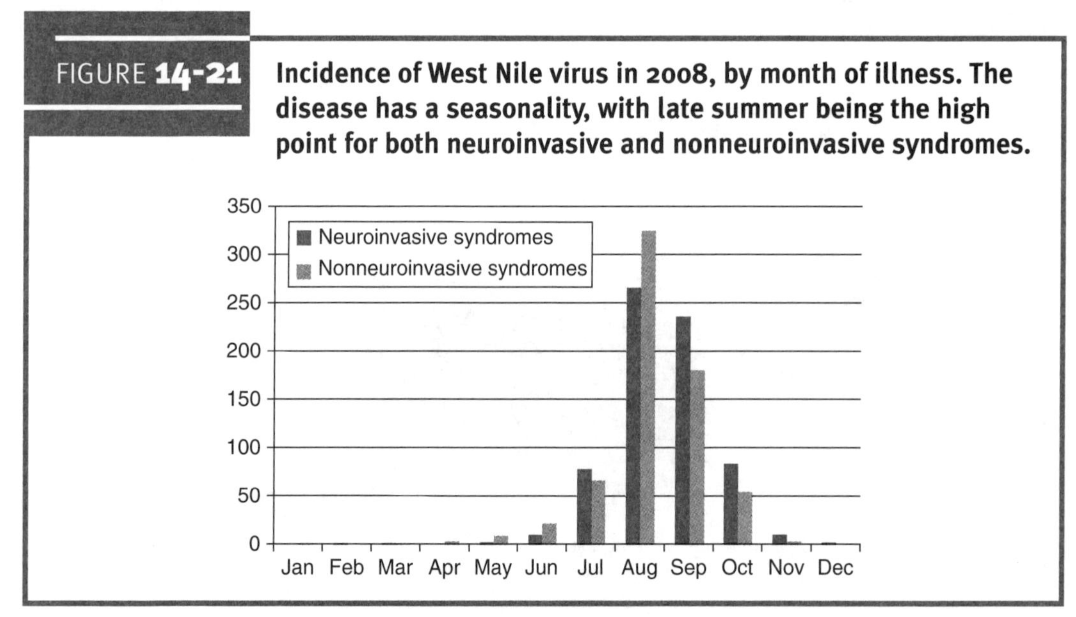

FIGURE 14-21 **Incidence of West Nile virus in 2008, by month of illness. The disease has a seasonality, with late summer being the high point for both neuroinvasive and nonneuroinvasive syndromes.**

SOURCE: Based on data from the Centers for Disease Control and Prevention.

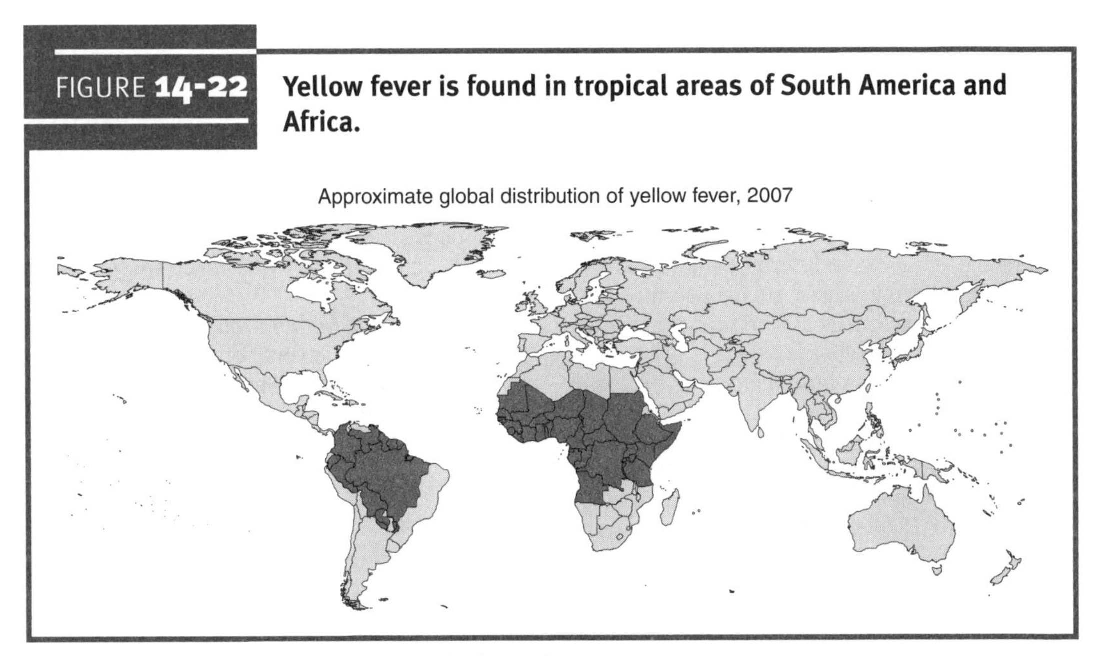

FIGURE 14-22 **Yellow fever is found in tropical areas of South America and Africa.**

SOURCE: Courtesy of the Centers for Disease Control and Prevention.

jaundice in skin and eyes that gives yellow fever its name. In some outbreaks, the fatality rate may reach 15%.

Another flavivirus is Kyasanur forest disease, which is limited to parts of India. This virus infects monkeys and a few other mammals, and humans are an accidental host. In outbreaks, the fatality rate is 3–5%.

Dengue Fever

Dengue is one of the most important viral diseases worldwide. It does not have a great impact in the United States, although it is endemic in Puerto Rico. It is transmitted by the *Aedes* mosquito, a very common insect. Some estimates postulate that 100 million cases occur each year. In some cases, this becomes a hemorrhagic fever with severe consequences.

Dengue fever is evident with fever, headache, pain behind the eyes (i.e., retroorbital), and mild bleeding or bruising episodes. This last symptom is due to damage to the capillaries that makes them more porous. In Dengue hemorrhagic fever, this damage is more pronounced and leads to hypovolemic shock. All symptoms of Dengue become much more pronounced. A decrease in platelets means that damage to blood vessels will not be quickly or easily repaired. Under these conditions, death is not uncommon.

A number of viruses are yet to be discovered. Some of these produce diseases that are similar to other known diseases. The Zika virus (Hayes, 2009) was originally described in Africa, where it is spread by the *Aedes africanus* mosquito. Surprisingly, it has now been found in other tropical areas of the world, where it is often mistaken for Dengue fever (Duffy et al., 2009). If effective vaccines are to be made, more precise identification of the disease will be needed to determine whether the vaccination program has been effective.

Obscure and Rare Viruses

Reoviruses. The Reoviruses have double-stranded RNA genomes. They include the coltiviruses, of which the Colorado tick fever virus is notable. In most cases, this infection leads to a fever and malaise and then resolves without complication, but in a number of cases, it may lead to meningitis or encephalitis. These cases can be life threatening. The virus is, of course, spread by the tick and has been found in rabbits and rodents.

A relatively new group within the Reoviruses is the Seadornaviruses (for Southeast Asian dodeca RNA virus, referring to the 12 segments of RNA in the genome). These viruses include Banna virus, Kadipiro virus, and Liao Ning virus. Of interest is the fact that they are transmitted through mosquitoes of several types: *Culex, Anopheles*, and *Aedes* (Attoui et al., 2005). While large outbreaks have not been reported, these merit further study. Some of these are localized illnesses, such as Barmah Forest virus, which is specific to certain parts of Australia (Lindsay et al., 1995).

Borna Viruses. An unusual group of viruses to watch for the future are the Borna viruses (Hatalski et al., 1997; Ikuta et al., 2002). These are from a new family of negative, single-stranded, RNA viruses that appear to be rather common in the animal kingdom. The finding that makes them interesting is their apparent association with neuropsychiatric diseases. Although they are not found in every patient with a psychiatric disorder, they are found in high percentages in people with psychiatric disorders, whereas controls have very low incidence of infection. It has also been suggested that they play a role in chronic fatigue syndrome, although many other pathogens have been considered for that role as well (Ortega-Hernandez and Shoenfeld, 2009).

Compounding the problem is that Borna viruses appear to have low copy numbers in human cells and appear to bind tightly to the cell membranes. This makes them difficult to isolate and enumerate. Very little is known about these viruses, including the mode of transmission and which species might act as reservoirs.

Crimean-Congo Hemmorrhagic Fever. Crimean-Congo hemorrhagic fever is one of the viral hemorrhagic fever viruses (see previous section), but is spread by ticks, with a variety of wild and domesticated animals as potential reservoirs.

Parasites

PLASMODIUM

Four species of *Plasmodium* are responsible for the disease malaria: *Plasmodium falciparum, P. vivax, P. ovale*, and *P. malariae*. Of these, *P. falciparum* is the most dangerous. Together these four cause at least 100 million illnesses every year. The World Health Organization (WHO) estimates that more than a million people die from malaria every year (**Fig-**

FIGURE **14-23** **Malaria is a worldwide problem.**

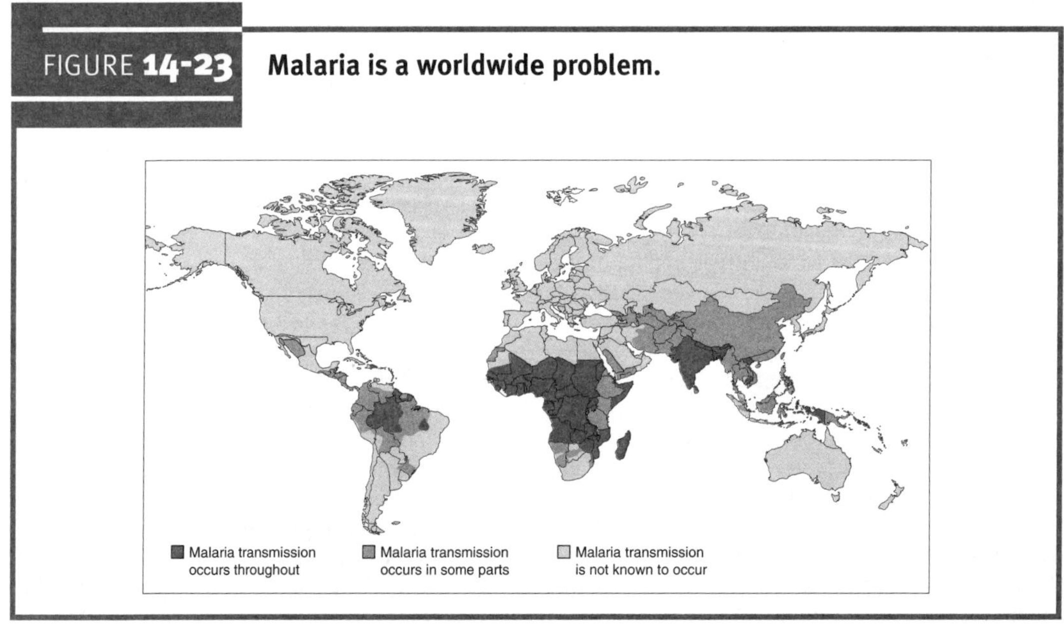

SOURCE: Courtesy of the Centers for Disease Control and Prevention.

ure 14-23). It is rare in the United States; most cases occur in visitors from tropical countries.

The vector is the *Anopheles* mosquito (**Figure 14-24**). The illness results in high fever and shaking chills. The illness can be treated and cured. The use of quinine drugs allows people to avoid infection. The quinine blocks receptor sites on the red blood cells so that the *Plasmodium* cells do not recognize it. Unfortunately, there is no vaccine for malaria, although great efforts to produce one are under way. Such a vaccine would have a huge impact on public health in many countries around the world.

Trypanosoma brucei

This is the causative agent of African sleeping sickness, which is transmitted through the tsetse fly. Of course, the victim does not actually sleep; after infection, the parasite crosses the blood/brain barrier and causes neurological symptoms, including extreme lethargy, confusion, and mental deterioration. The victim's normal pattern of sleeping is also affected. Gradually this leads to coma and death.

The key to success for this trypanosome is its regular switching of antigenic markers on its surface. It has hundreds of variant surface glycoproteins for this purpose. The immune system has a difficult time mounting an effective immune response to a constantly changing target (Stanne and Rudenko, 2010). It has been suggested that the switching of the variant genes might be a good target for a future therapy. With treatment the disease is now curable, but there is no vaccine. Some areas of Africa where the tsetse fly is endemic are probably best avoided.

FIGURE **14-24** ***Anopheles* mosquitoes, the carriers of malaria.**

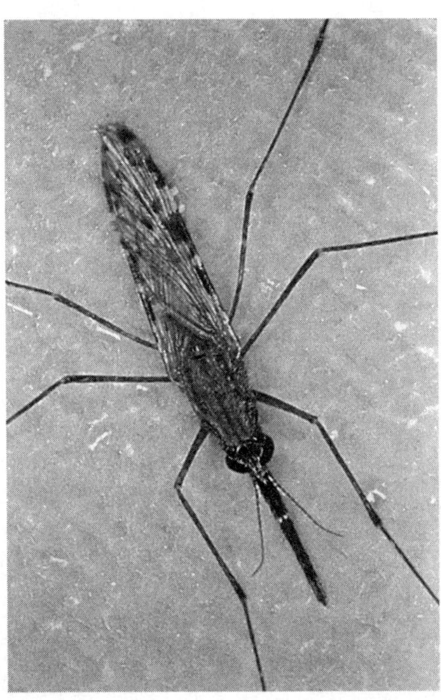

SOURCE: Courtesy of the Centers for Disease Control and Prevention.

Trypanosoma cruzi

This is the causative agent of Chagas disease, which is confined to the Western hemisphere, particularly Central America. It is spread by an insect called the kissing bug, which is an inapt name because the insect is actually a blood-sucking parasite. In its blood meal, it passes the trypanosome, typically through its fecal contamination of the new wound (**Figure 14-25**).

At its worst, the disease causes heart inflammation and associated poor cardiac function (e.g., arrhythmias and inadequate cardiac volume). In some patients, there is gastrointestinal involvement, with enlarged bowel and bleeding in the intestines.

Babesia

This is a parasite that is spread by ticks and also infects small mammals. It causes a flu-like illness (babesiosis) in some infected people. Most of the time the disease is not a problem, but occasionally it causes anemia, including a low platelet count. Blood clots may form from damaged blood cells, and these may plug capillaries, leading to organ damage. In rare cases, death results.

Leishmania

These parasites cause leishmaniasis, a disease that is largely confined to the tropics. It is transmitted by the sand fly. The disease involves skin lesions as

FIGURE **14-25** **The "kissing bug," which transfers the trypanosome for Chagas disease.**

SOURCE: Courtesy of the Centers for Disease Control and Prevention.

well as involvement of the major organs. In this latter case, a fatal outcome is not uncommon.

THE PUBLIC HEALTH RESPONSE

In many ways, these diseases are the most frustrating. Many of them cannot be cured, and the patient is left with only supportive therapy. The diseases are unlikely to be eradicated because they have a reservoir in some type of wildlife. And many of them strike sporadically, making them difficult to predict or monitor. The role of the public health authority is thus limited, although in these limited areas a vigorous activity can have major benefits. One aspect is the education of the public. In many areas of the world, this education concerns the reduction of conditions for breeding mosquitoes. Containers that may have standing (stagnant) water are emptied on a regular basis. The public is also educated to get their pets vaccinated and to stay away from wild animals.

Another aspect is monitoring, and this may be the most difficult. Budgets for monitoring are constrained, and the number of samples that must be examined can be very high. In addition, many of the diseases listed here are viruses, which will only be enumerated by an expensive molecular technique such as PCR. There is a great opportunity here for clever techniques in improved monitoring.

The last aspect to consider in relation to vector-borne and zoonotic diseases is the development of vaccines and the promotion of their use. The rabies vaccine has been mentioned, and additional vaccines for West Nile virus and *Borrelia burgdorferi* would be very welcome. Convincing people to get the vaccination after it has been created and tested can be a challenge; these diseases are still rather rare, and many people will ignore warnings. The public health authority must advocate vaccination programs.

QUESTIONS FOR DISCUSSION

1. Of the diseases mentioned in this chapter, which are known in your area?
2. Why must cats and dogs be vaccinated against rabies on a yearly basis?
3. How might the mosquito population be controlled?
4. What host factors might account for susceptibility to these illnesses?
5. If you were a veterinarian, what vaccinations would you want to receive?
6. Why are there many insect vectors and relatively few mammalian vectors?
7. Should the DNA sequence for smallpox have been published?

References

Attoui, H., F.M. Jaafar, P. de Micco, and X. de Lamballerie. 2005. Coltiviruses and seadornaviruses in North America, Europe, and Asia. Emerg. Infect. Dis. 11: 1673–1679.

Burgdorfer, W., A.G. Barbour, S.F. Hayes, J.L. Benach, E. Grunwaldt, and J.P. Davis. 1982. Lyme disease—a tick-borne spirochetosis? Science 216: 1317–1319.

Duffy, M.R., T. Chen, W.T. Hancock, A.M. Powers, J.L. Kool et al. 2009. Zika virus outbreak on Yap island, federated States of Micronesia. New England J. Med. 360: 2536–2543.

English, C.K., D.J. Wear, A.M. Margileth, C.R. Lissner, and G.P. Walsh. 1988. Cat scratch disease: isolation and culture of the bacterial agent. JAMA 259: 1347–1352.

Hatalski, C.G., A.J. Lewis, and W.I. Lipkin. 1997. Borna disease. Emerg. Infect. Dis. 3: 129–135.

Hayes, E.B. 2009. Zika virus outside Africa. Emerg. Inf. Dis. 15: 1347–1350.

Ikuta, K., M.S. Ibrahim, T. Kobayashi, and K. Tomonaga. 2002. Borna disease virus and infection in humans. Front Biosci. 7: 470–495.

Lindsay, M., C. Johansen, A.K. Broom, D.W. Smith, and J.S. Mackenzie. 1995. Emergence of Barmah Forest virus in Western Australia. Emerg. Infect. Dis. 1: 22–6.

Mangieri, N., F. Sorhage, C. Campbell, G. Fratz, C. Amari et al. 2008. Rabies in a dog imported from Iraq—New Jersey, June 2008. Morbid. Mortal. Weekly Rep. 57: 1076-1078.

Ortega-Hernandez, O.D., and Y. Shoenfeld. 2009. Infection, vaccination, and autoantibodies in chronic fatigue syndrome, cause or coincidence? Ann. N. Y. Acad. Sci. 1173: 600–609.

Pue, H.L., G. Turabelidze, S. Patrick, A. Grim, C. Bell, et al. 2009. Human arabies—Missouri, 2008. Morbid. Mortal. Weekly Rep. 58: 1207–1209.

Reboli, A.C., and W.E. Farrar 1989. Erysipelothrix rhusiopathiae: an occupational pathogen. Clin. Microbiol. Rev. 2: 354–359.

Regnery, R., and J. Tappero. 1995. Unraveling mysteries associated with cat-scratch disease, bacillary angiomatosis, and related syndromes. Emerg. Inf. Dis. 1: 16–21.

Stanne, T.M., and G. Rudenko. 2010. Active VSG expression sites in *Trypanosoma brucei* are depleted of nucleosomes. Eukaryot. Cell 9: 136–147.

Vilcarromero, S., P.V. Aguilar, E.S. Halsey, V. Alberto Laguna-Torres, H. Razuri, et al. 2010. Venezuelan equine encephalitis and 2 human deaths, Peru. Emerg. Infect. Dis. 16: 553–556.

Willoughby, R.E., M.M. Rotar, H.L. Dhonau, K.M. Ericksen, D.L. Cappozzo, et al. 2004. Recovery of a patient from clinical rabies—Wisconsin, 2004. Morbid. Mortal Weekly Rep. 53: 1171–1173.

CHAPTER 15

Biological Warfare

LEARNING OBJECTIVES

- List the pathogens that are most likely to be used in biowarfare.
- Describe how a biowarfare attack would be recognized.
- Describe the components of a biowarfare device and why they are difficult to construct.
- Describe the appropriate responses of public health authorities in the event of an attack.

KEY TERMS

- Bubonic form of plague
- Delivery system
- Dispersion system
- Munition
- Payload
- Pneumonic form of plague
- Septicemic form of plague

INTRODUCTION

Throughout the chapters of this book, microorganisms have been presumed to encounter humans accidentally. If a person is infected by another person, it is incidental to a sneeze or a cough or sexual contact. But this is not always the case. For example, isolated cases occur in which a person who has a sexually transmitted disease deliberately has sexual relations with another person for the express purpose of infecting that other person. These cases, however, are rare and can be ascribed to a mental disturbance or illness.

Another, more disturbing act of deliberate infection is the use of microorganisms in biological warfare. Despite the seeming technological nature of biological warfare, it has been utilized for many hundreds of years. In antiquity, diseased humans or diseased corpses were used to spread a contagion to an enemy population. This could be advantageous because the contagion would do the work of the warrior without the uncertainties of battle.

In the 1300s, the Tartar army besieged the city of Kaffa (in modern-day Ukraine). As often happened with siege armies, illness broke out in the camp, in this case due to bubonic plague. The Tartars, sensing an opportunity, catapulted the plague-ridden corpses over the city walls in an attempt to start the plague among the defenders. Plague did break out in Kaffa. The plague, however, may have been caused by rats from the Tartar camp that infiltrated the city rather than the corpses. Regardless, the situation was severe enough that the defenders were forced to abandon the city. As we will see, this was not the last use of plague as a weapon of war.

In many cases, illness and death by infectious disease were decisive factors in wars, though not by human design. In 1520, the conquistador Hernando Cortez led a relatively small force of Spaniards (probably about 500 men) against the Aztec empire. Due to a combination of superstition and Spanish technological superiority, the Spaniards were able to enter the Aztec capital briefly but eventually were driven out. However, one of the Spanish soldiers had smallpox (some historians speculate that it was measles), and this illness spread quickly through the Aztec population. By the time the epidemic had subsided, approximately 90% of the population was dead. Under these conditions, Cortez was able to conquer the city and eventually destroy the Aztec empire.

CHEMICAL AND BIOLOGICAL WEAPONS OF WAR

The use of chemical warfare during World War I resulted in approximately one million casualties, including perhaps 100,000 deaths. This had a profound effect on the perception of warfare and on which methods of warfare would be regarded as "allowable." Chemical warfare (often called "gas warfare" at the time) was singled out as particularly heinous. Article 171 of the Treaty of Versailles specified, "The use of asphyxiating, poisonous or other gases and all analogous liquids, materials or devices being prohibited, their manufacture and importation are strictly forbidden in Germany." No such prohibition was created for the Allied countries, although the Geneva Protocol of 1925 was more comprehensive in its prohibition of chemical weapons. Significantly for the time, the Geneva Protocol also outlawed "bacteriological methods of warfare." The inclusion of this language may have been prompted by allegations that Germany tried biological warfare during World War I, specifically by spreading glanders (melioidosis) among Russian horses and mules. Glanders is caused by *Burkholderia* species, typically through ingestion of contaminated food, and produces lesions in the respiratory system. It is frequently fatal in infected animals.

World War II and the Present Day

Biological warfare was carried out in World War II by the notorious Water Purification Unit 731 of the Imperial Japanese Army. Under the guise of a water purification unit, this group conducted some of the most hideous experiments on human beings that have ever been recorded. They deliberately infected people with deadly diseases and then kept careful notes about the progression of disease. In some cases, they performed vivisection on the victims during stages of illness. They used this information to design biological weapons for the Imperial Japanese Army. Several biowarfare attacks were carried out using a variety of pathogens. The plague bacillus seemed to work best (see "Possible Biowarfare Agents").

The Allies were also prepared to conduct biological warfare, although they never carried out any attacks or experimented in the way the Japanese did. The British conducted extensive testing of anthrax at their facility at Porton Down. Their field release site was located on Gruinard Island off the coast of Scotland. They placed flocks of sheep there and then exploded anthrax bombs to determine the efficacy of the weapons. Contamination with anthrax spores was so extensive that the island remained uninhabitable for decades. Remediation was conducted in the late 1990s and involved huge amounts of disinfectant and the removal of most of the topsoil. Even today, it would probably be unwise to visit the island.

Allegations have been put forward that the Soviet Union employed *Francisella tularensis*, the causative agent of tularemia, against the German army during the Battle of Stalingrad in 1942 (Alibek, 1999). The outbreak started in the German army but soon spread to the Soviet army, highlighting the unpredictable nature of biowarfare. The allegation is intriguing because tularemia, which is very infectious, is considered a possible biowarfare agent. However, the report remains unsubstantiated. Diseases often accompany military campaigns due to generally poor hygiene, poor food quality, poor sanitation, and lack of pest control.

The United States unilaterally renounced the use of all methods of biological warfare (including toxins) by order of President Nixon in 1970. The only activities that were allowed were research and development of vaccines, treatments, and other peaceful uses. Shortly thereafter, in 1972, the Convention on the Prohibition of the Development, Production and Stockpiling of Bacteriological (Biological) and Toxin Weapons and on their Destruction was created by international agreement. Most of the world's nations have since signed and ratified the agreement, which prohibits development, production, stockpiling, or acquisition of biowarfare agents or toxins. Again, exceptions were made for peaceful uses of biological agents. All signatories to the agreement agreed to quickly destroy whatever stockpiles they had already acquired.

As was true for many other countries during World War II, the Soviet Union developed a substantial biological weapons program (Caudle, 1997). Research continued during the Cold War; both the United States and the Soviet Union had access to the Japanese scientists involved with unit 731 (Alibek, 1999). Biowarfare was never seriously considered as a strategic or tactical weapon in the U.S. military; the military planners could not agree on a reliable way to deploy the weapons in a battlefield scenario. However, the Soviet Union considered biowarfare a valuable part of its military planning and weaponized many bacterial, rickettsial, and viral strains, as well as toxins derived from plants and microorganisms. Significantly, they were able to stabilize bulk viral cultures and to add dispersants for effective airborne distribution. These activities continued long after the Soviet Union signed the 1972 Convention. Biowarfare agents included anthrax, plague, and smallpox. The latter was available at any time in the amount of 20 tons (Alibek, 1999). This caused great concern because the weaponized virus, the seed stocks, and the archived strains might be subject to theft.

Biological weapons still pose a threat because they have distinct advantages. If effectively used, the potential for significant loss of life is high. Bioweapons can be deployed stealthily, allowing the perpetrator to sneak away before the attack becomes evident. Bioweapons are cheap; they cost an estimated 0.05% of the cost of a conventional weapon when potential casualties are calculated (Hawley and Eitzen, 2001). Many of the bioweapons are quite easy to obtain and produce, and production facilities are easy to hide. The same type of fermentation reactors and laboratories that can be used for legitimate purposes can be quickly converted to bioweapon production. It is conceivable that a small manufacturing facility could be located in an average home, making detection nearly impossible.

The U.S. government maintains a list of "select agents," which refer to either toxins or microorganisms that are considered potential weapons. This is a task for both the Department of Health and Human Services and the Department of Agriculture (see http://www.selectagents.gov/ for the latest information). While legitimate uses of these select agents are acknowledged, there is also a need to maintain a level of security around them. For example, a researcher may possess 100 mg of ricin without triggering any concern. An amount in excess of this level would be scrutinized. A mechanism for approval of attenuated strains of select agents is also in place, so that research can be performed safely.

Spreading Disease Sounds Easy—It Is Not

Historical anecdotes like these might tempt a modern-day, would-be conqueror to try something similar. After all, we are now much more sophisticated about microbiology and weapons design, and we have far more pathogens from which to choose. Obviously, however, not all pathogens are equally useful as weapons (**Box 15-1**). Many, such as the cold viruses (e.g., rhinoviruses) are associated with very low mortality, and have little overall impact on a population. Others (i.e., "exotics") are very difficult to obtain and/or propagate.

Pathogens, by definition, cause disease and therefore would seem to be effective weapons. However, this is an oversimplification. Pathogens have evolved over time to be transmitted to new hosts in certain, often specific, ways, and simply slopping a huge quantity of pathogens at a target is insufficient to cause infection. A pathogen is said to be "weaponized" if it can be delivered to a target in an active form in order to infect people and cause disease. To achieve this outcome, the weapon is thought to be composed of four components. The **payload** is the pathogen itself, which must remain alive long enough after dispersal that it can infect

BOX 15-1 Delivery and Damage

Delivery

- The pathogen must be easy to disperse.
- The pathogen must remain alive in the delivery vehicle and during dispersal.
- The pathogen must enter the human body in a manner that allows infection.

Damage

- The pathogen must cause significant illness or death.
- The pathogen should work quickly.
- The pathogen should not be easily defeated by treatment (e.g., antibiotics).

the targets. The **munition** refers to the package that the payload is in. The **delivery system** is the means to get the munition to the targets. The **dispersion system** is used to distribute the payload over a certain area (Hawley and Eitzen, 2001). The entire device can be sophisticated, such as the use of a missile that carries a warhead packed with anthrax spores and a dispersal agent. It may be quite mundane: smallpox is the payload, a man infected with it is both the munition and the delivery system, and the dispersion system is simply coughing on other people.

Even if an individual suffering from the pathogen (and therefore a source of the pathogen) were available, isolating the pathogen and bringing it to a weapons facility for further propagation and weaponizing it is no simple matter. Growth of viruses requires a tissue culture or some reservoir animal. While some bacterial pathogens are fairly easy to grow on agar or broth media, many others have specific growth requirements. For example, obligate intracellular pathogens, such as *Rickettsia,* require either a tissue culture medium or embryonated eggs for growth. Even in these conditions, they grow slowly. The danger that the bioterrorist may infect himself during any step of the weaponizing procedure is also imminent.

Given the many difficulties inherent in producing a bioweapon, it is not surprising that only wealthy countries are able to afford the technology to produce weaponized versions of these pathogens. However, crude preparations of pathogens and unsophisticated means of distribution have been used by smaller groups. And the range of delivery mechanisms is impressive: microbes can be added to food or water for ingestion; they can be released in the air for inhalation exposure; they can be spread person to person (typically by airborne droplet); and they can be loaded into small projectiles to break through skin and start an infection.

Keeping in mind the limitations of useful biological weapons, a list of possible pathogens is provided in **Box 15-2**. The list includes plant and animal pathogens because a country may be attacked through its food source. While the list includes many deadly pathogens, it also includes a number of pathogens that have a high morbidity rate but a low mortality rate. A country can be debilitated by sickening a number of people who then must be tended by the living. One of the harsh truths about illness is that once a person is dead, he or she no longer needs medical care. (For information about the disease progression and diagnosis of biowarfare agents against humans, see Oren, 2009).

BOX 15-2 Australia Group

An international meeting was held in 1985 to discuss measures that countries might take to limit the spread of biological and chemical weapons. Because the suggestion came from Australia, this body came to be known as the Australia Group. The Australia Group created an informal agreement regarding which pathogens should be recognized as potential biowarfare agents. Individual countries are then able to formulate their own measures to restrict the access to these pathogens. Their list is the most comprehensive of potential biowarfare agents for use against humans, animals, and plants.

The Australia Group

Plant Pathogens

Bacteria

- *Xanthomonas albilineans*
- *Xanthomonas campestris* pv. *citri*
- *Xanthomonas oryzae* pv. *oryzae* (*Pseudomonas campestris* pv. *oryzae*)
- *Clavibacter michiganensis* subsp. *sepedonicus* (*Corynebacterium michiganensis* subsp. *sepedonicum* or *Corynebacterium sepedonicum*)
- *Ralstonia solanacearum* races 2 and 3 (*Pseudomonas solanacearum* races 2 and 3 or *Burkholderia solanacearum* races 2 and 3)

Fungi

- *Colletotrichum coffeanum* var. *virulans* (*Colletotrichum kahawae*)
- *Cochliobolus miyabeanus* (*Helminthosporium oryzae*)
- *Microcyclus ulei* (syn. *Dothidella ulei*)
- *Puccinia graminis* (syn. *Puccinia graminis* f. sp. *tritici*)
- *Puccinia striiformis* (syn. *Puccinia glumarum*)
- *Pyricularia grisea/Pyricularia oryzae*

Viruses

- Potato Andean latent tymovirus
- Potato spindle tuber viroid

Animal Pathogens

Viruses

- African swine fever virus
- Avian influenza virus
- Bluetongue virus
- Foot and mouth disease virus
- Goat pox virus
- Herpes virus (Aujeszky's disease)
- Hog cholera virus (synonym: swine fever virus)
- Lyssa virus
- Newcastle disease virus
- Peste des petits ruminants virus
- Porcine enterovirus type 9 (synonym: swine vesicular disease virus)
- Rinderpest virus
- Sheep pox virus
- Teschen disease virus
- Vesicular stomatitis virus
- Lumpy skin disease virus
- African horse sickness virus

Bacteria

- *Mycoplasma mycoides* subsp. *mycoides* SC (small colony)
- *Mycoplasma capricolum* subsp. *capripneumoniae* ("strain F38")

Human Pathogens

Viruses

- Chikungunya virus
- Congo-Crimean haemorrhagic fever virus
- Dengue fever virus
- Eastern equine encephalitis virus
- Ebola virus
- Hantavirus
- Junin virus
- Lassa fever virus
- Lymphocytic choriomeningitis virus
- Machupo virus
- Marburg virus
- Monkeypox virus
- Rift Valley fever virus
- Tick-borne encephalitis virus (Russian spring-summer encephalitis virus)
- Variola virus
- Venezuelan equine encephalitis virus
- Western equine encephalitis virus
- White pox
- Yellow fever virus
- Japanese encephalitis virus
- Kyasanur Forest virus
- Louping ill virus
- Murray Valley encephalitis virus

- Omsk haemorrhagic fever virus
- Oropouche virus
- Powassan virus
- Rocio virus
- St. Louis encephalitis virus
- Hendra virus (equine morbillivirus)
- South American haemorrhagic fever (Sabia, Flexal, Guanarito)
- Pulmonary and renal syndrome–haemorrhagic fever viruses (Seoul, Dobrava, Puumala, sin nombre)
- Nipah virus

Rickettsiae

- *Coxiella burnetii*
- *Bartonella quintana* (*Rochalimea quintana, Rickettsia quintana*)
- *Rickettsia prowazeki*
- *Rickettsia rickettsii*

Bacteria

- *Bacillus anthracis*
- *Brucella abortus*
- *Brucella melitensis*
- *Brucella suis*
- *Chlamydia psittaci*
- *Clostridium botulinum*
- *Francisella tularensis*
- *Burkholderia mallei* (*Pseudomonas mallei*)
- *Burkholderia pseudomallei* (*Pseudomonas pseudomallei*)
- *Salmonella typhi*
- *Shigella dysenteriae*
- *Vibrio cholerae*
- *Yersinia pestis*
- *Clostridium perfringens*, epsilon toxin-producing types
- Enterohaemorrhagic *Escherichia coli*, serotype O157 and other verotoxin producing serotypes

Biotoxins

- *Botulinum* toxins
- *Clostridium perfringens* toxins
- Conotoxin
- Ricin
- Saxitoxin
- Shiga toxin
- *Staphylococcus aureus* toxins
- Tetrodotoxin
- Verotoxin and shiga-like ribosome inactivating proteins
- Microcystin (Cyanginosin)
- Aflatoxins
- Abrin
- *Cholera* toxin
- Diacetoxyscirpenol toxin
- T-2 toxin
- HT-2 toxin
- Modeccin toxin
- Volkensin toxin
- Viscum Album Lectin 1 (Viscumin)

Fungi

- *Coccidioides immitis*
- *Coccidioides posadasii*

In reviewing the Australia Group's list of pathogens, it is immediately obvious that most agents are not very common. These include exotic viruses that would be difficult to isolate and usually are difficult to propagate in the laboratory. The likelihood that they might be weaponized is low, although industrialized countries could conceivably achieve such a thing. A major consideration for unsophisticated terrorists is availability: it is much easier to obtain *Bacillus anthracis* than Kyasanur Forest virus.

The Centers for Disease Control and Prevention (CDC) maintains a list of possible and probable bioterrorism agents (see http://www.bt.cdc.gov/agent/agentlist-category.asp). These are grouped by category (i.e., A, B, C); the first group consists of those microorganisms that are most likely to be used. Specifically, they have the following characteristics:

- They can be easily disseminated or transmitted from person to person.

- They result in high mortality rates and have the potential for major public health impact.
- They might cause public panic and social disruption.
- They require special action for public health preparedness.

Several of these have already been used in a biowarfare or bioterrorism attack, and therefore more is known about their efficacy as weapons.

EVIDENCE OF AN ATTACK

How will we know if there has been a biowarfare attack? Often, terrorists brag about their attacks, and this may give authorities enough advance warning to blunt the worst of an attack. However, if this does not happen, other signs may surface. The appearance of a rare disease always suggests a biowarfare attack, especially if the pathogen is one found on these pages. For example, a single case of anthrax might not be unusual, especially if a patient history shows contact with farm animals. However, four or five seemingly unrelated cases would be alarming. In such cases, it is prudent to aggressively pursue the matter with local physicians. They will not panic simply because of a suggested concern, and they need to be primed to look for additional cases.

Another indicator is the presence of multiple unusual infections in victims. This could indicate the use of a mixed pathogen weapon. Other species, principally mammals like dogs and cats, may also have died. Sometimes these other species are more susceptible to pathogens. Alternatively, certain animals may not have died. Typically in cases of plague, the rats die first, and then the fleas jump onto nearby humans. If you do not hear about dead rats, for example, an attack may be suspected.

In an attack on a particular building (e.g., through its air-conditioning system), the victims will have only that factor in common. Perhaps the outbreak has an innocent explanation, such as a chemical spill, but it must be investigated. The Department of Homeland Security prepares for biowarfare attacks by running simulation exercises such as the TOPOFF 2 and TOPOFF 3 projects. These exercises allow local authorities to respond to simulated attacks in a realistic scenario, and they reveal deficiencies in training and operations (**Figure 15-1**). While no training can prepare completely for an actual attack, these exercises increase confidence among first responders that they can handle a real emergency.

POSSIBLE BIOWARFARE AGENTS

Anthrax

Bacillus anthracis is the causative agent of anthrax. Although it is usually considered an illness of sheep and cattle, it can certainly infect humans, sometimes with fatal results. The pathogenicity of this microorganism has been studied in great detail (Mock and Fouet, 2001). *Bacillus anthracis* is a Gram-positive bacillus, notable for its spore-forming ability. The spores are considered the weapon because they can be dispersed easily and can survive environmental conditions that would kill other bacterial agents.

How is *B. anthracis* likely to be used? The former Soviet Union believed it had great potential as a battlefield weapon and manufactured tremendous amounts of it at its facility in Sverdlovsk. In 1979, this facility had an unfortunate accident. The interior of the facility had a directed flow of air that swept along any loose spores. The air exited the facility through a filter at one end of the building. Occasionally, this filter had to be replaced. In this instance, the filter was removed but was not replaced, and the *B. anthracis* spores blew across the city (Meselson et al., 1994).

The official death count was 68, although many experts believe the number was far greater. Cattle died from the disease many miles away. It is not possible to estimate the concentration of spores that the victims received, but the extent of death from a simple accident demonstrates the potential of *B. anthracis* as an agent of biowarfare.

How can anthrax be detected? Anthrax is generally considered a disease of cattle and sheep, which occasionally infects a human in close contact with an infected animal. For that reason, it has been referred to as wool sorter's disease. This is also the reason that *B. anthracis* is not especially difficult to obtain; it can be found in barnyards in which an outbreak has occurred. Selective media have been described for the isolation of this microorganism, and anyone with a basic knowledge of microbiology

FIGURE 15-1 **Practice sessions such as these are important to maintain readiness in the event of a biowarfare attack.**

SOURCE: Courtesy of the U.S. Department of Defense.

can do so (Luna et al., 2009). There are three means of anthrax infection: cutaneous, gastrointestinal, and inhalation. Cutaneous anthrax is the form most commonly seen in humans (perhaps 95% of cases), typically those with close contact to an infected animal. The spores can enter through a small break in the skin and then start a localized infection (**Figure 15-2**). This is the least dangerous form; it is usually noticed and treated by a physician before the bacteria have a chance to spread. However, if the wound goes untreated, this form can be lethal.

Gastrointestinal cases of anthrax occur after consumption of food (i.e., meat) that is infected with spores and has not been adequately cooked. In bioterrorism cases, other types of food might be adulterated. Once they are swallowed, the spores can cause ulcerative lesions anywhere in the gastrointestinal tract, from the oropharynx to the rectum (Sirisanthana and Brown, 2002). The lesions may result in a fatal hemorrhage or a perforation of the bowel. Left untreated, this form is likely to be lethal; even with treatment, the outcome is often uncertain. It has been suggested that gastrointestinal anthrax is far more common than is generally believed, especially in developing countries (Sirisanthana and Brown, 2002).

The inhalation method of infection is considered the most likely for a biowarfare attack (**Figure 15-3**). It has been estimated that a dose of anywhere between 8,000 and 50,000 spores is an infectious dose. Two major obstacles block effective dispersal of anthrax spores. The first is the ability to mill the spores to a size that is no more than 2–3 microns. The spores are slightly smaller than 2–3 microns,

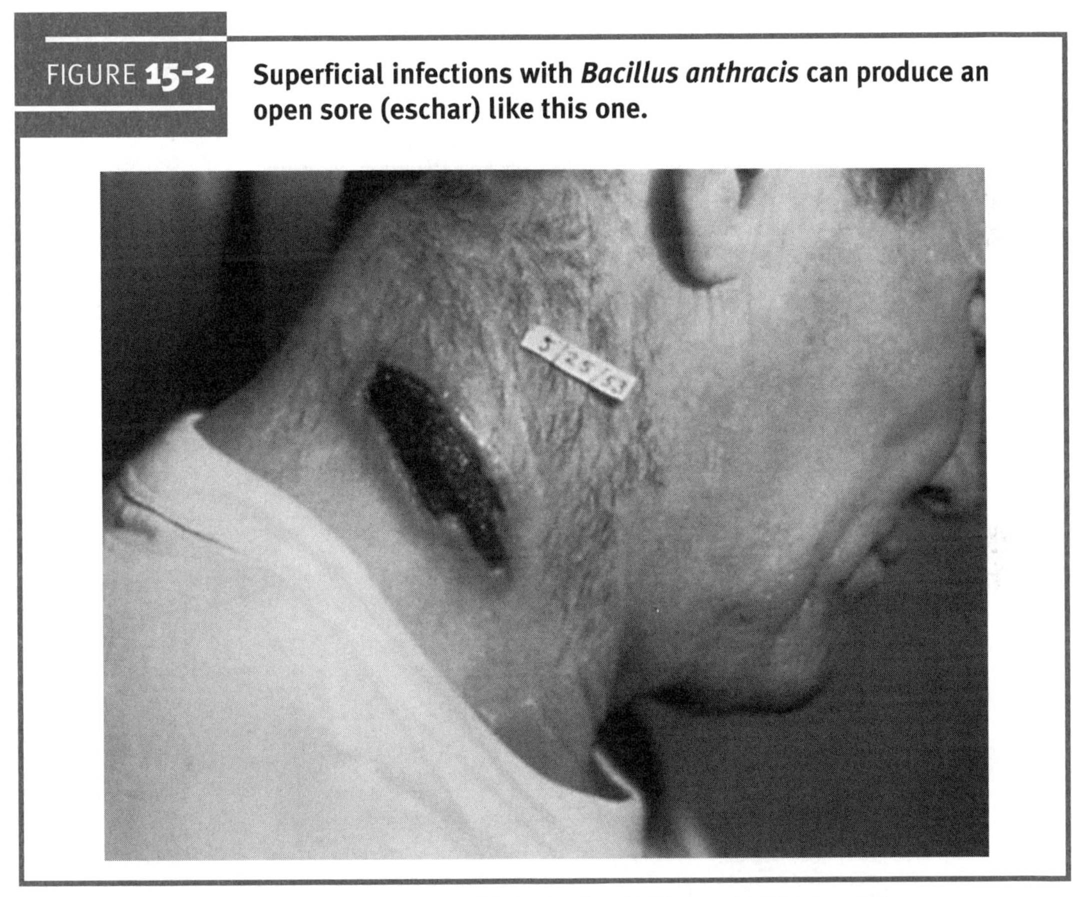

FIGURE 15-2 **Superficial infections with *Bacillus anthracis* can produce an open sore (eschar) like this one.**

SOURCE: Courtesy of the Centers for Disease Control and Prevention. (http://phil.cdc.gov/phil/details.asp). #1934.

but the associated medium causes crude preparations of spores to have a much larger diameter. At the 2–3 micron size, the dispersed spores can enter the bronchi and then the alveoli, where they can then invade the bloodstream, lodge in the lymph nodes, and germinate.

The second obstacle is related to the first. Once milled, the spores may tend to stick together in bunches. A suitable dispersant is needed so that the spores will dissociate from each other and behave as individual particles once they are released into the air. The difference that this can make is illustrated by another attack with anthrax. Shortly after the terrorist attacks of September 11, 2001, letters began arriving at the offices of several notable people in which anthrax spores were enclosed. At first the preparations were rather crude, but over time, the milling improved substantially. The last fatality was an elderly woman who had no notoriety or link to the other victims. It was hypothesized that the spores were so well crafted that one of her letters had become contaminated when it crossed in the mail with an anthrax letter, and that, along with her frail health, had been sufficient to produce an infectious dose. After a while the letters stopped, and investigators eventually linked them to a disgruntled scientist at a military research center.

Once infected in this manner, the patient can be treated with antibiotics such as ciprofloxacin. In many cases (such as the 2001 letter attacks), the

FIGURE 15-3 **During a case of inhalation anthrax, an X-ray of the lungs will show a mediastinal widening.**

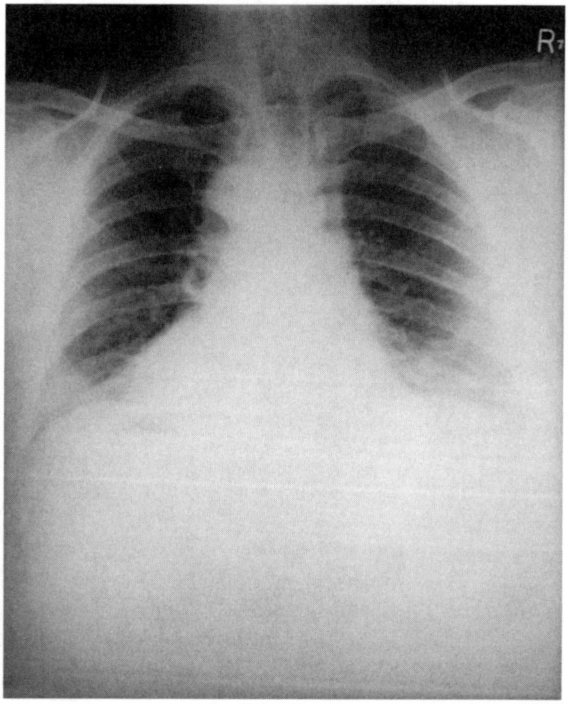

SOURCE: Courtesy of Arthur E. Kaye, the Centers for Disease Control and Prevention. (http://phil.cdc.gov/phil/details.asp). #5146.

strain is effectively treated with penicillin. However, it is not too difficult to introduce some antibiotic-resistance genes into *Bacillus* species, so prudence dictates using the antibiotic that is most likely to be effective. Prolonged therapy is required, in some cases as long as 60 days. Speed is of the essence because the survival rate associated with anthrax decreases markedly after respiratory symptoms appear. There is no need for quarantine of infected people because person-to-person transmission is unlikely (Oren, 2009).

Anthrax attacks are occasionally suspected when a "white powder" is found in an unusual location. First responders require a rapid method to determine whether the material is biological. Other substances, such as talcum powder, are often used as part of a hoax. Antibody-based, rapid tests have been developed as a presumptive test for *Bacillus* spores. Positive identification of the microorganism is performed using polymerase chain reaction (PCR) analysis.

Salmonellosis

Salmonella typhimurium (a subdesignation of *S. enterica*) causes severe gastroenteritis. It is not as serious as infection with *S. typhi*, the causative agent of typhoid fever. However, *S. typhi* has not been used for biowarfare, whereas *S. typhimurium* has. This may be due to the ease with which *S. typhimurium* can be obtained. In one notorious case

(**Box 15-3**), the bacteria were easily obtained from a commercial culture collection. *Salmonella*, as has been noted in this text, frequently contaminate chicken. Anyone with even a rudimentary knowledge of microbiology should be able to isolate and cultivate *Salmonella* (**Figure 15-4**). Distribution would be probable in either food or water. In an industrialized society with good medical care, an attack of this type would make many people sick, but the mortality would likely be low. Nevertheless, such an attack would create panic.

Plague

Yersinia pestis is the causative agent of the plague (or "The Black Plague" or "The Black Death"). It is a Gram-negative bacillus that can be easily cultivated in a microbiology laboratory (**Figure 15-5**). While often identified with the Middle Ages and the huge mortality that it caused at that time, it is still around today, although it no longer causes pandemics.

In the United States, the plague is a rare disease (perhaps 10–15 cases in a typical year) but is not unknown. The areas where the disease is endemic

BOX 15-3 Salmonella in Wasco County

The largest bioterrorist attack in the United States in terms of the number of people sickened occurred in Wasco County, Oregon, in 1984. About 3 years earlier, followers of the Bhagwan Shree Rajneesh, a cult leader, started a community on a ranch in Wasco County. The population grew quickly, and their political influence also increased. Some of the changes they proposed were met with suspicion by their neighbors, and an antagonistic attitude prevailed. The Rajneesh cult focused on winning the November 1984 county elections in order to consolidate its political power. Fearing that they would not have enough votes, some members of the cult conspired to sicken opposition voters to prevent them from going to the polls.

The group obtained a strain of *Salmonella typhimurium* and grew batches in a makeshift laboratory. They tested the plan by first poisoning some local officials with glasses of water spiked with *Salmonella*. This worked well and resulted in both becoming ill and one being hospitalized. They also tried to contaminate produce in grocery stores, and they applied bacteria to the door handles in public buildings, but these measures were ineffective. During September and October of that year, they visited ten restaurants in The Dalles, the largest town in the area. *Salmonella* was surreptitiously sprinkled over ingredients in salad bars and introduced into bottles of coffee creamer and salad dressing. (Investigators later determined that *Salmonella* was in the salad dressing, but not in the dry mix used to make it.) This last attack worked well, resulting in 751 cases and 45 hospitalizations. Fortunately, no one died (Torok et al., 1997).

The ultimate plan was, apparently, to poison the area water supply. However, the magnitude of the food poisoning episode attracted state authorities, and the final plan was not carried out. Initial explanations for the outbreak focused on food handlers as the source, although the outbreak was confined to Wasco County, and there was no clear link among the 10 restaurants. A survey of the victims found that the salad bars were the most likely common source of infection (relative risk = 7.5). Local citizens blamed the Rajneesh cult, and continued scrutiny by media and law enforcement finally resulted in a search warrant.

Investigators found the crude laboratory, the *Salmonella typhimurium* strain that had been used, and documentation of the attacks. The strain was identical by biotype and genotype with the isolates from the victims. Members of the cult also testified to investigators about the role of cult leaders in the plot. Significantly, investigators also found a vial of *Salmonella typhi* (causative agent of typhoid fever) that they had obtained from the American Type Culture Collection. According to information supplied by cult members, this strain was also to have been used in attacks.

The cult leaders were arrested and convicted on charges of assault (for the initial attacks) and for product tampering (for the restaurant poisonings) as well as other serious charges. Notably, the mass poisoning did not achieve the desired effect because area voters turned out in high numbers on Election Day and defeated the cult's candidates. The cult disbanded soon after. This attack and the anthrax attacks of 2001 are the only documented cases of bioterrorism in the United States. It demonstrates that individuals with very little specialized training can successfully propagate pathogenic bacteria and use them in a crude, yet effective, attack.

FIGURE **15-4** **Many biowarfare agents are uncomplicated to obtain and to cultivate. *Salmonella* can be easily cultivated to high concentrations using relatively simple media and equipment, demonstrating that even novices can utilize it for bioterrorism.**

are known (**Figure 15-6**). In these typically desert or mountainous regions, the microorganism is harbored by rodents such as desert rats, prairie dogs, or chipmunks. Fleas from these animals can then bite a human and pass along the disease. Occasionally the fleas infest a pet dog or cat and then the human owner. Therefore, most cases are typically among campers, hikers, and hunters who enter the habitats of these wild animals. It is a worldwide disease, although the vast majority of the cases (and the associated deaths) occur in Africa. In 2003, more than 2,100 cases and 182 deaths were recorded there.

Because the endemic areas are well known, it is not especially difficult to isolate and cultivate *Y. pestis*. Weaponizing the bacteria is, however, more difficult. The bacteria normally enter the bloodstream through a flea bite, and in this manner of introduction it is very infectious (**Figure 15-7**). But growing swarms of fleas infected with plague is a difficult task, and distributing them is equally difficult. During World War II, the Imperial Japanese Army created an army unit (the infamous Unit 731) to develop biological warfare agents and techniques. Their main weaponized system was plague, and to produce weapons, they created large rat

FIGURE **15-5** ***Yersinia pestis*** **bacilli in a blood smear. Note the peculiar "safety pin" appearance of the cells. This is not due to a spore, since they are not spore formers. Instead, it is a stain-opaque portion of the cell.**

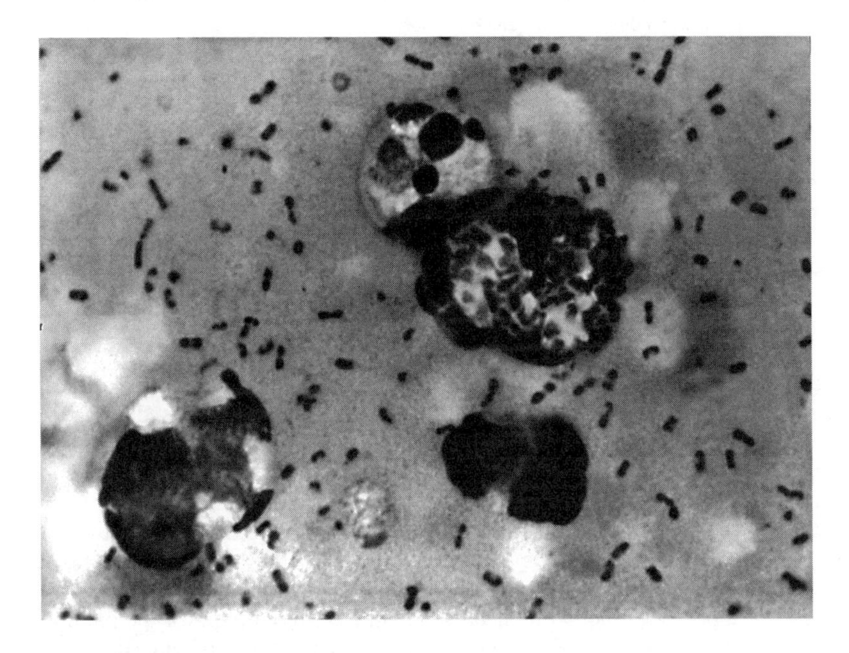

SOURCE: Courtesy of Margaret Parsons and Dr. Karl F. Meyer, the Centers for Disease Control and Prevention. (http://phil.cdc.gov/phil/details.asp). #2063.

FIGURE 15-6 ***Yersinia pestis* can be found in many countries around the world.**

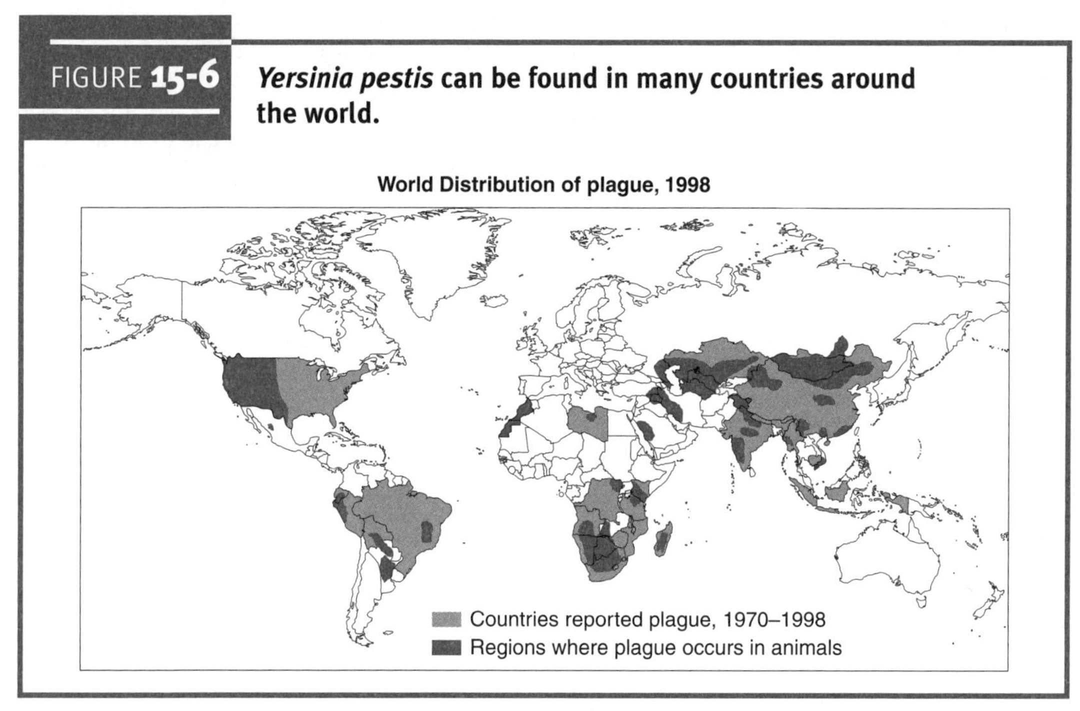

SOURCE: Courtesy of the Centers for Disease Control and Prevention.

farms holding infected rats. At certain times, they would vacuum the fleas off the rats and concentrate the fleas into a bomb casing. The bomb could then be dropped over an enemy village and disperse the fleas in midair, allowing them to drift down onto the populace. Despite the obvious planning and preparation of this weapon system, the actual attacks were apparently less than successful, killing very few people.

The most common means of acquiring a plague infection is through the bite of an infected flea. The fleas are typically found on rats and other rodents, and after the rodent dies, the carcass is disposed of by a human. During this time, it is possible for the fleas to jump off the dead host onto the human host and continue a blood meal. This leads to the **bubonic form of plague**, in which the plague bacillus travels to a lymph node where it replicates at high levels and causes such a severe inflammation that the node becomes greatly enlarged. This is termed a bubo, and provides the name of bubonic plague (**Figure 15-8**). In some cases, the bubo becomes so distended that it ruptures and creates an open sore.

The incubation period for the illness is between 2 and 6 days, after which the infected person experiences fever, chills, headache, and extreme exhaustion. These symptoms are not particularly diagnostic, but the development of a bubo is a key finding. It is possible for the bacillus to spread directly through the bloodstream without forming a bubo. This is called the **septicemic form of plague**. Again, it is caused by flea bites or by contact with infected materials (as in a hospital setting) in which the bacillus enters through breaks in the skin.

The pathogen is carried by the bloodstream throughout the body. If it colonizes the lungs, a severe pneumonia occurs. This is an especially dangerous condition for the patient and creates a **pneumonic form of plague** that is highly contagious. Under these conditions, an epidemic might start because it does not depend on the rodent/flea vector but passes easily from person to person.

FIGURE 15-7 **Fleas were used by the notorious Unit 731 to carry plague bacteria (*Yersinia pestis*).**

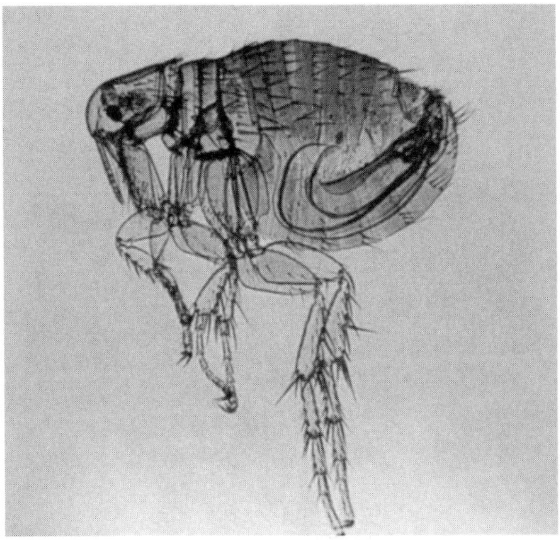

SOURCE: Courtesy of the Centers for Disease Control and Prevention. (http://phil.cdc.gov/phil/details.asp). #5451.

A patient history that includes travel to an endemic area is also important, especially if the case occurs in an area where plague is never seen. Sometimes a domestic cat brings the fleas that infect the humans. Cats are natural hunters of rodents and will kill them, causing the fleas to jump onto the cat. People who enjoy hunting, trapping, camping, and other outdoor activities are at a greater risk.

Death can be rapid if the plague is untreated, and even in the United States the fatality rate is about one in seven cases (14%). Fortunately, rapid tests for plague antigens are available, and many useful antibiotics are effective, such as gentamicin and streptomycin. Pneumonic plague patients should be isolated until at least 4 days of antibiotic therapy have been completed (Oren, 2009).

One of the diagnostic features of *Y. pestis* is its peculiar staining pattern. Both ends of the bacillus stain well, but not the center, giving the cells a "safety pin" appearance. A direct fluorescent antibody test is available for rapid diagnosis of the infection. Serological conversion confirms the disease.

Smallpox

Variola major virus is the causative agent of smallpox, a disease that has been eradicated after a concerted international effort (**Figure 15-9**). In its original form, it exists in only two repositories, one in the United States at the Centers for Disease Control and Prevention (CDC) in Atlanta and one in the State Research Center of Virology and Biotechnology VECTOR in Koltsovo, Russia. The last acknowledged case of smallpox to have been acquired from a human-to-human transmission was in Somalia in 1977. The disease was pronounced eradicated worldwide by the World Health Organization in 1980 (Barquet and Domingo, 1997). The Variola genome has been sequenced completely, and the sequence has been published (Shchelkunov et al., 1995). Concern remains, however, that a determined terrorist could rebuild the genome, perhaps using a related virus as a starting point and

FIGURE 15-8 **Bubonic plague takes its name from the appearance of buboes, in this case on the leg of this victim.**

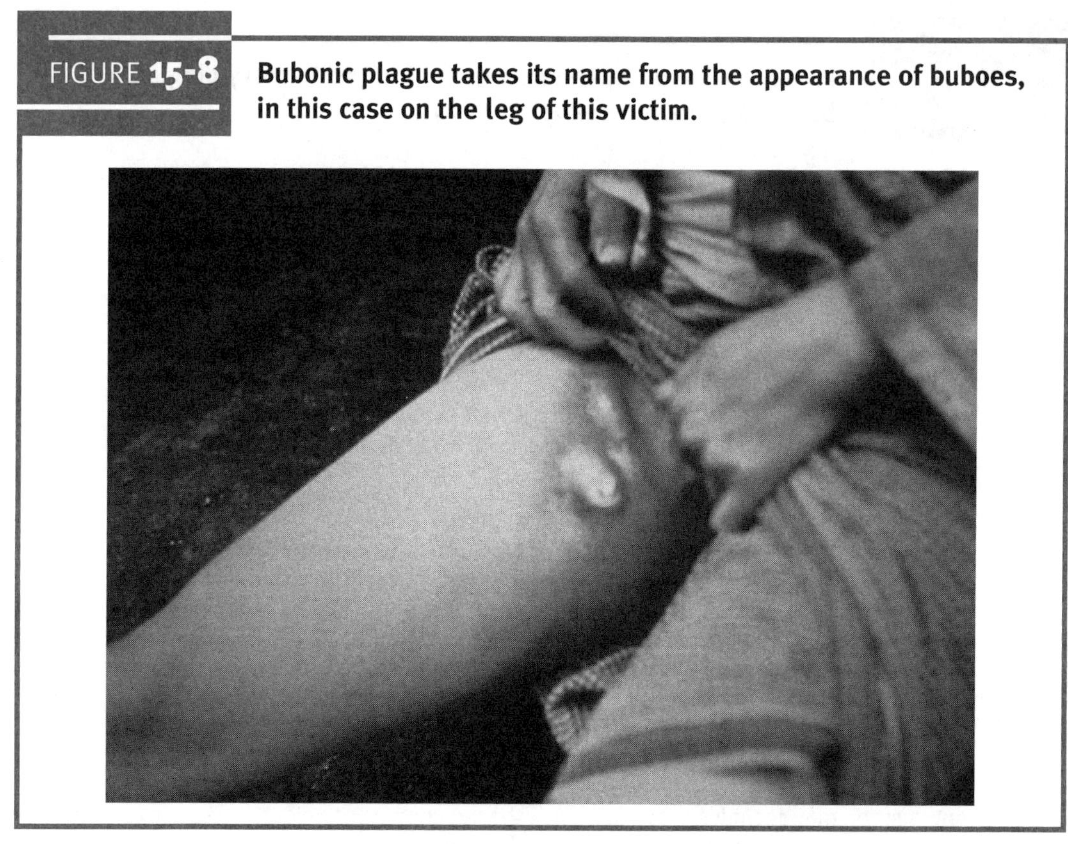

SOURCE: Courtesy of the Centers for Disease Control and Prevention. (http://phil.cdc.gov/phil/details.asp). #2047.

developing a potent weapon. Molecular technology is becoming so much more advanced and less expensive that this scenario is not out of the question.

A high risk of person-to-person transmission of smallpox exists during the first week of illness, when a rash appears on the body. Transmission takes place through aerosol droplets, and relatively close contact is all that is required. For this reason, infected people should be kept in isolation until they are no longer infectious. In an outbreak, this requirement might be impossible to achieve, or it might require immense resources. Fortunately, if the smallpox vaccine is administered within 3 days of exposure to the virus, the illness may be avoided or the symptoms may be significantly less severe.

Fortunately, an excellent vaccine for smallpox is available. Although it is no longer routinely used, stockpiles are accessible if the pathogen ever reappears. It is useful to note that the lethality of smallpox among unvaccinated people can be as high as 30%. Among vaccinated people, the lethality falls to 3% (Oren, 2009). Notably, vaccination does not confer absolute protection against the disease, but it does provide a large measure of protection.

Tularemia

Francisella tularensis is the causative agent of tularemia, also known as rabbit fever (**Figure 15-10**). This microbe, a small, Gram-negative bacillus, was discussed in Chapter 14 as a vectorborne pathogen. However, it is extremely infectious; the infectious dose is probably as low as 10 microorganisms. This makes it attractive as a weapon. The bacillus can enter through mucous membranes, and the infec-

FIGURE 15-9 **Extreme case of smallpox.**

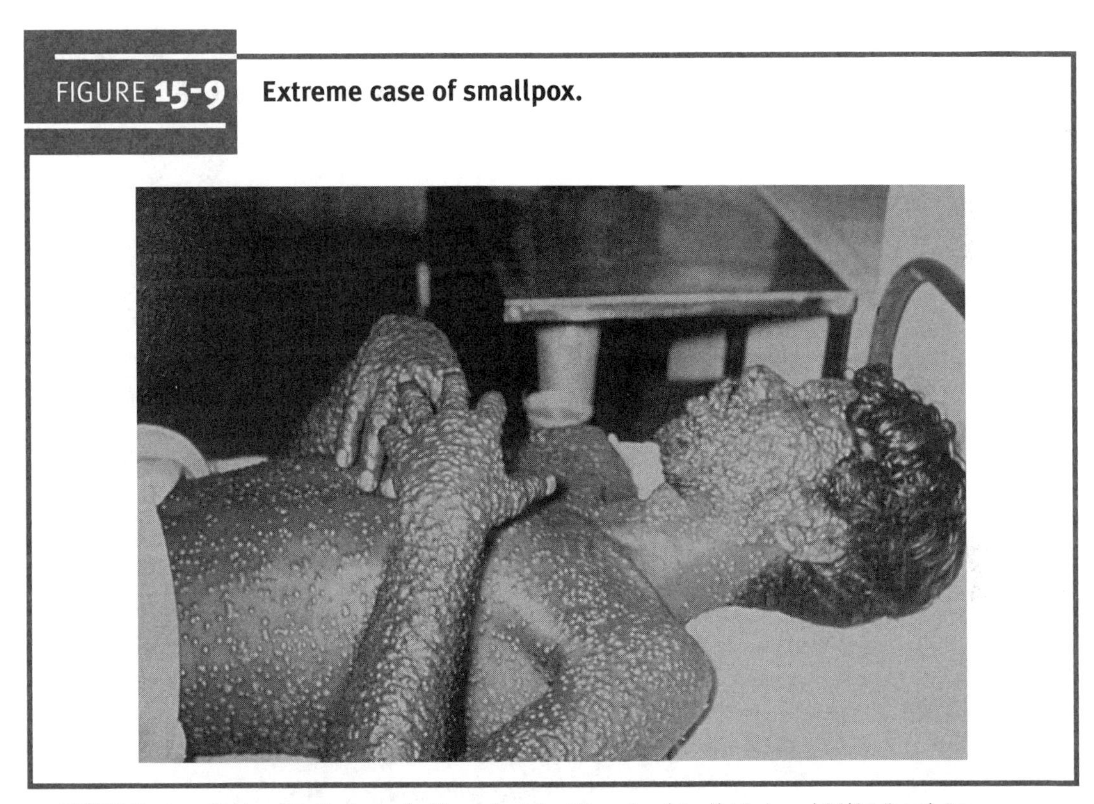

SOURCE: Courtesy of Barbara Rice, the Centers for Disease Control and Prevention. (http://phil.cdc.gov/phil/details.asp). #131.

tious dose is so low that an aerosolized droplet can easily contain enough cells to cause an infection. This illness is very rare and is usually associated with specific occupations, such as animal husbandry (e.g., raising rabbits) or butchering animals for food.

To determine whether tularemia has originated from a biowarfare attack, two key features must be considered. The first is the extent of the illness, because it is usually very rare. The second is the absence of any occupational exposure. If even a few people present with this illness without an obvious link to potentially infected animals, the probability of a bioattack is very high. Fortunately, this illness can be treated with antibiotics like gentamicin and streptomycin.

Viral Hemorrhagic Fever

The viral hemorrhagic fever viruses (e.g., Ebola, Lassa) are considered exotic and are unlikely to be used as biowarfare agents by small terrorist groups. Weaponization is possible if a government wanted to commit resources to their development. These are all RNA viruses. The newest member of this family is the Lujo virus (i.e., similar to Lassa fever virus) and has a reported lethality of 80% (Briese et al., 2009). Person-to-person transmission is possible in densely populated settings, such as hospital workers attending to a sick patient, but otherwise exposure is unlikely.

BIOTOXINS

Ricin

Ricin is a protein toxin obtained from the castor bean. There are actually two ricin proteins, I and II, and II is more dangerous. Ricin is fairly stable, although temperatures above 80°C destroy it. Ricin specifically attacks the 28S rRNA molecule of the 60S ribosomal subunit, which effectively stops protein synthesis.

FIGURE 15-10 **Smear of *Francisella tularensis* under the microscope.**

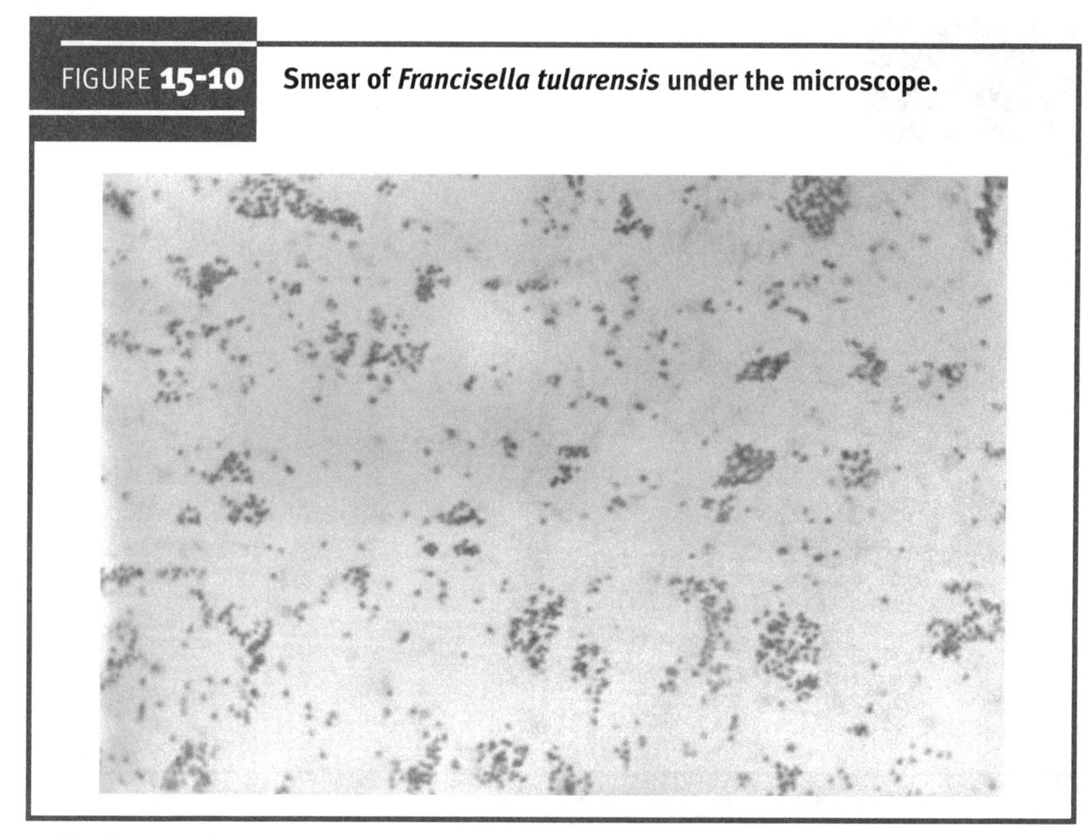

SOURCE: Courtesy of Dr. P.B. Smith, the Centers for Disease Control and Prevention. (http://phil.cdc.gov/phil/details.asp). #10526.

While castor beans have legitimate uses, ricin does not. Unfortunately, it is easy to separate ricin from the bean, requiring only chromatography to obtain a crude but effective preparation.

Reports of people chewing or eating the beans of the castor plant and then becoming ill or dying have been documented. The gastrointestinal LD50 (the dose required to kill 50% of a typical population) is estimated at 30 μg/kg of body weight. As few as 20 seeds may be enough to kill an adult (Norton, 2001). The gastrointestinal tract is heavily damaged by this illness, with foci of necrosis found throughout the bowel. As an aerosol, ricin is potentially more effective, with a LD50 of 3 μg/kg of body weight.

Ricin has developed a reputation as the bioweapon of choice for homicidal loners and disgruntled antigovernment groups (**Box 15-4**). The

BOX 15-4 **When Ricin Is Used by an Expert**

Vladimir Kostov was a Bulgarian dissident who lived in Paris in 1978. While riding an escalator in a subway station, he felt the sting of an object hitting him in the back. He later became feverish and had to be admitted to a hospital, but he recovered. The cause of his illness remained a mystery, but only for a short time. Ten days later, another Bulgarian exile, Georgi Markov, was attacked in a similar manner in London, where he worked as a writer. This time, a man with an umbrella pressed it against his leg, and a small pellet was fired into his thigh. Markov died 3 days later, but not before informing the authorities of his suspicions that he was targeted by the Bulgarian secret police and that they had poisoned him.

An autopsy revealed a 1.5-mm diameter sphere lodged in the wound on his leg. It revealed trace amounts of ricin. The volume of the partially hollow sphere was large enough to hold a toxic amount of ricin. When ricin enters a body parenterally (through a break in the skin or through mucous membranes), the LD50 is 3 μg/kg of body weight. In Markov's case, a dose of 250 μg would probably have been sufficient.

During an autopsy, Vladimir Kostov was examined, and a small projectile was also found in his back. He survived because the projectile was damaged, probably by his heavy coat, and the ricin leaked out of the little sphere before it could penetrate his skin. The relatively thin cloth of Markov's pant leg was not heavy enough to do the same for him.

Authorities speculate that the mysterious assailant used an "umbrella gun," which was essentially a compressed gas gun ("pellet gun"), and the toxic, ricin-containing spheres. Both the Bulgarian secret police and the Soviet KGB were suspected in these attacks, although definitive proof could not be established.

castor bean is easily cultivated, making it available to these individuals. In 1991, the Patriots Council, an unusual collection of aggrieved tax protestors in Minnesota, planned to kill federal officials with ricin. They theorized that ricin could be dissolved in dimethyl sulfoxide (DMSO) and then painted on surfaces that the targets would touch, allowing the ricin to enter their bloodstreams through contact. Dermal exposure is not the most effective route of intoxication for ricin, and the outcome is uncertain. Fortunately, the group was caught before it could act on its plans.

Botulism

Clostridium botulinum is itself not considered as a biowarfare agent, but the toxin it produces—botulinum toxin—is so powerful that it must be considered as a potential threat (**Figure 15-11**). The toxin causes the intoxication known as botulism and is usually associated with improperly canned food items (see Chapter 10). This is because *C. botulinum* is an obligate anaerobe and a spore former. The spores germinate in a closed container like a can and grow on the nutrients inside, producing a potent toxin.

Clostridium botulinum is relatively easy to find and not difficult to cultivate, although anaerobic conditions must be maintained. When conditions become disadvantageous, it forms a subterminal spore that is resistant to environmental insult. The spores of *C. botulinum* might be used to contaminate a food or water source (ingestion route). In these cases, person-to-person transmission is unlikely, and modern sanitation facilities should remove any spores that exit in the feces. Botulism can also be successfully treated with antibiotics, and the disease is relatively easy to diagnose.

The purified botulinum toxin concerns public health authorities more than the microbial infection. It is so powerful that the LD50 has been estimated at 0.003 μg/kg of body weight. That is, for a 150-lb (68-kg) person, a dose of only 0.2 μg of toxin is likely to be lethal. It would not be difficult to place 10 times as much toxin in a large number of small projectiles and disperse them in a crowded setting. However, even the addition of the toxin to food or water would result in a high death rate. The toxin is very stable. If it were used as a biowarfare agent, botulinum toxin exposure would be difficult to diagnose because a large group of affected people would not have the classic history of eating a home-canned product. It would probably be diagnosed as an intoxication, but the origin might remain elusive until too much time passed to find it. Once the toxin gets into the body, it damages nerves and leads to a flaccid paralysis, often starting in the head. The treatment involves the use of antitoxin.

THE PUBLIC HEALTH RESPONSE

In the wake of the anthrax letter attack of 2001, the federal government created the Department of

FIGURE 15-11 **Colonies of *Clostridium botulinum*. This is an obligate anaerobe and is somewhat difficult to cultivate. Its toxin is extremely dangerous.**

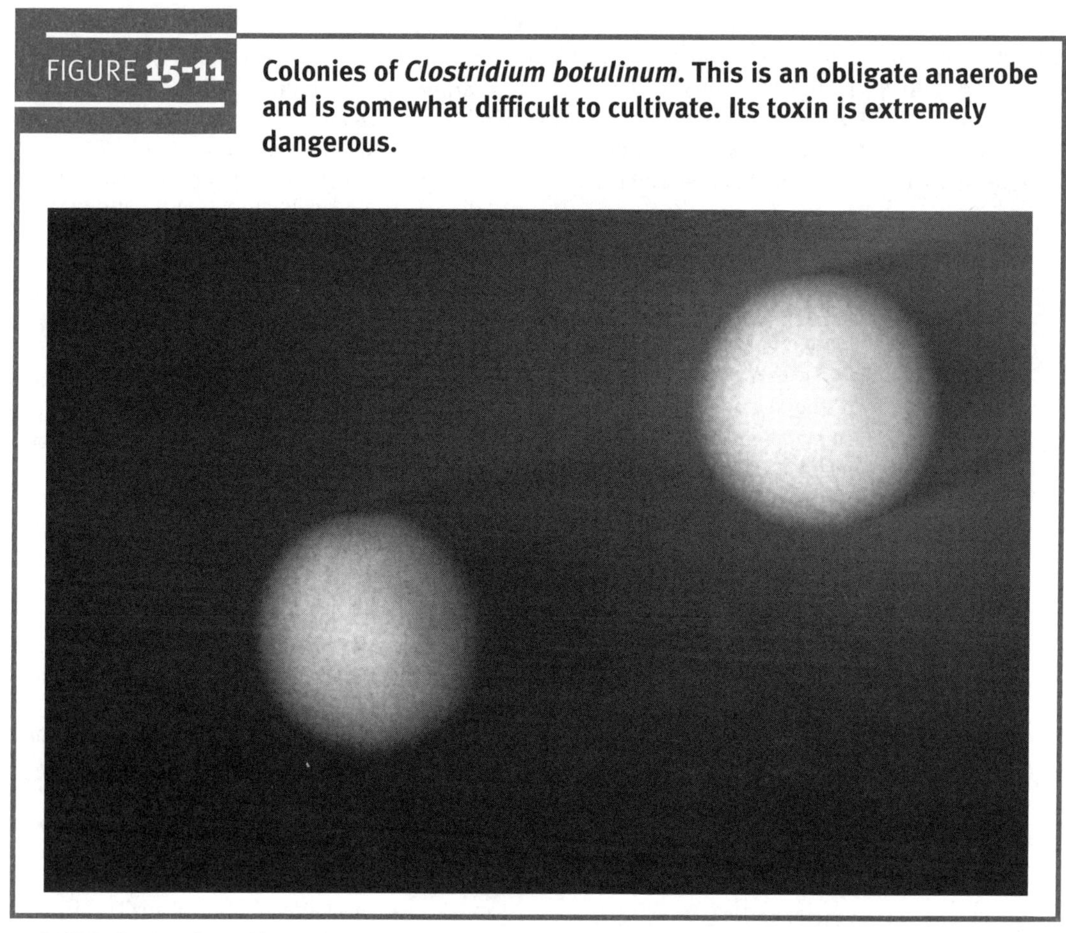

SOURCE: Courtesy of Dr. Holdeman, the Centers for Disease Control and Prevention. (http://phil.cdc.gov/phil/details.asp). #3868.

Homeland Security. They provided funds for local health departments to enable them to respond to local biological problems. Local health departments are responsible for surveying their populations for monitoring incidence of infectious disease. In particular, a major burden rests with the local medical examiner to identify cause of death and whether that death constitutes a murder (Nolte et al., 2004).

In the event of a biowarfare attack, public health department surveillance should both detect and identify the pathogen quickly. If a bacterial pathogen is used, the description of its antibiotic resistance profile should be rapidly distributed. Surveillance and good epidemiological analysis are key, because a small yet significant rise in a specific illness may herald a biowarfare event.

In one example, two individuals, a husband and wife, entered a hospital in New York City in late 2002 and were diagnosed with an infection with *Yersinia pestis* (plague). Following soon after the terrorist attacks of 2001, including the anthrax letter attacks, the public health authorities were especially cautious (Perlman et al., 2003). After all, some of the anthrax letters went to addresses in the city.

The problem was not that people in the United States contracted plague; there are a few cases every year. The problem was that the cases were "out of place." Plague does not occur in New York City or anywhere east of the Mississippi River. Therefore the discovery of only two cases was cause for alarm. The victims were tourists visiting from New Mexico, where plague is not unknown. In fact, New Mexico has the highest number of reported cases of plague. A search of the couple's property yielded rats with fleas. This is not entirely surprising because rodents are attracted to homes where food scraps might be found. *Yersinia pestis* was cultured from both the rats and the fleas, and the mystery of plague in New York City was solved.

The response to a biowarfare attack depends on the specifics of the attack. The response may include immunization of the population at risk or distribution of antibiotics if a bacterial agent is used. Identification of the carrier of the bioattack agent (e.g., airborne, waterborne, etc.) is necessary to educate the public on risk avoidance and to focus remediation efforts effectively.

While most of the examples presented in this chapter involve relatively small attacks undertaken by scientifically unsophisticated groups or individuals, an attack could certainly be conducted by a more developed country. Therefore, it is important to determine whether biowarfare agents have been genetically modified to increase their pathogenicity, to mask them from existing vaccines, or to give them antibiotic resistance.

It has been suggested that the Soviet Union tried to create a hybrid virus from smallpox and Ebola virus, two of the most feared viruses (Alibek, 1999). Engineering of antibiotic-resistant bacterial strains and other genetically engineered pathogens has also been suspected. The technology necessary to genetically engineer microorganisms is now commonplace, and it is conceivable that other countries might produce strains with enhanced pathogenicity or that are unrecognized by the immune system, even after vaccination.

A technological race between the opportunity to use a bioweapon and the means to detect and defeat that weapon will remain a grave concern for public health officials. A technological advantage in science is imperative to provide a credible deterrent to their use.

QUESTIONS FOR DISCUSSION

1. Assuming you were a terrorist, which of the biowarfare weapons would you use? Why?
2. In your own community, do you believe that the public health authorities are prepared for a biowarfare attack?
3. Why would you check other animals if you were expecting a biowarfare attack?
4. Which traits could be genetically engineered into a pathogen to make it a more effective biowarfare agent?
5. What are the advantages of using a biowarfare agent versus a conventional explosive device?
6. Why would you use a high morbidity weapon instead of a high mortality weapon?
7. Which of the biowarfare agents mentioned here could be distributed in the water supply?

References

Alibek, K. 1999. *Biohazard*. Random House, New York.

Barquet, N., and P. Domingo. 1997. Smallpox: the triumph over the most terrible of the ministers of death. Ann. Internal Med. 127: 635–642.

Briese, T., J.T. Paweska, L.K. McMullen, S.K. Hutchison, C. Street, G. Palacios, M.L. Khristova, J. Weyer, R. Swanepoel, M. Egholm, S.T. Nichol, and W.I. Lipkin. 2009. Genetic detection and characterization of Lujo virus, a new hemorrhagic fever-associated arenavirus from southern Africa. PLoS Pathog. 5: e1000455. doi:10.1371/journal.ppat.1000455.

Caudle, L.C. 1997. The biological warfare threat, chapter 21, p.451–466. *In* Medical aspects of chemical and biological warfare. Department of the Army, Office of The Surgeon General, Borden Institute.

Hawley, R.J., and E.M. Eitzen. 2001. Biological weapons—a primer for microbiologists. Annu. Rev. Microbiol. 55: 235–253.

Luna, V.A., J. Gulledge, A.C. Cannons, and P.T. Amuso. 2009. Improvement of a selective media for the isolation of *B. anthracis* from soils. J. Microbiol. Methods 79: 301–306.

Meselson, M., J. Guillemin, M. Hugh-Jones, A. Langmuir, I. Popova, A. Shelokov, and O. Yampolskaya. 1994. The Sverdlovsk anthrax outbreak of 1979. Science 266: 1202–1208.

Mock, M., and A. Fouet. 2001. Anthrax. Annu. Rev. Microbiol. 55: 647–671.

Nolte, K.B., R.L. Hanzlick, D.C. Payne, A.T. Kroger, W.R. Oliver, et al. 2004. Medical examiners, coroners, and biologic terrorism. Morbid. Mortal. Weekly Rep. 53: 1–27.

Norton, S. Toxic effects of Plants, in: *Casarett & Doull's Toxicology–the basic science of poisons*. C.D. Klaassen (ed.). McGraw-Hill, New York. 2001. (Ch 27) pp. 965–976.

Oren, M. 2009. Biological agents and terror medicine, chapter 12, p. 195–221. In S.C. Shapira, J.S. Hammond, and L.A. Cole (eds.), Essentials of terror medicine. Springer, New York.

Perlman, D.C., R. Primas, B. Raucher, R. Lis, B. Weinberg, et al. 2003. Imported plague—New York City, 2002. MMWR 52: 725–728.

Shchelkunov, S.N., R.F. Massung, and J.J. Esposito, 1995. Comparison of the genome DNA sequences of Bangladesh, 1975, and India, 1967, variola viruses. Virus Res. 36: 107–118.

Sirisanthana, T., and A.E. Brown. 2002. Anthrax of the gastrointestinal tract. Emerging Inf. Dis. 8: 649–651.

Torok, T.J., R.V. Tauxe, R.P. Wise, J.R. Livingood, et al. 1997. A large community outbreak of Salmonellosis caused by intentional contamination of restaurant salad bars. J. Amer. Med. Assoc. 278: 389–395.

CHAPTER 16

SURVEILLANCE, MONITORING, AND COMMUNICATION

LEARNING OBJECTIVES

- Define monitoring, surveillance, and epidemiology as they relate to public health.
- Define incidence and prevalence.
- Show how odds ratios and relative risks are calculated and when they are appropriate to use.
- Distinguish between active and passive surveillance, and give examples of each.
- List the types of disease outbreaks and the critical differences between them.

KEY TERMS

- Active surveillance
- Attack rate
- Case
- Cohort study
- Controls
- Endemic
- Epidemic
- Epidemiology
- Event
- Incidence
- Monitoring
- Mortality or lethality rate
- Nonevent
- Odds
- Pandemic
- Passive surveillance
- Prevalence
- Primary cases
- Prospective study
- Secondary cases
- Sporadic
- Surveillance

INTRODUCTION

When people become ill, they eventually go to a physician. Hopefully they are treated effectively, and they resume their normal lives. The disease process by which microorganisms invade the body, multiply, and cause symptoms can be studied, and these worthy efforts have been responsible for relieving a great deal of human suffering. But this does not provide an explanation of what caused that person to become ill in the first place, how many more people might be at risk, or what can be done to avoid this microbe in the future (**Box 16-1**).

To accomplish these other goals, reliable data are needed, and that is the objective of monitoring, surveillance, and epidemiology. **Monitoring** refers to the examination of inanimate objects for the presence of pathogens. For instance, food and water are routinely examined for the presence of microbes. Trace-back analysis is also a form of monitoring. In this case, an outbreak has already occurred, and it is the task of the public health authority to find out where things went wrong and who else is at risk from this episode. This method

BOX 16-1 The Monkeypox Outbreak of 2003

There are a number of viruses in the pox virus family. The most feared is the dreaded smallpox, which has now been eradicated. Cowpox virus is the virus that Jenner used to create his vaccine against smallpox; the closely related vaccinia virus is now used to genetically engineer other therapeutic aids. During the regime of Saddam Hussein in Iraq, his biowarfare program included work with camelpox virus. This caused concern because smallpox and camelpox are genetically similar (Gubser et al., 2004).

It is not unusual for these viruses to infect other mammalian species; this tendency provided the basis of the cowpox vaccine. Cowpox also infects a number of other wild and domesticated animals, although the likelihood of subsequent spread of infection from these other animals is probably not high. Unfortunately, data are lacking on most of these potential sources of infection.

Monkeypox was first described in the 1970s. As the name indicates, it infects monkeys, which are primates. Considering that fact, there was great concern that monkeypox could be used by nefarious regimes to recreate smallpox (Douglass and Dumbell, 1992). The two viruses are not especially close phylogenetically (Gubser et al., 2004; Shchelkunov et al., 2001); the most homologous region is only 96.3% identical to smallpox. However, a concerted research project could conceivably recreate smallpox using the sequence data.

In 2003, some 53 people in Indiana, Illinois, and Wisconsin became ill with symptoms of a high fever and a papular rash (Melski et al., 2003). This outbreak of monkeypox was unusual because none of the patients had contact with monkeys or other primates or had traveled to Africa. Their patient histories, collected in great detail, showed that the majority had had contact with prairie dogs, an exotic pet. A few clearly contracted the illness through contact with one of the sick people, demonstrating person-to-person transmission. Another person became ill from a pet rabbit that had had contact with a prairie dog at a veterinary clinic, demonstrating the infection of other mammalian species. Trace-back analysis found that the prairie dogs came from a distributor that housed the animals close to a shipment of Gambian giant rats. The giant rats were imported by a company in Texas that received them from Ghana, and there was concern that many other animals had been exposed to the virus. The giant rats had probably become infected in Ghana and had subsequently infected prairie dogs in the nearby enclosures. These animals had passed the virus to humans. The risk from exotic pets is clear.

The surveillance of this outbreak allowed the public health authorities to move decisively to end the problem. Based on the severity of the outbreak and the risk of further infection, the Wisconsin Department of Health and Family Services issued an order in June 2003 that stopped the importation or sale of prairie dogs as well as any animal known to have had contact with prairie dogs. The order included many other exotic species that might be kept as exotic pets as well: tree squirrels, rope squirrels, Gambian giant pouched rats, and so on.

was discussed in regard to sexually transmitted diseases in Chapter 12, but other diseases are traced back as well. **Surveillance** refers to an examination of a human population for signs of disease. For example, random phone interviews may be made after an outbreak of disease to determine the extent of the outbreak. **Epidemiology** is the science of health data analysis. The data that are collected by monitoring and surveillance must be analyzed rigorously to determine whether patterns can be discerned and whether any particular condition is a significant factor in the spread of disease. For example, if a number of people become ill at a party, the epidemiologist would want to know what they ate. If one dish was eaten by all the sick people, that would be a likely place to look for a cause of disease. Notably, this does not prove that a food caused the outbreak; it just suggests the best places to look for causes. Finding a high concentration of *Staphylococcus aureus* enterotoxin in the food dish as well as in the sick people provides proof of the source.

MONITORING AND SURVEILLANCE

Monitoring has been described in several chapters in this book. Food and water are constantly monitored for the presence of harmful microorganisms or for the presence of bioindicators that suggest contamination. In each case, a threshold value of tolerable contamination has been determined. For drinking water, the threshold value is less than 1 coliform bacterium per 100 ml. The threshold for *Clostridium botulinum* spores in canned food is zero. In most cases where these thresholds are in place, scheduling of carefully detailed sampling and

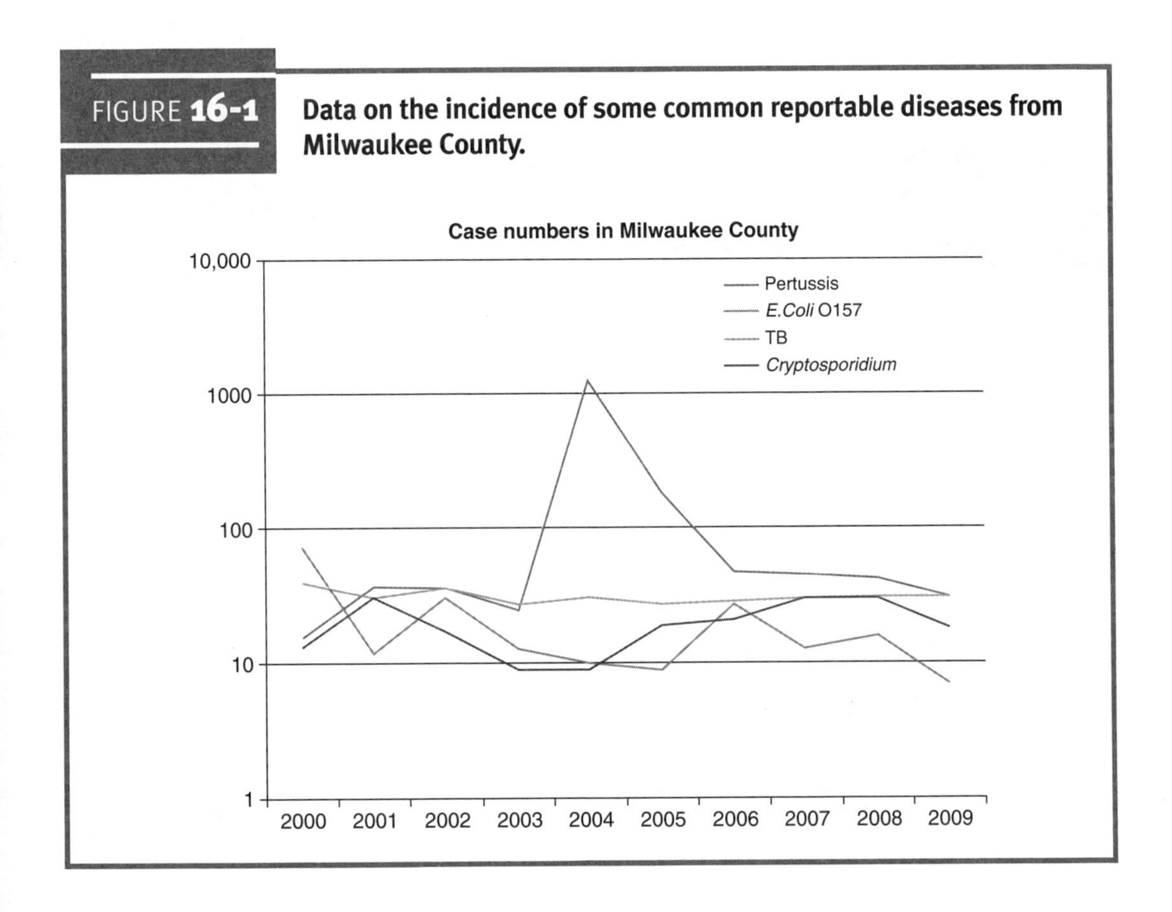

FIGURE **16-1** **Data on the incidence of some common reportable diseases from Milwaukee County.**

analysis is in place. The cost of monitoring is passed on to the consumer, but the benefits of monitoring outweigh this expense.

Surveillance can be a little more difficult because it involves a human population. Many, many polls and questionnaires attempt to discover attitudes and habits of a population. Much more information could be collected that would be of value to public health officials, but collecting these data would be far too intrusive.

Public health surveillance is concerned only with the incidence of disease and the human factors (e.g., environment, occupation, or personal habits) that lead to those diseases. Surveillance is divided into two types: passive and active. **Passive surveillance** refers to a means of data collection in which health authorities report specific diseases to a central collection site on a regular schedule. The notifiable disease program is an example of passive surveillance (**Figures 16-1** and **16-2**). Physicians are required by law to report certain diseases to the health authorities, who in turn report them to the Centers for Disease Control and Prevention (CDC). Some serious diseases must be reported immediately, such as an unlikely case of Ebola virus. Other diseases are reported weekly or monthly. Standard forms are used for this task, adding a necessary burden to the physician's paperwork.

Active surveillance involves a more proactive collection of data. Health authorities may respond to an outbreak of disease by searching for more cases or by collecting data from the affected population. The questions that are asked depend on the nature of the outbreak and the suspected vector for transmission (i.e., food, water, sexual contact, other contact, etc.). Active surveillance often also involves the collection and analysis of samples. For example, blood samples can be collected to determine how many people have seroconverted for Lyme disease. This means of data collection is far more expensive than questionnaire assays. Precise questions framed around a reasonable hypothesis usu-

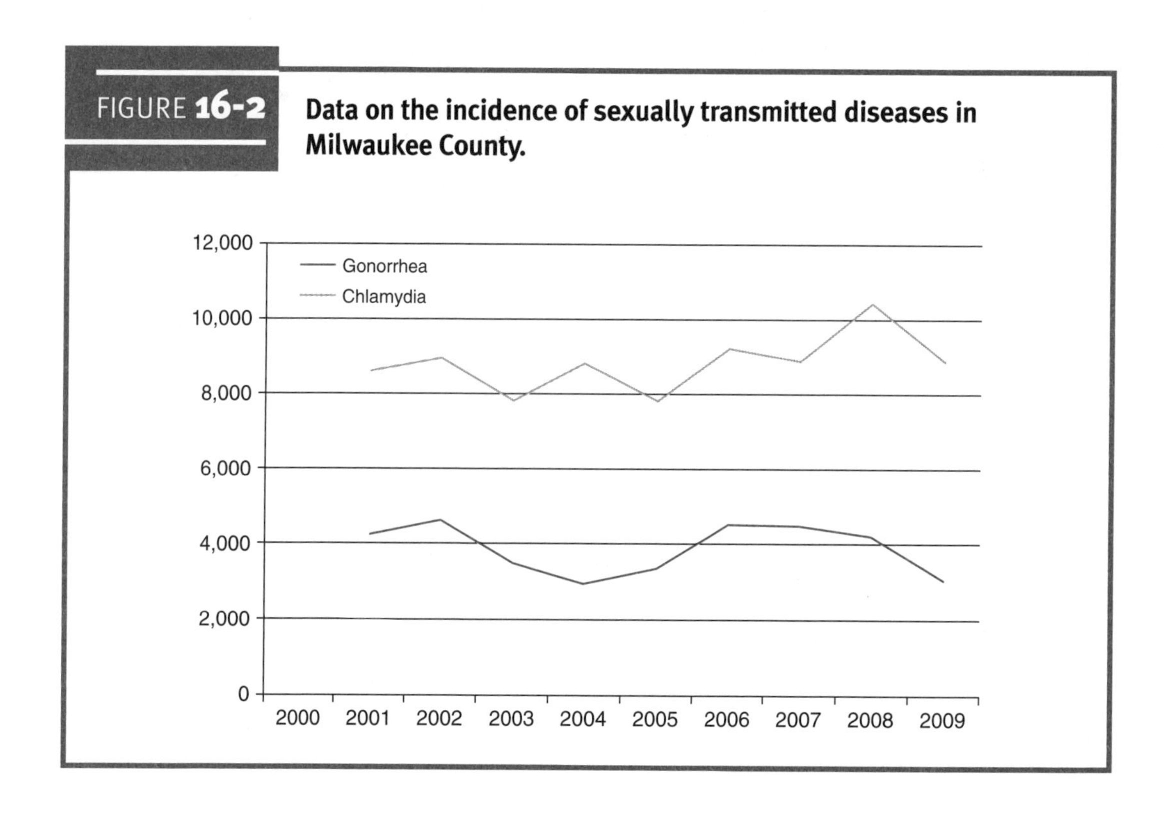

FIGURE **16-2** Data on the incidence of sexually transmitted diseases in Milwaukee County.

ally provide high-quality data, and the outcome of these measures is often fruitful.

EXAMPLES OF SURVEILLANCE SYSTEMS

Surveillance systems for disease operate at a number of government levels. Some are passive systems that largely depend on local physicians reporting data on a timely basis. The Milwaukee Health Department runs an online system called SurvNet (http://www.ci.mil.wi.us/DiseaseControlandEnv417/SurvNet.htm) that lists the total infection numbers for various diseases of interest. Other cities and counties have health departments that provide similar data.

State boards of health (which may use several different names) have their own data collection systems. The State of Wisconsin runs the Wisconsin Electronic Disease Surveillance System (WEDSS; http://dhs.wisconsin.gov/wiphin/WEDSS.htm).

By far the largest sources of information are the World Health Organization (WHO) and the CDC. WHO maintains a vast library of data on countries and diseases, including both infectious and noninfectious diseases. They maintain surveillance systems for a number of important diseases.

The CDC has an extensive library and arguably the best surveillance system for infectious disease in the world. Information is compiled for online reports and for its journals, *Morbidity and Mortality Weekly Report* (MMWR) and *Emerging Infectious Diseases* (EID). Collectively, the U.S. government runs a large number of specialized surveillance programs, some of which are very large. The Foodborne Diseases Active Surveillance Network (FoodNet) is run by the CDC for the purpose of determining the extent of foodborne illness in the United States. It uses data from 650 clinical laboratories in "catchment areas," in 10 states where foodborne diseases are actively investigated. This represents about 15% of the population and is a reliable indicator of the nation's health as well. Cases of diarrhea are examined for a potential foodborne cause, then the pathogen is identified using standard laboratory guidelines. These investigations have given us a better understanding of the scope of foodborne disease (Vugia et al., 2006).

PulseNet is another CDC initiative in which isolated pathogens are examined by pulsed-field gel electrophoresis (PFGE) to type the strain that caused a specific case of illness. Genomic typing of the pathogens has a great advantage because genotypes can be used to spot trends in disease transmission and virulence (von Belkum et al., 2001). The PFGE provides a "fingerprint" pattern of DNA bands that can be used to identify a specific strain, which may be important in identifying common sources of pathogens during outbreaks.

The National Electronic Disease Surveillance System (NEDSS) is the national model for the WEDSS system mentioned above. When complete, the system will integrate reporting systems at all levels of government and will enable faster reporting of outbreaks and better comparison of results among municipalities.

The Waterborne Disease and Outbreak Surveillance System (WBDOSS) is a repository for data on illnesses that result from all types of water-related activities, including drinking water, washing water, and recreational activities. These data also include illnesses that are not related to infectious agents, such as chemical poisonings.

The Cholera and Other *Vibrio* Illness Surveillance System (COVISS) is largely concerned with the Gulf Coast states, for good reason: *Vibrio* species occur there. This system collects data that include the relatively few cases of cholera in the United States, as well as the pathogenic species of *Vibrio parahaemolyticus* and *V. vulnificus*. Because these cases are waterborne, they are reported directly to the WBDOSS.

The National Antimicrobial Resistance Monitoring System (NARMS) is unlike other systems; it examines antibiotic resistance among bacterial isolates that are specifically obtained from domesticated animals. The idea here is that antibiotic resistance may appear first in animals and then be transferred to bacterial species that cause illness in humans. This system is managed jointly by the CDC, the U.S. Food and Drug Administration (FDA), and the U.S. Department of Agriculture (USDA).

The National Health and Nutrition Examination Survey (NHANES) is another ongoing activity that measures specific aspects of the health of the U.S. population. These reports are quite extensive and are run for a specific number of years, after which the data are compiled and analyzed. The NHANES is not directly concerned with infectious disease but rather the overall health of the population. It is an active surveillance system that uses both surveys and physical examinations.

The National HIV Behavioral Surveillance System (NHBS) was created to monitor behaviors that increase the risk of becoming infected with HIV (Lansky et al., 2009). This is particularly true for the IDU (injectible drug user) population. Needle sharing among drug users carries the risk of transmitting the HIV virus. Modifying the behavior of drug addicts to reduce needle sharing theoretically reduces the overall problem of infection. The accurate collection of information with regard to this behavior can affect health policies and can improve public health. There are many, many other systems that collect valuable data about the population and that may prove useful to specific studies.

EPIDEMIOLOGY—CRITICAL ELEMENTS

For public health microbiology, the aims of epidemiology can be summarized as follows:

- Identify the focus of infection.
- Reduce the spread of illness.
- Identify community risk factors.
- Identify new pathogens and new variants of old pathogens.
- Evaluate the efficacy of treatments and preventive measures.

The accurate and complete collection of data allows these goals to be met. Their "completeness," however, is a difficult concept. A complete data set does not necessarily include thorough examination of every possible sample. Sufficient sampling and analysis provides suitable explanations of disease. If three possible causes for an outbreak are indicated at the end of an analysis, the epidemiological analysis has been a success as long as further study finds the correct answer among these three possibilities.

The first objective of an epidemiological study is to determine exactly what the problem is, that is, what constitutes a **case**. This is not a trivial matter. When there is an outbreak of diarrhea and vomiting in town, the case might be defined as someone with vomiting and diarrhea. This seems inclusive enough. Or a case might be defined as suffering from vomiting and/or diarrhea, which would include more of the sick people. If early evidence has suggested *Salmonella* contamination, the case is defined as someone with a proven case of *Salmonella*. This definition has the advantage of eliminating people who have an ordinary gastroenteritis, but it has the disadvantages of extensive testing (i.e., culturing) and increased costs. If the preliminary data are incorrect, and the pathogen is really *E. coli*, searching only for *Salmonella* causes all of the relevant data to be missed. When data and/or resources are limited, broader definitions are preferred, and local medical records are searched later. Too much data is better than none at all. If possible, samples are obtained and analyzed for the presence of pathogens. Several pathogens may concurrently infect the population, which is typical when sewage infiltrates a drinking water system.

The International Classification of Disease (ICD) is a list of all known diseases categorized with numerical codes. The list includes infectious diseases and every other medical condition known: poisoning, cancer, heart disease, arthritis, and so on. Currently, the tenth version of the ICD is being used (it must be updated on a regular basis); it is referred to as the ICD-10 (**Table 16-1**). Numerical codes are useful for a number of reasons. First, they force the physician to make a decisive diagnosis so that it can be coded. Once a numerical code is determined, the data are easily manipulated in a computerized system. Numbers are precise, and written descriptions can be less so. For example, saying that a patient has a "high fever" is different than saying that a patient has "a fever of 39°." The computer cannot distinguish these two phrases and instead selects the only common word between them: "fever." On the other hand, a chart note stating that the patient "did not have a fever" would be recognized the same way. Although numerical codes force the diagnostic decision, inaccuracy sometimes occurs when a symptom is not quite right for the code. Overall, however, the numerical system provides a better format for data entry and comparison. This system is also vital for insurance companies and for epidemiologists because it gives precise numbers for precise definitions.

TABLE 16-1 ICD-10 Examples

Code	Disease
A00	Cholera (*Vibrio cholerae*)
A05.1	Botulism (*Clostridium botulinum*)
A36.1	Nasopharyngeal diphtheria (*Corynebacterium diphtheriae*)
A54.3	Gonococcal infection of eye (*Neisseria gonorrheae*)
A60.9	Anogenital herpes viral infection, unspecified
A78	Q fever (*Coxiella burnetti*)
A96.2	Lassa fever
B17.1	Acute hepatitis C
B21.1	HIV disease resulting in Burkitt's lymphoma
B35.6	*Tinea cruris* ("jock itch")
B50.0	*Plasmodium falciparum* malaria with cerebral complications
B68.0	*Taenia solium* (pork tapeworm)

A related classification system known as the Reason for Visit Classification (RVC) has been developed (**Table 16-2**). As the name indicates, numerical values are applied to common complaints by patients (e.g., headache, constipation, malaise, etc.). Again, data collection combines what people actually are feeling with the diseases with which they are eventually diagnosed. This system provides an excellent overview of the medical concerns and problems of a population.

TABLE 16-2 Reason for Visit Classification—Examples

Code	Explanation
1015.0	Tiredness, exhaustion
1035.3	Excessive thirst
1130.1	Antisocial behavior
1205.0	Convulsions
1410.2	Sinus inflammation, infection
1455.1	Sore throat
1470.1	Coughing up blood
1890.1	Too little hair (baldness, losing hair)
2015.1	HIV
2365.0	Migraine headache
2540.0	Varicose veins
3310.0	Glucose level determination

ANALYSIS

In any description of disease, one key question is how widespread the disease is in a population. For that measurement, an **attack rate** is determined. The attack rate is simply the number of people who become ill divided by the number of people who could potentially become ill. This is not always as simple as it sounds. The number of people who are ill might only be the clinical cases—the cases that come to the attention of a physician or other health provider. However, the subclinical cases—people who have the illness but are not sick enough to see a physician—should be counted as well. Typically, the subclinical cases greatly outnumber the clinical cases, so the numerator of this fraction could change dramatically based on available data. Finding the subclinical cases is more challenging, but research methods exist to do so.

The population at risk for a disease can also be difficult to define. If it is an easily communicable disease, like the seasonal flu, then a large population can be assumed, such as an entire city or county. Census data are very helpful here. However, if it is an occupational disease, such as hospital workers who may be exposed to *Coxiella burnetti*, then only the lab workers and perhaps a few other hospital personnel should be counted in the population at risk. The vast majority of people living in that area would have no realistic opportunity to become exposed to the disease. Finding acceptable definitions of affected populations can be a contentious issue.

The **mortality or lethality rate** is defined in a similar manner: the number of people who died from the disease divided by the number who could have become infected with the disease.

Distinguishing the primary cases from the secondary cases can also be useful. **Primary cases** are attributed to a point of contamination, such as a contaminated food or the source of a biowarfare attack. **Secondary cases** result from a healthy person becoming ill after contact with a primary case, assuming that the disease is contagious. Some diseases are certainly contagious (e.g., when the flu is spread from person to person). Some diseases are only marginally contagious (e.g., when an infant with diarrhea spreads the pathogen to the mother during diaper changing). Some are really not contagious at all, such as anthrax.

By itself, the attack rate is helpful, but other measures are more useful. The **incidence** of a disease is defined as the number of new cases within a specified time period as a fraction of people at risk for catching the disease. The emphasis here is on new cases, so a time period must be specified (e.g., the incidence of tuberculosis *in 2007*, or the incidence of influenza *in December*). Determining the number of people at risk for catching the disease can be problematic. An incidence rate of 115 cases in a population of 5,145 individuals may be accurate, but expressing the number as 22.4 cases per 1,000 people is much more convenient and is still accurate. However, if 2 cases occur in a population of 267 people, for example, and this is expressed as 7.5 cases per 1,000, there were not actaully 7.5 cases, only 2. Sometimes more data should be listed so that the extent of the outbreak is clear. Finally, only people who have a reasonable chance of getting the disease should be included in the denominator. The incidence of Lyme disease in Baraboo, Wisconsin (pop. 11,000), is much different than the incidence of Lyme disease in the United States (pop. 300 million). But Baraboo residents have a realistic chance of becoming infected with Lyme disease, while the vast majority of U.S. residents do not.

The **prevalence** of a disease is slightly different. Prevalence is a measure of the active cases of disease at a single moment in time. Again, the number of infected people is divided by the population at risk of becoming infected. Prevalence thus includes the new cases as well as the older cases that are still active. A good example of this is tuberculosis, which can be latent for many years and still count as an infection, and which takes a long time to treat effectively. An old case may appear in the prevalence statistic even though it was acquired years ago and treatment started a year ago. It may be helpful to ask for the prevalence at a point in time (e.g., today) or to specify a time period (e.g., the number of cases of disease in 2008). In the latter example, prevalence differs from incidence because cases that were acquired during the previous year but are still being treated would be counted in the prevalence.

OUTBREAK DESCRIPTIONS

In describing outbreaks of disease, several terms are used. People use these terms colloquially, but they have more precise meanings.

Sporadic infections are incidences of disease that usually occur in one individual and not in a group with a common focus of infection. Very often it is unclear why one person became ill while many others around him did not. The equine encephalitis viruses (i.e., EEE, WEE, and VEE) are good examples of this phenomenon. In these cases, the ill person is thought to have certain host factors that dispose him or her to catching the illness, perhaps genetic factors or immune characteristics.

Endemic infections are illnesses that are expected in a community, and they occur in a regular and continuing pattern. There is no way to know who will become ill, but a fraction of a community is certain to become ill within a given period of time. Some people use the term "endemic" to mean that the disease is confined to one geographic area, the way cholera is found in Bangladesh but not in Vermont. In fact, cholera is endemic to Bangladesh because every year a certain number of cases can be expected there. It is not endemic in Vermont because cases never occur there.

Epidemic infections refers to the incidence of disease that is significantly higher than normally expected. This is a key distinction. An example is seasonal influenza. In a typical community, a thousand cases may occur during the winter months and no cases during the summer. That is the endemic level—very predictable. If there were 2,000 cases in a particular winter, for example, that would be 100% more than expected and would certainly qualify as an epidemic. However, 20 cases during the summer would also be epidemic because the expected number is zero. In both cases, public health authorities would suspect that something was unusual about this outbreak. Perhaps a new variant of the flu is indicated, or perhaps a new pathogen is present whose symptoms mimic those of the flu. Absolute numbers do not determine an epidemic; the variance from the expected incidence defines an epidemic.

What, then, is a "significant" increase? The above example used 100%, which certainly seems significant. But perhaps in some cases an increase of only 10% would be significant. This is a biostatistical question.

A **pandemic** infection is an epidemic that covers large areas (i.e., continents). Pandemices are rare, and the term should not be used lightly. Before the advent of modern science and medicine, pandemics of many diseases, such as typhoid, smallpox, and cholera crisscrossed the globe. Today mechanisms are in place to quickly respond to these threats and to ensure the safety of the population. However, the airborne diseases, particularly the viruses, are much more difficult to control. Another pandemic of influenza, SARS, or avian flu is entirely possible. The last great pandemic was the 1918 influenza outbreak, in which millions of people died in countries all over the world. To stop such an event from occurring again, substantial resources have been used to survey populations and to study potential pathogens. The support of vaccine development programs has also proved vital to preventing pandemics.

HOW IS RISK CALCULATED?

Sophisticated models of risk have been developed for specific pathogens. These models make assumptions about exposure to the pathogen, health of the pathogen (infectivity), numbers of the pathogen in the medium when it contacts humans, and so on (Hurst, 2007). These models are valuable for giving perspective to a problem that is very complex. Often they depend on previous outbreaks of disease and the information gained from studying the conditions at the time of the outbreak.

Populations can be studied in several ways that are both ethical and effective. One is the **cohort study**, which is also called the **prospective study**. This study makes use of a population, called a cohort. The cohort is sometimes defined in a general manner (e.g., people who live in Chicago) and sometimes very specifically (e.g., single white males from age 25 to 44). An outcome of interest is also defined, such as becoming infected with tuberculosis. The cohort is divided according to different characteristics that are of interest in causing the disease (e.g., people who drink alcohol excessively and those who do not). The health of the cohort can be followed over time. The questions then become what is the rate of disease in each group, and is there a significant difference between the two?

The incidence rate, which is simply the ratio of the sick people in the exposed group to the total number in the exposed group, is the first statistic of interest. The same calculation is performed for the unexposed group. When the two figures are compared, the result may show that the test group and the control group have the same incidence of disease, in which case the test factor has no effect on the disease. Or the incidence in the test group may be significantly higher than in the control, which would indicate that the test factor does have an effect on the disease. Or in the opposite case, the test factor would appear to have a protective effect against the disease.

Comparisons between the test group and the control group are made using the relative risk (RR) equation. The RR is calculated using

$$RR = \frac{\text{Risk (test factor)}}{\text{Risk (control)}}$$

where "risk" is the incidence rate.

For example, suppose that the test-factor group consists of 803 people, and 47 get sick during the study. The control group consists of 1,215 people, and 12 get sick. The incidence rate (risk) for the test group is 47/803, or 0.059; the incidence rate for the control group is 12/1,215, or 0.009. Then the RR is 0.059/0.009, or 6.5, indicating that this test factor does have an effect on acquiring the disease.

Cohort studies can be effective at providing evidence for hypotheses about causes and effects. They cannot by themselves prove the hypothesis, but they can provide good evidence for it.

Another way to study a population is through the **case-control** study. This is a purely retrospective study, which is dependent on something going wrong somewhere. Simply put, a group of people become ill and the opportunity arises to study it. This group comprises the cases. The **controls** are selected from people who are unrelated to the case group. The case population became ill because of some event, like eating or drinking something contaminated or being exposed to some condition (e.g., aerosolized bacteria or proximity to a sick animal). The condition is unknown, but the right questioning and data analysis are expected to reveal it. The control population may include some members who were also exposed to the contaminated thing or condition but for some reason did not become ill. Therefore some members of the control population may be associated with the infectious condition, but the case group should have a much higher incidence rate.

The appropriate calculation for this study is the odds ratio (OR). The odds ratio depends on the **odds** of an **event** happening compared to the odds of a **nonevent** happening. These terms are in bold type because they have specific definitions. The event is whatever is being studied, for example, cases of hepatitis A illness. The odds refer to the probability of something happening divided by the probability that it will not happen. Adding together the probabilities of happening and not happening always equal 1 (unity), by definition. After all, either a thing happens or it does not happen. There are no fuzzy outcomes.

The example of a coin flip is instructive. What are the odds (risk) of flipping a coin and getting heads? A coin flipped 100 times comes up heads about 50 times, and about 50 times it comes up tails. In this example, the event happening is "heads," and the event not happening is "not heads" (i.e., tails). Those are the only two choices; the coin is not going to land on its side. So the probability of the events happening is 50/100 = 0.5, and the probability of the event not happening is also 50/100 = 0.5. The sum of these two probabilities is one. But the odds of getting heads has become 0.5/0.5 = 1.

At first this does not seem to make sense. If someone asks the odds of getting heads in a coin flip, that answer is not one. But in fact, the first task has been performed correctly, and the control group of the odds ratio has been defined. After all, this was a "fair" coin. But now suppose that the task is to test a coin that has been altered to be unfair. Once again, it is flipped 100 times; heads has resulted 75 times, and tails has resulted 25 times. Calculating the odds for this coin yields 0.75/0.25, or 3. The odds ratio is the odds for the unknown coin divided by the odds for the fair (control) coin, or 3/1 = 3.

If the result had been an odds ratio of 1 or something close to it, the conclusion would be that the unknown coin is essentially identical to the fair coin. But with an OR of 3, it is correct to suspect that something has been done to the unknown coin that makes it no longer fair.

This is fine for coin flips, but what about disease? What happens when a potential risk factor is compared with a control group? Suppose that a number of people at a hotel became ill with diarrhea and vomiting. Contaminated food from the hotel restaurant is suspected. Two groups are considered: the sick people and a group of people staying at the hotel that did not get sick. The sick group included 97 cases, and 73 had eaten in the restaurant. The "not sick" group included 147 people, and 23 got sick. The odds of a sick person eating in the restaurant are 73/97, or 0.75. The odds of the "not sick" people eating in the restaurant are 23/147, or 0.16. Therefore the OR is 0.75/0.16, or 4.7, which certainly suggests that eating in the restaurant was a factor in illness.

An elevated OR is not conclusive, however. Perhaps it is coincidental. To evaluate the OR, the 95% confidence interval (CI) for the OR is calculated. Some people became sick who did not eat in the restaurant. This may be due to several factors. Perhaps some were sick with diarrhea and vomiting from an entirely different source. Perhaps some forgot that they ate in the restaurant, or for some reason did not want to admit that fact. When humans are involved, many explanations for unexpected data may be in play.

The OR is comparable to RR when the numbers of individuals being studied is high. With low cohort numbers, the two figures can diverge substantially. Consider a population of only 10 people in which these positive numbers went from 1 to 2; this would constitute a major increase. Just one extra positive result has a major impact on the result. On the other hand, if 1,000 people are in the study and the positive number goes from 1 to 2 by mere coincidence, the overall effect is much less. Studying a large population is almost always a good thing. Although it is more expensive and the record-keeping is certainly more complex, the statistical accuracy of the results more than offsets these disadvantages.

A large number of test subjects may not be readily available. For example, when studying an outbreak of hepatitis A virus, perhaps only 10 people are affected. The risk factors can still be analyzed, and an OR for various factors can still be produced, but the result will not be entirely definitive. Because the OR tells you which factors are *likely* to be important, a smaller study population means more uncertainty in the analysis phase of the study.

COMMUNICATION WITH THE PUBLIC

Most of the time, the public is content to remain unaware of the work done by public health professionals. The public is concerned that water is purified, that wastewater is treated, that food is fresh and wholesome, and that someone is watching out for disease in the community. The public is content as long as nothing goes wrong. In fact, too much information from public health authorities can pose a problem. People can become indifferent to the messages of public officials if they hear them too often or if they relate to trivial matters. Information overload should be avoided.

There are extraordinary times in which public utilities are damaged and the risk of infectious disease to the public increases greatly. During natural catastrophes (e.g., earthquake, flood, power outages, etc.) the safety of the food and water supply cannot be guaranteed. During Hurricane Katrina (2005), a large population was displaced. Finding fresh water and food for so many people while housing them in tight spaces that could breed infectious disease was a major challenge. At times like this, the worth of public health services becomes evident to all.

There is certainly an element of judgment in a public health official's priorities. New or emerging pathogens are regularly cited as threats. In the recent past, there have been suggestions that a danger exists from Borna virus (Ikuta et al., 2002), the effect of vaccination on chronic fatigue syndrome (Ortega-Hernandez and Shoenfeld, 2009), and the XMRV retrovirus on both prostate cancer (Schlaberg et al., 2009) and chronic fatigue syndrome (Lombardi et al., 2009). Clearly, public health officials cannot react publicly to all of these potential threats; they must weigh the evidence and decide whether any credible threat exists. Then they must decide how to respond. Sound judgment and a calm demeanor are important.

At times, the public must be informed that a problem has arisen. The public health authorities then need the cooperation of the public to address the problem and to protect themselves from harm. Communicating clearly and effectively with the public is also a task of the public health professional. This is particularly true for infectious disease because everyone is concerned about their risk or the risk to their families and children.

A few simple rules apply to communicating with the public:

1. Clear language should be used rather than scientific jargon. Scientific terms may sound impressive, but they may not have meaning to the public. For example, most people do not use the word "cohort." Effective communication creates greater understanding.
2. Put risk in terms that people can understand. Again, most people have no idea what an odds ratio is. In many health presentations, information is sensationalized. For example, it might be said that a certain drug will "double your risk of cancer." What is not said is that the risk is usually one in a million cases, and that with the drug the risk becomes two in a million. Big deal. Meanwhile, that new drug may be helping people with another medical condition.

   Certain behaviors or activities that increase risk should be clearly stated; however, risk need not be quantified at this point. Information should be presented calmly and clearly without scare tactics. Most people respond well to useful information. Unfortunately, however, a fraction of the population always remains unreachable or unwilling to follow advice.
3. Emphasize risk assessment. The seriousness and the character of the problem should be clearly communicated. Is this a seasonal problem, like influenza? Is it a behavioral problem, such as with STDs? Comments should be framed simply so the public understands the scope of the problem.
4. Emphasize prevention. The public can often do much to help with an outbreak of disease, if they are given useful information in an understandable format. An advertising campaign may be useful at some point, and knowing what works and what does not is very helpful.

Sometimes information must be provided to a legislative body about a current health problem, about health care alternatives, or about public health policy. In these cases, the rules of discussion are much the same as above. Testimony should be clear and jargon free. Although listeners may be very bright and attentive, they probably are not scientists. Because cost is a major factor that affects alternatives for health policy, balancing cost with efficacy is necessary when tax dollars are spent. Therefore, the best suggestion may not be accepted despite its high cost. The reality is that public health must compete with many other causes worthy of funding e.g., police and fire protection and education.

QUESTIONS FOR DISCUSSION

1. Every 10 years the U.S. census is taken. Is this active or passive surveillance?
2. Think of a disease or illness (it does not have to be infectious). Can you find a surveillance system that searches for it?
3. Review the characteristics of hepatitis B virus infection. How would you define a case of HBV?
4. Appendix 1 provides the URL for the ICD-10 codes. What is the code for mumps?
5. What is the difference between incidence and prevalence?
6. How much of an increase in illness would you require before you declared an epidemic?
7. In a case-control study, the control population (906 people) included 102 people who ate potato salad at the picnic. The case population (427 people) included 97 people who ate the potato salad. What is the OR for this study? Should the potato salad be further examined for contamination?
8. Your community is experiencing an outbreak of an exotic disease. What are three things you want to know about it before you speak to the public, and what are three things that you will tell the public?

References

Douglass, N., and K. Dumbell. 1992. Independent evolution of monkeypox and Variola viruses. J. Virol. 66: 7565–7567.

Gubser, C., S. Hue, P. Kellam, and G.L. Smith. 2004. Poxvirus genomes: a phylogenetic analysis. J. Gen. Virol. 85: 105–117.

Hurst, C.J. 2007. Estimating the risk of infectious disease associated with pathogens in drinking water, p. 365–377. In *Manual of environmental microbiology*. American Society for Microbiology Press, Washington, DC.

Ikuta, K., M.S. Ibrahim, T. Kobayashi, and K. Tomonaga. 2002. Borna disease virus and infection in humans. Front. Biosci. 7: d470–d495.

Lansky, A., A. Drake, and H.T. Pham. 2009. HIV-associated behaviors among injecting-drug users—23 cities, United States, May 2005–February 2006. Morbid. Mortal. Weekly Rep. 58: 329–332.

Lombardi, V.C., F.W. Ruscetti, J. Da Gupta, M.A. Pfost, K.S. Hagen, D.L. Peterson, S.K. Ruscetti, R.K. Bagni, C. Petrow-Sadowski, B. Gold, M. Dean, R.H. Silverman, and J.A. Mikovits. 2009. Detection of an infectious retrovirus, XMRV, in blood cells of patients with chronic fatigue syndrome. Science 326: 585–589.

Melski, J., K. Reed, E. Stratman, M.B. Graham, J. Fairley, et al. 2003. Multistate outbreak of monkeypox—Illinois, Indiana, and Wisconsin, 2003. Morbid. Mortal. Weekly Rep. 52: 537–540.

Ortega-Hernandez, O.D., and Y. Shoenfeld. 2009. Infection, vaccination, and autoantibodies in chronic fatigue syndrome, cause or coincidence? Ann. N.Y. Acad. Sci. 1173: 600–609.

Schlaberg, R., D.J. Choe, K.R. Brown, H.M. Thanker, and I.R. Singh. 2009. XMRV is present in malignant prostatic epithelium and is associated with prostate cancer, especially high-grade tumors. Proc. Natl. Acad. Sci. USA 106: 16351–16356.

Shchelkunov, S.N., A.V. Totmenin, I.V. Babkin, P.F. Safronov, O.I. Ryazankina, et al. 2001. Human monkeypox and smallpox viruses: genomic comparison. FEBS Lett. 509: 66–70.

Von Belkum, A., M. Struelens, A. de Visser, H. Verbrugh, and M. Tibayrenc. 2001. Role of genomic typing in taxonomy, evolutionary genetics, and microbial epidemiology. Clin. Microbial. Rev. 14: 547–560.

Vugia, D., A. Cronquist, J. Hadler, M. Tobin-D'Angelo, D. Blythe, et al. 2006. Preliminary FoodNet data on the incidence of infection with pathogens transmitted commonly through food—10 states, United States, 2005. Morbid. Mortal. Weekly Rep. 55: 392–395.

APPENDIX 1

USEFUL SOURCES OF INFORMATION

Watershed
The Environmental Protection Agency offers a wealth of information on watershed protection, as well as a "Find your watershed" feature that allows you to see where your water comes from http://www.epa.gov/watershed/

Environmental Protection Agency
The EPA's Office of Water has its home page here. http://www.epa.gov/ow/

The World Health Organization (WHO)
http://www.who.int/en/

National Pollutant Discharge Elimination System (NPDES)
The EPA describes information regarding its discharge permitting on this site. http://cfpub.epa.gov/npdes/index.cfm

Biosolids
The ultimate disposal of sludge and other biosolids is described here. http://www.epa.gov/owm/mtb/biosolids/

Drinking Water
The EPA's site for drinking water databases, regulations, and contaminant limits. http://www.epa.gov/safewater/dwh/

The EPA's drinking water research program is described here. http://www.epa.gov/ord/npd/dwresearch-intro.htm

Vaccine Information
http://www.cdc.gov/vaccines/pubs/vis/default.htm#je

Biowarfare ABC list
http://emergency.cdc.gov/agent/agentlist-category.asp#catdef

Water and Wastewater Issues
The Water Environment Federation is one of the most knowledgeable organizations about water and wastewater issues. http://www.wef.org

Food pH
The U.S. Food and Drug Administration has compiled an extensive list of food pH values. http://www.foodscience.caes.uga.edu/extension/documents/FDAapproximatepHoffoodslacf-phs.pdf

Publications and Additional Organizations

Morbidity and Mortality Weekly Report
The home page for the CDC's premier online journal MMWR. http://www.cdc.gov/mmwr/

Emerging Infectious Diseases
The home page for the CDC's online journal devoted to disease outbreaks on a worldwide scale. http://www.cdc.gov/ncidod/EID/index.htm

Bulletin of the World Health Organization
http://www.who.int/bulletin/en/

American Industrial Hygiene Association
Section on preventing infectious disease in the workplace. http://www.aiha.org/Pages/default.aspx

Asthma and Allergy Foundation of America
Section on mold allergy. http://www.aafa.org/

National Institute of Allergy and Infectious Diseases (NIAID), the premier branch of the National Institute of Health dealing with infectious disease. http://www3.niaid.nih.gov/

National Institute for Occupational Safety and Health (NIOSH), the agency charged with promoting and regulating health in the workplace. Sections on infectious disease, mold, allergies. http://www.cdc.gov/NIOSH/

Food and Drug Administration Current Good Manufacturing Practices http://www.fda.gov/Food/GuidanceComplianceRegulatoryInformation/CurrentGoodManufacturingPracticesCGMPs/default.htm

Bacteriological Analytical Manual (BAM) http://www.fda.gov/Food/ScienceResearch/LaboratoryMethods/BacteriologicalAnalyticalManualBAM/default.htm

International Classification of Disease ICD-10 site http://www.who.int/classifications/icd/en/

Reason for Visit Classifications (RVC) http://www.cdc.gov/nchs/data/ahcd/rvc97.pdf

Hazard Analysis and Critical Control Points (HACCP) http://www.fda.gov/Food/FoodSafety/HazardAnalysisCriticalControlPointsHACCP/default.htm

APPENDIX 2

WISCONSIN ROUTINE ENTERIC FOLLOW-UP WORKSHEET

WEDSS ID:__________

WISCONSIN ROUTINE ENTERIC FOLLOW-UP WORKSHEET

PATIENT TAB

Patient name: ______________________ DOB: _____ / _____ / _____ Age: _________

Gender: M F Unknown

Address: __________________________ City: ________________ Zip: ________________________

County: __________________________ Jurisdiction: ____

Parent's name (If child): ____

Telephone number(s): ____

Ethnicity: _________________ Race: _______________ Occupation/School: _____________________

WEDSS DISEASE BEING REPORTED:

☐ CAMPYLOBACTEROSIS [*Campylobacter* _______________ (species)]
☐ CRYPTOSPORIDIOSIS
☐ E. COLI O157:H7— without HUS
☐ E. COLI, NON-O157 SHIGA TOXIN-PRODUCING(STEC) — without HUS
☐ GIARDIASIS
☐ HEMOLYTIC UREMIC SYNDROME, E. coli non-O157:H7
☐ HEMOLYTIC UREMIC SYNDROME, E. coli O157:H7
☐ HEMOLYTIC UREMIC SYNDROME, OTHER OR UNSPECIFIED
☐ SALMONELLOSIS [*Salmonella* ____________________ (serotype)]
☐ SHIGELLOSIS [*Shigella* ______________ (serogroup)]
☐ YERSINIOSIS

Interview date: _____ / _____ / _____ Interviewer: _____________________________________

Person interviewed: ☐patient ☐surrogate (specify): _____________________________________

WEDSS ID:__________

REQUESTED SECTIONS TO BE COMPLETED DURING ROUTINE INTERVIEWS

Expsoure Period	7 days				14 days		4 days	
Disease	E.coli O157/ STECs/HUS	Salmonella	Campy	Yersina	Crypto	Giardia	Shigella	Norovirus
Enteric - Travel	Yes	Yes	Yes	Yes	Yes	Yes	Yes	Yes
Enteric - Animal contact	Yes	Yes	Yes	Yes	Yes	Yes	No	No
Enteric - Soil-manure exposure	Yes	Yes	Yes	Yes	Yes	Yes	No	No
Enteric - Potential fecal exposure	Yes	Yes	Yes	Yes	Yes	Yes	Yes	Yes
Enteric - Water	Yes	Yes	Yes	Yes	Yes	Yes	Yes	Yes
Enteric - Food source	Yes	Yes	Yes	Yes	Yes	Yes	Yes	Yes
Enteric - Commercial food establishments	Yes	Yes	Yes	Yes	Yes	Yes	Yes	Yes
Enteric - Large gatherings	Yes	Yes	Yes	Yes	Yes	Yes	Yes	Yes
Enteric - Unpasteurized / Food General	Yes	Yes	Yes	Yes	Yes	Yes	No	No
Enteric - Food history	Yes	Yes	Yes	Yes	No	No	No	No
Enteric - Dairy	Yes	Yes	Yes	No	No	No	No	No
Enteric - Meat, Poultry, Fish	Yes	Yes	Yes	Yes	No	No	No	No
Enteric - Eggs	Yes	Yes	Yes	No	No	No	No	No
Enteric - Fruit	Yes	Yes	No	No	No	No	No	No
Enteric - Vegetables	Yes	Yes	No	No	No	No	No	No
Enteric - Other foods	Yes	Yes	No	No	No	No	No	No
Enteric - Beverages	Yes	Yes	No	No	No	No	No	No

LAB – CLINICAL TAB. ***This tab should be completed for all patients. Disease-specific laboratory information is not included on this worksheet because it is unique to each agent.***

Enteric - Symptoms

What was the first symptom? ______________________

Date of symptom onset: _____ / _____ / _____ Time of symptom onset: _____ : _____ AM / PM

Date first well: _____ / _____ / _____

Symptoms:

Nausea	Y	N	Unk	Vomiting	Y	N	Unk
Diarrhea	Y	N	Unk	Maximum number of stools in 24-hour period:_______			
(Defined as 3 or more loose stools in a 24-hour period)							
Bloody diarrhea	Y	N	Unk	Watery diarrhea	Y	N	Unk
Abdominal cramps	Y	N	Unk	Chills	Y	N	Unk
Sweats	Y	N	Unk	Headache	Y	N	Unk
Body/muscle aches	Y	N	Unk	Fatigue	Y	N	Unk
Fever	Y	N	Unk	Highest measured temperature:___________			
Other	Y	N	Unk	Describe: ____			

Enteric - Medical

Did the patient have:

Hemolytic Uremic Syndrome (HUS)? Y N Unk

Thrombocytopenic Purpura (TTP)? Y N Unk

Did patient see a physician/medical provider? Y N Unk If yes, provider name: _________________

Was patient hospitalized overnight? Y N Unk If yes, where: ________________________________

Admission date: _____ / _____ / _____ Discharge date: _____ / _____ / _____

WEDSS ID:__________

Did patient die as a result of this illness? Y N Unk

Does patient have any underlying medical conditions? Y N Unk
If yes, describe: ______________________________

Did patient take any medications in 30 days PRIOR to illness onset (prescription medication, over-the-counter medication, herbal preparations, vitamins, or other supplements)? Y N Unk
If yes, list medications? ______________________________

Were antibiotics prescribed FOR the illness? Y N Unk
If yes, date of first dose _____ / _____ / _____ List antibiotic(s)______________________________

Anti-diarrheal medication taken FOR this illness? Y N Unk
If yes, date of first dose _____ / _____ / _____ List medication(s) ______________________________

Other medication(s) taken FOR this illness? Y N Unk
If yes, list medication(s) ______________________________

Enteric – Other ill persons

Provide details—including common exposures—for household contacts, co-workers, classmates, and other associates that have a similar illness. Were others ill? Y N Unk
If yes, please describe who, when, relationship, and common exposures:______________________________
____

ENTERIC RISK TAB – *See table on page 1 for additional guidance on sections to complete*

Enteric Exposure period

The exposure period should be determined based on the incubation period of the etiologic agent under investigation and the illness onset date.

Campylobacter – 7 days
E. coli (Shiga toxin producing) – 7 days
Hemolytic uremic syndrome – 7 days
Salmonella – 7 days
Yersinia – 7 days

Cryptosporidium – 14 days
Giardia – 14 days

Shigella – 4 days
Viral agents (norovirus) – 4 days

If a specific agent is not known, the exposure period should include the 7 days prior to illness onset.
Earliest exposure date: _____ / _____ / _____ Onset date (end exposure): _____ / _____ / _____

Enteric – Travel (*All patients*)

During the exposure period, did the patient travel or visit a place outside of usual activities? Y N Unk
If yes: Where: ______________________________
Departure: _____ / _____ / _____ Return: ____ /____/____ Airline/Flight No.: __________
Activities: ______________________________
Foods consumed/Restaurants/Additional info: ______________________________
____

WEDSS ID:___________

Enteric – Animal Contact *(this section generally not necessary for Shigellosis and Viral Gastroenteritis)*
Indicate if the patient has any pets, or had contact with pets in the homes of others, during the exposure period (including reptiles, pocket pets, and fish). Y N Unk
If yes, list pets: ___________
What brand of pet food was used and where was it purchased?: ___________
Is pet fed raw meat? Y N Unk

Does patient live or work on a farm? Y N Unk

During the exposure period, did the patient have exposure to any of the following animals or their environment (including at home or visiting a farm, fair, petting zoo, school, etc)? Y N Unk

Indicate all animals (or their environments) that patient had contact with during the exposure period.

- ☐ Dog/puppies *(circle)*
- ☐ Cats/kittens *(circle)*
- ☐ Cattle
- ☐ Horses
- ☐ Sheep
- ☐ Goats
- ☐ Pigs
- ☐ Poultry
- ☐ Fish
- ☐ Rodents, specify: ___________
- ☐ Reptiles/amphibians, specify: ___________
- ☐ Other, specify: ___________

Animal contact details (Type/location/date/type of exposure): ___________
____

Enteric – Soil/Manure exposure *(Generally not necessary for Shigellosis and Viral Gastroenteritis)*
Did patient apply manure or compost? Y N Maybe Unk
If yes, please specify date and type of compost/manure: ___________

Enteric – Potential Fecal Exposure (*All patients*)
During the exposure period, was the patient exposed to:
Adults or children using diapers? Y N M Unk If yes, did the person have diarrhea? Y N M Unk
Describe the nature of the exposure (date, what type of contact with the person or diaper, etc.): ___________
____

Enteric – Water (*All patients*)
During the exposure period, was the patient exposed to the following sources of water (include drinking and recreational use):

Municipal water supply	Y	N	M	Unk	Private well	Y	N	M	Unk
Common well/Rural system	Y	N	M	Unk	Bottled water	Y	N	M	Unk
River/Lake/Pond	Y	N	M	Unk	Ocean	Y	N	M	Unk
Chlorinated pool	Y	N	M	Unk	Wading pool	Y	N	M	Unk
Water/Splash park	Y	N	M	Unk	Standing water	Y	N	M	Unk

Please provide details related to your exposures (dates, locations, etc.) and any additional source not covered above: ___________
____

WEDSS ID:__________

Enteric – Food Source (*All patients*)

List grocery stores and supermarkets where patient purchased the food eaten during the exposure period.

Name: ______________________________ Address: ____________________________________

Name: ______________________________ Address: ____________________________________

Name: ______________________________ Address: ____________________________________

During your exposure period, did you eat food obtained from any of the following sources?

Hunting/fishing/trapping Y N M Unk   Butcher shop Y N M Unk

Private kill Y N M Unk   Farmers market Y N M Unk

Own garden Y N M Unk   Friend/relative Y N M Unk

Home delivery (Meals on Wheels/Schwan's/Other) Y N M Unk

Provide details regarding any "Yes" answers above (What was consumed, where and when purchased, etc.):

____

____

Enteric – Food, unpasteurized/general diet (*All patients*)

Tendency to buy organic or natural foods? Y N M Unk

During the exposure period, did the patient consume items or dishes containing the following items?

Unpasteurized or raw dairy products (milk, cheese, etc) Y N M Unk

If yes, specify product type/brand: ______________________________________

Where and when it was purchased or obtained: ______________________________

Date(s) consumed: ____

Unpasteurized, raw, or freshly squeezed juices? Y N M Unk

If yes, specify product type/brand: ______________________________________

Where and when it was purchased or obtained: ______________________________

Date(s) consumed: ____

Does the patient eat a special or restricted diet? Y N Unk

(vegetarian, vegan, diabetic, gluten free, dairy free, infant formula and foods, etc.)

If yes, please specify: ____

Enteric – Commercial food establishments (*All patients*)

Indicate the commercial establishments where food was eaten or obtained during the exposure period. Include restaurants, catered events, fast food, cafeterias, delis, supermarkets, street vendors, concession stands, snack bars, and gas stations.

Did patient eat in any establishment? Y N Unk

1. Name:__________________________ Address/location: ________________________

Meal date: _____ / _____ / _____ Time: _____ : _____ AM / PM

Did you eat at a salad bar or buffet at the establishment? Y N M Unk

Foods consumed:____

____

2. Name:__________________________ Address/location: ________________________

Meal date: _____ / _____ / _____ Time: _____ : _____ AM / PM

Did you eat at a salad bar or buffet at the establishment? Y N M Unk

Foods consumed:____

____

WEDSS ID:__________

3. Name:____________________ Address/location: ____________________
Meal date: _____ / _____ / _____ Time: _____ : _____ AM / PM
Did you eat at a salad bar or buffet at the establishment? Y N M Unk
Foods consumed:____

4. Name:____________________ Address/location: ____________________
Meal date: _____ / _____ / _____ Time: _____ : _____ AM / PM
Did you eat at a salad bar or buffet at the establishment? Y N M Unk
Foods consumed:____

5. Name:____________________ Address/location: ____________________
Meal date: _____ / _____ / _____ Time: _____ : _____ AM / PM
Did you eat at a salad bar or buffet at the establishment? Y N M Unk
Foods consumed:____

Enteric – Large Gatherings (*All patients*)

Large gathering: Wedding, shower, parties, sports events, picnics during the exposure period? Y N Unk

1. Event description:____
Event date: _____ / _____ / _____ Time: _____ : _____ AM / PM
Address/location: ____
Food eaten at event: ____

2. Event description:____
Event date: _____ / _____ / _____ Time: _____ : _____ AM / PM
Address/location: ____
Food eaten at event: ____

3. Event description:____
Event date: _____ / _____ / _____ Time: _____ : _____ AM / PM
Address/location: ____
Food eaten at event: ____

Enteric – Food History (*E. coli O157, non-O157 STECs, HUS, Salmonella, Campylobacter, Yersinia*)

Day 1 prior to illness – Date: _____ / _____ / _____

Meal	Ate at Home	Ate outside of home	Outside location	Foods eaten
Breakfast	☐	☐	__________	__________
Lunch	☐	☐	__________	__________
Dinner	☐	☐	__________	__________
Other (snack, etc.)	☐	☐	__________	__________

WEDSS ID:__________

Day 2 prior to illness – Date: _____ / _____ / _____

Meal	Ate at Home	Ate outside of home	Outside location	Foods eaten
Breakfast	☐	☐	____________	____________
Lunch	☐	☐	____________	____________
Dinner	☐	☐	____________	____________
Other (snack, etc.)	☐	☐	____________	____________

Day 3 prior to illness – Date: _____ / _____ / _____

Meal	Ate at Home	Ate outside of home	Outside location	Foods eaten
Breakfast	☐	☐	____________	____________
Lunch	☐	☐	____________	____________
Dinner	☐	☐	____________	____________
Other (snack, etc.)	☐	☐	____________	____________

Day 4 prior to illness – Date: _____ / _____ / _____

Meal	Ate at Home	Ate outside of home	Outside location	Foods eaten
Breakfast	☐	☐	____________	____________
Lunch	☐	☐	____________	____________
Dinner	☐	☐	____________	____________
Other (snack, etc.)	☐	☐	____________	____________

Day 5 prior to illness – Date: _____ / _____ / _____

Meal	Ate at Home	Ate outside of home	Outside location	Foods eaten
Breakfast	☐	☐	____________	____________
Lunch	☐	☐	____________	____________
Dinner	☐	☐	____________	____________
Other (snack, etc.)	☐	☐	____________	____________

WEDSS ID:___________

The following sections are a comprehensive food history. For each item below, ask the patient to answer "yes," "no," or "maybe" if they remember eating the item, **either at their home or outside the home**, during their exposure period. Please remind the patient of the time frame of interest and try to obtain specific details when possible related to food items consumed.

Earliest exposure date: _____ / _____ / _____ **Onset date (end exposure):** _____ / _____ / _____

Enteric – Dairy (*E. coli O157, non-O157 STECs, HUS, Salmonella, Campylobacter*)

Item	Y	N	M	Unk	Additional information: variety or brand, purchase location, how prepared, when and where consumed, etc.
Milk (pasteurized)	Y	N	M	Unk	
Sour cream	Y	N	M	Unk	
Cream cheese	Y	N	M	Unk	
Cheese	Y	N	M	Unk	Specify type:
Queso fresco (Mexican style soft cheese)	Y	N	M	Unk	Product details:
Cheese curds	Y	N	M	Unk	
Cottage cheese	Y	N	M	Unk	
Ice cream / frozen yogurt / etc.	Y	N	M	Unk	
Yogurt	Y	N	M	Unk	
Any other dairy	Y	N	M	Unk	

Enteric – Meat, Poultry, Fish (*E. coli O157, non-O157 STECs, HUS, Salmonella, Campylobacter, Yersinia*)

Item	Y	N	M	Unk	Additional information: variety or brand, purchase location, how prepared, when and where consumed, etc.
Ground beef / hamburger	Y	N	M	Unk	Date(s) consumed: ____________ How prepared: ____________ Was it raw, bloody, or pink when eaten? Y / N Where purchased/consumed:____________ Date purchased: ____________ Type: (lean, %fat, chuck, round, etc.): ____________ Size: ______lb; Purchased as preformed patties? Y / N Was it purchased: Fresh / Frozen; Any left over? Y / N
Other beef (steak, roast, etc.)	Y	N	M	Unk	
Chicken	Y	N	M	Unk	Date(s) consumed: ____________ Describe product: ____________ Where purchased/consumed:____________ Was it purchased? Fresh / Frozen Date purchased: ____________
Turkey	Y	N	M	Unk	
Pork (including ham, bacon, etc.)	Y	N	M	Unk	
Lamb	Y	N	M	Unk	

WEDSS ID:__________

Fish	Y	N	M	Unk	
Shellfish / Seafood *(circle)*	Y	N	M	Unk	
Wild game	Y	N	M	Unk	
Hot dogs / Bratwurst *(circle)*	Y	N	M	Unk	
Sausage (other: breakfast salami, sausage, pepperoni, etc.)	Y	N	M	Unk	Specify:
Lunch meat (prepackaged or deli)	Y	N	M	Unk	Specify:
Any other meat	Y	N	M	Unk	
Any above raw, rare, or undercooked?	Y	N	M	Unk	Specify:

Enteric – Eggs (*E. coli O157, non-O157 STECs, HUS, Salmonella, Campylobacter*)

Item	**Y**	**N**	**M**	**Unk**	**Additional information: variety or brand, purchase location, how prepared, when and where consumed, etc.**
Eggs eaten	Y	N	M	Unk	How prepared (fried, scrambled, etc.): __________ Where purchased/eaten: __________ Date consumed: __________
Eggs used in cooking/baking	Y	N	M	Unk	
Uncooked batter or dough eaten	Y	N	M	Unk	
Egg products/substitutes used	Y	N	M	Unk	

Enteric – Fruit (*E. coli O157, non-O157 STECs, HUS, Salmonella*)

Item	**Y**	**N**	**M**	**Unk**	**Additional information: variety or brand, purchase location, how prepared, when and where consumed, etc. If other than FRESH please specify.**
Apples	Y	N	M	Unk	
Bananas	Y	N	M	Unk	
Cantaloupe	Y	N	M	Unk	
Cherries	Y	N	M	Unk	
Grapefruit	Y	N	M	Unk	
Grapes	Y	N	M	Unk	Red / Green
Honeydew	Y	N	M	Unk	
Oranges	Y	N	M	Unk	
Pears	Y	N	M	Unk	
Strawberries	Y	N	M	Unk	
Other berries	Y	N	M	Unk	Specify:
Watermelon	Y	N	M	Unk	
Other tree fruit (nectarines, plums, peaches, apricots)	Y	N	M	Unk	Specify:
Exotic/tropical fruit (mango, kiwi, papaya, pineapple, etc.)	Y	N	M	Unk	Specify:
Other citrus (lime, lemon, tangerine, Clementine)	Y	N	M	Unk	Specify:
Any other fruit	Y	N	M	Unk	Specify:

WEDSS ID:__________

Enteric – Vegetables (*E. coli O157, non-O157 STECs, HUS, Salmonella*)

Item	**Y**	**N**	**M**	**Unk**	**Additional information: variety or brand, purchase location, how prepared, when and where consumed, etc. If other than FRESH please specify.**
Packaged or bagged salads/lettuces/greens	Y	N	M	Unk	Brand: ____________ Description/type: ____________ Where purchased: ____________ Date purchased: ____________ Date(s) consumed: ____________
Lettuce (that was not bagged/ prepackaged)	Y	N	M	Unk	Iceberg / Leaf / Romaine / other: ________
Spinach (loose or bagged)	Y	N	M	Unk	Describe:
Cabbage	Y	N	M	Unk	
Other salad greens	Y	N	M	Unk	
Asparagus	Y	N	M	Unk	
Broccoli	Y	N	M	Unk	
Carrots	Y	N	M	Unk	Baby / Regular / Precut
Cauliflower	Y	N	M	Unk	
Celery	Y	N	M	Unk	
Cucumbers	Y	N	M	Unk	
Eggplant	Y	N	M	Unk	
Green onions / scallions	Y	N	M	Unk	
Other onions	Y	N	M	Unk	Specify:
Mushrooms	Y	N	M	Unk	
Pea pods	Y	N	M	Unk	
Peppers (sweet or hot)	Y	N	M	Unk	Specify:
Radishes / Jicama *(circle)*	Y	N	M	Unk	
Sprouts (alfalfa, bean, etc.)	Y	N	M	Unk	Specify:
Squash / Zucchini *(circle)*	Y	N	M	Unk	
Tomatoes	Y	N	M	Unk	Specify:
Fresh parsley	Y	N	M	Unk	
Fresh cilantro	Y	N	M	Unk	
Other fresh herbs	Y	N	M	Unk	Specify:
Any other vegetable(s)	Y	N	M	Unk	Specify:

WEDSS ID:__________

Enteric – Other Foods (*E. coli O157, non-O157 STECs, HUS, Salmonella*)

Item	Y	N	M	Unk	Additional information: variety or brand, purchase location, how prepared, when and where consumed, etc.
Peanut butter	Y	N	M	Unk	
Peanuts	Y	N	M	Unk	
Any other nuts (almonds/ walnuts / pecans / etc.)	Y	N	M	Unk	Specify:
Dips or spreads (hummus, Tahini, etc.)	Y	N	M	Unk	Specify:
Fresh salsa	Y	N	M	Unk	
Tofu	Y	N	M	Unk	
Chocolate	Y	N	M	Unk	
Other candy	Y	N	M	Unk	Specify:
Convenience foods (i.e., frozen, boxed, ready to eat)	Y	N	M	Unk	Specify:
Coleslaw	Y	N	M	Unk	
Tossed salad	Y	N	M	Unk	
Fruit salad	Y	N	M	Unk	
Potato salad	Y	N	M	Unk	
Other deli salads	Y	N	M	Unk	Specify:
Any other foods eaten	Y	N	M	Unk	

Enteric – Beverages Enteric – Other Foods (*E. coli O157, non-O157 STECs, HUS, Salmonella*)

Item	Y	N	M	Unk	Additional information: variety or brand, purchase location, how prepared, when and where consumed, etc.
Juice (commercial/pasteurized)	Y	N	M	Unk	Specify:
Smoothies/blended drinks	Y	N	M	Unk	
Health drinks or supplements	Y	N	M	Unk	

Any additional comments or items not already covered? ______________________________

____

____

____

APPENDIX 3

SAMPLE MUMPS SURVEILLANCE FORM

Patient Identification and Demographics

Name (Last, First, MI)			**Birth date**	**Age**	**Sex** ☐ M ☐ F
Address			**Phone**		
City	**State**	**Zip**	**County**	**LPHA**	
Race: ☐ White ☐ Black ☐ Asian or Pacific Islander ☐ Native American				**Ethnicity:** ☐ Hispanic ☐ Non-Hispanic	

Reporting and Case Management

Agency Reporting (Name and Location):	
Interviewer Name:	**Interview Date:**
Physician (Name, Location, and Phone):	
Date Reported:	**Investigation Initiation Date:**
Case Manager:	**Investigation Completion Date:**

Morbidity Information

Initial Case Status: ☐ Confirmed – Lab ☐ Confirmed – Epi ☐ Probable ☐ Suspect ☐ Unknown		
Final Case Status: ☐ Confirmed – Lab ☐ Confirmed – Epi ☐ Probable ☐ Suspect ☐ Ruled Out		
Pregnant: ☐ Yes ☐ No	**Due Date:**	**Miscarriage:** ☐ Yes ☐ No
Underlying Conditions: ☐ Yes ☐ No	**Specify:**	
Hospitalized: ☐ Yes ☐ No	**Hospital Name and Location:**	
Hospitalization Dates: to		**Total Days Hospitalized:**
Outcome: ☐ Survived ☐ Died	**Date of Death:**	
Origin: ☐ In-State ☐ Out of State ☐ Outside US	**Outbreak Related:** ☐ Yes ☐ No	

Case ______________________ Page ____ Case Manager Initials ______ Date ________

Lab Testing

Lab Testing: ☐ Yes ☐ No		Lab (Name, Location, Phone):			
Collected	**Test**	**Specimen**	**Collection Date**	**Results**	**Value**
☐ Yes ☐ No	IgG Acute	Blood			
☐ Yes ☐ No	IgG Convalescent	Blood			
☐ Yes ☐ No	IgM	Blood			
☐ Yes ☐ No	Viral Culture	Buccal Swab			
☐ Yes ☐ No	Viral Culture	Urine			
☐ Yes ☐ No	Rt PCR	Throat Swab			
☐ Yes ☐ No	Rt PCR	Urine			
☐ Yes ☐ No	Rt PCR	CSF			

Signs and Symptoms

Onset Date:		**Date Reported:**	
Fever	☐ Yes ☐ No ☐ Unk	**Muscle/Joint Pain**	☐ Yes ☐ No ☐ Unk
Highest Fever Measured: ________ F		**Headache**	☐ Yes ☐ No ☐ Unk
		Hearing Loss	☐ Yes ☐ No ☐ Unk
Pain with Chewing	☐ Yes ☐ No ☐ Unk	**Loss of Appetite**	☐ Yes ☐ No ☐ Unk
Facial or Salivary Gland Swelling	☐ Yes ☐ No ☐ Unk	**Nausea**	☐ Yes ☐ No ☐ Unk
		Vomiting	☐ Yes ☐ No ☐ Unk
Onset:	**to:**	**Convulsions**	☐ Yes ☐ No ☐ Unk
		Complications (*Consult Medical Record*)	
Total Days:		**Encephalitis**	☐ Yes ☐ No ☐ Unk
		Meningitis	☐ Yes ☐ No ☐ Unk
Location: ☐ **Unilateral** ☐ **Bilateral**		**Myocarditis**	☐ Yes ☐ No ☐ Unk
Testicular Swelling or Tenderness	☐ Yes ☐ No ☐ Unk	**Orchitis**	☐ Yes ☐ No ☐ Unk
Cough	☐ Yes ☐ No ☐ Unk	**CNS Involvement**	☐ Yes ☐ No ☐ Unk
Earache	☐ Yes ☐ No ☐ Unk	**Pancreatitis**	☐ Yes ☐ No ☐ Unk
Fatigue	☐ Yes ☐ No ☐ Unk	**Oophoritis**	☐ Yes ☐ No ☐ Unk
Other, Specify:			

Case ____________________ Page ____ Case Manager Initials ______ Date ________

Vaccination History

Vaccinated: ☐ Yes ☐ No ☐ Unknown	Total Number of Valid Doses:
In WIR: ☐ Yes ☐ No	
Previous History of Disease: ☐ Yes ☐ No ☐ Unknown	Date of Previous Disease:
If Unvaccinated, Reason: ☐ Religious Exemption ☐ Medical Contraindication ☐ Philosophical Objection ☐ Lab. Evidence of Previous Disease ☐ MD Diagnosis of Previous Disease ☐ Other ☐ Under Age for Vaccination ☐ Parental Refusal ☐ Unknown	

Vaccination Date	Administered By	Vaccine Type	Manufacturer	Lot Number
		☐ MMR ☐ Mumps ☐ Other ☐ Unknown	☐ Merck ☐ Other ☐ Unknown	
		☐ MMR ☐ Mumps ☐ Other ☐ Unknown	☐ Merck ☐ Other ☐ Unknown	
		☐ MMR ☐ Mumps ☐ Other ☐ Unknown	☐ Merck ☐ Other ☐ Unknown	

Source/Spread Investigation *(use additional pages as needed)*

Onset Date: ____________________
Source: 25 days prior to symptom onset ___________ to 12 days prior to symptom onset __________
Spread: 7 days prior to symptom onset ___________ to 9 days after symptom onset __________

Individuals *(Complete for close contacts only)*

	Contact ___	Contact ___	Contact ___
Name			
Contact Information			
Relationship	☐ Parent/Guardian ☐ Other family member ☐ Other	☐ Parent/Guardian ☐ Other family member ☐ Other	☐ Parent/Guardian ☐ Other family member ☐ Other
Household Member	☐ Yes ☐ No ☐ Unknown	☐ Yes ☐ No ☐ Unknown	☐ Yes ☐ No ☐ Unknown
Symptomatic	☐ Yes ☐ No ☐ Unknown If yes, Onset date: Salivary gland swelling or tenderness? ☐ Yes ☐ No ☐ Unk Tested: ☐ Yes ☐ No ☐ Unk If tested, where/when/lab	☐ Yes ☐ No ☐ Unknown If yes, Onset date: Salivary gland swelling or tenderness? ☐ Yes ☐ No ☐ Unk Tested: ☐ Yes ☐ No ☐ Unk If tested, where/when/lab	☐ Yes ☐ No ☐ Unknown If yes, Onset date: Salivary gland swelling or tenderness? ☐ Yes ☐ No ☐ Unk Tested: ☐ Yes ☐ No ☐ Unk If tested, where/when/lab

Case ______________________ Page ____ Case Manager Initials ______ Date ________

Vaccinated	☐ **Yes** ☐ **No** ☐ **Unknown** If yes, # of doses: In WIR ☐ Yes ☐ No	☐ **Yes** ☐ **No** ☐ **Unknown** If yes, # of doses: In WIR ☐ Yes ☐ No	☐ **Yes** ☐ **No** ☐ **Unknown** If yes, # of doses: In WIR ☐ Yes ☐ No
Source/Spread	☐ **Possible Source** ☐ **Possible Spread** ☐ **Unknown**	☐ **Possible Source** ☐ **Possible Spread** ☐ **Unknown**	☐ **Possible Source** ☐ **Possible Spread** ☐ **Unknown**
Contacted	☐ **Yes** ☐ **No** ☐ **Unknown**	☐ **Yes** ☐ **No** ☐ **Unknown**	☐ **Yes** ☐ **No** ☐ **Unknown**
High Risk	☐ **Yes** ☐ **No** ☐ **Unknown**	☐ **Yes** ☐ **No** ☐ **Unknown**	☐ **Yes** ☐ **No** ☐ **Unknown**
Intervention	☐ **Isolation** ☐ **Quarantine** ☐ **Vaccination** ☐ **Education** ☐ **Symptom Monitoring** ☐ **Testing**	☐ **Isolation** ☐ **Quarantine** ☐ **Vaccination** ☐ **Education** ☐ **Symptom Monitoring** ☐ **Testing**	☐ **Isolation** ☐ **Quarantine** ☐ **Vaccination** ☐ **Education** ☐ **Symptom Monitoring** ☐ **Testing**
Notes			

Organizations/Groups (Employer, School, Teams, Churches, etc.)

Use additional pages as necessary

	Organization ____	**Organization ____**
Name, Location, and Description		
Relationship	☐ Employer ☐ Worksite Location ☐ School or Child Care Provider ☐ Patron, Customer, Visitor ☐ Member/Participant ☐ Healthcare Provider ☐ Other	☐ Employer ☐ Worksite Location ☐ School or Child Care Provider ☐ Patron, Customer, Visitor ☐ Member/Participant ☐ Healthcare Provider ☐ Other
Possible Source of Exposure	☐ Yes ☐ No ☐ Unk If yes, possible exposure dates: Source Case(s)	☐ Yes ☐ No ☐ Unk If yes, possible exposure dates: Source Case(s)
Possible Spread and Dates	☐ Yes ☐ No ☐ Unk If yes, dates of possible spread: Subsequent Case(s):	☐ Yes ☐ No ☐ Unk If yes, dates of possible spread: Subsequent Case(s):
Contacted/Date	☐ Yes ☐ No ☐ Unk If yes, date contacted:	
Org. Contact (Name and Number)		

Case ____________________ Page ____ Case Manager Initials ______ Date ________

Other Affiliated Cases and Names	☐ Yes ☐ No ☐ Unk If yes, names and case status:	☐ Yes ☐ No ☐ Unk If yes, names and case status:
High-Risk Setting	☐ Yes ☐ No ☐ Unk	☐ Yes ☐ No ☐ Unk
Intervention	☐ Notification Letter Date Sent: ____________ ☐ Isolation ☐ Quarantine ☐ Exclusion ☐ Symptom Monitoring ☐ Vaccination ☐ Verify Immunity ☐ Education ☐ Infection Control (PPE) ☐ Contact List Obtained	☐ Notification Letter Date Sent: ____________ ☐ Isolation ☐ Quarantine ☐ Exclusion ☐ Symptom Monitoring ☐ Vaccination ☐ Verify Immunity ☐ Education ☐ Infection Control (PPE) ☐ Contact List Obtained
Notes:		

Case ______________________ Page ____ Case Manager Initials ______ Date ________

Index

Boxes, figures, and tables are indicated with b, f, and t following the page number.

C

D

E

F

G

H

I

J

K

L

N

O

P

Q

R

S

T

U

V

W

X

Y

Z